Chi Tat Chong, Liang Yu

Recursion Theory

De Gruyter Series in Logic and Its Applications

Volume 8

Chi Tat Chong, Liang Yu

Recursion Theory

Computational Aspects of Definability

DE GRUYTER

AMS subject classification (2010)
03-01, 03-02, 03D, 03E

Authors
Prof. Dr. Chi Tat Chong
National University of Singapore
Department of Mathematics
10 Lower Kent Ridge Road
119076 Singapore
chongct@math.nus.edu.sg

Prof. Dr. Liang Yu
Nanjing University
Institute of Mathematical Science
210093 Nanjing
People's Republic of China
yuliang.nju@gmail.com

ISBN 978-3-11-027555-1
e-ISBN (PDF) 978-3-11-027564-3
e-ISBN (EPUB) 978-3-11-038129-0
Set-ISBN 978-3-11-027565-0
ISSN 1438-1893

Library of Congress Cataloging-in-Publication Data
A CIP catalog record for this book has been applied for at the Library of Congress.

Bibliographic information published by the Deutsche Nationalbibliothek
The Deutsche Nationalbibliothek lists this publication in the Deutsche Nationalbibliografie; detailed bibliographic data are available on the Internet at http://dnb.dnb.de.

Typesetting: PTP-Berlin, Protago TEX-Produktion GmbH
Printing and binding: CPI books GmbH, Leck
♾ Printed on acid-free paper
Printed in Germany

www.degruyter.com

Chi Tat Chong dedicates this book to the memory of his parents;
Liang Yu dedicates it to Li Zhang.

Preface

... logic is about definability.
– Gerald E. Sacks [124]

The above quote underlines the authors' view that recursion theory is anchored on the notion of definability. Although classically the theory began with Turing's conception of effective computability (of various types of objects), it soon took on a much broader perspective. A fundamental theorem states that a set of natural numbers is computable if and only if it is Δ^0_1-definable in first order arithmetic. This sets the stage for a general study of sets of natural numbers and their syntactic classes.

The view on definability does not conflict with the fact that computation is the heart of recursion theory. Indeed a definability problem naturally translates into one about (possibly relativized) computation and vice versa. Kleene's characterization of the Δ^1_1-sets of numbers as precisely those that are hyperarithmetic, and the corresponding theorem of Spector and Gandy that these are all the reals in $L_{\omega_1^{\mathrm{CK}}}$, is a case in point. Once the syntax (the language of first and second order arithmetic) and the structure in which to interpret members of ω and its subsets are specified, definability naturally takes its place. Tools and construction techniques have been introduced to study problems concerning computability for syntactic classes over ω, 2^ω and even the power set of 2^ω or ω^ω. Post's problem for the recursively enumerable sets, the Gandy basis theorem for Σ^1_1-sets of reals, characterization of Δ^1_1-sets in terms of hyperarithmeticity, the definability of the jump operator, are immediate examples. Indeed it is hard to find a problem in the area that has little to do with definability. The title of this book is chosen to reflect this point.

The idea of writing a book in recursion theory, focusing on the computational aspects of the set of reals and its subsets, evolved over a period of time. The subject was established by Kleene who introduced the notion of hyperarithmetic hierarchy relative to a real and its associated ordinals. It grew in the hands of Spector, Gandy, Kreisel and Sacks. These were major developments in what one now calls higher recursion theory. The theory was a subject of intensive study for almost thirty years beginning in the 1950s, but seemed to lose momentum in the 1980s. In an interview reproduced in the Appendix, Sacks remarked: "I tended over the years to keep predicting the immediate end of recursion theory, but I have been wrong every time." Despite the rather pessimistic pronouncement by one of the founders of the field, recursion theory is alive and well. However, higher recursion theory in the grand tradition of the pioneers received only occasional attention since, even though a number of highly significant contributions to the subject, such as Slaman and Woodin's theorem on the rigidity of the hyperdegrees, were made towards the end of the last century, and many interesting as well as challenging problems remain. Indeed Sacks himself bade farewell to higher recursion theory in 1998. Today the subject is not pursued with as much vigor or zeal as it deserves. We view this as rather unfortunate since in it one sees a beautiful illustration of the unity of mathematical logic, where ideas and tools from set theory and model theory are fruitfully applied to study problems about computability. It was a tempting proposition to write another book (following Sacks's *Higher Recursion Theory* [123]) to exhibit this interplay. Recent advances in the study of the global theory of Turing and hyperdegrees, and the emergence of higher randomness as a subject of investigation, provided the impetus for us to turn the idea into reality. It is hoped that the reader would agree, at the end of it, that the enterprise of generalizing recursion theory that began more than half a century ago has been mathematically rewarding, and that the subject is both attractive and highly challenging, with many deep and elegant theorems.

This book consists of four main parts. The first six chapters constitute Part I and develop the fundamental theory of Π^1_1-sets and fine structure theory of the constructible universe, the ramified analytical hierarchy, and set theory. They lay the groundwork for the rest of the book. Part II focuses on the structure of Turing degrees. Here we discuss classification of the jump operator in relation to Martin's conjecture and the construction of Π^1_1-sets, as well as degree-theoretic independence results. We turn our attention to hyperdegrees in Part III. In particular, we prove the rigidity of the structure of the hyperdegrees and its biinterpretability with second order arithmetic. We also include a chapter on basis/antibasis theorems as well as perfect subsets in L. Part IV is devoted to the study of higher randomness, a subject which is naturally linked to the topics covered in the first three parts. Exercises are sprinkled throughout the chapters. Many of them are intended to supplement the topics covered. The most challenging ones are labelled with a *.

The book ends with an interview with Gerald Sacks (in Appendix B) conducted in 2009. We think that it is fitting to conclude the book with a tribute to one whose work has arguably substantially shaped the development of recursion theory, both classical and higher, in the last century.

A word about prerequisites: While this book attempts to cover as much background material as possible, it is not intended to be encyclopedic. The reader is assumed to be familiar with basic logic and set theory, including classical recursion theory, basic set theory (including forcing) and algorithmic randomness. A number of the results are used in proofs, sometimes without explicit mention. In line with our preference and taste, some of the proofs of theorems in set theory presented are discernibly recursion-theoretic in flavor. These proofs are not necessarily the simplest. Our aim is to offer the reader a different perspective to these theorems by highlighting the effective content that is not immediately apparent. Hopefully these provide additional insights to the ideas that went behind the proofs.

The materials in the first two chapters closely follow Sacks's *Higher Recursion Theory* [123]. Mansfield and Weitkamp's monograph *Recursive Aspects of Descriptive Set Theory* [88] was also one of our key references. Of course one should not fail to mention Hartley Rogers's classic *Theory of Recursive Functions and Effective Computabil-*

ity [116], the book that first made recursion theory accessible to several generations of students in logic and which largely set the style for books on this subject.

In this book we give preference to calling the subject recursion theory rather than computability theory (as it is otherwise known) in deference to its historical origin, although both terms are used interchangeably throughout.

During the preparation of the manuscript, the authors benefited from comments and suggestions of colleagues and students, some of whom read through various portions of the book. Certain chapters were used by the second author in a graduate course given at the National University of Singapore. We thank Laurent Bienvenu, Peter Cholak, Decheng Ding, Qi Feng, Su Gao, Junle Goh, Noam Greenberg, Wei Li, Yiqun Liu, Yong Liu, Benoit Monin, André Nies, Dilip Raghavan, Gerald Sacks, Richard Shore, Stephen Simpson, Theodore A. Slaman, Frank Stephan, Wei Wang, Yue Yang and Jing Zhang for their interest, ideas, suggestions and/or support of this project.

Finally, for this work Chong was partially supported by NUS grant WBS C140-000-025-00, while Yu was supported in part by the National University of Singapore, the Humboldt Foundation and the National Science Foundation of China grant no. 11071114 and 11322112. Their generosity is gratefully acknowledged.

Chitat Chong and Liang Yu
15 June 2015

Contents

Part I: **Fundamental theory**

1 An introduction to higher recursion theory

1.1 Projective predicates

1.1.1 Recursive and arithmetical predicates

A *recursive predicate* R on ω is a total recursive function $\Phi : \omega \to 2$ such that $R(n) \Leftrightarrow \Phi(n) = 1$. With a standard coding of $\omega^{<\omega}$ in ω, one may define recursive predicates on $\omega^{<\omega}$. Similarly one may code ω^n in ω to obtain recursive n-ary predicates on ω, as well as recursive predicates on $(\omega^{<\omega})^m \times \omega^n$. The class of arithmetical predicates on $\omega^\omega \times \omega$ is now defined as follows:

Definition 1.1.1. A predicate $R(x, n)$ on $\omega^\omega \times \omega$ is *partial recursive* if there is a partial recursive function $\Phi(\sigma, n)$ on $\omega^{<\omega} \times \omega$ such that
(i) $(\forall\sigma)(\forall\tau)(\forall n)[\Phi(\sigma, n)\downarrow \wedge\, \sigma \preceq \tau \to (\Phi(\tau, n)\downarrow \wedge\, \Phi(\sigma, n) = \Phi(\tau, n))]$.
(ii) For any $x \in \omega^\omega$ and $n \in \omega$, $R(x, n)$ if and only if $(\exists m)(\Phi(x \upharpoonright m, n)\downarrow= 0)$.
A predicate $R(x, n)$ on $\omega^\omega \times \omega$ is *recursive* if the partial recursive function Φ above also satisfies the following property:
(iii) $(\forall x)(\forall n)(\exists m)(\Phi(x \upharpoonright m, n)\downarrow)$.

Remark 1.1.2. We may similarly define recursive predicates on $(\omega^\omega)^n \times (\omega^{<\omega})^m \times \omega^k$. In general, there is no effective enumeration of recursive predicates on $\omega^\omega \times \omega$.

The hierarchy Σ^0_n of arithmetical predicates is defined by induction on $n < \omega$.

Definition 1.1.3.
(i) $\Sigma^0_1 = \{P \mid P \text{ is a partial recursive predicate}\}$.
(ii) P is Π^0_n if $\{(x, m) | \neg P(x, m)\}$ is Σ^0_n.
(iii) P is Σ^0_{n+1} if there is a Π^0_n-predicate $R(x, \langle m, k\rangle)$ such that

$$P(x, m) \Leftrightarrow (\exists k)R(x, \langle m, k\rangle)\,.$$

(iv) P is Δ^0_n if it is Σ^0_n and Π^0_n.

We say that a predicate P is *arithmetical* if $P \in \Sigma^0_n$ for some n.

It follows from results in classical recursion theory that a predicate $P(x, m)$ is Σ^0_1 if and only if there is a recursive predicate $R(x, m, n)$ such that $P(x, m) \Leftrightarrow (\exists n)R(x, m, n)$.

Fix an effective enumeration of partial recursive functions $\{\Phi_e(\sigma, n)\}_{e<\omega}$ which satisfies property (i) in Definition 1.1.1. For each real x, $\Phi_e(x, n)$ denotes $\Phi_e(x \upharpoonright m, n)$ for the least m such that $\Phi_e(x \upharpoonright m, n)\downarrow$ at stage m. If no such m exists, then $\Phi_e(x, n)$ is undefined. We say that $\Phi_e(x, n)$ is an x-partial recursive function. Hence there is a uniform enumeration of partial recursive functions $\{\Phi_e(x, n)\}_{e<\omega}$ with oracle x. In other words, the predicate $P(e, x, n, m) \Leftrightarrow \Phi_e(x, n)\downarrow= m$ is Σ^0_1.

There is an obvious generalization of the above to allow parameters in ω^ω, yielding the class of Σ^0_n-predicates relative to a real or a finite set of reals.

Definition 1.1.4. Let $a \in \omega^\omega$.
(i) A predicate $P^a(y, m)$ is $\Sigma^0_1(a)$ if there is a partial recursive predicate $R(a, y, m)$ such that for all y and m, $P^a(y, m) \Leftrightarrow R(a, y, m)$.
(ii) $P^a(y, m)$ is $\Pi^0_n(a)$ if $\neg P^a(y, m)$ is $\Sigma^0_n(a)$.
(iii) $P^a(y, m)$ is $\Sigma^0_{n+1}(a)$ if there is a $\Pi^0_n(a)$-predicate $R^a(y, m, k)$ such that $P^a(y, m)$ if $(\exists k)R^a(y, m, k)$.
(iv) $P^a(y, m)$ is $\Delta^0_{n+1}(a)$ if it is $\Sigma^0_{n+1}(a)$ and $\Pi^0_{n+1}(a)$.

Given a collection of predicates Γ, we say that a set $a \subseteq \omega$ (resp. $P^* \subseteq \omega^\omega$) is a Γ-set, or a member of Γ, if $m \in a \Leftrightarrow P(\emptyset, m)$ (resp. $x \in P^* \Leftrightarrow P(x, 0)$) for some predicate $P(x, m) \in \Gamma$. For each $n < \omega$, let Σ^0_n, Π^0_n and Δ^0_n denote respectively the class of Σ^0_n, Π^0_n and Δ^0_n-predicates. The following lemma is well known. A proof may be found in [53].

Lemma 1.1.5.
(i) *Let $n < \omega$. There is a Σ^0_n-set $U \subseteq \omega^\omega \times \omega^\omega$ such that for each $P \in \Sigma^0_n$, $P = \{x \mid (x, y) \in U\}$ for some y.*
(ii) *For each n, $\Sigma^0_n \subset \Delta^0_{n+1} \subset \Sigma^0_{n+1}$ and $\Pi^0_n \subset \Delta^0_{n+1} \subset \Pi^0_{n+1}$.*

In classical recursion theory, a set $a \subseteq \omega$ is Σ^0_{n+1} if and only if a is $\Sigma^0_1(\emptyset^n)$. However, it is not generally true for sets of reals. For example, the set $\{\emptyset^{(\omega)}\}$ is Σ^0_3 but is not $\Sigma^0_1(x)$ (as a singleton) for any $x \in \omega^\omega$ (see Exercise 1.1.5).

1.1.2 Analytical predicates

Definition 1.1.6.
(i) A predicate $P(x, m)$ is Σ^1_0 if it is Σ^0_1;
(ii) A predicate $P(x, m)$ is Π^1_n if $\neg P(x, m)$ is Σ^1_n.
(iii) A predicate $P(x, m)$ is Σ^1_{n+1} if there is a Π^1_n-predicate $R(x, y, m) \subseteq \omega^\omega \times \omega^\omega \times \omega$ such that $P(x, m)$ if and only if $(\exists y)R(x, y, m)$.
(iv) A predicate $P(x, m)$ is Δ^1_n if it is both Π^1_n and Σ^1_n.
(v) A predicate P is *analytical* if $P \in \Sigma^1_n$ for some n.

As in Lemma 1.1.5, we have the following facts which are straightforward to verify.

Lemma 1.1.7.
(i) *Let $n < \omega$. There is a Σ^1_n-set $U \subseteq \omega^\omega \times \omega^\omega$ such that for each $P \in \Sigma^1_n$, $P = \{x \mid (x, y) \in U\}$ for some y.*
(ii) *For each n, $\Sigma^1_n \subset \Delta^1_{n+1} \subset \Sigma^1_{n+1}$ and $\Pi^1_n \subset \Delta^1_{n+1} \subset \Pi^1_{n+1}$.*
(iii) *Every arithmetical set is Δ^1_1.*

More basic facts about analytical sets may be found in Jech [53]. For $\sigma \in \omega^{<\omega}$, let $[\sigma]$ denote the set of functions in ω^ω which extend σ. We say that a function $f : \omega^\omega \to \omega^\omega$ is in the class Γ if $\{(x, \sigma) \in \omega^\omega \times \omega^{<\omega} \mid x \in f^{-1}([\sigma])\} \in \Gamma$. Note that every Σ^0_1-function is continuous.

The following proposition is immediate.

Proposition 1.1.8.
(i) *If $f : \omega \to \omega$ is total, then $f \in \Sigma^0_n \Leftrightarrow f \in \Delta^0_n$.*
(ii) *A function $f : \omega^\omega \to \omega^\omega$ is Σ^1_n if and only if its* graph $G^f = \{(x, y) \mid f(x) = y\}$ *is Σ^1_n; if f is Π^1_n then G^f is Π^1_n; if f is Δ^1_n then G^f is Δ^1_n.*
(iii) *If f is total, then G^f is Δ^1_n if and only if f is Δ^1_n.*
(iv) *If $f : \omega^\omega \to \omega^\omega$ is total, then $f \in \Sigma^1_n \Leftrightarrow f \in \Pi^1_n \Leftrightarrow f \in \Delta^1_n$.*

Proof. (i) is obvious. For (ii), note that $(x, y) \in G^f$ if and only if $(\forall\sigma)(y \succ \sigma \to x \in f^{-1}([\sigma]))$. Hence if f is Σ^1_n (resp. Π^1_n) then G^f is Σ^1_n (resp. Π^1_n). Moreover, if G^f is Σ^1_n, then $x \in f^{-1}([\sigma])$ if and only if $(\exists y)(y \succ \sigma \wedge (x, y) \in G^f)$. So f is Σ^1_n if and only if its graph $G^f = \{(x, y) \mid f(x) = y\}$ is Σ^1_n.

(iii). If f is total, then $x \in f^{-1}([\sigma])$ if and only if $(\forall\tau)(\tau \mid \sigma \to x \notin f^{-1}([\tau]))$.

(iv). If f is total, then $x \in f^{-1}([\sigma])$ if and only if $(\forall y)(y \not\succ \sigma \to (x, y) \notin G^f)$. So if G^f is Σ^1_n, then f is Π^1_n. By (iii), f is Δ^1_n. □

Given σ_i and σ'_i in $\omega^{<\omega}$, where $i \leq n$, we say that $(\sigma'_0, \dots, \sigma'_n)$ is a substring of $(\sigma_0, \dots, \sigma_n)$ if $\sigma'_i \preceq \sigma_i$ for each i. A tree $T \subseteq \omega^{<\omega} \times \cdots \times \omega^{<\omega}$ (n-fold product) is assumed to have the property that if $(\sigma_0, \dots, \sigma_n) \in T$ then all of its substrings are in T. For our purpose, all trees grow downwards with root at the top. The following tree representation theorem provides a useful picture of Σ^1_1-sets of reals.

Proposition 1.1.9. *A predicate $P(x, m)$ is Σ^1_1 if and only if there is a recursive sequence of recursive trees $\{T_m\}_{m<\omega}$ such that $T_m \subseteq \omega^{<\omega} \times \omega^{<\omega}$ and*

$$P(x, m) \Leftrightarrow (\exists y)(\forall n)((x \upharpoonright n, y \upharpoonright n) \in T_m)\,.$$

Proof. A predicate $P(x, m)$ is Σ^1_1 if and only if there is a partial recursive function $\Phi(\sigma, \tau, m)$ which satisfies properties (i)–(iii) of Definition 1.1.1, so that $P(x, m) \Leftrightarrow (\exists y)(\forall k)(\Phi(x \upharpoonright k, y \upharpoonright k, m)\downarrow \to \Phi(x \upharpoonright k, y \upharpoonright k, m) \neq 0)$. At stage $\max\{|\sigma|, |\tau|\}$, put (σ, τ) in T_m if and only if $\Phi(\sigma, \tau, m)\uparrow$ or $\Phi(\sigma, \tau, m) \neq 0$. We leave it to the reader to verify that $\{T_m\}_{m\in\omega}$ satisfies the requirement. □

1.1.3 Σ^1_1-sets of reals

A set of reals is Σ^1_1 if there is a Σ^1_1-predicate P (possibly with a number parameter m) such that $P(x, m)$ holds if and only if x belongs to the set. A set of reals is Π^1_1 if its complement is Σ^1_1. Both Σ^1_1 and Π^1_1-sets of reals play significant roles in higher recursion theory. In this section, we introduce some basic facts about these sets.

Recall that $P \subseteq \omega^\omega$ is *perfect* if it is a closed set with no isolated point. The next theorem shows that every uncountable Σ^1_1-set of reals has what may be described as the *perfect set property*.

Theorem 1.1.10 (Suslin [162]). *If P is Σ^1_1 and uncountable, then it contains a perfect subset.*

Proof. Suppose $P(x)$ is Σ^1_1 (for simplicity assume that there is no number variable defining P). We give two proofs of Theorem 1.1.10. One is set-theoretic while the other is recursion-theoretic.

Set-theoretic proof. By Proposition 1.1.9, there is a recursive tree $T \subseteq \omega^{<\omega} \times \omega^{<\omega}$ such that $x \in P$ if and only if $(\exists y)(\forall n)(x \upharpoonright n, y \upharpoonright n) \in T$. Let T' be a subtree of T such that

$$(\sigma, \tau) \in T' \Leftrightarrow |\{x \succ \sigma \mid (\exists y \succ \tau)(\forall n)(x \upharpoonright n, y \upharpoonright n) \in T\}| > \aleph_0.$$

Define

$$Q = \{x \mid (\exists y)(\forall n)(x \upharpoonright n, y \upharpoonright n) \in T'\}.$$

Since P is uncountable, Q is not empty. Obviously $Q \subseteq P$. Note that if $(\sigma, \tau) \in T'$ then there exist k and k' such that $(\sigma^\frown k, \tau^\frown k') \in T'$. We show that Q contains a perfect set. By induction on n, build a perfect tree S_0 and a tree S_1 with the following properties:

(i) $S_1 \subseteq T'$;

(ii) let $[S_0]$ denote the set of infinite paths in S_0. Then $[S_0]$ is contained in $\{x \mid (\exists y)(\forall n)(x \upharpoonright n, y \upharpoonright n) \in S_1\}$;

(iii) $(\forall(\sigma, \tau) \in S_1)(|\{x \succ \sigma \mid (\exists y \succ \tau)(\forall n)((x \upharpoonright n, y \upharpoonright n) \in T)\}| > \aleph_0)$.

We begin with setting S_0 and S_1 to be empty at level 0. Suppose we are at level $n+1$. Let $\sigma \in S_0$ be defined at level n with $|\sigma| = n$. By induction, there is a τ such that $(\sigma, \tau) \in S_1$ was defined at level n and $Q_{\sigma,\tau} = \{x \succ \sigma \mid (\exists y \succ \tau)(\forall n)((x \upharpoonright n, y \upharpoonright n) \in T)\}$ is uncountable.

Claim. *There exist incompatible extensions σ_0, σ_1 of σ such that $Q_{\sigma_0,\tau}$ and $Q_{\sigma_1,\tau}$ are both uncountable.*

If the Claim is false, then there is a unique $x \succ \sigma$ such that if $\sigma' \succ \sigma$ is incompatible with x, then $Q_{\sigma',\tau}$ is countable. But this would imply that $Q_{\sigma,\tau} = \{x\} \cup \bigcup_{\sigma' \succ \sigma \wedge \sigma' \nprec x} Q_{\sigma',\tau}$ is countable, which is a contradiction.

Fix (σ_0, σ_1) as in the Claim. Then there exist $\tau_0, \tau_1 \succ \tau$ such that $|\{x \succ \sigma_i \mid (\exists y \succ \tau_i)(\forall n)((x \upharpoonright n, y \upharpoonright n) \in T)\}| > \aleph_0$ for $i = 0, 1$. Put σ_0 and σ_1 into S_0 and (σ_i, τ_i), for $i = 0, 1$, into S_1. This procedure is carried out for every $\sigma \in S_0$. Clearly S_0 is a perfect tree and for each $x \in [S_0]$ there is a y such that $(\forall n)(x \upharpoonright n, y \upharpoonright n) \in S_1$. Since $S_1 \subseteq T' \subseteq T$, $[S_0] \subseteq Q \subseteq P$.

Recursion-theoretic proof. As above, we have a recursive tree $T \subseteq \omega^{<\omega} \times \omega^{<\omega}$ such that $x \in P$ if and only if $(\exists y)(\forall n)(x \upharpoonright n, y \upharpoonright n) \in T$. We perform a Cantor–Bendixon style operation on the tree as follows:

At stage 0, let $T_0 = T$.
At limit stage α, let $T_\alpha = \bigcap_{\beta<\alpha} T_\beta$.
At successor stage $\alpha + 1$, let

$$T_{\alpha+1} = \{(\sigma, \tau) \in T_\alpha \mid (\exists \sigma_0 \succ \sigma)(\exists \tau_0 \succ \tau)(\exists \sigma_1 \succ \sigma)(\exists \tau_1 \succ \tau)((\sigma_0, \tau_0) \in T_\alpha \wedge (\sigma_1, \tau_1) \in T_\alpha \wedge \sigma_0 | \sigma_1)\}.$$

Observe that $T_\alpha \subseteq T_\beta$ for all $\alpha \geq \beta$. Moreover, for any successor ordinal $\alpha + 1$, if $T_{\alpha+1} \neq T_\alpha$, then there is a $(\sigma_\alpha, \tau_\alpha) \in T_\alpha \setminus T_{\alpha+1}$. Note that $(\sigma_\alpha, \tau_\alpha) \neq (\sigma_\beta, \tau_\beta)$ for any $\alpha \neq \beta$ if these are defined. Since the set $\omega^{<\omega} \times \omega^{<\omega}$ is countable, there is a least countable ordinal γ such that $(\sigma_\gamma, \tau_\gamma)$ does not exist. Then $T_\gamma = T_{\gamma+1}$ and so $T_\gamma = T_\alpha$ for any $\alpha \geq \gamma$. By replacing T' in the set-theoretic proof above with T_γ, we conclude that P contains a perfect subset. □

Remark 1.1.11.

(i) The proofs above raise two interesting questions. First, what is the complexity of the perfect subset obtained? Since the statement, "P is uncountable" is not second-order definable in arithmetic, the perfect subset defined is generally neither Σ^1_1 nor Π^1_1. However, we will show later by a more powerful recursion-theoretic argument that its complexity can be controlled (see Exercise 1.5.5). Second, what is the least ordinal γ for which $T_\gamma = T_{\gamma+1}$? We will give a calculation of this ordinal later (see Proposition 3.6.10).

(ii) In general, not every uncountable set has the perfect set property. In fact, "Σ^1_1-ness" is the best possible since it is not provable in ZFC that every Π^1_1-set has the perfect set property (see Corollary 4.3.6).

Proposition 1.1.12. *Suppose $\{P_i\}_{i<\omega}$ is a sequence of sets of reals such that for any nonempty Σ^1_1-set A and $i < \omega$, $A \cap P_i$ contains a nonempty Σ^1_1-set. Let P be Σ^1_1 and nonempty. Then $P \cap \bigcap_{i<\omega} P_i \neq \emptyset$.*

Proof. By hypothesis, there is a nonempty Σ^1_1-set $F_0 \subseteq P \cap P_0$. By Proposition 1.1.9, there is a recursive tree $T \subseteq \omega^{<\omega} \times \omega^{<\omega}$ such that $x \in F_0 \Leftrightarrow (\exists y)(\forall n)((x \upharpoonright n, y \upharpoonright n) \in T)$. Fix such a pair (x, y) so that $(\forall n)(x \upharpoonright n, y \upharpoonright n) \in T$. Define $\sigma_0 = x \upharpoonright 1$ and $\tau_0 = y \upharpoonright 1$. Note that $G_0 = \{x \succ \sigma_0 \mid (\exists y \succ \tau_0)(\forall n)(x \upharpoonright n, y \upharpoonright n) \in T)\}$ is a nonempty Σ^1_1-set. Assume by induction that for all $i' \leq i$, a pair $(\sigma_{i'}, \tau_{i'}) \in T$ and a nonempty Σ^1_1-set $G_{i'} \subseteq P_{i'}$ are defined such that

(1) $(\forall i' < i)[(\sigma_{i'} \prec \sigma_{i'+1}) \wedge (\tau_{i'} \prec \tau_{i'+1})]$
(2) $(\forall i' \leq i)(G_{i'} \subseteq P_{i'})$
(3) $(\forall i' < i)[G_{i'} \supseteq G_{i'+1} \wedge (\forall x \in G_{i'})(x \succ \sigma_{i'} \wedge (\exists y \succ \tau_{i'})(\forall n)((x \upharpoonright n, y \upharpoonright n) \in T)]$.

Since G_i is Σ^1_1, there is a nonempty Σ^1_1-set $F_{i+1} \subseteq P_{i+1} \cap G_i$. Furthermore, by (3) every $x \in F_{i+1}$ extends σ_i and there is a $y \succ \tau_i$ such that for all n, $(x \upharpoonright n, y \upharpoonright n) \in T$. Fix such a pair (x, y). Define $\sigma_{i+1} = x \upharpoonright (|\sigma_i| + 1)$ and $\tau_{i+1} = y \upharpoonright (|\tau_i| + 1)$. Let

$$G_{i+1} = \{x \in F_{i+1} \mid x \succ \sigma_{i+1} \wedge (\exists y \succ \tau_{i+1})(\forall n)((x \upharpoonright n, y \upharpoonright n) \in T)\}.$$

Then G_{i+1} is a nonempty Σ^1_1-subset of F_{i+1} and hence of P_{i+1}. Let $x(i) = m$ if and only if $\sigma_{i+1}(i) = m$ and $y(i) = m$ if and only if $\tau_{i+1}(i) = m$. Then for all i, $x \in G_i$ and for all n, $(x \upharpoonright n, y \upharpoonright n) \in T$. Hence $x \in P \cap \bigcap_{i<\omega} P_i$. □

As a consequence of Proposition 1.1.12, we may define the Gandy topology as follows.

Definition 1.1.13. The *Gandy topology* on ω^ω is the topology generated by taking all Σ^1_1-sets as basic open sets.

Proposition 1.1.12 implies that the Gandy topology satisfies the Baire category theorem: The intersection of a countable collection of dense open sets is dense.

1.1.4 Π^1_1-sets of reals

By Proposition 1.1.9, we have the following:

Proposition 1.1.14. *A predicate $P(x, m)$ is Π^1_1 if and only if there is a recursive sequence of recursive trees $T_m \subseteq \omega^{<\omega} \times \omega^{<\omega}$ such that $P(x, m) \Leftrightarrow (\forall y)(\exists n)((x \upharpoonright n, y \upharpoonright n) \notin T_m)$.*

For each real x, define a binary relation $\leq_x$ on ω such that $m \leq_x n$ if and only if $x(\langle m, n\rangle) = 1$. Define

$$\mathsf{WF}_1 = \{x \in \omega^\omega \mid \leq_x \text{ is a wellfounded relation}\}.$$

Then $x \in \mathsf{WF}_1$ if and only if

$$(\forall f \in \omega^\omega)[(\forall n)(\forall m)(n \geq m \leftrightarrow f(n) \leq_x f(m)) \to$$
$$(\exists m)(\forall n)(\forall k)(n > m \wedge k > m \to f(n) = f(k))],$$

so that WF_1 is a Π^1_1-set of reals. Fix a standard effective enumeration $\{\sigma_n\}_{n<\omega}$ of $\omega^{<\omega}$. We say that x codes a tree relation if

$$m \leq_x n \Leftrightarrow \sigma_m \succeq \sigma_n.$$

Now let P be a Π^1_1-set of reals and let T be a recursive tree such that $x \in P \Leftrightarrow (\forall y)(\exists n)((x \upharpoonright n, y \upharpoonright n) \notin T)$. Define $T_x = \{\sigma \mid (x \upharpoonright |\sigma|, \sigma) \in T\}$. Then $x \in P$ if and only if T_x is wellfounded. Define the *Kleene–Brouwer ordering* $<_{\mathrm{KB}}$ on strings so that for any $\sigma, \tau \in \omega^{<\omega}$, $\sigma <_{\mathrm{KB}} \tau$ if and only if

$$\sigma \succ \tau \vee (\exists i \leq \min\{|\sigma|, |\tau|\})(\sigma \upharpoonright i - 1 = \tau \upharpoonright i - 1 \wedge \sigma(i) < \tau(i)).$$

In other words, $\sigma <_{\mathrm{KB}} \tau$ whenever either σ extends τ or σ is "to the left" of τ. Then for each x, x codes a wellfounded tree if and only if x codes a well-ordering under $<_{\mathrm{KB}}$.

Let $T \subseteq \omega^{<\omega} \times \omega^{<\omega}$ be a recursive tree which witnesses P being Π^1_1. Thus $x \in P$ if and only if $(\forall y)(\exists n)(x \restriction n, y \restriction n) \notin T$. Define $f(x) = z$ if and only if z is the unique real coding the tree T_x. Obviously f is a well-defined recursive function. Furthermore, $P(x)$ if and only if $f(x) \in \mathsf{WF}_1$.

Define

$$\mathsf{WO}_1 = \{x \mid x \text{ codes a well-ordering}\}.$$

Then $x \in \mathsf{WO}_1$ if and only if $x \in \mathsf{WF}_1 \wedge (\forall n)(\forall m)(n \leq_x m \vee m \leq_x n)$. Hence WO_1 is Π^1_1. Then for each Π^1_1-set P, there is a recursive function f such that $x \in P$ if and only if $f(x) \in \mathsf{WO}_1$.

Let $\{R_e\}_{e<\omega}$ be an effective enumeration of recursively enumerable (r.e.) binary relations. Then the set

$$\mathsf{WF}_0 = \{e \mid R_e \text{ is wellfounded}\}$$

is Π^1_1. Let $x \subseteq \omega$ be a Π^1_1-set and $T \subseteq \omega^{<\omega} \times \omega$ be a recursive tree such that $n \in x$ if and only if $(\forall y)(\exists m)(y \restriction m, n) \notin T$. Clearly there is a recursive function f such that for any n, $T_n = \{\sigma \mid (\sigma, n) \in T\} = R_{f(n)}$. Then $n \in x$ if and only if $R_{f(n)}$ is a wellfounded binary relation. In other words, x is many-one reducible to WF_0. By the same argument above, every Π^1_1-set of numbers is many-one reducible to the Π^1_1-set $\mathsf{WO}_0 = \{e \mid R_e \text{ is a well-ordering}\}$.

Note that the ordinal height of WF_1 may be as high as ω_1. An interesting question is the ordinal height of WF_0. In other words, what is the least ordinal which is not the ordinal of a wellfounded recursive tree? We shall see that this ordinal, denoted ω_1^{CK}, is a countable ordinal.

The following proposition provides an interesting connection between Σ^1_1-sets of reals and Π^1_1-subsets of ω.

Proposition 1.1.15 (Spector [153]). *Suppose P is a Σ^1_n-set of reals. Then*
(i) *$\bigcap_{x \in P} x$ is a Π^1_n-set of integers.*
(ii) *If P has a unique element x, then x is Δ^1_n.*

Proof. For (i): $n \in \bigcap_{x \in P} x \Leftrightarrow (\forall x)(P(x) \to n \in x)$.
For (ii): $n \in x \Leftrightarrow (\forall y)(P(y) \to n \in y) \leftrightarrow (\exists y)(P(y) \wedge n \in y)$. □

Proposition 1.1.15 is very useful. First, (i) says that one obtains a Π^1_1-set by taking the intersection of a Σ^1_1-condition. Second, (ii) provides a link between sets of numbers and sets of reals. The unique real in (ii) is called a Σ^1_n-singleton. By (ii), each Π^1_n-singleton is Δ^1_{n+1}. As we will see (cf. Proposition 1.2.5), this is the best possible. In general, the link between sets of numbers and sets of reals in terms of singleton sets is weak. For example, it will be shown that a Π^0_2-singleton may itself have a definition much more complex than Π^0_2 (see also Exercise 1.1.5).

1.1.5 Relativization

As in classical recursion theory, all the results above may be relativized to a real x. We use boldface symbols with tilde $\undertilde{\Sigma}^0_n$, $\undertilde{\Pi}^0_n$, $\undertilde{\Delta}^0_n$, $\undertilde{\Sigma}^1_n$, $\undertilde{\Pi}^1_n$ and $\undertilde{\Delta}^1_n$ to denote the corresponding sets (of numbers or reals) relativized to a real. A set is *projective* if it belongs to $\undertilde{\Pi}^1_n$ for some n.

1.1.6 Exercises

Exercise 1.1.1. *Prove that a set $P \subseteq 2^\omega$ is Δ^0_1 if and only if it is defined by a recursive predicate.*

Exercise 1.1.2. *Prove that there is a Π^0_1-total function which is not Δ^0_1.*

Exercise 1.1.3. *Prove that every closed uncountable set $P \subseteq \omega^\omega$ can be decomposed into the union of a countable set and a perfect set.*

Exercise 1.1.4. *Prove that every uncountable set $P \subseteq \omega^\omega$ can be decomposed into the union of a countable set and a set without isolated members.*

Exercise 1.1.5. *Prove that $\emptyset^{(\omega)} \in 2^\omega$ is a Π^0_2-singleton but not a $\Pi^0_1(x)$-singleton for any $x \not\geq_T \emptyset^{(\omega)}$.*

1.2 Ordinal notations

A key goal in descriptive set theory is to investigate properties of real numbers and sets of real numbers within the system ZFC. However, many statements about complexity classes above the Π^1_1-level are independent of ZFC. Hence it is natural to study in greater depth properties of Π^1_1-sets. Higher recursion theory provides an ideal platform as well as powerful tools for this purpose. In this section, we introduce the notion of a recursive ordinal and an ordinal notation which is at the heart of analysis of Π^1_1-sets. The recursive ordinals allow one to carry out effective transfinite induction over the ordinal ω_1^{CK} similar to what is done over ω_1 for transfinite induction.

1.2.1 Kleene's $\mathcal{O}$

Let $\langle , \rangle : \omega \times \omega \to \omega$ be a recursive bijection. Define an arithmetical set $P \subseteq 2^\omega$ such that $x \in P$ if and only if x satisfies the following conditions:

(i) $\langle 1, 2\rangle \in x$;

(ii) $(\forall m)(\forall n)(\langle m, n\rangle \in x \to \langle n, 2^n\rangle \in x)$;

(iii) $(\forall e)((\forall n)\Phi_e(n)\downarrow \wedge (\forall n)\langle \Phi_e(n), \Phi_e(n+1)\rangle \in x) \to (\forall n)(\langle \Phi_e(n), 3\cdot 5^e\rangle \in x))$;

(iv) $(\forall n)(\forall m)(\forall k)(\langle n, m\rangle \in x \wedge \langle m, k\rangle \in x \to \langle n, k\rangle \in x)$.

Thus (i) implies that x is nonempty; (ii) says that x is "closed under successor", and (iii) that it is "closed under limit"; (iv) asserts that x defines a transitive relation. We introduce the following notations.

Definition 1.2.1. $<_o = \bigcap_{x \in P} x$.

$\mathcal{O} = \{n \mid (\exists m)(n <_o m)\}$.

By Proposition 1.1.15, $<_o$ is a Π^1_1-relation and $\mathcal{O}$ is Π^1_1. We define a set a by induction:

Stage 0. Put $\langle 1, 2\rangle$ into a_0.

Stage $\beta + 1$. Put $\langle n, 2^n\rangle$ and $\langle m, 2^n\rangle$ into $a_{\beta+1}$ if $\langle m, n\rangle \in a_\beta$ at stage β.

Stage λ, for λ a limit ordinal. For each e such that Φ_e is total and $(\forall n)(\exists \beta < \lambda)(\langle \Phi_e(n), \Phi_e(n+1)\rangle \in a_\beta)$, put $\langle \Phi_e(n), 3\cdot 5^e\rangle$ into a_λ for each n. Furthermore, if $\langle m, \Phi_e(n)\rangle \in a_\beta$ for some $\beta < \lambda$, then $\langle m, 3 \cdot 5^e\rangle \in a_\lambda$.

Let $a = \bigcup_\beta a_\beta$.

It is straightforward to verify that $a \in P$ and for each $x \in P$, $a \subseteq x$. Hence $a = <_o$.

Define a function $|\cdot| : \mathcal{O} \mapsto \omega_1$ such that $|1|=0$ and $|n|$ is the least ordinal β for which there is a number m with $\langle m, n\rangle \in a_\beta$. If $n \in \mathcal{O}$, then n is said to be a notation for the ordinal $|n|$. An ordinal β is *constructive* if $\beta = |n|$ for some $n \in \mathcal{O}$. Obviously $m <_o n$ implies $|m| < |n|$. Hence $<_o$ is a wellfounded partial ordering.

Definition 1.2.2. ω_1^{CK} = the least ordinal α greater than $|n|$ for every $n \in \mathcal{O}$.

By induction on ordinals $\beta < \omega_1^{\mathrm{CK}}$, it is easy to see that for each $n \in \mathcal{O}$, $\{m \mid m <_o n\}$ is a linear ordering under $<_o$.

Remark 1.2.3.

(i) A path $x \subseteq \mathcal{O}$ is a set such that

$$(\forall m, n \in x)(n <_o m \vee m <_o n \vee m = n) \wedge$$
$$(\forall m)(\forall n \in x)(m <_o n \to m \in x).$$

An interesting question is whether every path can be extended to one whose ordinal height is ω_1^{CK}. Such a path is necessary for carrying out effective transfinite induction. However, this is in general not true. Note that there are many possible choices of notations at a limit stage β below ω_1^{CK}. If every path up to β were extendible to one of ordinal height ω_1^{CK}, then there would be $2^{\aleph_0}$ many sets of notations. Hence to obtain a path through $<_o$, one must choose the notations judiciously. We will show that there exists a "well-behaved" Π^1_1-path through $<_o$.

(ii) It is not difficult to see that the set of constructive ordinals forms an initial segment, and hence is equal to ω_1^{CK}.

1.2.2 Relativized $\mathcal{O}$

As in § 1.2.1, one may relativize the notion of ordinal notations to any real $x \in \omega^\omega$. Let $\{\Phi_e(x,n)\}_{e<\omega}$ be an effective enumeration of partial recursive functions with oracle x (cf. § 1.1.1). Define an arithmetical set $P \subseteq \omega^\omega \times 2^\omega$ such that $(x,y) \in P$ if and only if y satisfies the following conditions:

(i) $\langle 1,2\rangle \in y$;

(ii) $(\forall m)(\forall n)(\langle m,n\rangle \in y \to \langle n,2^n\rangle \in y)$;

(iii) $(\forall e)((\Phi_e(x,\cdot)$ is total $\wedge (\forall n)\langle \Phi_e(x,n),\Phi_e(x,n+1)\rangle \in y) \to$
$(\forall n)(\langle \Phi_e(x,n),3\cdot 5^e\rangle \in y))$, and

(iv) $(\forall n)(\forall m)(\forall k)(\langle n,m\rangle \in y \wedge \langle m,k\rangle \in y \to \langle n,k\rangle \in y)$.

Definition 1.2.4. $<_{\mathcal{O}^x} = \bigcap_{(x,y)\in P} y$.
$\mathcal{O}^x = \{n \mid (\exists m)(n <_{\mathcal{O}^x} m)\}$.

One may also let relativized $a^x = \bigcup_\beta a^x_\beta$. Then $a^x = <_{\mathcal{O}^x}$ and the function $|\cdot|_x : \mathcal{O}^x \mapsto \omega_1$ is a well-defined rank function so that $|1|_x=0$ and $|n|_x$ is the least ordinal β for which there is a number m with $\langle m,n\rangle \in a^x_\alpha$.

Note that there is a subtle difference between $\mathcal{O}$ and $\mathcal{O}^\emptyset$: An enumeration of oracle functionals is used to define $\mathcal{O}^\emptyset$. Hence the set of indices used in the definition of $\mathcal{O}^\emptyset$ may be different from that of $\mathcal{O}$, resulting in possibly different notations at limit stage. However, such a difference does not affect the overall properties of $\mathcal{O}$. Indeed, one may redefine $\mathcal{O}$ as $\mathcal{O}^\emptyset$ if desired.

By the relativized version of Proposition 1.1.15, both $<_{\mathcal{O}^x}$ and $\mathcal{O}^x$ are $\Pi^1_1(x)$. The following proposition is immediate.

Proposition 1.2.5.
(i) *The sets* $\{(n,x) \mid n \in \mathcal{O}^x\}$ *and* $\{(\langle m,n\rangle,x) \mid m <_{\mathcal{O}^x} n\}$ *are* Π^1_1.
(ii) *The sets* $\{(x,y) \mid y = <_{\mathcal{O}^x}\}$ *and* $\{(x,y) \mid y = \mathcal{O}^x\}$ *are* Π^1_1.

Proof. Let P be the set of reals defined before Definition 1.2.4. For (i), $n \in \mathcal{O}^x$ if and only if $(\exists m)\langle n,m\rangle \in <_{\mathcal{O}^x}$ if and only if $(\exists m)(\forall y)((x,y) \in P \to \langle n,m\rangle \in y)$.

For (ii), note that $y = <_{\mathcal{O}^x}$ if and only if $(x,y) \in P$ and $(\forall z)((x,z) \in P \to (\forall n)(n \in y \to n \in z))$ which is Π^1_1. We leave the other part of (ii) as an exercise. □

Proposition 1.2.5 has some interesting consequences. Consider for example the set $O = \{\mathcal{O}\}$. Then $x \in O$ if and only if $(\emptyset,x) \in \{(x,y) \mid y = \mathcal{O}^x\}$. Hence O is a Π^1_1-singleton whose unique member is also Π^1_1.

As in § 1.2.1, one may define the notion of an x-constructive ordinal and denote the least non-x-constructive ordinal as ω^x_1. Note that at limit stage, different oracle sets x may use different indices e for the notation. Hence $\mathcal{O}^x$ may be different for different x's. Moreover, for x's with “sufficient computational power”, one may have $\omega^x_1 > \omega^{\mathrm{CK}}_1$. We will present a characterization of this property in due course.

1.2.3 Exercise

Exercise 1.2.1. *Complete the proof of Proposition* 1.2.5.

1.3 Effective transfinite induction

1.3.1 Recursion theorem

The motivation for introducing recursive ordinals is to perform effective transfinite induction. We make it precise in this section, by first recalling the following theorem famously known as the Recursion Theorem (also called the Kleene Fixed Point Theorem).

Theorem 1.3.1 (Kleene). *Suppose $f : \omega \to \omega$ is recursive. Then $\Phi_{f(e)} = \Phi_e$ for some e.*

Proof. By the s-m-n Theorem, there is a recursive function g such that $\Phi_{g(e)} = \Phi_{\Phi_e(e)}$ for each e. Let i be the index of the recursive function $f \circ g$. Then $\Phi_{f(g(i))} = \Phi_{\Phi_i(i)} = \Phi_{g(i)}$. □

The Recursion Theorem is a powerful tool in the study of effective computability in classical recursion theory. In higher recursion theory, it is used for effective transfinite induction. The following proposition is the first illustration of such an application.

Proposition 1.3.2 (Sacks [123]). *Let $<_R$ be an ordering relation on ω with a minimal element. Suppose f is a recursive function such that for all $e \in \omega$:*

(i) *For each n in the field of $<_R$, $\Phi_e(n)$ is defined whenever $\Phi_{f(e)}(m)$ is defined for all $m <_R n$;*

(ii) *$\Phi_{f(e)}$ is defined on every minimal element of $<_R$.*

Then there is an index i such that $\Phi_i = \Phi_{f(i)}$ and Φ_i is defined on every wellfounded initial segment of R.

Proof. By Theorem 1.3.1, there is an i such that $\Phi_i = \Phi_{f(i)}$. By (i), if n is an $<_R$-minimal element, then $\Phi_i(n)$ is defined. By an easy induction and an application of (ii), Φ_i is defined on every wellfounded initial segment of R. □

In particular, if R is wellfounded then Φ_i is defined on the field of R. Note that no assumption is made on the complexity of R in Proposition 1.3.2. As long as the recursive function f exists, we may find such an index i. In general, some nice properties concerning R are required to guarantee the existence of f.

Intuitively, $\Phi_{f(e)}$ is a "one-step effective induction" operator over Φ_e. For effective transfinite induction, one needs the assumption that Φ_e itself is produced by f through another function $\Phi_{e'}$, i.e. $\Phi_e = \Phi_{f(e')}$. The Recursion Theorem says that this is possible. We shall see many applications of Proposition 1.3.2 to effective transfinite induction. In this chapter we give detailed verifications for some of these. In subsequent chapters, such details will be suppressed and left to the reader.

1.3.2 Operators on ordinal notations

We now define an operator $+_o$ on $\mathcal{O}$. In addition to requiring it to be closed in $\mathcal{O}$ (i.e. $m \in \mathcal{O}$ and $n \in \mathcal{O}$ implies $m +_o n \in \mathcal{O}$), we also require $|m| + |n| = |m +_o n|$.

Let g be a a recursive function such that

$$\Phi_{g(e,m,n)}(k) = \Phi_e(m, \Phi_n(k))$$

for all e, k, m, n. Intuitively, g is a function to be used to handle induction at limit stage.

By the s-m-n Theorem, there is a recursive function h such that

$$\Phi_{h(e)}(m,n) = \begin{cases} m & n = 1, \\ 2^{\Phi_e(m,k)} & n = 2^k, \\ 3 \cdot 5^{g(e,m,k)} & n = 3 \cdot 5^k, \\ 7 & \text{Otherwise.} \end{cases}$$

By the Recursion Theorem, there is a c such that $\Phi_{h(c)} = \Phi_c$. Define

$$m +_o n = \Phi_c(m,n).$$

It is straightforward to verify that

$$m +_o n = \begin{cases} m & n = 1, \\ 2^{m +_o k} & n = 2^k, \\ 3 \cdot 5^{g(c,m,k)} & n = 3 \cdot 5^k, \\ 7 & \text{Otherwise.} \end{cases}$$

Note that

$$\Phi_{g(c,m,k)}(n) = \Phi_c(m, \Phi_k(n)) = m +_o \Phi_k(n)$$

for all n. Obviously $+_o$ is a recursive function on $\omega \times \omega$. The following is proved by induction on $<_o$.

Proposition 1.3.3. *The function $+_o$ satisfies the following properties: For all m, n,*
(i) $m, n \in \mathcal{O} \Leftrightarrow m +_o n \in \mathcal{O}$.
(ii) $m, n \in \mathcal{O} \implies |m +_o n| = |m| + |n|$.
(iii) $m, n \in \mathcal{O}$ *and* $n \neq 1 \implies m <_o m +_o n$.
(iv) $m \in \mathcal{O}$ *and* $k <_o n \Leftrightarrow m +_o k <_o m +_o n$.
(v) $m \in \mathcal{O} \wedge n = k \in \mathcal{O} \Leftrightarrow m +_o n = m +_o k \in \mathcal{O}$.

An interesting application of effective transfinite induction is the following basic result.

Theorem 1.3.4 (Kleene [71]). *There are two recursive functions f and g such that for all* $n \in \mathcal{O}$,

(i) $W_{f(n)} = \{m \mid m <_o n\}$,
(ii) $W_{g(n)} = \{\langle i, j\rangle \mid i <_o j <_o n\}$.

Proof. (i) By the *s-m-n* Theorem, there are two recursive functions p, q such that

$$\begin{aligned} W_{p(e,m)} &= \{m\} \cup W_{\Phi_e(m)}, \\ W_{q(e,m)} &= \bigcup\{W_{\Phi_e(\Phi_m(n))} \mid n \in \omega\}. \end{aligned}$$

Fix e_0 such that $W_{e_0} = \emptyset$. Define a recursive function h as follows:

$$\Phi_{h(e)}(n) = \begin{cases} e_0 & n = 1, \\ p(e,k) & n = 2^k, \\ q(e,k) & n = 3 \cdot 5^k, \\ 0 & \text{Otherwise.} \end{cases}$$

By the Recursion Theorem, there is a constant c such that $\Phi_{h(c)} = \Phi_c$. Set $f = \Phi_c$. Then

$$f(n) = \begin{cases} e_0 & n = 1, \\ p(c,k) & n = 2^k, \\ q(c,k) & n = 3 \cdot 5^k, \\ 0 & \text{Otherwise.} \end{cases}$$

Obviously f is total since both p and q are total. Then

$$W_{f(n)} = \begin{cases} \emptyset & n = 1, \\ \{k\} \cup W_{f(k)} & n = 2^k, \\ \bigcup\{W_{f(\Phi_k(m))} \mid m \in \omega\} & n = 3 \cdot 5^k, \\ W_0 & \text{Otherwise.} \end{cases}$$

An easy induction shows that $W_{f(n)} = \{m \mid m <_o n\}$ for each $n \in \mathcal{O}$.

(ii) By the *s-m-n* Theorem, there are two functions p and q such that

$$\begin{aligned} W_{p(e,m)} &= \{\langle i, m\rangle \mid i \in W_{f(m)}\} \cup W_{\Phi_e(m)}, \\ W_{q(e,m)} &= \bigcup\{W_{\Phi_e(\Phi_m(n))} \mid n \in \omega\}, \end{aligned}$$

where f was defined in (i). Fix e_0 so that $W_{e_0} = \emptyset$. Let

$$\Phi_{h(e)}(n) = \begin{cases} e_0 & n = 1, \\ p(e,k) & n = 2^k, \\ q(e,k) & n = 3 \cdot 5^k, \\ 0 & \text{Otherwise.} \end{cases}$$

By the Recursion Theorem, there is a constant c such that $\Phi_{h(c)} = \Phi_c$. Set $g = \Phi_c$. Hence

$$g(n) = \begin{cases} e_0 & n = 1, \\ p(c,k) & n = 2^k, \\ q(c,k) & n = 3 \cdot 5^k, \\ 0 & \text{Otherwise.} \end{cases}$$

Note that g is total since both p and q are total. Then

$$W_{g(n)} = \begin{cases} \emptyset & n = 1, \\ \{\langle i, k\rangle | i <_o k\} \cup W_{g(k)} & n = 2^k, \\ \bigcup\{W_{g(\Phi_k(m))} \mid m \in \omega\} & n = 3 \cdot 5^k, \\ W_0 & \text{Otherwise.} \end{cases}$$

An easy induction shows that $W_{g(n)} = \{\langle i, j\rangle \mid i <_o j <_o n\}$ for $n \in \mathcal{O}$. □

Theorem 1.3.4 allows one to carry out transfinite induction uniformly for $e \in \mathcal{O}$. Here we give two applications of Proposition 1.3.3.

Lemma 1.3.5. *There is a recursive function g such that for all e:*
(i) $g(e) \in \mathcal{O} \Leftrightarrow W_e \subseteq \mathcal{O}$.
(ii) $g(e) \in \mathcal{O} \Leftrightarrow |n| < |g(e)|$ *for every* $n \in W_e$.

Proof. By the *s-m-n* Theorem, there is a recursive function r such that $\Phi_{r(e)}(0) = 1$ and the range of $\Phi_{r(e)}$ is $(W_e \setminus \{0\}) \cup \{1\}$. By the *s-m-n* Theorem again, there is a recursive function s such that $\Phi_{s(e)}(0) = 1$ and $(\forall n)(\Phi_{s(e)}(n+1) = \Phi_{s(e)}(n) +_o 2^{\Phi_{r(e)}(n+1)})$. Define

$$g(e) = 3 \cdot 5^{s(e)}.$$

Note that if $W_e \subseteq \mathcal{O}$, then $g(e) \in \mathcal{O}$ by Proposition 1.3.3. If $g(e) \in \mathcal{O}$, then $\Phi_{s(e)}(n) \in \mathcal{O}$ for each n. Hence $W_e \subseteq \mathcal{O}$ by Proposition 1.3.3.

Assume $g(e) \in \mathcal{O}$ and $n \in W_e$. Then $n <_o \Phi_{s(e)}(m) +_o 2^n = \Phi_{s(e)}(m+1) <_o 3 \cdot 5^{s(e)} = g(e)$ where m is the least number such that $\Phi_{r(e)}(m + 1) = n$. □

As a consequence of the proof of Lemma 1.3.5, we have

Corollary 1.3.6. *There is a recursive function g such that for all e, if* $W_e \subseteq \mathcal{O}$ *and* $<_o$ *on* W_e *is linear, then* $g(e) \in \mathcal{O}$ *and* $(\forall n)(n \in W_e \to n <_o g(e))$.

Corollary 1.3.6 may be viewed as a Σ^0_1-boundedness principle, i.e. there is an effective way to find an upper bound for any r.e. subset of $\mathcal{O}$. Lemma 1.3.5 will be significantly strengthened in Theorem 1.5.5.

1.4 Recursive ordinals

What are the possible order types of a recursive well-ordering? What is the connection between a recursive well-ordering and a constructive ordinal? We answer these questions in this section.

Definition 1.4.1. An ordinal β is *recursive* if there is a recursive well-ordering R whose order type is β.

Lemma 1.4.2. *Let* $\{R_e\}_{e<\omega}$ *be an effective list of partial recursive binary relations. There exists a recursive function f such that for all e,*

(i) R_e *is wellfounded if and only if* $f(e) \in \mathcal{O}$.
(ii) *If* R_e *is wellfounded, then* $|R_e| \leq |f(e)|$, *where* $|R_e|$ *is the height of* R_e.

Proof. By the *s-m-n* Theorem, there is a recursive function h such that

$$R_{h(e,k)}(m, n) \Leftrightarrow R_e(m, n) \wedge R_e(m, k) \wedge R_e(n, k)$$

for all e, k, m and n. $R_{h(e,k)}$ is the initial segment of R_e preceding k. The idea is to define a one-one, order-preserving map from the field of R_e into $\mathcal{O}$. Fix the recursive function g in Lemma 1.3.5. If k is a minimal element of R_e, then k is mapped to an element in $\mathcal{O}$ determined by applying $W_{h(e,k)} = \emptyset$ in Lemma 1.3.5. Otherwise, by induction $R_{h(e,k)}$ is mapped to an r.e. set $W \subset \mathcal{O}$ with index e'. By Lemma 1.3.5, $g(e') \in \mathcal{O}$ and $|n| < g(e')$ for all $n \in W$. Hence we may define $f(k) = g(e')$.

We now turn to the formal proof. First of all, there is a recursive function p such that

$$W_{p(i,e)} = \begin{cases} \emptyset & R_e = \emptyset \\ \{\Phi_i(h(e,k)) \mid k \in \omega\} & \text{Otherwise.} \end{cases}$$

Hence there is a recursive function q such that $\Phi_{q(i)}(e) = g(p(i, e))$. By the Recursion Theorem, there is a fixed point c such that $\Phi_{q(c)} = \Phi_c$. Define

$$f(e) = \Phi_c(e)$$

and

$$p(e) = p(c, e).$$

Then

$$W_{p(e)} \text{ wellfounded} = \begin{cases} \emptyset & R_e = \emptyset \\ \{\Phi_c(h(e,k)) \mid k \in \omega\} = \{f(h(e,k)) \mid k \in \omega\} & \text{Otherwise.} \end{cases}$$

Note that $f(e) = \Phi_c(e) = \Phi_{q(c)}(e) = g(p(c, e)) = g(p(e))$.

Now suppose R_e is wellfounded. If R_e is empty, then by Lemma 1.3.5, $W_{p(e)} = \emptyset \subset \mathcal{O}$ and $f(e) \in \mathcal{O}$. Assume for all wellfounded relations R_e with $|R_e| < \alpha$, (i) and (ii) hold. Fix a wellfounded relation R_e with $|R_e| = \alpha$. Then for each k in the field of R_e, $|R_{h(e,k)}| < \alpha$. Hence $W_{p(h(e,k))} \subset \mathcal{O}$ and $f(h(e,k)) \in \mathcal{O}$. Thus $W_{p(e)} = \{f(h(e,k)) \mid k \in \omega\} \subset \mathcal{O}$. By Lemma 1.3.5, $f(e) = g(p(e)) \in \mathcal{O}$ and $|f(e)| > |n|$ for all $n \in W_{p(e)}$. Thus $|f(e)| \geq |R_e|$.

Suppose $f(e) \in \mathcal{O}$. By Lemma 1.3.5, $W_{p(e)} \subset \mathcal{O}$ and $|f(e)| > |f(h(e,n))|$ for all n. By induction on $<_o$, $R_{h(e,n)}$ is wellfounded for all n. Hence

$$R_e = \bigcup_n (R_{h(e,n)} \cup \{(m, n) \mid R_e(m, n) \wedge m \in \omega\})$$

is wellfounded. □

We leave the proof of the next lemma to the reader (Exercise 1.5.1):

Lemma 1.4.3. *Every partial recursive well-ordering R is isomorphic to a recursive well-ordering.*

The main result of this section states:

Theorem 1.4.4 (Kleene [71], Markwald [89]). *An ordinal is recursive if and only if it is constructive.*

Proof. Suppose R is a recursive well-ordering with $|R| = \beta$. By Lemma 1.4.2, there is a constructive ordinal $\gamma \geq \beta$. Hence β is constructive. Conversely, if β is constructive, then by Theorem 1.3.4, there is a partial recursive well-ordering R with $|R| = \beta$. By Lemma 1.4.3, β is recursive. □

1.4.1 Exercise

Exercise 1.4.1. *Prove that there is a set $x \subseteq \mathcal{O}$ restricted to which $<_o$ is a well-ordering of order type ω^2, and a partial recursive function $p : \omega^2 \to \omega$ such that*
(i) *For any $n \in x$, the function p_n such that $p_n(i) = p(n, i)$ is total;*
(ii) *for any $n <_o m \in x$, $(\exists i)(\forall j > i)(p_n(j) < p_m(j))$, and*
(iii) *for any recursive function f, there exists an $n \in x$ such that $(\exists i)(\forall j > i)(f(j) < p_n(j))$.*

1.5 Π^1_1-completeness and Σ^1_1-boundedness

1.5.1 Π^1_1-completeness

Theorem 1.5.1. *$\mathcal{O}$ is Π^1_1-complete. Hence $\mathcal{O}$ is not Σ^1_1.*

Proof. By Lemma 1.4.2, WF_0 is many-one reducible to $\mathcal{O}$. As discussed in § 1.1.4, WF_0 is Π^1_1-complete, hence $\mathcal{O}$ as well. □

We now consider sets of reals. Suppose $P(x, n)$ is Π^1_1. By relativizing Theorem 1.5.1, there is a recursive function f_P such that $P(x, n) \Leftrightarrow f_P(n) \in \mathcal{O}^x$ for all n and x. In particular, a Π^1_1-predicate $Q(x)$ is $P(x, 0)$ for some Π^1_1-predicate P. Hence $Q(x)$ if and only if $f_Q(0) \in \mathcal{O}^x$ for some recursive function f_Q. It follows that the set $C = \{(x, n) \mid n \in \mathcal{O}^x\}$ is Π^1_1-universal. In other words, for each Π^1_1-set Q, there is a number n such that $Q = \{x \mid (n, x) \in C\}$. It is not difficult to see that C is not Σ^1_1. We may also code C into a subset of 2^ω by defining $D = \{0^n 1^\frown x \mid n \in \mathcal{O}^x\}$. Then D is a Π^1_1-complete set.

1.5.2 Σ^1_1-boundedness

The next corollary is known as the Σ^1_1-bounding principle. We will see a number of applications of this principle later.

Corollary 1.5.2 (Spector [153]). *Suppose $x \subseteq \mathcal{O}$ and $x \in \Sigma^1_1$. Then there exists an $n \in \mathcal{O}$ such that $|m| \leq |n|$ for all $m \in x$.*

Proof. Otherwise, for each $\beta < \omega_1^{\mathrm{CK}}$ there is an element $n \in x$ with $|n| > \beta$.

Since $\mathcal{O}$ is Π_1^1, there is a recursive function f such that $n \in \mathcal{O}$ implies $f(n) \in \mathsf{WF}_0$. Then $n \in \mathcal{O}$ if and only if

$$(\exists m)(m \in x \wedge (\exists h \in \omega^\omega)(\forall i)(\forall j)(R_{f(n)}(i,j) \rightarrow \langle h(i), h(j)\rangle \in W_{g(m)})),$$

where g is as defined in Theorem 1.3.4. So $\mathcal{O}$ is Σ_1^1, a contradiction. □

We now turn to Theorem 1.5.5. This may be viewed as an effective version of Corollary 1.5.2.

Lemma 1.5.3. *There is a recursive function f such that $n \in \mathcal{O} \Leftrightarrow f(n) \in \mathsf{WF}_0 \wedge |n| < |R_{f(n)}|$.*

Proof. Since $\mathcal{O}$ is Π_1^1, there is a recursive function g such that

$$n \in \mathcal{O} \Leftrightarrow g(n) \in \mathsf{WF}_0.$$

By Theorem 1.3.4, there is a function h such that $n \in \mathcal{O}$ implies $R_{h(n)} = \{(i,j) \mid \langle i,j\rangle \in W_{k(n)}\}$, $h(n) \in \mathsf{WF}_0$ and $|n| = |R_{h(n)}|$, where $k(n)$ is the second function in Theorem 1.3.4. Define a recursive function s such that $R_{s(m,n)} =$

$$R_m \oplus R_n = \{(i,j) \mid (i,j \in \text{Even} \wedge \left(\frac{i}{2},\frac{j}{2}\right) \in R_m) \vee (i,j \in \text{Odd} \wedge \left(\frac{i-1}{2},\frac{j-1}{2}\right) \in R_n)\}.$$

Then

$$(m \in \mathsf{WF}_0 \wedge n \in \mathsf{WF}_0) \Leftrightarrow (s(m,n) \in \mathsf{WF}_0 \wedge \max\{|R_m|, |R_n|\} \leq |R_{s(m,n)}|).$$

So $n \in \mathcal{O} \Leftrightarrow (g(n) \in \mathsf{WF}_0 \wedge h(n) \in \mathsf{WF}_0) \Leftrightarrow s(g(n),h(n)) \in \mathsf{WF}_0$ and $|n| \leq |R_{s(g(n),h(n))}|$. Now let $f(n) = s(g(n),h(n))$. □

Given two binary r.e. relations R_i, R_j on ω, define a recursive function t such that $R_{t(i,j)} = R_i \otimes R_j$ where

$$R_i \otimes R_j - \{(\langle n_1, m_1\rangle, \langle n_2, m_2\rangle) | (n_1, n_2) \in R_i \wedge (m_1, m_2) \in R_j\}.$$

It is easy to verify that $t(i,j) \in \mathsf{WF}_0 \Leftrightarrow i \in \mathsf{WF}_0 \vee j \in \mathsf{WF}_0$ and that $\min\{|R_i|, |R_j|\} \leq |R_{t(i,j)}|$. We have the following corollary.

Corollary 1.5.4. *There is a recursive function p such that*

$$(\forall n)(\forall m)(n \in \mathcal{O} \vee m \in \mathcal{O}) \Leftrightarrow (p(m,n) \in \mathcal{O}),$$

and if $p(m,n) \in \mathcal{O}$ then

$$\min\{|n|, |m|\} \leq |p(m,n)|,$$

where $|m| = \infty$ for $m \notin \mathcal{O}$.

Proof. Let f_0 be the recursive function "f" in Lemma 1.4.2, f_1 be the recursive function "f" in Lemma 1.5.3 and let t be as above. Then the function $p(i,j) = f_0(t(f_1(i), f_1(j)))$ is exactly what is required. □

Given a standard enumeration $\Phi_e(n, m, y)$ of partial recursive functions, there is an enumeration $\{x_e\}_{e<\omega}$ of Σ^1_1-reals such that $n \in x_e \Leftrightarrow (\exists y)(\forall m)(\Phi_e(n, m, y)\downarrow\rightarrow \Phi_e(n, m, y) \neq 0)$. We have the following effective version of Spector's Σ^1_1-boundedness principle.

Theorem 1.5.5 (Sacks). *There is a recursive function $\hat{g}$ such that*

$$(\forall n)(x_n \subseteq \mathcal{O} \leftrightarrow \hat{g}(n) \in \mathcal{O} \wedge x_n \subseteq \mathcal{O}_{\hat{g}(n)}),$$

where $\mathcal{O}_{\hat{g}(n)} = \{m \in \mathcal{O} \mid |m| < |\hat{g}(n)|\}$.

Proof. Since the set $\{(m, n) \mid m \notin x_n\}$ is Π^1_1, there is a recursive function h such that $m \in x_n \Leftrightarrow h(m, n) \notin \mathcal{O}$. Let p be the function in Corollary 1.5.4. Define a recursive function f such that $W_{f(n)} = \{p(h(m, n), m) \mid m \in \omega\}$. If $x_n \subseteq \mathcal{O}$, then $W_{f(n)} \subseteq \mathcal{O}$. Let $\hat{g}(n) = 2^{g(f(n))}$ where g is the function in Lemma 1.3.5. By Lemma 1.3.5, $x_n \subseteq \mathcal{O} \implies \hat{g}(n) \in \mathcal{O}$. If $\hat{g}(n) \in \mathcal{O}$, then by Lemma 1.3.5 again, $W_{f(n)} \subseteq \mathcal{O}$. But $h(m, n) \notin \mathcal{O}$ for all $m \in x_n$. Then by Corollary 1.5.4, $m \in \mathcal{O}$ for all $m \in x_n$ and hence $x_n \subseteq \mathcal{O}$. Moreover, if $x_n \subseteq \mathcal{O}$, then $m \in x_n \implies h(m, n) \notin \mathcal{O} \implies |m| \leq |p(h(m, n), m)|$. □

By Theorem 1.5.1, every Π^1_1-set is many-one reducible to $\mathcal{O}$. Hence we may view $\mathcal{O}$ as providing the stages for enumerating a Π^1_1-set (this view will become clearer after the set $\mathcal{O}_1$ is introduced). Hence it is natural to think of a Π^1_1-set as analog of a Σ^0_1 (i.e. recursively enumerable) set. Now if one considers a "finite set" to be one whose members are all enumerated by some stage $\beta < \omega_1^{\mathrm{CK}}$, then Spector's boundedness principle says that any Δ^1_1-set is "finite". We shall make this precise with the introduction of admissible set theory.

In terms of definability, one would expect a big gap between possible order types of a recursive ordering and a Σ^1_1-ordering. The following observation says otherwise.

Proposition 1.5.6. *Suppose R is a Σ^1_1 well-ordering relation on ω. Then $|R| < \omega_1^{\mathrm{CK}}$.*

Proof. Suppose $|R| \geq \omega_1^{\mathrm{CK}}$. Then $n \in \mathsf{WF}_0 \Leftrightarrow (\exists h)(\forall i)(\forall j)(R_n(i, j) \rightarrow R(h(i), h(j)))$. This implies that WF_0 is Σ^1_1 which is a contradiction. □

Remark 1.5.7. For sets of reals, there is a boundedness principle for Δ^1_1-sets. As discussed in §1.1.4, for each Π^1_1-set A of reals, there is a recursive function $f : \omega^\omega \rightarrow \omega^\omega$ such that $x \in A \Leftrightarrow f(x) \in \mathsf{WO}_1$. Then if A is Δ^1_1, by the same argument as in the proof of Proposition 1.5.6, the supremum $\beta = \sup\{|f(x)| \mid x \in A\}$ of $f(A)$ is less than ω_1^{CK}.

For Σ^1_1 well-orderings of reals, by the same argument as in the proof of Corollary 1.5.2, each such well-ordering has order type less than ω_1^{CK}.

1.5.3 Exercises

Exercise 1.5.1. *Show that a partial recursive well-ordering is isomorphic to a recursive well-ordering.*

Exercise 1.5.2. *Prove that if $x \in \omega^\omega$ is a nonrecursive Π^0_1-singleton, then there is a $z \equiv_T x$ and a Π^0_1-set $P \subseteq 2^\omega$ such that z is the unique nonrecursive real in P.*

Exercise 1.5.3 (Jockusch and Soare [60]). *Prove that if $P \subseteq 2^\omega$ is a nonempty Π^0_1-set, then there is an $x \in P$ such that $x' \leq_T \emptyset'$.*

Exercise 1.5.4. *Prove that every uncountable Π^0_1-set $P \subseteq \omega^\omega$ contains a perfect set $[T] = \{x \in \omega^\omega \mid (\forall n)(x \upharpoonright n \in T)\}$, where T is a Σ^1_1-tree.*

Exercise 1.5.5. *Prove that every uncountable Σ^1_1-set $P \subseteq \omega^\omega$ contains a perfect set $[T]$, where T is a tree recursive in $\mathcal{O}$.*

2 Hyperarithmetic theory

In this chapter, we introduce the hyperarithmetic hierarchy and prove that the class of hyperarithmetic sets is precisely that of the Δ_1^1-sets. Hence the hyperarithmetic hierarchy provides the right framework for analyzing Δ_1^1 and Π_1^1-sets. This equivalence also allows one to perform transfinite induction over Δ_1^1-sets.

2.1 H-sets and Π_2^0-singletons

2.1.1 H-sets

Definition 2.1.1.

- The sequence $\{H_n\}_{n \in \mathcal{O}}$ is defined by transfinite induction over $\mathcal{O}$ as follows.

$$\begin{aligned} H_1 &= \emptyset, \\ H_{2^n} &= (H_n)', \text{ the Turing jump of } H_n \\ H_{3 \cdot 5^e} &= \{\langle m, i \rangle \mid m \in H_{\Phi_e(i)}\}. \end{aligned}$$

- A member of $\{H_n\}_{n \in \mathcal{O}}$ is called an H-set. A real y is *hyperarithmetic* if $y \leq_T H_n$ for some $n \in \mathcal{O}$. We use HYP to denote the collection of all hyperarithmetic sets.
- Similarly, for any real x, the set H_n^x is defined for all $n \in \mathcal{O}^x$:

$$\begin{aligned} H_1 &= x, \\ H_{2^n}^x &= (H_n^x)', \\ H_{3 \cdot 5^e}^x &= \{\langle m, i \rangle \mid m \in H_{\Phi_e^x(i)}^x\}. \end{aligned}$$

Clearly $n <_o m$ implies $H_n <_T H_m$. In fact,

Lemma 2.1.2 (Kleene). *There is a recursive function f such that for all n, m, $n <_o m$ implies $H_n = \Phi_{f(n,m)}^{H_m}$.*

Proof. Let e_0 be an index such that $x = \Phi_{e_0}^{x'}$ and let p, q be recursive functions such that $x = \Phi_n^y \wedge y = \Phi_m^z$ implies $x = \Phi_{p(n,m)}^z$ and $\Phi_{q(n)}^x(m) = m$ where $\langle m, n \rangle \in x$. Fix a function g as in (i) of Theorem 1.3.4.

Define f by induction on $<_o$. By the Recursion Theorem, we may assume that if $n <_o m$ then $H_n = \Phi_{f(n,m')}^{H_{m'}}$ for all $m' <_o m$. If $m = 2^{m'}$ for some $m' \in \mathcal{O}$, then $\Phi_{f(n,m)}^{H_m} = \Phi_{f(n,m')}^{\Phi_{e_0}^{H_m}}$. If $m = 3 \cdot 5^i$ for some i, then a recursive search for a pair (m', k) such that $m' = \Phi_i(k)$ and $n \in W_{g(m')}$ allows one to define $\Phi_{f(n,m)}^{H_m} = \Phi_{f(n,m')}^{H_{m'}}$. □

Lemma 2.1.2 remains valid for relativized H-sets. Thus there is a recursive function f such that for all x, if $n <_{o^x} m$ then $H_n^x = \Phi_{f(n,m)}^{H_m^x}$.

2.1.2 Π^0_2-singletons

Let x be the unique real satisfying a predicate in the class Γ. Then x is said to be a *Γ-singleton*. We have seen that every Σ^1_1-singleton is Δ^1_1 and $\mathcal{O}$ is a Π^1_1-singleton. In this section we will show that every H-set is a Π^0_2-singleton.

Lemma 2.1.3. *If x is a Π^0_2-singleton, then x' and any $y \equiv_T x$ is a Π^0_2-singleton.*

Proof. Let x be a Π^0_2-singleton for the predicate R. Note that the set $\{y \mid R(\Phi^y_e)\}$ is Π^0_2 for each e. Let e_0 be an index such that $\Phi^{y'}_{e_0} = y$ for all y. Then the predicate

$$Q(y) \Leftrightarrow R(\Phi^y_{e_0}) \wedge (\forall e)(e \in y \leftrightarrow (\exists s)(\exists \sigma)(\sigma \prec y \wedge \Phi_e^{\Phi^\sigma_{e_0}[s]}(e)[s]\downarrow))$$

is Π^0_2. Note that x' is the unique y such that $Q(y)$ holds. If $x = \Phi^y_{e_1}$ and $y = \Phi^x_{e_2}$ then the predicate

$$S(y) \Leftrightarrow R(\Phi^y_{e_1}) \wedge (\forall e)(e \in y \leftrightarrow (\exists s)(\exists \sigma)(\sigma \prec y \wedge \Phi_{e_2}^{\Phi^\sigma_{e_1}[s]}(e)[s]\downarrow= 1))$$

is Π^0_2 and witnesses the uniqueness of y. □

Theorem 2.1.4. *There exists a Π^0_2-predicate H such that for all reals y and $n \in \mathcal{O}^y$,*

$$(\exists! x)(H(n, x, y)) \wedge H(n, H^y_n, y).$$

Hence every H^y-set is a $\Pi^0_2(y)$-singleton.

Proof. By the Recursion Theorem, we may assume that there exists a Π^0_2-predicate $H(n, x, y)$ such that for each $m' <_{\mathcal{O}^x} m \in \mathcal{O}^x$, $H(m', H^x_{m'}, x)$ is defined and $H^x_{m'}$ is the unique real y satisfying $H(m', y, x)$. If $m = 2^{m'}$, by Lemma 2.1.3 we may define $H(m, H^x_m, x)$ if and only if $(H^x_{m'})' = H^x_m \wedge H(m', H^x_{m'}, x)$. If $m = 3 \cdot 5^k$ for some k, then define $H(m, H^x_m, x)$ if and only if $(\forall n)(H(\Phi^x_k(n), H^x_{\Phi^x_k(n)}, x))$. □

We remark that the condition Π^0_2 in Theorem 2.1.4 cannot be replaced by Σ^0_2. It is not difficult to show that every Σ^0_2-singleton in 2^ω is recursive. On the other hand, every Π^0_2-singleton in 2^ω is Turing reducible to a Π^0_1-singleton in ω^ω.

Proposition 2.1.5. *If $x \in 2^\omega$ is a Π^0_2-singleton, then there is a Π^0_1-singleton $T \subseteq \omega^\omega$ whose unique member y is Turing equivalent to x.*

Proof. Suppose that $x \in 2^\omega$ is a Π^0_2-singleton. Then there is a recursive predicate $R(n, \sigma)$ such that x is the unique real z satisfying $(\forall n)(\exists m)R(n, z \upharpoonright m)$. We may assume that $(\forall n)(\forall \sigma)(R(n, \sigma) \to |\sigma| \geq n)$. Let T be the set of y's such that

- If $y(n) = \sigma$ then $R(n, \sigma)$;
- $(\forall \tau \prec y(n))(\neg R(n, \tau))$, and
- $(\forall k \leq n)(y(k) \preceq y(n) \vee y(n) \preceq y(k))$.

Then T is a Π^0_1-set. Furthermore, if $y \in T$ then for each n, $y(n)$ is the shortest initial segment σ of x satisfying $R(n, \sigma)$. Hence if $y \in T$, then $\bigcup_{n\in\omega} y(n) = x$. It follows that y is the unique real in T. Clearly $y \equiv_T x$. □

2.1.3 Exercise

Exercise 2.1.1. *Prove that every Δ^0_2-real $x \in 2^\omega$ is a Π^0_2-singleton but there is a Δ^0_3-real in 2^ω which is not a Π^0_2-singleton.*

2.2 Δ^1_1-ness and hyperarithmeticity

In this section we prove the fundamental result that a set is Δ^1_1 if and only if it is hyperarithmetic.

2.2.1 Δ^1_1-subsets of ω

Lemma 2.2.1. *The following two predicates are Π^1_1:*
(i) $n \in \mathcal{O}^y \wedge k \in H^y_n$,
(ii) $n \in \mathcal{O}^y \wedge k \notin H^y_n$.

Proof. Note that $n \in \mathcal{O}^y \wedge k \in H^y_n$ if and only if $n \in \mathcal{O}^y \wedge (\forall z)(H(n,z,y) \to k \in z)$, and $n \in \mathcal{O}^y \wedge k \notin H^y_n$ if and only if $n \in \mathcal{O}^y \wedge (\forall z)(H(n,z,y) \to k \notin z)$. By Proposition 1.2.5 and Theorem 2.1.4, both are Π^1_1. □

Corollary 2.2.2 (Kleene [69]).
(i) *If x is hyperarithmetic in y, then x is $\Delta^1_1(y)$.*
(ii) *The set $\{(x,y) \mid x \in \mathsf{HYP}^y\}$ is Π^1_1.*

Proof. If x is hyperarithmetic in y, then there is an H^y-set H^y_n such that $x \leq_T H^y_n$ for some $n \in \mathcal{O}^y$. Hence $x = \Phi_e^{H^y_n}$ for some e. Then both the predicates

$$k \in x \Leftrightarrow (\exists\sigma)(\Phi^\sigma_e(k) = 1 \wedge \sigma \prec H^y_n)$$

and

$$k \notin x \Leftrightarrow (\exists\sigma)(\Phi^\sigma_e(k) = 0 \wedge \sigma \prec H^y_n)$$

are $\Pi^1_1(y)$ by Lemma 2.2.1.

Now $x \in \mathsf{HYP}^y \Leftrightarrow (\exists n)(\exists e)(n \in \mathcal{O}^y \wedge x = \Phi_e^{H^y_n}) \Leftrightarrow (\exists n)(\exists e)(\forall m)(\forall z)(n \in \mathcal{O}^y \wedge (H(n,z,y) \to x = \Phi^z_e))$ where H is as in Theorem 2.1.4. Thus the set $\{(x,y) \mid x \in \mathsf{HYP}^y\}$ is Π^1_1. □

For each $n \in \mathcal{O}^y$, define $\mathcal{O}^y_n = \{m \mid m \in \mathcal{O}^y \wedge |m| < |n|\}$.

Lemma 2.2.3. *The following predicates are Π^1_1:*
(i) $n \in \mathcal{O}^y \wedge m \in \mathcal{O}^y_n$,
(ii) $n \in \mathcal{O}^y \wedge m \notin \mathcal{O}^y_n$.

Proof. By Theorem 1.3.4, define a Σ^1_1-predicate $P(x, y)$ to be the conjunction of the following three sentences:

$$x^{[1]} = \emptyset,$$

$$(\forall e)(3 \cdot 5^e \in \mathcal{O}^y \to x^{[3 \cdot 5^e]} = \bigcup_n x^{[\Phi^y_e(n)]}),$$

and

$$(\forall n)(2^n \in \mathcal{O}^y \to x^{[2^n]} = \{1\} \cup \{k \mid (\exists j)(k = 2^j \wedge j \in x^{[n]})\} \cup \{k \mid (\exists e)(k = 3 \cdot 5^e \wedge (\forall j)(\Phi^y_e(j) <_{\mathcal{O}^y} \Phi^y_e(j+1) \in x^{[n]}))\}).$$

Define a Π^1_1-set $A = \{\langle m, n, y\rangle \mid (\forall z)(P(z, y) \to \langle m, n\rangle \in z)\}$. It is easy to see that

$$n \in \mathcal{O}^y \wedge m \in \mathcal{O}^y_n \Leftrightarrow \langle m, n, y\rangle \in A.$$

The argument for (ii) is similar and is left as an exercise. □

Lemma 2.2.4. *There is a recursive function g such that for all $n \in \mathcal{O}^y$, $\mathcal{O}^y_n = \Phi^{H^y_{2^n}}_{g(n)}$.*

Proof. We prove the unrelativized version. By the Recursion Theorem, we may assume that $\mathcal{O}_m = \Phi^{H_{2^m}}_{g(m)}$ for all $m <_o n$.

If $n = 2^m$, define

$$E = \{e \mid (\forall n)(\Phi_e(n) \downarrow \wedge \Phi_e(n) \in W_{f(\Phi_e(n+1))})\}$$

where f is the function in (i) of Theorem 1.3.4. Note $E \leq_T \emptyset''$. If $m \neq 1$ and $m \in \mathcal{O}$, then

$$k \in \mathcal{O}_{2^m} \Leftrightarrow k = 1 \vee (\exists j)(k = 2^j \wedge j \in \mathcal{O}_m) \vee (\exists e)(k = 3 \cdot 5^e \wedge e \in E \wedge (\forall j)(\Phi_e(j) \in \mathcal{O}_m)).$$

Thus $\mathcal{O}_{2^m}$ may be computed by H_{2^n} via an oracle Turing machine whose index can be effectively obtained from $g(m)$. Set the index to be $g(n)$. Then $\mathcal{O}_n = \Phi^{H_{2^n}}_{g(n)}$.

If $n = 3 \cdot 5^e$, then

$$i \in \mathcal{O}_n \Leftrightarrow (\exists k)(i \in \mathcal{O}_{\Phi_e(k)}) \Leftrightarrow (\exists k)(i \in \Phi^{H_{2^{\Phi_e(k)}}}_{g(\Phi_e(k))}).$$

By Lemma 2.1.2, $H_{2^{\Phi_e(k)}}$ may be computed by H_n, and hence by H_{2^n} uniformly. Thus $\mathcal{O}_n$ is Σ^0_1 over H_n and so recursive in H_{2^n}. Moreover, the computation from H_{2^n} is uniform. Hence the index i of the oracle Turing machine computing $\mathcal{O}_n$ from H_{2^n} can be obtained effectively. Set $g(n) = i$. Then $\mathcal{O}_n = \Phi^{H_{2^n}}_{g(n)}$. □

Corollary 2.2.5. *If x is $\Delta^1_1(y)$, then x is hyperarithmetic in y.*

Proof. If x is $\Delta^1_1(y)$, then there is recursive function f such that $n \in x \Leftrightarrow f(n) \in \mathcal{O}^y$. The set $A = \{f(n) \mid n \in x\}$ is $\Delta^1_1(y)$. By Spector's Boundedness Theorem (1.5.5), there is a number $m \in \mathcal{O}^y$ such that $A \subseteq \mathcal{O}^y_m$. Then $n \in x \Leftrightarrow f(n) \in \mathcal{O}^y_m$. Hence $x \leq_T H^y_{2^m}$ by Lemma 2.2.4. □

We have as a consequence the following characterization due to Kleene.

Theorem 2.2.6 (Kleene [69]). *A real x is $\Delta^1_1(y)$ if and only if it is hyperarithmetic in y.*

The following corollary says that Δ^1_1-sets are strongly reducible to H-sets.

Corollary 2.2.7. *Every Δ^1_1-real is many-one reducible to an H-set.*

Proof. By Theorem 2.2.6, if x is Δ^1_1 then it is Turing reducible to H_n for some $n \in \mathcal{O}$. Then x is many-one reducible to $H_{2^n} = (H_n)'$. □

2.2.2 Δ^1_1-sets of reals

There is a subtle point regarding uniformity when it comes to defining the notion of hyperarithmetic sets of reals. For this, we define $\hat{H}^x$-sets by induction on $\mathcal{O}$ rather than on $\mathcal{O}^x$:

$$\hat{H}^x_1 = x,$$
$$\hat{H}^x_{2^n} = (\hat{H}^x)'$$

and

$$\hat{H}^x_{3\cdot 5^e} = \{\langle m, i\rangle \mid m \in \hat{H}^x_{\Phi_e(i)}\}.$$

Definition 2.2.8. A set of reals A is *hyperarithmetic* if there is a number $n \in \mathcal{O}$ and an index e such that for all x, $x \in A \Leftrightarrow \Phi_e^{\hat{H}^x_n}(0)\downarrow$.

Hence if A is hyperarithmetic, then there exist n and e such that

$$x \in A \Leftrightarrow (\exists\sigma)(\sigma \prec \hat{H}^x_n \wedge \Phi^\sigma_e(0)\downarrow)$$

and

$$x \notin A \Leftrightarrow (\forall\sigma)(\sigma \prec \hat{H}^x_n \rightarrow \Phi^\sigma_e(0)\uparrow).$$

It follows that Lemma 2.2.1 remains true for such $\hat{H}^x$-sets and so A is Δ^1_1.

Conversely, assume A is Δ^1_1. According to the discussion in § 1.5.1, there are two numbers k_0 and k_1 such that $x \in A \Leftrightarrow k_0 \in \mathcal{O}^x \Leftrightarrow k_1 \notin \mathcal{O}^x$. By relativizing Corollary 1.5.4, there is a recursive function p such that $\min\{|k_0|_x, |k_1|_x\} \le |p(k_0, k_1)|_x$ for all x. Let $n = 2^{p(k_0,k_1)}$. Note that $n \in \mathcal{O}^x$ for all x and so $n \in \mathcal{O}^\emptyset = \mathcal{O}$. Since Lemma 2.2.4 remains true, $\mathcal{O}^x_n = \Phi^{\hat{H}^x_{2^n}}_{g(n)}$ for all x. Hence there is a recursive function h such that $k_0 \in \mathcal{O}^x_n \Leftrightarrow 0 \in W^{\hat{H}^x_{2^n}}_{h(n)}$ for all n, giving $x \in A \Leftrightarrow \Phi^{\hat{H}^x_{2^n}}_{h(n)}(0)\downarrow$. Thus we arrive at the following conclusion.

Theorem 2.2.9. *A set of reals is hyperarithmetic if and only if it is Δ^1_1.*

Hyperarithmeticity also provides a scheme for inductive analysis of Δ^1_1-sets of reals. We will show how this is implemented after introducing the ramified analytical hierarchy in § 3.8. The notion of a hyperarithmetic code, to be introduced in § 2.7, will also provide a powerful tool for the study of Δ^1_1-sets of reals. The next proposition sets an effective bound on $|n|_x$ if $n \in \mathcal{O}^x$ for all x.

Proposition 2.2.10. *If $n \in \mathcal{O}^x$ for all x, then there is a recursive ordinal β such that $|n|_x < \beta$ for all x.*

Proof. If not, $m \in \mathsf{WF}_0 \Leftrightarrow (\exists f)(\exists x)(\forall i)(\forall j)(R_m(i,j) \to \langle f(i), f(j)\rangle \in W^x_{g(n)})$ where g is the recursive function in (ii) of Theorem 1.3.4. So WF_0 is Σ^1_1, a contradiction. □

It also follows from the discussion above that $\omega^x_1 = \omega^{\mathrm{CK}}_1$ if and only if every $y \le_h x$ is recursive in $\hat{H}^x_n$ for some $n \in \mathcal{O}$.

2.2.3 Parametrization of Δ^1_1-sets

As a first application of the results above, we demonstrate a parametrization of Δ^1_1-reals.

Proposition 2.2.11. *There is a partial function $D : \omega \times \omega \to \omega$ with a Π^1_1-graph G^D such that for all x, $x \in \Delta^1_1 \Leftrightarrow (\exists m)(\forall i)(\forall j)(x(i) = j \leftrightarrow (m,i,j) \in D)$.*

Proof. Define $D = \{(\langle n,e\rangle, i, j) | n \in \mathcal{O} \wedge \Phi^{H_n}_e$ is total $\wedge\, \Phi^{H_n}_e(i) = j\}$. By Theorem 2.2.6, for all x, $x \in \Delta^1_1 \Leftrightarrow (\exists m)(\forall i)(\forall j)(x(i) = j \leftrightarrow (m,i,j) \in D))$. Moreover, $(\langle n,e\rangle, i, j) \in D \Leftrightarrow n \in \mathcal{O} \wedge (\forall z)(H(n,z) \wedge \Phi^z_e$ is total $\to \Phi^z_e(i) = j)$, where H is as defined in § 2.1.2. Hence G^D is Π^1_1. □

Proposition 2.2.11 may be applied to reduce certain Σ^1_2-predicates to Π^1_1.

Corollary 2.2.12. *For any Π^1_1-predicate $P(x,y)$, the predicate $Q(y) \Leftrightarrow (\exists x \in \Delta^1_1)(P(x,y))$ is Π^1_1.*

Proof. By Proposition 2.2.11, $Q(y) \Leftrightarrow (\exists x \in \Delta^1_1)P(x,y)$ is equivalent to

$$(\exists m)(\forall i)(\exists j)((m,i,j) \in D) \wedge \\ (\forall x)(\forall i)(\forall j)(x(i) = j \leftrightarrow (m,i,j) \in D) \to P(x,y))).$$

Note that for each m, if $(\forall i)(\exists j)((m,i,j) \in D)$, then the predicate $(m,i,j) \in D$ is Δ^1_1. Hence Q is Π^1_1. □

One may relativize the above proposition to the statement "for any Π^1_1-predicate $P(x,y)$, the predicate $Q(y) \Leftrightarrow (\exists x \in \Delta^1_1(y))P(x,y)$ is Π^1_1", which is used to derive a classical result in descriptive set theory:

Corollary 2.2.13. *If the function $f : \omega^\omega \to \omega^\omega$ is Δ^1_1 and one-one, then $f(A)$ is Δ^1_1 for any Δ^1_1-set A.*

Proof. If f is Δ^1_1 and total, by Proposition 1.1.8, G^f is Δ^1_1. Hence $f^{-1}(\{y\})$ is $\Delta^1_1(y)$ for any y. Since f is one-one, by Proposition 1.1.15, the unique real $f^{-1}(y)$ is $\Delta^1_1(y)$. So $y \in$

$f(A) \Leftrightarrow (\exists x \in \Delta^1_1(y))(y \in A \wedge (x, y) \in G^f) \Leftrightarrow (\exists x)(x \in A \wedge (x, y) \in G^f)$. By relativizing Corollary 2.2.12 to y, one concludes that $f(A)$ is Δ^1_1. □

Thus we obtain a characterization of Δ^1_1-sets of reals:

Corollary 2.2.14. *A set of reals A is Δ^1_1 if and only if there is a Π^0_1-set B and a recursive function $f : \omega^\omega \to \omega^\omega$ such that $f(B) = A$ and f is one-one on B.*

Proof. If there is a Π^0_1-set B and a one-one recursive function $f : \omega^\omega \to \omega^\omega$ such that $f(B) = A$, then, by Corollary 2.2.13, one easily verifies that A is Δ^1_1.

If $A \subset \omega^\omega$ is Δ^1_1, then there is a recursive function g such that $x \in A \Leftrightarrow g(x) \in \mathrm{WO}_1$ and, by the discussion in §1.5.2, there is a recursive well-ordering z such that $x \in A \Leftrightarrow |g(x)| \leq |z|$. Define $P(x, y)$ if and only if y is an isomorphism from $\leq_{g(x)}$ to an initial segment of $\leq_z$. Since z is recursive, $P(x, y)$ is Π^0_2 and $x \in A \Leftrightarrow (\exists! y)P(x, y)$. Hence there is a recursive predicate R such that $P(x, y) \Leftrightarrow (\forall n)(\exists m)R(x, y, n, m)$. Uniformize P by defining a Π^0_1-set $B = \{x_0 \oplus x_1 \oplus x_2 \mid (\forall n)(R(x_1, x_2, n, x_0(n)) \wedge (\forall m < x_0(n))\neg R(x_1, x_2, n, m))\}$. Define a recursive function f such that $f(x) = x_1$ if $x = x_0 \oplus x_1 \oplus x_2$. Obviously f is one-one on B. □

In certain situations, Corollary 2.2.14 enables one to reduce a construction over a Δ^1_1-set to one that is over a Π^0_1-set. Additionally, Theorem 2.2.9 provides a parametrization of Δ^1_1-sets of reals.

Proposition 2.2.15. *There is a Π^1_1-set $A \subseteq \omega \times \omega^\omega$ with the property that for any set B of reals, B is Δ^1_1 if and only if there is an n such that $B = A_n = \{x \mid (n, x) \in A\}$.*

Proof. Define $A = \{(\langle e, m\rangle, x) \mid m \in \mathcal{O} \wedge \Phi_e^{\hat{H}^x_m}(0)\downarrow\}$. By Theorem 2.2.9, A is as required. □

2.2.4 Exercises

Exercise 2.2.1. *Complete the proof of Lemma 2.2.3.*

Exercise 2.2.2. *Prove that for any $n \in \mathcal{O}$, there exists an uncountable Π^0_1-set $P \subseteq 2^\omega$ in which every real recursive in H_n is recursive.*

Exercise 2.2.3. *Prove that there is a $B \subseteq \mathcal{O}$ such that $<_o$ is a well-ordering on B of order type ω_1^{CK}.*

2.3 Spector's Uniqueness Theorem

Since the H-sets are generated by a Π^0_2-singleton process (cf. Theorem 2.1.4), they provide a way of carrying out transfinite induction over hyperarithmetic sets. We make this observation precise by noting Corollary 2.3.2, the Spector Uniqueness Theorem, which implies that induction over H-sets is degree-theoretic invariant.

Theorem 2.3.1 (Spector [153]). *There is a recursive function h such that*

$$n, m \in \mathcal{O} \wedge |n| \leq |m| \Rightarrow H_n = \Phi_{h(n,m)}^{H_m}.$$

Proof. We prove this by induction on the well-ordering $<$ of $\mathcal{O}^2$:

$$(n_1, m_1) < (n_2, m_2) \Leftrightarrow |n_1| < |n_2| \vee (|n_1| = |n_2| \wedge |m_1| < |m_2|).$$

There are four cases to consider.

(1) $n = 1$. Set $h(1, m) = c_1$ so that $\Phi_{c_1}^x = H_1 = \emptyset$ for all x.

(2) $n = 2^i \wedge m = 2^j$. Fix a recursive function g so that for any e, x and y, $x = \Phi_e^y$ implies $x' = \Phi_{g(e)}^{y'}$. Set $h(n, m) = g(h(i, j))$.

(3) $n = 3 \cdot 5^k$. Set $h(n, m) = c_3$ so that $\Phi_{c_3}^{H_m}$ is the characteristic function of $\{\langle i, j\rangle \mid i \in \Phi_{h(\Phi_k(j),m)}^{H_m}\}$.

(4) $n = 2^k \wedge m = 3 \cdot 5^l$. This is the most difficult case. By Lemma 2.2.4, we may H_m-recursively search for an i such that $n \in \mathcal{O}_{\Phi_l(i)}$ and $\mathcal{O}_{\Phi_l(i)} = \Phi_{g(\Phi_l(i))}^{H_{2^{\Phi_l(i)}}}$. By Lemma 2.1.2, there is a recursive function f so that $H_{2^{\Phi_l(i)}} = \Phi_{f(2^{\Phi_l(i)},m)}^{H_m}$. By induction, $H_n = \Phi_{h(n,2^{\Phi_l(i)})}^{H_{2^{\Phi_l(i)}}}$. Then there exists a c_4 with $H_n = \Phi_{h(n,2^{\Phi_l(i)})}^{\Phi_{f(2^{\Phi_l(i)},m)}^{H_m}} = \Phi_{c_4}^{H_m}$. Set $h(n, m) = c_4$.

Note that these cases can be identified recursively. Hence h is a recursive function. □

Thus we have the following uniqueness theorem.

Corollary 2.3.2 (Spector). *If $n, m \in \mathcal{O}$, then $|n| = |m|$ implies $H_n \equiv_T H_m$.*

In particular, for any recursive ordinal β, the "β-th Turing jump", denoted $\emptyset^{(\beta)}$, may be taken to be H_n for any $n \in \mathcal{O}$ with $|n| = \beta$. There is a parallel result for $\hat{H}^x$-sets with exactly the same proof as that of Theorem 2.3.1. We leave it to the reader as an exercise.

Theorem 2.3.3. *There is a recursive function h such that for any real x,*

$$n \in \mathcal{O}^x \wedge m \in \mathcal{O} \wedge |n| \leq |m| \implies H_n^x = \Phi_{h(n,m)}^{\hat{H}_m^x}.$$

2.3.1 Exercise

Exercise 2.3.1. *Prove Theorem* 2.3.3.

2.4 Hyperarithmetic reducibility

2.4.1 Δ_n^1-reducibility

Definition 2.4.1. $x \leq_{\Delta_n^1} y$ if x is $\Delta_n^1(y)$. In particular, $x \leq_h y$ if $x \leq_{\Delta_1^1} y$.

Proposition 2.4.2 (Shoenfield). *If $x \leq_{\Delta_n^1} y$ and $y \leq_{\Delta_n^1} z$, then $x \leq_{\Delta_n^1} z$.*

Proof. If $x \leq_{\Delta_n^1} y$, then there is a Δ_n^1-predicate R such that $n \in x \Leftrightarrow R(n, y)$.
If $y \leq_{\Delta_n^1} z$, then there is a Δ_n^1-predicate S such that $n \in y \Leftrightarrow S(n, z)$.
Then

$$\begin{aligned} n \in x &\Leftrightarrow (\exists r)(R(n,r) \wedge (\forall m)((m \in r \leftrightarrow S(m,z))) \\ &\Leftrightarrow (\forall r)(\forall m)((m \in r \leftrightarrow S(m,z)) \to R(n,r)). \end{aligned}$$

Thus $x \leq_{\Delta_n^1} z$. □

Reals x, y are said to have the same Δ_n^1-degree if $x \leq_{\Delta_n^1} y$ and $y \leq_{\Delta_n^1} x$. If $n = 1$, then we also say that they have the same hyperdegree. By Theorem 2.2.6, $x \leq_h y$ if and only if there is an $n \in \mathcal{O}^y$ and an e such that $x = \Phi_e^{H_n^y}$. The *hyperjump* of x is defined to be the set $\mathcal{O}^x$.

The h-reduction relation $x \leq_h y$ suggests another reason for introducing the $\hat{H}^x$-sets. One of the problems with h-reducibility is that, unlike the case of Turing reducibility where one examines a fixed recursive list of possible reduction algorithms (independently of the sets x and y being considered), there is no uniform procedure to search for an algorithm that h-reduces x to y. The reduction depends on what can be Turing computed by the H^y-sets and this involves ordinal notations for y. Different reals y have different ordinal notations. In certain cases the notations may be markedly different since $\{\omega_1^y \mid y \in 2^\omega\}$ is unbounded in ω_1. The use of $\hat{H}^y$-sets circumvents this difficulty. If $\omega_1^y = \omega_1^{\mathrm{CK}}$, then the sets Turing computable by the $\hat{H}^y$-sets are precisely the $\Delta_1^1(y)$-sets. Hence while it is not true in general, for reals y with $\omega_1^y = \omega_1^{\mathrm{CK}}$, one does have a uniform reduction.

2.4.2 The partial order of hyperdegrees

The Turing jump operator is order preserving on the Turing degrees. This remains true for the hyperjump operator which preserves order on the hyperdegrees.

Proposition 2.4.3. $x \leq_h y \Leftrightarrow \mathcal{O}^x \leq_m \mathcal{O}^y$.

Proof. First of all $\mathcal{O}^x$ is $\Pi_1^1(x)$. If $x \leq_h y$, then as in the proof of Proposition 2.4.2 $\mathcal{O}^x$ is $\Pi_1^1(y)$. By Theorem 1.5.1, $\mathcal{O}^x \leq_m \mathcal{O}^y$.

If $\mathcal{O}^x \leq_m \mathcal{O}^y$, then $x \leq_m \mathcal{O}^x \leq_m \mathcal{O}^y$. So x is $\Pi_1^1(y)$. Since $\mathcal{O}^{\omega \setminus x}$ is $\Pi_1^1(x)$, $\omega \setminus x \leq_m \mathcal{O}^{\omega - x} \leq_m \mathcal{O}^x \leq_m \mathcal{O}^y$. Then $\omega \setminus x$ is $\Pi_1^1(y)$ and this implies that $x \leq_h y$. □

One of the most basic theorems in classical recursion theory is the Friedberg–Mučnik Theorem ([28] and [109]) which asserts the existence of two incomparable r.e. Turing degrees. Since Π^1_1-sets are seen as analogs of r.e. sets in the higher setting, the following result shows that there is a fundamental difference between the two notions of reducibility.

Theorem 2.4.4 (Spector [153]). *If $x, y \in \Pi^1_1$ and $y \notin \mathrm{HYP}$, then $x \leq_h y$.*

Proof. We show that $\mathcal{O} \leq_h y$. By Theorem 1.5.1, there is a recursive function f such that $n \in y \Leftrightarrow f(n) \in \mathcal{O}$. Since $y \notin \mathrm{HYP}$, by Lemma 2.2.4, $\sup\{|f(n)| \mid n \in y\}$ is ω_1^{CK}. Then

$$n \in \mathcal{O} \Leftrightarrow (\exists m)(m \in y \wedge n \in \mathcal{O}_{f(m)})$$

and

$$n \notin \mathcal{O} \Leftrightarrow (\forall m)(m \in y \rightarrow n \notin \mathcal{O}_{f(m)}).$$

By Lemma 2.2.3, $\mathcal{O} \leq_h y$. □

Remark 2.4.5. Theorem 2.4.4 prompted Kreisel to suggest that h-reduction was not the "correct" notion of reducibility for higher recursion theory. This led to the introduction of the notion of metarecursive reducibility in metarecursion theory. It is shown in Sacks [118] that there exist two Π^1_1-sets with incomparable metadegrees.

There is a tight link between ω_1^x and the h-degree of x, as illustrated in the following proposition:

Proposition 2.4.6 (Spector [153]). $x \leq_h y \implies \omega_1^x \leq \omega_1^y$.

Proof. Since $x \leq_h y$, every x-recursive well-ordering relation $R \subseteq \omega^2$ is Δ^1_1 in y. By relativizing Proposition 1.5.6 to y, R has order type $< \omega_1^y$. Hence $\omega_1^x \leq \omega_1^y$. □

The converse of Proposition 2.4.6 does not hold in general. However, there is an important partial result which characterizes ω_1^{CK}.

Theorem 2.4.7 (Sacks). *$\omega_1^x > \omega_1^{\mathrm{CK}}$ if and only if $x \geq_h \mathcal{O}$.*

We decompose the proof into two lemmas, with the following left as an exercise:

Lemma 2.4.8 (Sacks). *Suppose $x \leq_T y$. Then there is a recursive function f such that*
(i) $(\forall n)(\forall m)(n <_{\mathcal{O}^x} m \rightarrow f(n) <_{\mathcal{O}^y} f(m))$,
(ii) $(\forall n)(n \in \mathcal{O}^x \rightarrow |n|_x = |f(n)|_y)$.

Lemma 2.4.9 (Spector [153]).
(i) $\mathcal{O}^x \leq_h y \implies \omega_1^x < \omega_1^y$.
(ii) $\omega_1^x < \omega_1^y \wedge x \leq_h y \implies \mathcal{O}^x \leq_h y$.

Proof. (i) $<_{\mathcal{O}^x}$ is $\Pi^1_1(x)$ and hence is recursive in $\mathcal{O}^x$. Thus by Proposition 2.4.6, $\omega_1^x < \omega_1^{\mathcal{O}^x} \leq \omega_1^y$.

(ii) If $x \le_h y$, then $x \le_T H^y_n$ for some $n \in \mathcal{O}^y$. Let f be as in Lemma 2.4.8. Since $\omega_1^x < \omega_1^y$, by Proposition 2.4.6, $\omega_1^x < \omega_1^{H^y_n}$. Hence there is a number $m \in \mathcal{O}^{H^y_n}$ such that for all i, $i \in \mathcal{O}^x \Leftrightarrow f(i) \in \mathcal{O}^{H^y_n}_m$. Thus $\mathcal{O}^x \le_h H^y_n \le_h y$. □

Proof of Theorem 2.4.7. If $\omega_1^x > \omega_1^{\mathrm{CK}} = \omega_1^{\emptyset}$, then by (ii) of Lemma 2.4.9, $\mathcal{O} \le_h x$.

If $\mathcal{O} \le_h x$, then by (i) of Lemma 2.4.9, $\omega_1^{\mathrm{CK}} < \omega_1^x$. □

The following corollary says that each h-degree not above the hyperjump is GL_1. This is in sharp contrast to Turing reducibility where there exist reals $x \not\ge_T \emptyset'$ such that $x' \not\le_T x \oplus \emptyset'$:

Corollary 2.4.10. $x \not\ge_h \mathcal{O} \Leftrightarrow \mathcal{O}^x \le_h x \oplus \mathcal{O}$.

Proof. If $\mathcal{O}^x \le_h x \oplus \mathcal{O}$, then $x <_h \mathcal{O}^x \le_h x \oplus \mathcal{O}$. Hence $x \not\ge_h \mathcal{O}$. Conversely, if $x \not\ge_h \mathcal{O}$, then by Theorem 2.4.7, $\omega_1^x = \omega_1^{\mathrm{CK}} < \omega_1^{x\oplus\mathcal{O}}$. By (ii) of Lemma 2.4.9, $\mathcal{O}^x \le_h x \oplus \mathcal{O}$. □

Corollary 2.4.11. *The set* $\{x \mid \omega_1^x = \omega_1^{\mathrm{CK}}\}$ *is* Σ^1_1.

Proof. By Proposition 1.2.5, $\mathcal{O}$ is a Π^1_1-singleton. Then $x \ge_h \mathcal{O}$ if and only if $(\exists n \in \mathcal{O}^x)(\exists e)(\Phi_e^{H^x_n} \in \{\mathcal{O}\})$, which is a Π^1_1-statement. By Theorem 2.4.7, the set $\{x \mid \omega_1^x = \omega_1^{\mathrm{CK}}\}$ is Σ^1_1. □

2.4.3 Exercises

Exercise 2.4.1. *Prove Lemma* 2.4.8.

Exercise 2.4.2. *Prove that the set* $\{x \mid \omega_1^x = \omega_1^{\mathrm{CK}}\}$ *is* $\Delta^1_1(\mathcal{O})$.

2.5 Some basis theorems and their applications

2.5.1 Gandy's Basis Theorem

A set of reals A is a *basis* for a collection of sets $\mathbf{P}$ of reals if for any nonempty $P \in \mathbf{P}$, $P \cap A \neq \emptyset$. Basis theorems are very important in recursion theory. They are used among other things to pin down the complexity of predicates. For example, let $\{T_i\}_{i<\omega}$ be an effective enumeration of recursive trees in $2^{<\omega}$. The index set $N = \{i \mid T_i \text{ contains an infinite path}\}$ appears to be Σ^1_1. But one can show that every recursive tree in $2^{<\omega}$ containing an infinite path must contain an infinite path recursive in $\emptyset'$ (see Exercise 1.5.3). So N is an arithmetical set.

Lemma 2.5.1. *Suppose that* $T \subseteq \omega^{<\omega}$ *is a recursive tree with an infinite path. There is an infinite path* $x \in [T]$ *recursive in* $\mathcal{O}$.

Proof. Suppose that $T \subseteq \omega^{<\omega}$ is a recursive tree with an infinite path. Note that the set $S = \{\sigma \in T \mid (\exists y \succ \sigma)(\forall n)(y \upharpoonright n \in T)\} \subseteq T$ is a nonempty Σ^1_1-tree without dead ends and so recursive in $\mathcal{O}$. Since T has an infinite path, so does S. Hence, the leftmost infinite path x in S is recursive in $\mathcal{O}$. □

Thus the collection of reals recursive in $\mathcal{O}$ is a basis for recursive trees in $\omega^{<\omega}$, while the set of hyperarithmetic reals is not:

Proposition 2.5.2. *There is a recursive tree $S \subseteq \omega^{<\omega}$ with uncountably many infinite paths but without a hyperarithmetic infinite path.*

Proof. By Corollary 2.2.2, the set A of non-hyperarithmetic reals is Σ^1_1. Then there is a recursive tree $T \subseteq 2^{<\omega} \times \omega^{<\omega}$ such that for any $x \in 2^\omega$, $x \in A$ if and only if there exists a y satisfying $(\forall n)((x \upharpoonright n, y \upharpoonright n) \in T)$. Let $S \subseteq \omega^{<\omega}$ be a recursive tree such that $\sigma \in S$ if and only if $(\sigma_0, \sigma_1) \in T$ where $\sigma_0(n) = \sigma(2n)$ and $\sigma_1(n) = \sigma(2n+1)$ for all n which makes sense for the σ_i's. Then for any infinite path $x \in [S]$, $x = x_0 \oplus x_1$ so that $x_0 \in A$. In other words, every infinite path in S Turing computes a nonhyperarithmetic real. Moreover, by the definition of A, every nonhyperarithmetic real is recursive in an infinite path in S. Hence S contains uncountably many infinite paths. □

Theorem 2.5.3 (Gandy [33]). *If A is Σ^1_1 and nonempty, then A contains a real x such that $\omega_1^x = \omega_1^{\mathrm{CK}}$ and $x \leq_T \mathcal{O}$.*

Proof. By Corollary 2.2.2, the set $B = \{x \oplus y \mid y \not\leq_h x \wedge x \in A\}$ is also a Σ^1_1 nonempty set. By Lemma 2.5.1, there is a real $x \oplus y \in B$ and $x \oplus y \leq_T \mathcal{O}$. Hence $x \in A$ is recursive in $\mathcal{O}$. Since $y \not\leq_h x$, we have $x \not\geq_h \mathcal{O}$. By Theorem 2.4.7, $\omega_1^x = \omega_1^{\mathrm{CK}}$. □

Theorem 2.5.3 is known as the Gandy Basis Theorem.

2.5.2 Incomparable hyperdegrees

We apply the Gandy Basis Theorem to construct a pair of incomparable hyperdegrees.

Lemma 2.5.4. *If A is a Σ^1_1-set containing a nonhyperarithmetic real, then A is uncountable.*

Proof. Suppose that A is a countable Σ^1_1-set containing a nonhyperarithmetic real. Then the set $B = \{x \in A \mid x \not\leq_h \emptyset\}$ is a nonempty Σ^1_1-subset of A. Hence there is a recursive tree T such that $x \in B$ if and only if there exists a y satisfying for every n, $(x \upharpoonright n, y \upharpoonright n) \in T$. Let $T' = \{(\sigma, \tau) \mid (\exists x \succ \sigma)(\exists y \succ \tau)(\forall n)((x \upharpoonright n, y \upharpoonright n) \in T)\}$ be a subtree of T. Obviously T' is a Σ^1_1-tree. Moreover, for any $(\sigma, \tau) \in T'$, there are $\sigma_0, \sigma_1 \succ \sigma$ and $\tau_0, \tau_1 \succ \tau$ with $\sigma_0 | \sigma_1$ so that $(\sigma_i, \tau_i) \in T'$ for $i \leq 1$. Otherwise, the Σ^1_1-set $\{x \mid (\exists y)(\forall n)(x \succ \sigma \wedge y \succ \tau \wedge (x, y) \in T')\} \subseteq B$ contains a unique real which must be hyperarithmetic, contradicting the definition of B. Thus imitating the proof of Theorem 1.1.10, one shows that B contains a perfect subset. □

Theorem 2.5.5 (Spector [155]). *There exist reals x and y such that $x \not\leq_h y$ and $y \not\leq_h x$. In fact, there is a real $x \not\leq_h \mathcal{O}$ with $\omega_1^x = \omega_1^{\mathrm{CK}}$.*

Proof. Let $A = \{x \mid x \not\leq_h \emptyset\}$ be a nonempty Σ_1^1-set. By Theorem 2.5.3 and Corollary 2.4.11, the set $B = \{x \in A \mid \omega_1^x = \omega_1^{\mathrm{CK}}\}$ is a nonempty Σ_1^1-set. By Lemma 2.5.4, B is uncountable. Since there are at most countably many reals hyperarithmetically reducible to $\mathcal{O}$, there is a real $x \in B$ such that $x \not\leq_h \mathcal{O}$. Furthermore, $x \not\geq_h \mathcal{O}$ since $\omega_1^x = \omega_1^{\mathrm{CK}}$. □

It should be pointed out that Spector's original proof of Theorem 2.5.5 was quite different from the one given above. His proof involved a measure-theoretic argument and was nonconstructive. That idea will be explored in Chapter 14 on higher randomness.

2.5.3 Kreisel's Basis Theorem

Theorem 2.5.6 (Kreisel [35]). *If x is nonhyperarithmetic and $A \subseteq \omega^\omega$ is a nonempty Σ_1^1-set, then there is a $y \in A$ such that $x \not\leq_h y$.*

Proof. Suppose that x is nonhyperarithmetic and $A \subset \omega^\omega$ is a nonempty Σ_1^1-set. By Gandy's Basis Theorem 2.5.3, the set $B = \{z \mid \omega_1^z = \omega_1^{\mathrm{CK}} \wedge z \in A\}$ is a nonempty Σ_1^1-set.

For each pair (e, n) with $n \in \mathcal{O}$, let

$$P_{(e,n)} = \{z \mid (\exists i)(\Phi_e^{\hat{H}_n^z}(i) \uparrow \vee \Phi_e^{\hat{H}_n^z}(i) \downarrow \neq x(i))\}.$$

We claim that for any nonempty Σ_1^1-set C, there is a nonempty Σ_1^1-set $Q_{(e,n)} \subseteq C \cap P_{(e,n)}$. There are three cases to consider.

(1) There exist $z \in C$ and $i \in \omega$ such that $\Phi_e^{\hat{H}_n^z}(i) \uparrow$. Then let $Q_{(e,n)} = \{z \mid (\Phi_e^{\hat{H}_n^z}(i) \uparrow \wedge z \in C)\}$.

(2) Not Case (1) but there exist $z_0, z_1 \in C$ and $i \in \omega$ such that $\Phi_e^{\hat{H}_n^{z_0}}(i) \neq \Phi_e^{\hat{H}_n^{z_1}}(i)$. Then there is a $k \leq 1$ such that $\Phi_e^{\hat{H}_n^{z_k}}(i) \neq x(i)$. Let $Q_{(e,n)} = \{z \mid (\Phi_e^{\hat{H}_n^z}(i) = \Phi_e^{\hat{H}_n^{z_k}}(i) \wedge z \in C)\}$.

(3) Neither (1) nor (2). Then the Σ_1^1-set $\{u \mid (\exists z)(\forall i)(z \in C \wedge u(i) = \Phi_e^{\hat{H}_n^z}(i))\}$ contains exactly one real which must be hyperarithmetic. Let $Q_{(e,n)} = C$.

By Proposition 1.1.12, there is a $y \in B \cap \bigcap_{e<\omega, n\in\mathcal{O}} P_{(e,n)} \subseteq A$. By the definition of $P_{(e,n)}$, $x \neq \Phi_e^{\hat{H}_n^y}$. Note that $\omega_1^y = \omega_1^{\mathrm{CK}}$. Hence by Theorem 2.3.3, we conclude $x \not\leq_h y$. □

Theorem 2.5.6 says that the cone of sets hyperarithmetically above one that is nonhyperarithmetic does not contain a nonempty Σ_1^1-subset. In particular, no cone above a nonhyperarithmetic set is Σ_1^1. From the model-theoretic point of view, the argument used in the proof of the theorem is a type-omitting argument. This will be used again later to construct nonstandard models for characterizing hyperarithmetic sets via the constructible hierarchy L.

2.5.4 Exercises

Exercise 2.5.1. *Prove that for any real z, the set* $\{(x, \mathcal{O}^x) \mid x \in 2^\omega\}$ *is not* $\Sigma^1_1(z)$.

Exercise 2.5.2. *Prove that there exist* $x, y \leq_h \mathcal{O}$ *such that* $x \nleq_h y$ *and* $y \nleq_h x$.

Exercise 2.5.3. *Prove that there is a sequence of reals* $\{x_i\}_{i<\omega}$ *such that* $x_i <_h x_{i+1} <_h \mathcal{O}$.

***Exercise 2.5.4** (Slaman [140]). *Prove that there is no* $\underset{\sim}{\Sigma}^1_1$*-set* $A \subseteq \omega^\omega$ *such that every* $\underset{\sim}{\Sigma}^1_1$*-set is a continuous injective image of A.*

***Exercise 2.5.5** (Hjorth [49]). *Prove that every* $\underset{\sim}{\Sigma}^1_1$*-set is a continuous injective image of the* Π^1_1*-set* $\{\mathcal{O}^x \mid x \in \omega^\omega\}$.

Exercise 2.5.6. *Prove that any nonempty* Σ^1_1*-set A contains an element x such that* $\mathcal{O}^x \leq_T \mathcal{O}$.

***Exercise 2.5.7** (Solovay [152]). *Given functions* $f, g : \omega \to \omega$, *we say that g dominates f if* $g(i) \geq f(i)$ *for* $i < \omega$. *Prove that a real x is hyperarithmetic if and only if there is a function* $f \geq_T x$ *such that* $g \geq_T f$ *whenever g dominates f.*

2.6 More on $\mathcal{O}$

2.6.1 A nonstandard $\mathcal{O}$

We construct a nonstandard $\mathcal{O}$ which shares many properties with $\mathcal{O}$. The following lemma says that there is an r.e. set whose field is an end extension of $\mathcal{O}$.

Lemma 2.6.1. *There is an r.e. set x and an r.e. partial ordering relation* $<_x$ *defined on x such that*
(i) *The conditions (i)–(iv) in § 1.2.1 hold for* $<_x$;
(ii) $\mathcal{O} \subseteq x$;
(iii) $<_o$ *is the restriction of* $<_x$ *to* $\mathcal{O}$; *and*
(iv) *if* $m <_x n$ *and* $n \in \mathcal{O}$, *then* $m \in \mathcal{O}$.

Proof. We build $<_x$ by effective transfinite induction. Define a recursive function h as follows:

$$\Phi_{h(e)}(\langle m, n\rangle) = \begin{cases} 1 & m = 1 \wedge n = 2, \\ 1 & (\exists k)(\Phi_e(\langle m, k\rangle) = 1 \wedge \Phi_e(\langle k, n\rangle) = 1), \\ 1 & (\exists k)(\Phi_e(\langle k, m\rangle) = 1 \wedge n = 2^m), \\ 1 & m \neq 1 \wedge n = 3 \cdot 5^d \wedge (\exists k)(\exists j)(\Phi_e(\langle j, m\rangle) = 1 \wedge \\ & m = \Phi_d(k)), \\ \uparrow & \text{Otherwise.} \end{cases}$$

By the Recursion Theorem, there is a constant c such that $\Phi_{h(c)} = \Phi_c$. Set $g = \Phi_c$ to be a partial recursive function. Hence

$$g(\langle m,n\rangle) = \begin{cases} 1 & m = 1 \wedge n = 2, \\ 1 & (\exists k)(g(\langle m,k\rangle) = 1 \wedge \langle g(k,n) = 1), \\ 1 & (\exists k)(g(\langle k,m\rangle) = 1 \wedge n = 2^m), \\ 1 & m \neq 1 \wedge\ n = 3\cdot 5^d \wedge (\exists k)(\exists j)(g(\langle j,m\rangle) = 1 \wedge, \\ & m = \Phi_d(k)), \\ \uparrow & \text{Otherwise.} \end{cases}$$

Define $m <_x n$ if and only if $g(\langle m,n\rangle) = 1$. Set $x = \{m \mid (\exists n)(m <_x n\}$. It is straightforward to verify that the field of $<_x$ satisfies (i)–(iv). □

Note that given $n \in x$, $<_x$ restricted to the set $\{m \mid m <_x n\}$ is not necessarily linear. We introduce the following modification. Define $z \subseteq x$ to be such that $i \in z$ if and only if $i \in x$ and $<_x$ restricted to $\{m \mid m <_x i\}$ is linear. Then z is an arithmetical set. It is obvious that $<_x$ restricted to z is a partial ordering, $\mathcal{O} \subseteq z$ and the conditions (i)–(iv) hold for $<_x$ restricted to z.

Now for any $n \in \mathcal{O}$, let z_n be a hyperarithmetic set such that $i \in z_n$ if and only if $i \in \mathcal{O}_{2^{2^n}}$ or there exists $j \in \mathcal{O}_{2^{2^n}} \setminus \mathcal{O}_{2^n}$ satisfying $j <_x i$ and $i \in z$. In other words, z_n is obtained by "cutting off" those nodes in z not above $\mathcal{O}_{2^{2^n}}$. Hence z_n has $\mathcal{O}_{2^{2^n}}$ as an "initial segment". Note that $<_x$ restricted to z_n also satisfies conditions (i)–(iv).

Definition 2.6.2. Let $\mathcal{O}^* = \bigcap A$ where

$$A = \{y \mid y \subseteq x \text{ is hyperarithmetic} \wedge \text{Conditions (i)–(iv) in § 1.2.1} \\ \text{hold for } <_x \text{ restricted to } y\}.$$

Denote by $<_{o^*}$ the ordering $<_x$ restricted to $\mathcal{O}^*$. Note that $z_n \in A$ for all $n \in \mathcal{O}$. Moreover $\mathcal{O} \subseteq \mathcal{O}^*$ by the definition of $\mathcal{O}$. Since conditions (i)–(iv) in § 1.2.1 are arithmetical statements, by Corollary 2.2.2 it follows that $\mathcal{O}^*$ is Σ^1_1.

2.6.2 A Π^1_1-path through $\mathcal{O}$

Since $\mathcal{O}^*$ is Σ^1_1, there must be an $i \in \mathcal{O}^* \setminus \mathcal{O}$. Note that for every $n \in \mathcal{O}$, $i \in z_n \setminus \mathcal{O}_n$. Hence for every $n \in \mathcal{O}$, there is an $i_n \in \mathcal{O}_{2^{2^n}} \setminus \mathcal{O}_{2^n}$ such that $i_n <_x i$ and thus $i_n <_{o^*} i$. Let $\mathcal{O}^*_i = \{j \mid j <_{o^*} i \wedge j \in \mathcal{O}\}$. Then $j \in \mathcal{O}^*_i$ if and only if $j \in \mathcal{O} \wedge j <_x i$. Hence $\mathcal{O}^*_i \subseteq \mathcal{O}$ is a Π^1_1-set.

Lemma 2.6.3. *$<_o$ restricted to $\mathcal{O}^*_i$ is a well-ordering of order type ω_1^{CK}.*

Proof. Obviously $<_o$ restricted to $\mathcal{O}^*_i$ is a well-ordering. Suppose that the order type of $<_o$ restricted to $\mathcal{O}^*_i$ is less than ω_1^{CK}. Then there is an $n \in \mathcal{O}$ such that $\mathcal{O}^*_i \subseteq \mathcal{O}_n$. Then for any $k \in \mathcal{O}_{2^{2^n}} \setminus \mathcal{O}_{2^n}$, $k \not<_{o^*} j$. Hence $j \notin z_n$, a contradiction. □

Thus $\mathcal{O}_i^* \subseteq \mathcal{O}$ is a Π_1^1-path through $\mathcal{O}$, yielding the following theorem.

Theorem 2.6.4 (Feferman and Spector [26]). *There is a Π_1^1-set $\mathcal{O}_1 \subsetneq \mathcal{O}$ such that $<_o$ restricted to $\mathcal{O}_1$ is a well-ordering of order type ω_1^{CK}.*

The proof of Theorem 2.6.4 is essentially model-theoretic. A purely recursion-theoretic proof can be found in Sacks [123]. The results in this section may also be uniformly relativized. Namely, there is a Σ_1^1-set $\{(x, \mathcal{O}^{x,*}) \mid x \in 2^\omega\}$ such that for each x, $\mathcal{O}^{x,*}$ is a proper end extension of $\mathcal{O}^x$ (a nonstandard $\mathcal{O}^x$).

2.6.3 A regular path through $\mathcal{O}$

Despite the apparent complexity of the partial ordering on $\mathcal{O}$, there is a Π_1^1-path through it in which every proper initial segment is recursive. This indicates that there exists certain regularity hidden within $<_o$.

Theorem 2.6.5 (Jockusch [56]). *There is set $O \subseteq \mathcal{O}$ such that*
- *$<_o$ (Definition 1.2.1) is a well-ordering of order type ω_1^{CK} on O;*
- *O is $<_o$-downward closed, and*
- *for any $n \in O$, $\{m \mid m <_o n\}$ is recursive.*

Proof. First, observe that by the Padding Lemma in classical recursion theory, we may assume that

$$(\forall m \in \mathcal{O})(\forall n \in \mathcal{O})(n +_o m > \max\{m, n\}).$$

By the *s-m-n* Theorem, there is a recursive function g_0 such that for every m,

$$\Phi_{g_0(e,m)}(0) = \Phi_e(\Phi_m(0)) \wedge (\forall n)(\Phi_{g_0(e,m)}(n+1) = \Phi_{g_0(e,m)}(n) +_o \Phi_e(\Phi_m(n+1)).$$

Define a recursive function h_0 as follows:

$$\Phi_{h_0(e)}(n) = \begin{cases} 1 & n = 1, \\ 2^{\Phi_e(k)} & n = 2^k, \\ 3 \cdot 5^{g_0(e,k)} & n = 3 \cdot 5^k, \\ \uparrow & \text{Otherwise.} \end{cases}$$

By the Recursion Theorem, there is a constant c such that $\Phi_{h_0(c)} = \Phi_c$. Then Φ_c is total on $\mathcal{O}$. Note that $\Phi_c(1) = 1$, $\Phi_c(2^n) = 2^{\Phi_c(n)}$, and if $n = 3 \cdot 5^k$, then $\Phi_c(n) = 3 \cdot 5^w$ where $\Phi_w(0) = \Phi_c(\Phi_k(0))$ and $\Phi_w(j+1) = \Phi_w(j) +_o \Phi_c(\Phi_k(j+1))$. Φ_c follows the format set in (i)–(iv) of § 1.2.1 with the additional property that for $n = 3 \cdot 5^k$, $\Phi_c(n)$ gives an index w of a recursive function that is recursively defined from Φ_k. Thus intuitively one may view this as a version of $\mathcal{O}$ on the range of Φ_c. Let $i \in \mathcal{O}^* \setminus \mathcal{O}$. We claim that the set

$$O = \{m \mid (\exists n \in \mathcal{O})(m <_{o^*} \Phi_c(n) \wedge n <_{o^*} i)\}$$

is as required. We first leave it as an exercise to verify that

(1) Φ_c is strictly increasing in the sense of $<_o$;
(2) for each j_0 and m, if $\Phi_w(j_0) \le_o m <_o \Phi_w(j_0+1)$, then there is a $j_1 \in \mathcal{O}$ such that $m = \Phi_c(j_0) +_o j_1$.

Since $n <_o m$ implies $\Phi_c(n) <_o \Phi_c(m)$, $<_o$ is a well-ordering on O. Note that for any $n \in \mathcal{O}$, $|n| \le |\Phi_c(n)|$. Hence $\Phi_c : \{n \in \mathcal{O} \mid n <_{o^*} i\} \to O$ is an embedding and hence $<_o$ is a well-ordering of order type ω_1^{CK} on O. By performing induction over the standard ordinal notations $<_{o^*}$ below i, one easily sees that O is $<_o$-downward closed.

By Theorem 1.3.4, there is a recursive function g_1 such that $W_{g_1(n)} = \{j \mid j <_o n\}$ for any $n \in O$. We claim that $W_{g_1(n)}$ is a recursive set when restricted to $n \in O$. It will then follow that the set $\{\langle j,k\rangle \mid j <_o k <_o n\}$ is recursive as well.

By an effective transfinite induction together with an application of the Recursion Theorem, we will define a partial recursive function q such that $W_{q(n)} = \omega \setminus W_{g_1(n)}$ for $n \in O$. The case $n = 2^k$ and $\Phi_c(n) = 2^{\Phi_c(k)}$ is left to the reader as an exercise (Exercise 2.6.1). Suppose $n = 3 \cdot 5^k$ and $\Phi_c(3 \cdot 5^k) = 3 \cdot 5^w$, where w is as defined earlier, and e is given such that $\Phi_e(\Phi_w(j))$ is defined for each $j < \omega$. First note that an induction on j shows that $\Phi_w(j) \ge j$. If for each j, $W_{\Phi_e(\Phi_w(j))} = \omega \setminus W_{g_1(\Phi_w(j))}$, then we claim

$$m \in \bigcup_{j<\omega} W_{g_1(\Phi_w(j))} \Leftrightarrow (\exists j_0 \le m)(m \in W_{g_1(\Phi_w(j_0+1))}).$$

The proof of the claim from right to left is immediate. For the other direction, let j_0 be such that $\Phi_w(j_0) \le_o m <_o \Phi_w(j_0+1)$. Then by (2) there is a $j_1 \in \mathcal{O}$ such that $\Phi_w(j_0) +_o j_1 = m$. This implies that $j_0 \le \Phi_w(j_0) \le m$. Since $m \le_o \Phi_w(j_0+1)$, we have proved the Claim.

Hence $\bigcup_{j<\omega} W_{g_1(\Phi_w(j))} = \omega \setminus \bigcap_{j<\omega} W_{\Phi_e(\Phi_w(j))}$ is recursive. Now this calculation is uniform, and hence there is a recursive function g_2 such that if $W_{\Phi_e(\Phi_w(j))} = \omega \setminus W_{g_1(\Phi_w(j))}$ for every j, then $W_{g_2(e,w)} = \bigcap_{j<\omega} W_{\Phi_e(\Phi_w(j))}$. Let g_3 be a recursive function such that $W_{g_3(e,w)} = W_e \setminus \{w\}$. Recall that if $\Phi_c(3 \cdot 5^k) = 3 \cdot 5^w \in O$, then $\Phi_w(j+1) = \Phi_w(j) +_o \Phi_c(\Phi_k(j+1))$. Let $W_{c_1} = \omega$.

Define a recursive function h_1 as follows:

$$\Phi_{h_1(e)}(n) = \begin{cases} c_1 & n = 1, \\ g_3(\Phi_e(m), m) & n = 2^m, \\ g_2(e,w) & n = 3 \cdot 5^w, \\ \uparrow & \text{Otherwise.} \end{cases}$$

By the Recursion Theorem, there is a d such that $\Phi_{h_1(d)} = \Phi_d$. Let $q = \Phi_d$. Then q is as required. □

2.6.4 Exercises

Exercise 2.6.1. *Complete the inductive argument for the case* $n = 2^k$ *and* $\Phi_c(n) = 2^{\Phi_c(k)}$ *in Theorem* 2.6.5.

Exercise 2.6.2. *Prove that for any* $i \in \mathcal{O}^* \setminus \mathcal{O}$, $<_{o^*}$ *is not a well-ordering on the set* $\{j \mid j \in \mathcal{O}^* \wedge j <_{o^*} i\}$.

Exercise 2.6.3. *Prove (1) and (2) in Theorem* 2.6.5.

Exercise 2.6.4. *Prove that there are* $2^{\aleph_0}$*-many paths through* $\mathcal{O}$.

Exercise 2.6.5. *Prove that there is no partial recursive function* $p : \omega^2 \to \omega$ *satisfying properties (1)–(3) in Exercise* 1.4.1 *upon replacing* x *with* $\mathcal{O}_1$.

2.7 Codes for sets

In § 2.2.3, a method of coding Δ^1_1-sets was introduced. However, the method does not provide an effective way of tracking the construction of Δ^1_1-sets. Here we discuss the method of hyperarithmetic coding which serves as a useful tool for performing inductive analysis of Δ^1_1-sets of reals.

2.7.1 Borel codes

Throughout this section, fix an effective enumeration $\{\sigma_i\}_{i<\omega}$ *of* $\omega^{<\omega}$.

For each $x \in 2^\omega$, define $(x)_n = \{m \mid 2^n \cdot 3^m \in x\}$. Define a sequence $\{\Sigma_\beta\}_{\beta<\omega_1}$ of sets of reals as follows:

$\beta = 0$: Let $\Sigma_0 = \{x \mid x(0) = 0\}$.

$\beta > 0$: Let $\Sigma_\beta = \{x \in 2^\omega \setminus \Sigma_0 \mid (\forall n)(\exists \gamma < \beta)((x)_n \in \Sigma_\gamma)\}$.

Let $\mathrm{BC} = \bigcup_{\beta<\omega_1} \Sigma_\beta$ be the set of *Borel codes*. For each $x \in \mathrm{BC}$, define A_x to be:

$$x \in \Sigma_0\colon A_x = \bigcup_{\{i \mid i>0 \wedge x(i)=1\}} [\sigma_i].$$

$\beta > 0$ and $x \in \Sigma_\beta$: $A_x = \bigcup_n (\omega^\omega \setminus A_{(x)_n})$.

If σ is the empty string, let $(y)_\sigma = y$. For each $\sigma \in \omega^{<\omega}$, define

$$(y)_\sigma = (\cdots((y)_{\sigma(1)})_{\sigma(2)})\cdots)_{\sigma(|\sigma|)}.$$

Given a real y, we define a tree $T^y \leq_T y$ over $\omega^{<\omega}$ as follows:

$$\emptyset \in T^y;$$

For any $\sigma \in T^y$, if $(y)_\sigma \in \Sigma_0$ then σ will have no extension in T^y. Otherwise, $\sigma^\frown n \in T^y$ for every n.

Lemma 2.7.1. $y \in \mathsf{BC}$ *if and only if* T^y *is a wellfounded tree. Hence* BC *is* Π^1_1. *Moreover* $y \in \Sigma_\beta$ *if and only if* $|T^y| \leq \beta + 1$.

Proof. If T^y is wellfounded, then one may show by induction on the rank of σ, for $\sigma \in T^y$, that $(y)_\sigma$ belongs to BC.

If $y \in \mathsf{BC}$, then for any $\tau \succ \sigma$ in T^y there exists a $\gamma < \beta$ such that $(y)_\tau \in \Sigma_\gamma$ and $(y)_\sigma \in \Sigma_\beta$. Thus T^y is a wellfounded tree.

Since $T^y \leq_T y$ uniformly and $y \in \mathsf{BC}$ if and only if T^y is wellfounded, we conclude that BC is Π^1_1. We leave the verification of the last statement of the lemma to the reader. □

2.7.2 Hyperarithmetic codes

A *hyperarithmetic code* is a real in $\mathsf{HYP} \cap \mathsf{BC}$. Let $\mathsf{BC(HYP)}$ denote the collection of hyperarithmetic codes. By Corollary 2.2.2 and Lemma 2.7.1, BC(HYP) is Π^1_1.

Since $T^x \leq_T x$, if $x \in \mathsf{BC(HYP)}$ then by Lemma 2.7.1 and the discussion above, $|T^x| < \omega_1^{\mathrm{CK}}$. Thus $\mathsf{BC(HYP)} \subseteq \Sigma_{<\omega_1^{\mathrm{CK}}} = \bigcup_{\beta<\omega_1^{\mathrm{CK}}} \Sigma_\beta$.

Theorem 2.7.2 (Kleene [70]). *The following statements are equivalent:*
(i) $A = A_z$ *for some* $z \in \mathsf{BC} \cap \Delta^0_1$.
(ii) $A = A_z$ *for some* $z \in \mathsf{BC(HYP)}$.
(iii) A *is* Δ^1_1.

Proof. Obviously, (i) ⇒ (ii). For (ii) ⇒ (iii), suppose $A = A_z$ for some $z \in \mathsf{BC(HYP)}$. We define a Π^1_1-relation $R_z \subseteq \omega^{<\omega} \times \omega^\omega$ as follows. R_z will decode A_z from T^z, and retraces the steps that constructed T^z from A_z:

(1) $(\forall \tau \notin T^z)(\forall y)\neg R_z(\tau, y)$;
(2) $(\forall \tau \in T^z)(\forall y)(\tau \text{ has no extensions} \to (R_z(\tau, y) \leftrightarrow (\exists n)(y \succ \sigma_n \wedge (z)_\tau(n) = 1)))$;
(3) $(\forall \tau \in T^z)((\exists n)(\tau^\frown n \in T^z) \to (\forall y)(R_z(\tau, y) \leftrightarrow (\exists m)\neg R_z(\tau^\frown m, y)))$.

It is obvious that R_z is unique and well defined. Moreover, for any $z \in \mathsf{BC}$, $m \in \omega$, and $y \in \omega^\omega$, $R_z(\emptyset^\frown m, y)$ if and only if $R_{(z)_m}(\emptyset, y)$. We make the following claim.

Claim. $R_z(\emptyset, y)$ *if and only if* $y \in A_z$.

Proof of Claim. We prove this by induction. First assume that $z \in \Sigma_0$. Then T^z consists only of the empty string and so by (2) $R_z(\emptyset, y)$ if and only if $y \in A_z$. Suppose the Claim holds for all Σ_γ-codes where $\gamma < \beta$. Let $z \in \Sigma_\beta$. If $T_z(\emptyset, y)$ then by (3) for some m, $\neg R_z(\emptyset^\frown m, y)$. Then by induction $y \notin A_{(z)_m}$, hence $y \in A_z$. Conversely, if $y \in A_z$ then there is an m such that $y \notin A_{(z)_m}$. Now $(z)_m \in \Sigma_\gamma$ for some $\gamma < \beta$. By induction, $\neg R_{(z)_m}(\emptyset, y)$. Thus $R_z(\emptyset, y)$ by (3), proving the Claim.

The Claim implies that A_z is $\Pi^1_1(z)$. By a similar argument, one shows that the complement of A_z is also $\Pi^1_1(z)$. Hence A_z is $\Delta^1_1(z)$. Since $z \in \mathsf{HYP}$, A_z is Δ^1_1.

We now show (iii) ⇒ (i). Suppose that A is Δ^1_1. Then both A and $B = \omega^\omega \setminus A$ are Σ^1_1. So there are two recursive trees S and T such that

$$x \in A \Leftrightarrow (\exists y)(\forall n)(x \upharpoonright n, y \upharpoonright n) \in S$$

and

$$x \in B \Leftrightarrow (\exists y)(\forall n)(x \upharpoonright n, y \upharpoonright n) \in T.$$

Define $R = S \otimes T$ as follows:

$$(\sigma, \tau, \nu) \in R \Leftrightarrow (\sigma, \tau) \in S \wedge (\sigma, \nu) \in T.$$

Define $(\sigma, \tau, \nu) < (\sigma', \tau', \nu')$ if $\sigma \succ \sigma'$, $\tau \succ \tau'$ and $\nu \succ \nu'$. Since $A \cap B = \emptyset$, R is a wellfounded tree under $<$. We construct a hyperarithmetic code z so that $B \subseteq A_z$ and $A_z \cap A = \emptyset$. Then $A_z = B$.

We define a recursive function $h : \omega \to \omega$ such that for some c, $\Phi_{h(c)}(n, \emptyset, \emptyset, \emptyset)$ is a hyperarithmetic code for A.

For (σ, τ, ν) such that $|\sigma| = |\tau| = |\nu|$, there are three cases to consider:
(1) $(\sigma, \tau, \nu) \notin R$ and $(\sigma, \nu) \in T$. Define $\Phi_{h(e)}(0, \sigma, \tau, \nu) = 0$ and $\Phi_{h(e)}(n, \sigma, \tau, \nu) = 1$ for all $n > 0$.
(2) $(\sigma, \tau, \nu) \notin R$ and $(\sigma, \nu) \notin T$. Define $\Phi_{h(e)}(n, \sigma, \tau, \nu) = 0$ for all n.
(3) $(\sigma, \tau, \nu) \in R$. For (i, j, k, l), we define a recursive function g determined by the following four subcases:

3(i) $i = k$ and $(\sigma^\frown i, \tau^\frown j, \nu^\frown l) \in R$. Define $g(e, i, j, k, l, n) = \Phi_e(n, \sigma^\frown i, \tau^\frown j, \nu^\frown l)$.

3(ii) $i \neq k$. Define $g(e, i, j, k, l, n) = m$ for all $n > 0$ and $g(e, i, j, k, l, 0) = 0$, where m is a code such that $\sigma_m = \sigma^\frown k$.

3(iii) $i = k$ and $(\sigma^\frown i, \tau^\frown j) \notin S$. Define $g(e, i, j, k, l, n) = 1$ for all $n > 0$ and $g(e, i, j, k, l, 0) = 0$.

3(iv) $i = k$ and $(\sigma^\frown i, \tau^\frown j) \in S$ but $(\sigma^\frown i, \nu^\frown l) \notin T$. Define $g(e, i, j, k, l, n) = 0$ for all n.

Intuitively, g is a hyperarithmetic code for a set $A_{i,j,k,l}$ disjoint from $[S_{(\sigma^\frown i, \nu^\frown j)}]$ but covers $[T_{(\sigma^\frown k, \nu^\frown l)}]$, where

$$[S_{(\sigma^\frown i, \tau^\frown j)}] = \{x \succ \sigma^\frown i \mid (\exists y)(y \succ \tau^\frown j \wedge (\forall n)((x \upharpoonright n, y \upharpoonright n) \in S))\}$$

and

$$[T_{(\sigma^\frown k, \nu^\frown l)}] = \{x \succ \sigma^\frown k \mid (\exists y)(y \succ \nu^\frown l \wedge (\forall n)((x \upharpoonright n, y \upharpoonright n) \in T))\}.$$

The plan is to define $\Phi_{h(e)}(n, \sigma, \tau, \nu)$ to code the set $C = \bigcup_{k,l} \bigcap_{i,j} A_{i,j,k,l}$. Then $C \supseteq [T_{(\sigma,\nu)}]$ and $C \cap [S_{(\sigma,\tau)}] = \emptyset$. Define a recursive function $p^1_{k,l}$ such that $p^1_{k,l}(0) = 1$ and $p^1_{k,l}(2^{\langle i,j \rangle} \cdot 3^n) = 1$ if and only if $g(e, i, j, k, l, n) = 1$. Then $p^1_{k,l}$ will code the set $\bigcup_{i,j}(\omega^\omega \setminus A_{i,j,k,l})$.

Define a recursive function p^2 so that $p^2(0) = 1$ and $p^2(2^{\langle k,l \rangle} \cdot 3^n) = 1$ if and only if $p^1_{k,l}(n) = 1$. Then p^2 is a code for $\bigcup_{k,l}(\omega^\omega \setminus \bigcup_{i,j}(\omega^\omega \setminus A_{i,j,k,l})) = \bigcup_{k,l} \bigcap_{i,j} A_{i,j,k,l}$.

Let $\Phi_{h(e)}(n, \tau, \nu_0, \nu_1) = 1$ if and only if $p^2(n) = 1$. By the Recursion Theorem, there is a fixed point c such that $\Phi_{h(c)} = \Phi_c$.

Let $z(n) = 1 \Leftrightarrow \Phi_c(n, \emptyset, \emptyset, \emptyset) = 1$. By an easy transfinite induction along $<$ on R, one shows that for any $(\sigma, \tau, \nu) \in R$, $z_{(\sigma,\tau,\nu)} = \{n \mid \Phi_c(n, \sigma, \tau, \nu) = 1\}$ is a recursive Borel code such that $A_{z_{(\sigma,\tau,\nu)}} \supseteq [T_{(\sigma,\nu)}]$ and $A_{z_{(\sigma,\tau,\nu)}} \cap [S_{(\sigma,\tau)}] = \emptyset$. In particular, $z^* = z_{(\emptyset,\emptyset,\emptyset)}$ is a recursive Borel code such that $A \cap A_{z^*} = A \cap [S_{\emptyset,\emptyset}] = \emptyset$ and $[T_{(\emptyset,\emptyset)}] = B \subseteq A_{z^*}$. Thus z^* is a recursive Borel code for B. □

Note that more was proved in the theorem above. Recall (see § 1.5.1) that there is a Π^1_1-set $A \subseteq \omega \times \omega^\omega$ such that every Σ^1_1-set B is of the form $B = \neg A_n = \{x \mid (n, x) \notin A\}$ for some n. These n's are called Σ^1_1-codes. Then Theorem 2.7.2 may be strengthened to: There is a recursive function $f : \omega \times \omega \to \omega^\omega$ such that if n, m are Σ^1_1-codes and $(2^\omega \setminus A_n) \cap (2^\omega \setminus A_m) = \emptyset$, then $f(n, m) \in \mathsf{BC} \cap \Delta^0_1$, $(2^\omega \setminus A_{f(n,m)}) \cap (2^\omega \setminus A_n) = \emptyset$, and $(2^\omega \setminus A_m) \subseteq (2^\omega \setminus A_{f(n,m)})$. This is an *effective* separation theorem which is a central result in effective descriptive set theory. We will discuss related facts in the sequel.

Remark 2.7.3. One may wonder how recursive codes are able to code the same sets as those coded hyperarithmetically. The answer is that there is a trade-off in the number of induction steps used. For example, a hyperarithmetic open set can be coded by a hyperarithmetic code in one step but any recursive code for it will be in some Σ_β where β is a possibly large recursive ordinal.

We say that $A \subseteq \omega^{<\omega}$ is $\Sigma_\beta(\mathsf{HYP})$ if $A = A_x$ for some hyperarithmetic $x \in \Sigma_\beta$. Louveau [85], using Gandy forcing, proved the following result.

Theorem 2.7.4 (Louveau [85]). *If A, B are Σ^1_1 and separated by A_x for some $x \in \Sigma_\beta$, then A is separable from B by A_y for some $y \in \Sigma_\beta \cap \Delta^1_1$. Hence if A is Δ^1_1 and Σ_β, then A is $\Sigma_\beta(\mathsf{HYP})$.*

Theorem 2.7.2 yields another parameterization of Δ^1_1-sets of reals.

Corollary 2.7.5. *There is a Π^1_1-set $A \subseteq \omega \times \omega^\omega$ such that*
- $(\forall n)(A_n = \{y \mid (n, y) \in A\} \in \Delta^1_1)$,
- $(\forall C)(\exists n)(C \in \Delta^1_1 \to C = A_n)$.

Proof. By Proposition 2.2.11, there is a partial function $D : \omega \times \omega \to \omega$ with a Π^1_1-graph parameterizing Δ^1_1-reals. Define $(n, y) \in A$ if and only if

$$(\forall i)(\exists j)(n, i, j) \in D \wedge (\forall x)((\forall i)(\forall j)(x(i) = j \leftrightarrow (n, i, j) \in D) \to x \in \mathsf{BC(HYP)} \wedge y \in A_x).$$

Note that if $(\forall i)(\exists j)(n, i, j) \in D$, then the predicate $(n, i, j) \in D$ is Δ^1_1. So A is Π^1_1. Obviously each Δ^1_1-set C is A_n for some n, and A_n codes a real in $\mathsf{BC(HYP)}$. For each n, either $A_n = \emptyset$ or $A_n = A_x$ for some $x \in \mathsf{BC(HYP)}$. Hence A is as required. □

2.7.3 Exercises

Exercise 2.7.1. *Prove Lemma* 2.4.8.

Exercise 2.7.2. *Prove that* $y \in \Sigma_\beta$ *if and only if* $|T^y| \leq \beta + 1$.

3 Admissibility and constructibility

This chapter introduces the basic notions of effective computability from the set-theoretic perspective. While somewhat abstract at first sight, they are natural when viewed from the standpoint of definability.

An admissible structure is essentially an inner model of computation. The constructible universe L is defined via a transfinite iteration of the Gödel operators. By considering an appropriate initial segment of L which forms an admissible structure, we will see how a Π_1^1-set assumes the role of an r.e. set in higher recursion theory.

3.1 Kripke–Platek set theory

The Kripke–Platek (KP) theory of sets was developed by Kripke [75] and Platek [113]. KP is a weak version of the ZF axiom system and captures the basic features of effective computation in the setting of set theory. It offers a natural platform for the study in domains beyond ω.

3.1.1 KP

Definition 3.1.1. Let $\mathscr{L}$ denote the language of ZF. The collection of *Δ_0-formulas* in $\mathscr{L}$ is the smallest collection X consisting of the atomic formulas of $\mathscr{L}$ and closed under the following operations:
(i) If $\varphi \in X$ then $\neg\varphi \in X$.
(ii) If $\{\varphi, \psi\} \subseteq X$, then $\{\varphi \wedge \psi, \varphi \vee \psi\} \subseteq X$.
(iii) If $\varphi \in X$, then $\{(\forall u \in v)\varphi, (\exists u \in v)\varphi\} \subseteq X$.

In (iii) above, both of the formulas $(\forall u \in v)\varphi$, $(\exists u \in v)\varphi$ are called *bounded formulas*. A prefix of the form $\exists u \in v$ or $\forall u \in v$ is a *bounded quantifier. Δ_0-formulas are analogs of primitive recursive predicates in recursion theory.*

Definition 3.1.2. A Δ_0-formula is also known as a Σ_0-formula as well as a Π_0-formula. In general, a formula $\varphi(u)$ is Σ_{n+1} if it is of the form $(\exists v)\psi(u, v)$ where ψ is a Π_n-formula. A formula $\varphi(u)$ is Π_{n+1} if it is of the form $(\forall v)\psi(u, v)$ where ψ is a Σ_n-formula.

Definition 3.1.3. The class of *Σ-formulas* is the smallest class X consisting of the Δ_0-formulas and closed under conjunction, disjunction, bounded quantification, and if $\varphi \in X$ then $(\exists u)\varphi \in X$.

The class of *Π-formulas* is the smallest class X' consisting of the Δ_0-formulas, closed under conjunction, disjunction, bounded quantification, and if $\varphi \in X'$ then $(\forall u)\varphi \in X'$.

Σ-formulas are analogs of recursively enumerable predicates in classical recursion theory. Note that there is a difference between a Σ_1-formula and a Σ-formula. A Σ_1-formula is a Σ-formula but not necessarily conversely.

Definition 3.1.4. The theory KP consists of the universal closure of the following formulas:
- Extensionality: $(\forall x)(\forall y)((\forall u)(u \in x \leftrightarrow u \in y) \rightarrow x = y)$.
- Foundation: $(\exists x)\varphi(x) \rightarrow (\exists x)(\varphi(x) \wedge (\forall y \in x)\neg\varphi(y))$, for all formulas φ in which y does not occur free.
- Pair: $(\forall x)(\forall y)(\exists z)(x \in z \wedge y \in z)$.
- Union: $(\forall x)(\exists y)(\forall u \in x)(\forall v)(v \in u \rightarrow v \in y)$.
- Δ_0-separation: $(\forall x)(\exists y)(\forall u)(u \in y \leftrightarrow u \in x \wedge \varphi(u))$, for all Δ_0-formulas φ in which u does not occur free.
- Δ_0-collection: $(\forall x)((\forall u \in x)(\exists v)\varphi(u, v) \rightarrow (\exists y)(\forall u \in x)(\exists v \in y)\varphi(u, v))$, for all Δ_0-formulas φ in which y does not occur free.

Given a formula φ, we write $\varphi^{(w)}$ for the formula obtained upon replacing each unbounded quantifier in φ with a bounded quantifier restricted to w. Specifically, replace $\exists u$ with $\exists u \in w$ and $\forall u$ with $\forall u \in w$. Clearly if φ is Σ, then $\exists w \varphi^{(w)}$ is Σ. Under KP, each Σ-formula is equivalent to a Σ_1-formula, as given by the following theorem which is proved by induction on the complexity of formulas. We leave it as an exercise.

Theorem 3.1.5 (The Σ-reflection principle). *Let φ be a Σ-formula. Then*

$$\mathrm{KP} \vdash \varphi \leftrightarrow (\exists w)\varphi^{(w)}.$$

Hence the theorem asserts under KP the existence of a Δ_0-definable set for a given Σ-formula. In fact, it says more:

Theorem 3.1.6 (The Σ-collection principle). *Let φ be a Σ-formula and let a be a set, Then*

$$\mathrm{KP} \vdash (\forall x \in a)(\exists y)\varphi(x, y) \rightarrow (\exists w)((\forall x \in a)(\exists y \in w)\varphi(x, y) \wedge$$
$$(\forall y \in w)(\exists x \in a)\varphi(x, y)).$$

Proof. Assume that $(\forall x \in a)(\exists y)\varphi(x, y)$. Then the formula $(\forall x \in a)(\exists y)\varphi(x, y)$ is Σ. By Theorem 3.1.5, there is a set c such that $(\forall x \in a)(\exists y \in c)\varphi^{(c)}(x, y)$. Define $b = \{y \mid (\exists x \in a)\varphi^{(c)}(x, y)\}$. By Δ_0-separation, such a b exists. It is straightforward to verify that $\varphi^{(c)}(x, y)$ implies $\varphi(x, y)$ since $\varphi(x, y)$ is Σ. Hence $(\forall y \in b)(\exists x \in a)\varphi(x, y)$. Since $(\forall x \in a)(\exists y \in c)\varphi^{(c)}(x, y)$, we have in fact $(\forall x \in a)(\exists y)\varphi^{(c)}(x, y)$. It follows that $(\forall x \in a)(\exists y \in b)\varphi^{(c)}(x, y)$. □

Theorem 3.1.7 (The Δ-separation principle). *Let a be a set. For any Σ-formula $\varphi(x)$ and Π-formula $\psi(x)$,*

$$\mathrm{KP} \vdash (\forall x \in a)(\varphi(x) \leftrightarrow \psi(x)) \rightarrow (\exists w)(\forall x)(x \in w \leftrightarrow x \in a \wedge \varphi(x)).$$

Proof. Assume $(\forall x \in a)(\varphi(x) \leftrightarrow \psi(x))$. Then $(\forall x \in a)(\varphi(x) \vee \neg\psi(x))$. Note that the formula $\varphi(x) \vee \neg\psi(x)$ is Σ. By Theorem 3.1.5, there is a set c such that $(\forall x \in a)\varphi(x) \vee \neg\psi(x)$ if and only if $(\forall x \in a)\varphi^{(c)}(x) \vee \neg\psi^{(c)}(x)$. Define $b = \{y \mid y \in a \wedge \varphi^{(c)}(y)\}$. By Δ_0-separation, such a b exists.

If $y \in b$, then $y \in a \wedge \varphi^{(c)}(y)$ and so $y \in a \wedge \varphi(y)$ since $\varphi(y)$ is Σ. If $y \in a \wedge \varphi(y)$, then $y \in a \wedge \psi(y)$ and so $y \in a \wedge \psi^{(c)}(y)$ since ψ is Π. □

3.1.2 Definition by Σ-recursion

As seen in Chapters 1 and 2, definition by recursion using effective transfinite induction is a powerful tool in second order arithmetic. We now study definition by recursion in the setting of admissible set theory. This may be viewed as a generalization of recursion over ω_1^{CK}.

Recall that a set x is transitive if $(\forall y)(\forall z)(y \in x \wedge z \in y \to z \in x)$. Note that the formula "x is transitive" is Δ_0.

A function f is Σ if there is a Σ-formula φ such that $f(x) = y \Leftrightarrow \mathrm{KP} \vdash \varphi(x, y)$. By Theorem 3.1.5, f is defined by a Σ_1-formula. It is not difficult to see that if f is Σ, then for any set u, $\mathrm{KP} \vdash (\exists v)(v = f \restriction u)$, where $f \restriction u = \{(x, y) \mid f(x) = y \wedge x \in u\})$.

Theorem 3.1.8. *Let G be an $n+2$-ary Σ-function. There is an $n+1$-ary Σ-function F such that*

$$\mathrm{KP} \vdash F(y, x_1, \dots, x_n) = G(y, x_1, \dots, x_n, \{(z, F(z, x_1, \dots, x_n)) \mid z \in y\}).$$

Proof. Assume $n = 0$ for simplicity. Let $P(f)$ be the sentence:

$$\begin{aligned}&f \text{ is a function } \wedge \mathrm{Dom}(f) \text{ is transitive} \wedge \\ &(\forall y \in \mathrm{Dom}(f))(f(y) = G(y, \{(z, f(z)) \mid z \in y\})).\end{aligned}$$

Note that $P(f)$ is a Σ-formula. Define

$$Q(y, z) = (\exists f)(P(f) \wedge f(y) = z),$$

and

$$F(y) = z \Leftrightarrow Q(y, z).$$

It suffices to show that

$$\mathrm{KP} \vdash F \text{ is a function } \wedge (\forall y)(\exists z)Q(y, z).$$

We show that $\mathrm{KP} \vdash (\forall y)(\exists z)Q(y, z)$. The other conjunct may be proved using a similar argument. Suppose KP does not prove $(\forall y)(\exists z)Q(x, y, z)$. Then there is a model $\mathcal{M} = \langle M, \in\rangle$ of KP such that $\mathcal{M} \models (\exists y)(\forall z)\neg Q(x, y, z)$. By the Axiom of Foundation, choose y such that

$$\mathcal{M} \models (\forall z)\neg Q(y, z) \wedge (\forall y' \in y)(\exists z)Q(y', z).$$

Since $Q(y, z)$ is Σ_1, by Theorem 3.1.6, there is a $v \in M$ such that

$$\mathcal{M} \models (\forall y' \in y)(\exists f \in v)(\exists z)(f(y') = z \wedge P(f)).$$

Hence there is a $w \in M$ such that $w = \{f \mid f \in v \wedge P(f)\}$.

For any $f_1, f_2 \in w$ and $y' \in \mathrm{Dom}(f_1) \cap \mathrm{Dom}(f_2)$, we claim that $f_1(y') = f_2(y')$. Suppose not. Then by the Axiom of Foundation again, there is a y' such that $f_1(y') \neq f_2(y')$ but $f_1(y'') = f_2(y'')$ for all $y'' \in y'$. However,

$$f_1(y') = G(y', \{(y'', f_1(y'')) \mid y'' \in y'\}) = G(y', \{(y'', f_2(y'')) \mid y'' \in y'\}) = f_2(y'),$$

a contradiction.

Now take $f = \bigcup_{f \in w} f$. By Theorem 3.1.7, $f \in M$. Obviously, $\mathrm{Dom}(f)$ is transitive and so $P(f)$. Define $f' = f \cup \{(y, G(y, f \upharpoonright y))\}$. Then $\mathcal{M} \models f'(y) = G(y, f \upharpoonright y) \wedge P(f')$ and so $\mathcal{M} \models Q(y, G(y, f' \upharpoonright y))$. Hence $\mathcal{M} \models (\exists z)Q(y, z)$, a contradiction.

Thus $\mathrm{KP} \vdash (\forall y)(\exists z)Q(y, z)$. □

Remark 3.1.9. Defining a function such as G in Theorem 3.1.8 quite often requires one to go through a series of straightforward, though tedious, steps of Σ_1-induction. Henceforth we will leave the details of these steps to the reader and refer to them simply as "applying Theorem 3.1.8 by Σ_1-induction".

Theorem 3.1.8 is a central result in the theory of admissible sets. It plays the same role as that of effective transfinite induction in descriptive set theory. We illustrate this point with an application of the theorem. Define the transitive closure operator TC to be:

$$\mathrm{TC}(\emptyset) = \emptyset, \text{ and}$$
$$\mathrm{TC}(a) = a \cup \bigcup \{\mathrm{TC}(x) \mid x \in a\}.$$

By induction on the relation $\in$, one shows that for any set x, $\mathrm{TC}(x)$ is the smallest transitive set containing x. The question is whether this operator is well defined in KP.

Let G be a Σ-function such that $G(\emptyset, y) = \emptyset$ and $G(x, y) = x \cup \bigcup \{y_1 \mid (y_0, y_1) \in y\}$. Then by Theorem 3.1.8 there is a Σ-function F such that $F(x) = G(x, \{(y, F(y)) \mid y \in x\})$. It is not difficult to verify that for each x, $F(x) = \mathrm{TC}(x)$. We have the following corollary.

Corollary 3.1.10. $\mathrm{KP} \vdash \mathrm{TC}(x)$ *is a* Σ*-function.*

By Corollary 3.1.10, together with the proof of Theorem 3.1.8, we have the following theorem.

Theorem 3.1.11. *Let G be an $n+2$-ary Σ-function. There is an $n+1$-ary Σ-function F such that*

$$\mathrm{KP} \vdash F(y, x_1, \ldots, x_n) = G(y, x_1, \ldots, x_n, \{(z, F(z, x_1, \ldots, x_n)) \mid z \in \mathrm{TC}(y)\}).$$

Let rk be a function such that

$$\mathrm{rk}(\emptyset) = 0, \text{ and}$$
$$\mathrm{rk}(x) = \sup\{\mathrm{rk}(y) + 1 \mid y \in x\}.$$

By Theorem 3.1.8, rk is a Σ-function from sets to ordinals. For any set x, $\mathrm{rk}(x)$ is called the *rank* of x. The rank function is a convenient tool to use when performing transfinite induction.

3.1.3 The collapsing lemma

A set x is *transitive in* z if $y \cap z \subseteq x$ for all $y \in x$. A set x is *transitive* if $z = x$, i.e. $y \subseteq x$ for all $y \in x$. Using Theorem 3.1.8, we define a Σ-operation $C_x(y)$ as follows:

$$C_x(\emptyset) = \emptyset;$$
$$C_x(a) = \{C_x(y) \mid y \in a \cap x\}.$$

It is easy to verify that $C_b(b)$ is transitive for all b. Moreover if $a \subseteq b$ and a is transitive, then $C_b(x) = x$ for all $x \in a$. Define $c_b = \{(x, C_b(x)| x \in b\}$. One may view c_b as a map from b onto the transitive set $C_b(b)$. In general, c_b is not injective.

A set a is *extensional* if for all $b_1, b_2 \in a$, $(\forall x \in a)(x \in b_1 \leftrightarrow x \in b_2) \rightarrow b_1 = b_2)$. In other words, a is extensional if each set in a is determined by its membership in a. Obviously, any transitive set is extensional. The following theorem says that any extensional set is isomorphic to a transitive set.

Theorem 3.1.12 (Mostowski [108]). *If a is extensional then c_a is a bijection between a and $C_a(a)$. Furthermore, for all $x, y \in a$, $x \in y \leftrightarrow c_a(x) \in c_a(y)$.*

Proof. Obviously c_a is a surjective map from a to $C_a(a)$ and $x \in y$ implies $c_a(x) \in c_a(y)$. We show that for all $x, y \in a$, $c_a(x) \in c_a(y)$ implies $x \in y$. Then by the extensionality of a, c_a is injective.

Suppose there is a $y_0 \in a$ such that for all $y \in a$ with $\mathrm{rk}(y) < \mathrm{rk}(y_0)$ and $x \in a$, $x \in y \Leftrightarrow c_a(x) \in c_a(y)$ but there is an $x_0 \in a$ such that $x_0 \notin y_0$ and $c_a(x_0) \in c_a(y_0)$. Since $c_a(x_0) \in c_a(y_0)$, there is a $z \in a \cap y_0$ such that $c_a(x_0) = c_a(z)$. Then $\mathrm{rk}(z) < \mathrm{rk}(y_0)$. However, this implies $x_0 = z \in y_0$, a contradiction. □

3.1.4 Definability

Denote by M the domain of a structure $\mathcal{M}$.

Definition 3.1.13. Let $\mathcal{F}$ be a set of formulas in $\mathcal{L}$.

(i) Given a structure $\mathcal{M}$ and $A \subseteq M$, we say that A is *$\mathcal{F}$-definable* in $\mathcal{M}$ if there is a formula $\varphi(u, v_0, \dots, v_n) \in \mathcal{F}$ and a finite sequence $a_0, \dots, a_n$ in M such that $x \in A$ if and only if $\mathcal{M} \models \varphi(x, a_0, \dots, a_n)$;

(ii) if $\mathfrak{M}$ is a collection of structures and $A_M \subseteq M$ for $\mathcal{M} \in \mathfrak{M}$, then $\{A_M\}_{\mathcal{M}\in\mathfrak{M}}$ is $\mathcal{F}$-*uniformly definable* in $\mathfrak{M}$ if there is a formula $\varphi(u)$ in $\mathcal{F}$ without parameters such that for any $\mathcal{M} \in \mathfrak{M}$, $x \in A_{\mathcal{M}}$ if and only if $\mathcal{M} \models \varphi(x)$.

Note that for a collection of structures to be uniformly definable, there should be no parameters in the formula defining it.

3.1.5 Absoluteness

Definition 3.1.14. Given a set of sentences $\mathcal{F}$ in $\mathscr{L}$, we say that two $\mathscr{L}$-structures $\mathcal{M}$ and $\mathcal{N}$ are *$\mathcal{F}$-elementarily equivalent*, written $\mathcal{M} \equiv_{\mathcal{F}} \mathcal{N}$, if for every $\varphi \in \mathcal{F}$,

$$\mathcal{M} \models \varphi \Leftrightarrow \mathcal{N} \models \varphi.$$

If $\mathcal{F}$ is the set of all sentences, then we say that $\mathcal{M}$ and $\mathcal{N}$ are *elementarily equivalent* and write $\mathcal{M} \equiv \mathcal{N}$.

Definition 3.1.15. Given a set of sentences $\mathcal{F}$, we say that an embedding j from $\mathcal{M}$ into $\mathcal{N}$ is *$\mathcal{F}$-elementary* and write $\mathcal{M} \prec^{j}_{\mathcal{F}} \mathcal{N}$ if for every $\varphi(\vec{v}) \in \mathcal{F}$ and finite sequence $\vec{a} \in M^{<\omega}$,

$$\mathcal{M} \models \varphi(\vec{a}) \Leftrightarrow \mathcal{N} \models \varphi(j(\vec{a})).$$

If $\mathcal{F}$ is the set of all formulas, then we write $\mathcal{M} \prec^{j} \mathcal{N}$. If j is the inclusion map, then we write $\mathcal{M} \prec \mathcal{N}$ and say that $\mathcal{M}$ is an *elementary substructure* of $\mathcal{N}$, and $\mathcal{N}$ is an *elementary extension* of $\mathcal{M}$.

Definition 3.1.16. Given a class $\mathfrak{M}$ of structures and a formula $\varphi(\vec{v})$,
(i) φ is *$\mathfrak{M}$-upward absolute* if for every pair $\mathcal{M} \subseteq \mathcal{N}$ in $\mathfrak{M}$ and any finite sequence $\vec{a} \in M^{<\omega}$, $\mathcal{M} \models \varphi(\vec{a})$ implies $\mathcal{N} \models \varphi(\vec{a})$;
(ii) $\varphi(\vec{v})$ is *$\mathfrak{M}$-downward absolute* if for every pair $\mathcal{M} \subseteq \mathcal{N}$ in $\mathfrak{M}$ and any finite sequence $\vec{a} \in M^{<\omega}$, $\mathcal{N} \models \varphi(\vec{a})$ implies $\mathcal{M} \models \varphi(\vec{a})$;
(iii) $\varphi(\vec{v})$ is *$\mathfrak{M}$-absolute* if it is both $\mathfrak{M}$-upward and downward absolute.

Proposition 3.1.17. *Let $\mathfrak{M}$ be a class of transitive structures.*
(i) *Every Σ_0-formula is $\mathfrak{M}$-absolute;*
(ii) *every Σ_1-formula is $\mathfrak{M}$-upward absolute;*
(iii) *every Π_1-formula is $\mathfrak{M}$-downward absolute.*
Hence every Δ_1-formula is $\mathfrak{M}$-absolute.

Proof. We prove (i) by induction on the complexity of Σ_0-formulas and leave (ii) and (iii) to the reader. Suppose $\mathcal{M} \subseteq \mathcal{N}$ and $\varphi(a)$ has the form $(\exists u \in a)\psi(u, a)$ for some Σ_0-formula $\psi(u, a)$ with $a \in M$. If $\mathcal{M} \models \varphi(a)$, then there exists an $a' \in M$ such that $\mathcal{M} \models \psi(a', a)$. Since $\mathcal{M} \subseteq \mathcal{N}$, by induction $\mathcal{N} \models \psi(a', a)$. If $\mathcal{N} \models \varphi(a)$, then there exists an $a' \in N$ such that $\mathcal{N} \models \psi(a', a)$. Since $a' \in a$ and $a \in M$, by transitivity we have $a' \in M$. Then by induction, $\mathcal{M} \models \psi(a', a)$. □

Further facts concerning KP may be found in [3].

3.1.6 Exercises

Exercise 3.1.1. *Prove Theorem* 3.1.5.

Exercise 3.1.2. *Complete the proof of Proposition* 3.1.17.

3.2 Admissible sets

The universe of sets V is defined inductively on ordinals:

$$\begin{aligned} V_0 &= \emptyset, \\ V_{\beta+1} &= \mathcal{P}(V_\beta), \\ V_\lambda &= \bigcup_{\alpha<\lambda} V_\alpha \text{ when } \lambda \text{ is a limit ordinal, and} \\ V &= \bigcup_{\alpha} V_\alpha. \end{aligned}$$

Definition 3.2.1. An *admissible structure* (or an admissible set) is a model $\mathcal{M}$ of KP of the form $\mathcal{M} = \langle M, \in_M \rangle$ where M is a transitive set in V and $\in_M$ is the restriction of $\in$ to M. We will often write $\in_M$ simply as $\in$ if the context is clear.

Definition 3.2.2. The *ordinal of* a transitive structure $\mathcal{M}$, denoted $o(\mathcal{M})$, is the least ordinal not in M. An ordinal α is *admissible* if $\alpha = o(\mathcal{M})$ for some admissible structure $\mathcal{M}$.

A relation $S \subseteq M^n$ on $\mathcal{M}$ is Σ_n if there is a Σ_n-formula φ such that

$$S(a_1, a_2, \ldots, a_n) \Leftrightarrow \mathcal{M} \models \varphi(a_1, a_2, \ldots, a_n).$$

We allow parameters in the above definition. Clearly both a and $\{a\}$ are Σ_1 for any $a \in M$. Moreover, it follows from Theorem 3.1.5 that for $\mathcal{M}$ admissible, Σ_1-relations are closed under $\vee$, $\wedge$, $\exists x \in a$, $\forall x \in a$ and $\exists x$.

3.2.1 The smallest admissible set

Definition 3.2.3. A set $a \in V$ is *hereditarily finite*, written $a \in \mathrm{HF}$, if $\mathrm{TC}(a)$ is finite.

HF can be redefined by induction as follows:

$$\begin{aligned} \mathrm{HF}(0) &= \emptyset, \\ \mathrm{HF}(n+1) &= \{a \mid a \subseteq \mathrm{HF}(n)\}, \\ \mathrm{HF} &= \bigcup_{n<\omega} HF(n). \end{aligned}$$

Thus $\mathrm{HF} = V_\omega$. We leave the following theorem as an exercise.

Theorem 3.2.4. HF *is the smallest admissible set.*

$\langle \mathrm{HF}, \in \rangle$ is the first natural example of an admissible set we encounter. There is a natural and standard way of identifying n as a natural number (in arithmetic) and n as a finite ordinal (in set theory). We will follow the convention of identifying these two objects. The following theorem highlights an important connection between recursion theory and admissibility theory, and explains the reason for introducing admissible sets in the study of generalized recursion theory.

Theorem 3.2.5. *Let S be a relation on the natural numbers. Then S is r.e. if and only if S is Σ_1 over $\langle \mathrm{HF}, \in \rangle$.*

Proof. It is sufficient to assume that S is infinite. The class of r.e. relations is the class of relations which are Σ_1-definable over $\langle \omega, +, \cdot, 0, 1, < \rangle$ in the language of first order arithmetic. Thus for the "only if" part, it is sufficient to show that if $S \subseteq \omega$ is Σ_1 over $\langle \omega, +, \cdot, 0, 1, < \rangle$, then it is Σ_1 over $\langle \mathrm{HF}, \in \rangle$ in the language of set theory. It is obvious that the set of ordinals in HF, i.e. ω, is Δ_1-definable over $\langle \mathrm{HF}, \in \rangle$. Now it is routine to show that if S is defined by an open formula φ involving $+, \cdot$, then it is defined over $\langle \mathrm{HF}, \in \rangle$ by a Δ_1-formula φ^* in the language of set theory. Also "$u < v$" is definable over $\langle \mathrm{HF}, \in \rangle$ by "$u \in v$". Hence any open formula defining S involving $+, \cdot, <$ is definable by a corresponding Δ_1-formula over $\langle \mathrm{HF}, \in \rangle$ in the language of set theory.

For S defined by a bounded formula φ of the form $(\exists i < n)\psi$ where ψ is open, let φ^* be $(\exists u \in n)\psi^*(u)$, where ψ^* is the Δ_1-formula corresponding to ψ (note that the "n" in φ^* denotes the finite ordinal n). Define φ^* for φ of the form $(\forall i < n)\psi$ and ψ open, in the obvious way. By induction on the complexity of formulas, any $S \subseteq \omega$ defined by a bounded formula in the language of first order arithmetic has a corresponding definition via a Δ_1-formula in the language of set theory over $\langle \mathrm{HF}, \in \rangle$.

Now assume that $S = \{m | (\exists i)\psi(i, m)\}$ where ψ is a bounded formula in arithmetic. By induction, let ψ^* be a Δ_1 such that $\langle \omega, +, \cdot, 0, 1, < \rangle \models \psi(n, m)$ if and only if $\langle \mathrm{HF}, \in \rangle \models \psi^*(n, m)$. Then $m \in S$ if and only if

$$\langle \mathrm{HF}, \in \rangle \models (\exists u)(u \text{ is an ordinal} \wedge \psi^*(u, m)).$$

Hence S is Σ_1 over $\langle \mathrm{HF}, \in \rangle$.

For the other direction, we first define a function $f : \omega \to \mathrm{HF}$ as follows:

$$f(0) = 0,$$
$$f(1) = \{0\},$$
$$f(2^{n_1} + 2^{n_2} + \cdots . + 2^{n_k}) = \{f(n_1), f(n_2), \ldots, f(n_k)\}.$$

It is straightforward to verify that f is a bijection. Furthermore,

- f is Δ_1 over $\langle \mathrm{HF}, \in \rangle$.
- Define $f^*(n) = m$ if and only if m as an ordinal is the set $f(n)$. Then the relation "$f^*(n) = m$" is recursive.
- The set $T_\varphi = \{(n_1, n_2, \ldots, n_k) \mid \langle \mathrm{HF}, \in \rangle \models \varphi(f(n_1), f(n_2), \ldots, f(n_k))\}$ is recursive for each Δ_0-formula φ.

Now if S is Σ_1 over $\langle \mathrm{HF}, \in \rangle$, then there is a Δ_0-formula φ such that $S(m)$ if and only if there is a $u \in \mathrm{HF}$ with $\langle \mathrm{HF}, \in \rangle \models \varphi(u, m)$. Thus $S(m)$ if and only if $(\exists i)(\exists j)(f^*(i) = m \wedge (i, j) \in T_\varphi)$. We conclude that S is an r.e. relation. □

By the proof of Theorem 3.2.5, we have the following result.

Proposition 3.2.6. *For any $S \subseteq \mathrm{HF}$, there is an $x_S \subseteq \omega$ such that S is Δ_1-definable in $\langle \mathrm{HF}, \in, x_S \rangle$ and x_S is Δ_1-definable in $\langle \mathrm{HF}, \in, S \rangle$.*

Proof. By the proof of Theorem 3.2.5, there is a bijection $f : \omega \mapsto \mathrm{HF}$ which is Δ_1-definable in $\langle \mathrm{HF}, \in \rangle$. Define $n \in x_S$ if and only if $f(n) \in S$. □

Model-theoretically, Theorem 3.2.5 may be viewed as a biinterpretation of Σ_1-formulas between $\langle \mathrm{HF}, \in \rangle$ and $\langle \omega, \in \rangle$ (see [52]).

3.2.2 Exercise

Exercise 3.2.1. *Prove Theorem* 3.2.4.

3.3 Constructibility

The constructible universe L was introduced by Gödel [37] to prove the consistency of the Generalized Continuum Hypothesis and the Axiom of Choice. Jensen [55] developed the fine structure theory of L and the theory has had a deep impact on higher recursion theory.

3.3.1 The constructible universe

Expand the language of set theory $\mathscr{L}$ with a unary relation symbol $\dot{A}$ and denote it as $\mathscr{L}^{\dot{A}}$.

Definition 3.3.1. Let A be a set. Given a set X, we say that Y is *definable relative to A* over the structure $(X, \in, A \cap X)$, written $Y \in Def^A(X)$, if there is a formula $\varphi(v)$ in $\mathscr{L}^{\dot{A}}$ such that upon interpreting $\dot{A}$ to be $A \cap X$, we have $Y = \{y \mid \langle X, \in, A \cap X\rangle \models \varphi(y)\}$.

By recursion on ordinals, define

$$\begin{aligned} L_0[A] &= \emptyset, \\ L_{\beta+1}[A] &= Def^A(L_\beta[A]), \\ L_\lambda[A] &= \bigcup_{\beta<\lambda} L_\beta[A], \text{ if } \lambda \text{ is a limit ordinal.} \end{aligned}$$

Define $L[A] = \bigcup_\beta L_\beta[A]$, and $L = L[\emptyset]$. Obviously $L[A]$ is a proper class.

If $x \in L[A]$, then x is said to be *A-constructible*. In particular, x is *constructible* if $A = \emptyset$. The construction of L is uniformly Σ_1-definable over any admissible structure $\mathcal{M}$, in the sense of the following theorem whose proof we omit. It involves a tedious coding device that is Σ_1-definable by appealing to Theorem 3.1.8. For details, see [22].

Theorem 3.3.2. *There exists a Σ_1-formula $\varphi(u, v)$ such that for any admissible structure $\mathcal{M}$ and ordinal $\beta \in M$, $\mathcal{M} \models \varphi(\beta, x)$ if and only if $x = L_\beta$. Moreover, L_β is the unique set such that $\mathcal{M} \models \varphi(\beta, x)$. Hence L_β is uniformly Δ_1-definable over $\mathcal{M}$.*

There is a recursion-theoretic way of generating L via the following set of *primitive recursive operators* introduced by Gödel:
(i) $F_0(x, y) = \{x, y\}$.
(ii) $F_1(x, y) = x \setminus y$.
(iii) $F_2(x, y) = x \times y = \{x, \{x, y\}\}$.
(iv) $F_3(x, y) = \{(u, z, v) \mid z \in x \wedge (u, v) \in y\}$.
(v) $F_4(x, y) = \{(u, v, z) \mid z \in x \wedge (u, v) \in y\}$.
(vi) $F_5(x, y) = \bigcup x$.
(vii) $F_6(x, y) = \{y \mid \exists z(y, z) \in x\}$.
(viii) $F_7(x, y) = \in \cap\, x \times x$.
(ix) $F_8(x, y) = \{z_1 \mid \exists z_0((z_0, z_1) \in x \wedge z_0 \in y)\}$.
(x) $F_9(x) \quad = A \cap x$.

Let

$$\mathcal{S}^A(U) = U \cup \{U\} \cup \bigcup_{i\leq 9}\{z \mid (\exists x \in U \cup \{U\})(\exists y \in U \cup \{U\})(z = F_i(x, y)\}$$

and

$$\mathcal{R}^A(U) = \bigcup_{n<\omega} (\mathcal{S}^A)^{(n)}(U).$$

Define

$$J_0[A] = \emptyset,$$
$$J_{\beta+1}[A] = \mathcal{R}^A(J_\beta[A]),$$
$$J_\lambda[A] = \bigcup_{\beta<\lambda} J_\beta[A], \text{ if } \lambda \text{ is a limit ordinal.}$$
$$J[A] = \bigcup_{\beta} J_\beta[A]$$

If $A = \emptyset$, then we write J for $J[A]$. Intuitively and according to Jensen [55], $J_{\beta+1}$ is the result of "stretching" the collection of the definable subsets of J_β upwards by ω-levels in rank without adding new sets. The proof of the following proposition may be found in [22].

Proposition 3.3.3. *For all β,*
- *$J_\beta[A]$ is transitive.*
- *There is a Σ_1-formula $\varphi(u, w)$ such that for any two ordinals $\nu < \beta$, $\langle J_\beta, \in\rangle \models \varphi(\nu, x)$ if and only if $x = J_\nu$;*
- *There is a Σ_1-formula $\varphi_<(u, v)$ such that*
  (i) *for any β, $\varphi_<(u, v)$ defines a well-ordering $<_{J_\beta}$ on J_β;*
  (ii) *for any $\nu > \beta$, if $y \in J_\beta$ and $\langle J_\nu, \in\rangle \models \varphi_<(x, y)$, then $x \in J_\beta$ and $\langle J_\beta, \in\rangle \models \varphi_<(x, y)$.*
- *Theorem 3.3.2 remains true for J_β;*
- *$J_{\beta+1}[A] \cap \mathcal{P}(J_\beta[A]) = Def^A(J_\beta[A])$.*
- *$L_\beta[A] \subseteq J_\beta[A] \subseteq L_{\omega\beta}$.*
- *$J_\beta[A] = L_\beta[A]$ if and only if $\omega\beta = \beta$.*
- *$L[A] = \bigcup_\beta J_\beta[A]$;*
- *Theorem 3.3.2 remains valid for J.*

The Gödel operators provide an algorithmic approach to the construction of L. It reveals the "black box" hidden within the reference to definability. Based on Proposition 3.3.3, we may think of L as being obtained by an effective transfinite induction over the class of ordinals, analogous to the way hyperarithmetic sets are defined over ω_1^{CK}. This connection is useful as it suggests initial segments of L as models of computation in the general setting.

3.3.2 Condensation and Skolemization

If one works within a structure $\mathcal{M}$ and performs the stage-by-stage construction of L, then one arrives at the same L_β for each $\beta \in M$ if $\mathcal{M}$ satisfies a sufficiently strong condition such as admissibility. The following theorem formalizes this idea. Let $(L_\beta)^{\mathcal{M}}$ denote the L_β constructed in $\mathcal{M}$.

Theorem 3.3.4. *Let $\mathcal{M}$ be a transitive model of* KP. *Then for each $\beta < o(M)$, $L_\beta \in M$ and $(L_\beta)^{\mathcal{M}} = L_\beta$.*

Proof. By Theorem 3.3.2, the formula "$x = L_\beta$" is Δ_1-definable over $\mathcal{M}$. By the absoluteness of Δ_1-formulas, $L_\beta \in M$ and $(L_\beta)^{\mathcal{M}} = L_\beta$. □

Corollary 3.3.5. *Suppose $\mathcal{M}$ is transitive, admissible and $\mathcal{M} \models$ "V = L". Then $M = L_\alpha$ where $\alpha = o(M)$.*

Theorem 3.3.6. *α is admissible if and only if $\langle L_\alpha, \in\rangle$ is admissible.*

Proof. If $\langle L_\alpha, \in\rangle$ is admissible, then α is admissible.

If $\alpha = o(\mathcal{M})$ for some admissible $\mathcal{M}$, then $(L_\beta)^{\mathcal{M}} = L_\beta$ for each $\beta < \alpha$ by Theorem 3.3.4. Hence it suffices to show that $L_\alpha = \bigcup_{\beta<\alpha}(L_\beta)^{\mathcal{M}}$ is admissible. We verify that Δ_0-collection holds in L_α and leave the verification of other properties to the reader.

Suppose that $\langle L_\alpha, \in\rangle \models (\forall x \in a)(\exists y)\varphi(x, y)$. By the absoluteness of Δ_0-formulas, for each $x \in a$, there is a (least) $\beta < \alpha$ such that

$$\mathcal{M} \models (\exists y \in L_\beta)\varphi(x, y).$$

Then Theorem 3.3.2 and Theorem 3.1.6 on Σ-collection imply that there is an $\alpha_0 < \alpha$ such that $a \in L_{\alpha_0}$ and

$$\mathcal{M} \models (\forall x \in a)(\exists \beta < \alpha_0)(\exists y \in L_\beta)\varphi(x, y).$$

The absoluteness of Δ_1-formulas now gives

$$\langle L_{\alpha_0}, \in\rangle \models (\forall x \in a)(\exists y)\varphi(x, y).$$

Then $\{y \mid \langle L_{\alpha_0}, \in\rangle \models (\exists x \in a)\varphi(x, y)\} \in L_{\alpha_0+1}$ and is contained in L_α. □

Obviously if $\alpha = \omega$, then L_α = HF. Hence L_ω is the smallest admissible set (cf. Theorem 3.2.4). Note that if $\alpha > \omega$ is admissible, then $\omega\alpha = \alpha$.

A fundamental property of L is the following known as condensation:

Theorem 3.3.7. *Suppose that $\beta > 0$ and $\langle X, \in\rangle \prec_{\Sigma_1} \langle J_\beta, \in\rangle$. Then $\langle X, \in\rangle \cong \langle J_\gamma, \in\rangle$ for some $\gamma \le \beta$. Moreover $\pi(x) \le_{J_\beta} x$ for every $x \in X$.*

Proof. Suppose $\langle X, \in\rangle \prec_{\Sigma_1} \langle J_\beta, \in\rangle$. Then $\langle X, \in\rangle$ is wellfounded and satisfies the extensionality axiom. By Theorem 3.1.12, there is a transitive set M such that

$$\pi : \langle X, \in\rangle \cong \langle M, \in\rangle.$$

We claim that $M = J_\gamma$ where $\gamma = \sup\{\pi(\delta) \mid \delta \in \beta \cap X\}$.

Clearly $\langle M, \in\rangle \prec_{\Sigma_1}^{\pi^{-1}} \langle J_\beta, \in\rangle$ and M is transitive. By Proposition 3.3.3, there is a Σ_0-formula $\varphi(x, \delta, s)$ such that for any $x \in J_\beta$ and $\delta < \beta$, $\langle J_\beta, \in\rangle \models \exists s\varphi(x, \delta, s)$ if and only if $x = J_\delta$. Since M is transitive, by an induction on $\delta < \gamma$, we prove that $J_\delta \in M$ for any $\delta < \gamma$. Hence $J_\gamma \subseteq M$.

It is immediate that $J_0 = \emptyset \in M$. If $\delta > 0$, then $\pi^{-1}(\delta) < \beta$. Then $\langle J_\beta, \in\rangle \models (\exists x)(\exists s)\varphi(x, \pi^{-1}(\delta), s)$ and so $\langle M, \in\rangle \models (\exists x)(\exists s)\varphi(x, \delta, s)$. Let $x \in M$ and let $s \in M$ such that $\langle M, \in\rangle \models \varphi(x, \delta, s)$. By the absoluteness of $\varphi(x, \delta, s)$, any admissible structure $\mathcal{N}$ containing x, s and δ satisfies $\varphi(x, \delta, s)$. By Proposition 3.3.3, $J_\delta = x \in M$.

Furthermore, for any $x \in M$, there exists a $\delta < \beta$ such that $\pi^{-1}(x) \in J_\delta$. Hence $\langle J_\beta, \in\rangle \models (\exists\delta)(\exists y)(\exists s)(\pi^{-1}(x) \in y \wedge \varphi(y, \delta, s))$. Then $\langle M, \in\rangle \models (\exists\delta)(\exists y)(\exists s)(x \in y \wedge \varphi(y, \delta, s))$. By Proposition 3.3.3 again, $x \in J_{\delta'}$ for some $\delta' < \gamma$. We conclude that $J_\gamma \supseteq M$ and hence $J_\gamma = M$.

Suppose $x <_{J_\beta} \pi(x)$ for some $x \in X$. Take x_0 to be the $<_{J_\beta}$-least such set. Then $x_0 \in J_\gamma$ since $\pi(x_0) \in J_\gamma$ and so there is an $x_1 \in X$ such that $\pi(x_1) = x_0$. If $x_0 <_{J_\beta} x_1$, then $\pi(x_0) <_{J_\alpha} \pi(x_1) = x_0$ which contradicts the assumption that $x_0 <_{J_\beta} \pi(x_0)$. So $x_1 <_{J_\beta} \pi(x_1) = x_0$ which contradicts the choice of x_0 to be the J_β-least counterexample. □

An immediate consequence of the Condensation Theorem (3.3.7) is that V = L implies the Continuum Hypothesis. We leave the proof as an exercise.

Corollary 3.3.8. $2^\omega \cap L = 2^\omega \cap L_{\omega_1}$.

Another important tool in the theory of constructibility is the use of Skolem functions. A relation or function is $\Sigma_n(\mathcal{M})$-definable if it is Σ_n-definable over $\mathcal{M}$ (possibly with parameters).

Definition 3.3.9. Given a structure $\mathcal{M}$, a *Σ_n-Skolem function* for $\mathcal{M}$ is a $\Sigma_n(\mathcal{M})$-definable function $h : \omega \times M \to M$ with parameter $p \in M$ such that for any $A \subseteq M$ which is Σ_n-definable over $\mathcal{M}$ with parameters $p, x \in M$, there is an $i \in \omega$ satisfying $h(i, x) \in A$.

Thus a Σ_n-Skolem function is a Σ_n-definable choice function for Σ_n-sets.

Proposition 3.3.10. *Suppose that $\alpha > 0$ is admissible and h is a Σ_n-Skolem function for $\langle J_\alpha, \in\rangle$. Then*
(i) *for any $x \in J_\alpha$, $x \in \mathrm{Range}(h(\omega \times \{x\}))$ and $\langle \mathrm{Range}(h(\omega \times \{x\})), \in\rangle \prec_{\Sigma_n} \langle J_\alpha, \in\rangle$;*
(ii) *if $q \in J_\alpha$ and $X \subseteq J_\alpha$ is closed under ordered pairs, then $X \cup \{q\} \subseteq \mathrm{Range}(h(\omega \times (X \cup \{q\}))$ and $\langle \mathrm{Range}(h(\omega \times (X \cup \{q\})), \in\rangle \prec_{\Sigma_n} \langle J_\alpha, \in\rangle$.*

Proof. Note that (i) is straightforward to verify and is a standard fact in model theory.

For (ii), since X is closed under ordered pairs, any finite sequence from X used to define a finite sequence in $\mathrm{Range}(h(\omega \times (X \cup \{q\}))$ may be coded by a single parameter. Hence it is routine to show $\langle \mathrm{Range}(h(\omega \times (X \cup \{q\})), \in\rangle \prec_{\Sigma_n} \langle J_\alpha, \in\rangle$. □

We remark that the existence of a Σ_n-Skolem function is not a trivial fact. In model theory, defining a Skolem function for a structure $\mathcal{M}$ requires a well-ordering of M. In the constructible universe L, however, there is a uniformly Σ_1-definable Skolem function.

Proposition 3.3.11. *There is a Σ_1-formula $\varphi(x, y, i, z)$ such that for any ordinal α, the function*

$$h_\alpha(i, x) = y \Leftrightarrow \langle J_\alpha, \in\rangle \models (\exists z)\varphi(x, y, i, z)$$

is a Σ_1-Skolem function for $\langle J_\alpha, \in\rangle$.

Proof. By Proposition 3.3.3, there is a Σ_1-formula $\varphi_<(u, v)$ defining a well-ordering $<_J$ of J_α for any admissible α. Clearly there is a Σ_1-enumeration $\{\varphi_i(u, v, w)\}$ of all Σ_0-formulas. In other words, there is a Σ_1-formula $\varphi(x, y, i, z)$ so that $\langle J_\alpha, \in\rangle \models \varphi(x, y, i, z) \Leftrightarrow \langle J_\alpha, \in\rangle \models \varphi_i(x, y, z)$. Define $h_\alpha(i, x)$ to be y_0 such that (y_0, y_1) is the $<_J$-least pair with $\langle J_\alpha, \in\rangle \models \varphi_i(y_0, y_1, x)$. By Proposition 3.3.3, $h_\alpha(i, x)$ is Σ_1-definable.

By the admissibility of α, using some tedious coding argument which may be found in [22], one can show that the relation "$\langle J_\alpha, \in\rangle \models \varphi_i(y_0, y_1, x)$" is uniformly Δ_1. Let $h_\alpha(i, x) = y_0$ if and only if there exists a y_1 such that $\langle J_\alpha, \in\rangle \models \varphi_i(y_0, y_1, x)$ and there exists a $z = \{(z_0, z_1) \mid (z_0, z_1) <_J (y_0, y_1)\}$ such that for any pair $(z_0, z_1) \in z$, $\langle J_\alpha, \in\rangle \models \neg\varphi_i(y_0, y_1, x)$. Since z is uniformly Δ_1-definable for any admissible α, we have $h_\alpha(i, x)$ to be uniformly Σ_1-definable for any admissible α. □

If α is admissible, we call the Σ_1-Skolem function h_α the *canonical* Σ_1-Skolem function for α.

Proposition 3.3.12. *If $\alpha > \omega$ is admissible, then $\alpha \geq \omega_1^{\mathrm{CK}}$.*

Proof. Suppose α is admissible and $\omega < \alpha < \omega_1^{\mathrm{CK}}$. By Theorem 3.3.6, L_α is admissible. By Theorem 1.3.4, there is a recursive function f such that for each $n \in \mathcal{O}$, $W_{f(n)} = \{\langle i, j\rangle \mid i <_o j <_o n\}$. Fix a number $n \in \mathcal{O}$ with $|n| = \alpha$. Then the order type of $W_{f(n)}$ is α and $W_{f(n)} \in L_{\omega+1} \subseteq L_\alpha$ by Theorem 3.2.5. By Σ-induction there is a Σ_1 order-preserving bijection $g : W_{f(n)} \to \alpha$. Since $W_{f(n)} \in L_\alpha$, $\alpha \in L_\alpha$ by the admissibility of α. This is a contradiction. □

3.3.3 Exercises

Exercise 3.3.1. *Prove that $\langle L_{\omega_1}, \in\rangle \models \mathrm{KP} + \text{"V = L"}$.*

Exercise 3.3.2. *Prove Corollary* 3.3.8.

Exercise 3.3.3 (Putnam [115]). *Prove that for any countable ordinal α, there is a countable ordinal β such that $L_{\beta+\alpha} \setminus L_\beta \cap 2^\omega = \emptyset$.*

3.4 Projecta and master codes

In the interest of keeping the chapter within reasonable length and the proofs simpler, we present weaker versions of results in Jensen [55]. Nevertheless, the key ideas are extracted from [55] and the results are sufficient for recursion-theoretic applications.

3.4.1 Projecta

Definition 3.4.1. Let $\alpha > 0$ and $n \geq 1$. The *Σ_n-projectum ρ_α^n*, of α is the least ordinal ρ for which there is a $\Sigma_n(J_\alpha)$-definable function f mapping a subset of J_ρ onto J_α.

Clearly ρ_α^n exists for any $n \geq 1$. One may replace J_ρ with $\omega\rho$ in the above definition. To prove this, we need the following lemma whose proof is left to the reader:

Lemma 3.4.2. *Let $\alpha > 0$. For each $1 \leq \gamma \leq \omega\alpha$, there is a $\Sigma_1(J_\alpha)$-definable bijection between $\omega\alpha$ and $\gamma \times \omega\alpha$.*

Proposition 3.4.3. *Let $\alpha > 0$. There is a $\Sigma_1(J_\alpha)$-definable function from $\omega\alpha$ onto J_α.*

Proof. Let $f_0 = (f^0, f^1)$ be a bijection between $\omega\alpha$ and $(\omega\alpha)^2$ guaranteed by Lemma 3.4.2 (setting $\gamma = \omega\alpha$) defined via the $<_{J_\alpha}$-least parameter $p \in J_\alpha$). Define $f_{n+1}(\gamma) = (f^0(\gamma), f_n(f^1(\gamma)))$ for $\gamma < \omega\alpha$. It is easy to verify that f_n is $\Sigma_1(J_\alpha)$-definable with parameter p and is a bijection between $\omega\alpha$ and $(\omega\alpha)^{n+2}$. Let h_α be the canonical Σ_1-Skolem function for J_α. Set $X = h_\alpha(\omega \times (\omega\alpha \times \{p\}))$. Note that for each $n < \omega$ and $\gamma < \omega\alpha$, there is an i_n such that $f_n(\gamma) = h_\alpha(i_n, (\gamma, p))$. Thus $\gamma \in X$ for every $\gamma < \omega\alpha$.

By Proposition 3.3.10, $\langle X, \in\rangle \prec_{\Sigma_1} \langle J_\alpha, \in\rangle$. By Theorem 3.3.7, there is a $\beta \leq \alpha$ and an isomorphism $\pi : \langle X, \in\rangle \cong \langle J_\beta, \in\rangle$. Moreover, $\pi(p) \leq_{J_\alpha} p$. Since $\gamma \in X$ for every $\gamma < \omega\alpha$, we have $\beta = \alpha$ and $\pi(\gamma) = \gamma$ for every $\gamma < \omega\alpha$. Hence $\pi^{-1} : \langle J_\alpha, \in\rangle \prec_{\Sigma_1} \langle J_\alpha, \in\rangle$ and $\pi^{-1}(f_0) = f_0$. Thus f_0 is also defined by $\pi(p)$ in $\langle J_\alpha, \in\rangle$. By the assumption on p, $p \leq_{J_\alpha} \pi(p)$ and so $p = \pi(p)$. Then for any $x \in X$, there are $i < \omega$ and $\gamma < \omega\alpha$ such that

$$x = h_\alpha(i, (\gamma, p)) = h_\alpha(\pi(i), (\pi(\gamma), \pi(p))) = \pi(x),$$

implying that $X = J_\alpha$.

Note that h_α is not necessarily total. Let $\varphi(u, v, w, s)$ be a Σ_0-formula such that $h_\alpha(i, \gamma, p) = x$ if and only if $\langle J_\alpha, \in\rangle \models (\exists y)\varphi(i, \gamma, p, y)$. Then by the proof of Proposition 3.3.11, there is a total Σ_1-function

$$h'(i, \gamma_1, \gamma_2) = \begin{cases} x & \langle J_\alpha, \in\rangle \models (\exists n \in \omega)(\exists y \in J_{\gamma_2} \cup (\mathbb{S})^{(n)}(J_{\gamma_2}))\varphi(i, \gamma_1, p, y), \\ \emptyset & \text{Otherwise.} \end{cases}$$

Let $h(\gamma) = h'(f_1(\gamma))$ for $\gamma < \omega\alpha$. □

Applying Proposition 3.4.3, we have:

Proposition 3.4.4. *α is admissible if and only if there is no $\Sigma_1(J_\alpha)$-definable function mapping a $\gamma < \omega\alpha$ onto $\omega\alpha$.*

The next lemma is a characterization of ρ_α^1.

Lemma 3.4.5. *Suppose that $\alpha > 0$ and $\rho = \rho_\alpha^1$. Then $A \in J_\rho$ for any $\gamma < \rho$ and any $\Sigma_1(J_\alpha)$-definable $A \subseteq J_\gamma$.*

Proof. If $\rho = 1$ then the conclusion is immediate. Suppose $\rho > 1$. Then ρ is a limit ordinal. Let $\gamma < \rho$ and $A \subseteq J_\gamma$. Let h_α be the canonical Σ_1-Skolem function for J_α. Fix $p \in J_\alpha$ such that A is $\Sigma_1(J_\alpha)$-definable with parameter p. Let $X = h_\alpha(\omega \times (J_\gamma \times \{p\}))$. Then $\langle X, \in\rangle \prec_{\Sigma_1} \langle J_\alpha, \in\rangle$. By Theorem 3.3.7, there is a $\beta \leq \alpha$ such that $\pi : \langle X, \in\rangle \cong \langle J_\beta, \in\rangle$ is an isomorphism. Note that $\pi(x) = x$ for all $x \in J_\gamma$ since J_γ is transitive. Hence $\pi(x) = x$ for $x \in A$. Thus A is also Σ_1-definable in J_β with parameter $\pi(p) \in J_\beta$. We claim that $\beta < \rho$. Otherwise, let f be $\Sigma_1(J_\alpha)$ mapping a subset of J_ρ onto J_α. Then $f \circ \pi \circ h_\alpha$ is $\Sigma_1(J_\alpha)$ and maps a subset of $\omega \times (J_\gamma \times \{p\})$ onto J_α. But $\omega \times J_\gamma \in J_{\gamma+2} \in J_\rho$, and so it maps a subset of $J_{\gamma+2}$ onto J_α, contradicting the choice of ρ. □

Lemma 3.4.6. *Let $\alpha > 0$ and $\rho = \rho^1_\alpha$. Then there is an $A \subseteq J_\rho$ which is $\Sigma_1(J_\alpha)$-definable and $A \notin J_\alpha$.*

Proof. Let f be $\Sigma_1(J_\alpha)$-definable mapping a subset of J_ρ onto J_α. Then the domain of f restricted to J_ρ is $\Sigma_1(J_\alpha)$ and not an element of J_α by admissibility. □

Lemma 3.4.5 and Lemma 3.4.6 yield the following characterization of ρ^1_α:

Theorem 3.4.7. *Let $\alpha > 0$. Then ρ^1_α is the least ordinal ρ for which there is a $\Sigma_1(J_\alpha)$ set $A \subseteq J_\rho$ not in J_α.*

Now suppose that $x \subseteq \omega$ is $\Sigma_1(J_\alpha)$-definable with parameter p and $x \notin J_\alpha$. Let $X = h_\alpha(\omega \times (J_1 \times \{p\})))$. We may identify $\omega \times J_1$ with J_1. Hence $X = h(J_1 \times \{p\}))$ for some Σ_1-function h. Then $\langle X, \in\rangle \prec_{\Sigma_1} \langle J_\alpha, \in\rangle$. By transitive collapse there is a $\beta \leq \alpha$ such that $\pi : \langle X, \in\rangle \cong \langle J_\beta, \in\rangle$. Since $x \subseteq J_1$, $\pi(n) = n$ for all $n \in \omega$. Thus x is $\Sigma_1(J_\beta)$-definable with parameter $\pi(p)$. If $\beta < \alpha$, then $x \in J_\alpha$ which is a contradiction. Thus $\beta = \alpha$. Now π is $\Sigma_1(J_\alpha)$-definable and so $\pi \circ h$ is $\Sigma_1(J_\alpha)$-definable and maps a set $y \subseteq J_1$ onto J_α. Since $J_1 = \mathrm{HF}$, by Proposition 3.2.6, we may assume that y is a real. Hence for any admissible α, if $J_{\alpha+1} \setminus J_\alpha$ contains a $\Sigma_1(J_\alpha)$-definable real, then it contains a real "coding J_α".

Remark 3.4.8. Theorem 3.4.7 remains valid for $n \geq 1$. However, the proof requires more than a straightforward generalization of what was presented above. The major difficulty is that there is no canonical Σ_n-Skolem function for $n \geq 2$. To overcome this difficulty, a finer analysis of J_α is introduced. See Jensen [55] for details.

3.4.2 Master codes

Definition 3.4.9. Given an ordinal α and $n \geq 0$, a *Σ_n-master code* for J_α is a $\Sigma_n(J_\alpha)$-definable set $A \subseteq J_{\rho^n_\alpha}$ such that for any $m \geq 1$, any set $B \subseteq J_{\rho^n_\alpha}$ that is $\Sigma_{n+m}(J_\alpha)$-definable is Σ_m-definable in $\langle J_{\rho^n_\alpha}, \in, A\rangle$.

There is an interpretation of the notion of a master code in the context of recursion theory. First suppose $\alpha = 1$ and hence $\rho^1_\alpha = 1$. Then $J_\alpha = J_\rho = \mathrm{HF}$. Let $A \subseteq \mathrm{HF}$ be a Σ_1-master code. By Proposition 3.2.6, there is a real x such that x is Δ_1-definable in

$\langle \mathrm{HF}, \in, A\rangle$ and A is Δ_1-definable in $\langle \mathrm{HF}, \in, x\rangle$. By Theorem 3.2.5, every Σ_1-definable real in $\langle \mathrm{HF}, \in, x\rangle$ is r.e. in x. Hence every real Σ_2-definable in $\langle \mathrm{HF}, \in\rangle$ is Σ_1-definable in $\langle \mathrm{HF}, \in, A\rangle$ and so r.e. in x. Thus the Σ_1-master codes are precisely the reals that are r.e. and complete (hence Turing equivalent to $\emptyset'$). In general, a real which is a Σ_n-master code in $\langle J_1, \in\rangle$ is Σ_n-complete and hence Turing equivalent to $\emptyset^{(n)}$. We may therefore consider master codes to be Turing jumps in the set-theoretic sense. Second, suppose that $\rho_\alpha^1 = 1$. A Σ_1-master code for $\langle J_\alpha, \in\rangle$ is a set $A \subseteq \mathrm{HF}$ such that every $\Sigma_2(J_\alpha)$-definable real is Σ_1 in $\langle \mathrm{HF}, \in, A\rangle$. Again by Proposition 3.2.6 and Theorem 3.2.5, every $\Sigma_2(J_\alpha)$-definable real is r.e. in some real x which is $\Sigma_1(J_\alpha)$-definable. In particular, every real $z \in J_\alpha$ is recursive in x. Thus if α is much larger than ρ_α^1, the real x must be remarkably strong in computational power. Historically, the notion of a master code for $\rho = 1$ was first introduced by Boolos and Putnam [8] when they studied the Turing degrees of reals in the constructible universe. We now show that Σ_1-master codes exist for any admissible ordinal α.

Theorem 3.4.10. *Let $\alpha > 0$. There is a Σ_1-master code for $\langle J_\alpha, \in\rangle$, i.e. a $\Sigma_1(J_\alpha)$-definable $A \subseteq J_{\rho_\alpha^1}$ such that for any $B \subseteq J_{\rho_\alpha^1}$, if B is $\Sigma_{m+1}(J_\alpha)$-definable, then it is Σ_m-definable in $\langle J_{\rho_\alpha^1}, \in, A\rangle$.*

Proof. We only prove the case $m = 1$ for simplicity. Suppose that f is $\Sigma_1(J_\alpha)$-definable with parameter p mapping a subset of ρ_α^1 onto α. Fix a Δ_1-enumeration $\{\varphi_i(x, p)\}_{i<\omega}$ of Σ_1-formulas with parameter p. Let

$$A = \{(i, x) \mid i < \omega \wedge x \in J_{\rho_\alpha^1} \wedge \langle J_\alpha, \in\rangle \models \varphi_i(x, p)\}.$$

It is obvious that A is Σ_1-definable over $\langle J_\alpha, \in\rangle$. Now suppose that $B \subseteq J_{\rho_\alpha^1}$ is Σ_2-definable over $\langle J_\alpha, \in\rangle$. Then there is a $\Sigma_1(J_\alpha)$-formula $\varphi(u, v, w)$ such that

$$x \in B \Leftrightarrow \langle J_\alpha, \in\rangle \models (\exists y)\neg\varphi(x, y, s),$$

where s is a parameter in J_α. Since f is onto, there is an $s_0 \in J_{\rho_\alpha^1}$ such that $f(s_0) = s$. Then

$$x \in B \Leftrightarrow (\exists y \in J_{\rho_\alpha^1})(\langle J_\alpha, \in\rangle \models \neg\varphi(x, f(y), f(s_0))).$$

Since f is $\Sigma_1(J_\alpha)$-definable with parameter p, there is an index i such that for $x, y \in J_{\rho_\alpha^1}$,

$$\langle J_\alpha, \in\rangle \models \neg\varphi_i(x, y, s_0, p) \Leftrightarrow \langle J_\alpha, \in\rangle \models \neg\varphi(x, f(y), f(s_0)).$$

Thus

$$x \in B \Leftrightarrow (\exists y \in J_{\rho_\alpha^1})(i, x, y, s_0) \notin A \Leftrightarrow \langle J_{\rho_\alpha^1}, \in, A\rangle \models (\exists y)((i, (x, y, s_0)) \notin A).$$

Hence B is Σ_1-definable over $\langle J_{\rho_\alpha^1}, \in, A\rangle$. □

As an example, $\emptyset'$ is a $\Sigma_1(J_1)$-master code since it satisfies the property described in Theorem 3.4.10. Jensen [55] proved the following general result:

Theorem 3.4.11. *For any ordinal $\alpha > 0$ and $n > 0$, there is a $\Sigma_n(J_\alpha)$-master code $A \subset J_{\rho_\alpha^n}$. Moreover, ρ_α^n is the least ordinal γ for which there is a $\Sigma_n(J_\alpha)$ $A \subseteq \omega\gamma$ not in J_α.*

A Σ_n-master code corresponds to $\emptyset^{(n)}$ in recursion theory. The equality $\emptyset^{(n+m)} = (\emptyset^n)^{(m)}$ has a natural analog in fine structure theory. See [55] for details.

3.4.3 Exercises

Exercise 3.4.1. *Prove Lemma* 3.4.2.

Exercise 3.4.2. *Prove Proposition* 3.4.4.

Exercise 3.4.3. *Prove that there is a countable admissible ordinal $\alpha > \omega$ such that $\rho_\alpha^1 = \alpha$.*

3.5 ω-models

Definition 3.5.1. A structure $\mathcal{M}$ is an *ω-model* if $\omega^M = \omega$.

ω-models are widely used in recursion theory and descriptive set theory. A major point of interest in an ω-model is its standard part.

Definition 3.5.2. Let $\mathcal{M} = \langle M, E\rangle$ be a structure and $M_1 \subseteq M$. Then M_1 is *E-closed* if for every $x \in M_1$, yEx and $y \in M$ implies $y \in M_1$. The *standard part* of $\mathcal{M}$ is the maximal E-closed wellfounded subset of M.

Obviously if $M_1 \subseteq M$ is E-closed, then for any Σ_0-formula $\varphi(v)$ and $x \in M_1$, $\langle M, E\rangle \models \varphi(x)$ if and only if $\langle M_1, E\rangle \models \varphi(x)$.

Lemma 3.5.3. *Suppose that $\mathcal{M} \models$ KP. Then x belongs to the standard part of M if and only if* rk(x)*, the rank of x, belongs to the standard part of $\mathcal{M}$.*

Proof. Since $\mathcal{M} \models$ KP, by the Σ-induction Theorem (3.1.8) the rank function rk is Δ_1-definable.

If x belongs to the standard part of M, then by induction on the rank of x, it is straightforward to verify that rk(x) also belongs to the standard part.

If x does not belong to the standard part of M, then there is a descending sequence $\{x_n\}_{n<\omega}$ in M so that $x_{n+1}Ex_n$ for every n and x_0Ex. Then $\mathrm{rk}(x_{n+1})E\mathrm{rk}(x_n)$ for every n and $\mathrm{rk}(x_0)E\mathrm{rk}(x)$. So rk$(x)$ does not belong to the standard part of M either. □

Lemma 3.5.4. *Suppose that $\mathcal{M} \models$ KP and $M_1 \subseteq M$ is the standard part of M. Then $\langle M_1, E\rangle \models$ KP.*

Proof. Since $M_1 \subseteq M$ is E-closed, the extensionality axiom holds in $\mathcal{M}_1 = \langle M_1, E\rangle$. $\mathcal{M}_1$ satisfies the foundation axiom since M_1 is the standard part of M. Similarly it is straightforward to verify that pairing, union and the Δ_0-comprehension axioms hold in $\mathcal{M}_1$. We prove $\mathcal{M}_1 \models \Delta_0$-collection.

Suppose that $\varphi(u, v)$ is a Δ_0-formula and $a \in M_1$ such that

$$\langle M_1, E\rangle \models (\forall xEa)(\exists y)\varphi(x, y).$$

Since $\varphi(u, v)$ is Δ_0 and M_1 is an E-closed subset of M, we have

$$\langle M, E\rangle \models (\forall xEa)(\exists y)\varphi(x, y).$$

Now $\langle M, E\rangle \models$ KP implies that there is a $b \in M$ such that

$$\langle M, E\rangle \models (\forall xEa)(\exists yEb)\varphi(x, y).$$

By Lemma 3.5.3, there is a b_0 such that $\mathrm{rk}(b_0) = \alpha_0$ is the least ordinal in $\mathcal{M}$ to satisfy the collection axiom for φ. If $\alpha_0 \in M_1$, then by Lemma 3.5.3, $b_0 \in M_1$ and we are done. Otherwise, $\alpha_0 \in M \setminus M_1$. But then there is a nonstandard $\alpha_1 E\alpha_0$ such that for every xEa, there is a yEb_0 in M_1 with $\mathrm{rk}(y)E\alpha_1$ and $\langle M, E\rangle \models \varphi(x, y)$. Note that $\varphi(x, y) \wedge \mathrm{rk}(y)E\alpha_1$ is a Σ-formula and

$$\langle M, E\rangle \models (\forall xEa)(\exists y)(\varphi(x, y) \wedge \mathrm{rk}(y)E\alpha_1).$$

By the Σ-collection Theorem (3.1.6), there is a $b_1 \in M$ such that

$$\langle M, E\rangle \models (\forall xEa)(\exists yEb_1)\varphi(x, y) \wedge \mathrm{rk}(y)E\alpha_1).$$

Define

$$c = \{y \mid yEb_1 \wedge \langle M, E\rangle \models \mathrm{rk}(y)E\alpha_1 \wedge \exists xEa(\varphi(x, y))\}.$$

By Theorem 3.1.6, $c \in M$.

Then $\mathrm{rk}(c)E\alpha_0$ and $\alpha_0 = \mathrm{rk}(b_0)$. Since α_0 is the least ordinal in $\mathcal{M}$ to satisfy the collection axiom for φ, we have a contradiction. □

Proposition 3.5.5. *Let $\mathcal{M} \models$ KP be an ω-model. Then $L_{\omega_1^{\mathrm{CK}}} \subseteq M$.*

Proof. Let $M_1 \subseteq M$ be the standard part of M. By Lemma 3.5.4, $\langle M_1, E\rangle \models$ KP. Hence $\langle M_1, E\rangle$ is also an ω-model of KP. Since $\omega \cup \{\omega\} \subseteq M_1$, by Δ_0-separation every recursive set belongs to $\langle M_1, E\rangle$. By Σ-induction, $\omega_1^{\mathrm{CK}} \subseteq M_1$. Let $\pi : \langle M_1, E\rangle \to \langle N, \in\rangle$ be the collapsing map. Then $\langle N, \in\rangle$ is admissible so that $\omega_1^{\mathrm{CK}} \subseteq N$. By Theorem 3.3.4, $L_{\omega_1^{\mathrm{CK}}} \subseteq N$. We prove by induction on $\beta < \omega_1^{\mathrm{CK}}$ that $\pi^{-1} \restriction L_{\omega_1^{\mathrm{CK}}}$ is the identity map.

Note that by Proposition 3.3.3 there exists a Δ_1-formula $\varphi(u, v)$ such that for any ordinal $\beta < \omega_1^{\mathrm{CK}}$, $\langle N, \in\rangle \models \varphi(\beta, x)$ if and only if $x = J_\beta$. Moreover,

$$\langle N, \in\rangle \models \varphi(\beta, J_\beta) \wedge (\forall x \in J_\beta)(\exists \gamma \in \beta)(\exists y)(x \in y \wedge \varphi(\gamma, y)).$$

We prove by induction that $\pi^{-1}(J_\gamma) = J_\gamma$ for every $\gamma < \omega_1^{\mathrm{CK}}$.

Since $\omega_1^{\mathrm{CK}} \subseteq M_1$, we have $\pi(\beta) = \beta$ for any $\beta < \omega_1^{\mathrm{CK}}$. Thus $\pi(x) = x$ for $x \in J_1$. Since $\langle N, \in\rangle \models \varphi(1, x)$ if and only if $x = J_1$, we have $\langle M_1, E\rangle \models \varphi(1, x)$ if and only if $x = J_1$. So $\pi^{-1}(J_1) = J_1$.

For a successor ordinal $\beta + 1$. π clearly preserves the Gödel primitive recursive operators. Hence Range$(\pi^{-1} \restriction J_{\beta+1}) = J_{\beta+1}$. Since $\langle N, \in\rangle \models \varphi(\beta + 1, x)$ if and only if $x = J_{\beta+1}$, we have $\langle M_1, E\rangle \models \varphi(\beta + 1, x)$ if and only if $x = J_{\beta+1}$. So $\pi^{-1}(J_{\beta+1}) = J_{\beta+1}$.

Now suppose $\beta < \omega_1^{\mathrm{CK}}$ is a limit ordinal. By induction, if $\gamma < \beta$ then $\langle M_1, E\rangle \models \varphi(\gamma, x)$ if and only if $x = J_\gamma$. Let $z = \pi^{-1}(J_\beta)$. Since $\pi^{-1}(\beta) = \beta$, we have

$$\langle M_1, E\rangle \models \varphi(\beta, z) \wedge (\forall x E z)(\exists \gamma E \beta)(\exists y(xEy) \wedge \varphi(\gamma, y)).$$

Thus $z = J_\beta$. By the transitivity of $J_{\omega_1^{\mathrm{CK}}}$, we conclude that $\pi^{-1} \restriction L_{\omega_1^{\mathrm{CK}}}$ is the identity map so that $L_{\omega_1^{\mathrm{CK}}} \subseteq M$. □

3.5.1 Exercise

Exercise 3.5.1. *Prove that if $\langle M, E\rangle \models$ KP is a nonstandard ω-model, then for any nonstandard ordinal $\alpha \in M$, there is a set $A \subseteq \{\beta \mid \langle M, E\rangle \models \beta E \alpha\}$ such that $(\mathbb{Q}, <)$ is isomorphic to $\langle A, E\rangle$, where $\mathbb{Q}$ is the set of rational numbers.*

3.6 Coding structures

The method of coding a language or a structure by a set of natural numbers was introduced by Gödel in the proof of the Incompleteness Theorem. Coding is widely used today in mathematical logic. In this section, we will not discuss the formalism of coding. Instead, we present a general description of how the method works.

3.6.1 Coding a language

Given a countable language $\mathscr{L}$, there are many ways of coding it by a set of numbers. What we want is a "nice" coding scheme. In many cases, one may code a language $\mathscr{L}$ by a recursive set $\ulcorner\mathscr{L}\urcorner \subseteq \omega$. A language that may be thus coded is called a *recursive language*. For example, both the first order language of arithmetic and the language of Zermelo–Fraenkel set theory are recursive. Moreover, in a recursive language, the set of well-formed formulas and the set of sentences are recursive. We may effectively associate each formula φ with its *Gödel number* $\ulcorner\varphi\urcorner$. We say that a theory T is *recursive* if its corresponding set of Gödel numbers $\ulcorner T\urcorner = \{\ulcorner\varphi\urcorner \mid \varphi \in T\}$ is recursive. The theories considered in this book, such as KP, ZFC, are recursive.

Under coding with Gödel numbers, a proof sequence is identified with a finite sequence of numbers, and may therefore be considered as a finite subset of ω. Moreover, if both $\mathscr{L}$ and T are recursive, then so is the set of proofs in T. It follows that the set of sentences which are theorems of T is r.e.

3.6.2 Coding a structure

Given a recursive language $\mathscr{L}$, one may associate with it a recursive set $\hat{\mathscr{L}}$ and code a countable structure in $\mathscr{L}$ by the set of natural numbers ω with relations in $\hat{\mathscr{L}}$ defined over ω, so that the resulting structure is isomorphic to the original structure. If $\mathcal{M} = \langle M, R_0, \cdots, R_n\rangle$ is countable, a coding $\ulcorner\mathcal{M}\urcorner = \langle \omega, \hat{R}_0, \ldots, \hat{R}_n\rangle$ of $\mathcal{M}$ is a structure such that $\mathcal{M} \cong \ulcorner\mathcal{M}\urcorner$. If all the relations $\hat{R}_0 \ldots, \hat{R}_n$ are recursive, then the structure $\ulcorner\mathcal{M}\urcorner$ is said to be recursive. However, very few structures are recursive.

Let $\mathscr{L}$ be a recursive language and $\mathcal{M}$ a structure coded by $\ulcorner\mathcal{M}\urcorner$. For a formula $\hat{\varphi}(u)$ in $\hat{\mathscr{L}}$ corresponding to the formula $\varphi(u)$ in $\mathscr{L}$, we may compute the complexity of the set $A = \{m \in \omega \mid \ulcorner\mathcal{M}\urcorner \models \hat{\varphi}(m)\}$. The procedure to calculate the complexity goes as follows: Recursively decode $\hat{\varphi}(u)$ to obtain the formula $\varphi(u)$. Then the complexity of A is the complexity of $\varphi(u)$. For example, suppose $\mathscr{L}$ is the language of set theory and $\hat{\varphi}(u)$ corresponds to the formula $\varphi(w)$: $(\forall u)(\forall v)(u \in v \wedge v \in w \to u \in w)$ ("w is transitive"). Then the set

$$\{m \mid \langle \omega, \hat{\in}\rangle \models (\forall i)(\forall j)(i \hat{\in} j \wedge j \hat{\in} m \to i \hat{\in} m\}$$

is $\Pi_1(\hat{\in})$.

In general, given a language $\mathscr{L}$ with relation symbols $R_0, \ldots, R_n$, a structure and a formula $\varphi \in \mathscr{L}$, the set $\{m \in \omega \mid \ulcorner\mathcal{M}\urcorner \models \hat{\varphi}(m)\}$ is arithmetical in $\hat{R}_0 \oplus \cdots \oplus \hat{R}_n$. Moreover, this calculation is uniform. Hence if $\{\varphi_i\}_{i<\omega}$ is recursive, then the set $\{(i, m) \mid \ulcorner\mathcal{M}\urcorner \models \hat{\varphi}_i(m)\}$ is Turing computable in $(\hat{R}_0 \oplus \cdots \oplus \hat{R}_n)^{(\omega)}$ via an algorithm with index e. In particular, there is an e such that if $\{\varphi_i\}_{i<\omega}$ is a recursive set of sentences, then $\ulcorner\mathcal{M}\urcorner \models \hat{\varphi}_i$ if and only if $\Phi_e^{(\hat{R}_0 \oplus \cdots \oplus \hat{R}_n)^{(\omega)}}(i) = 1$. As a consequence, we have the following:

Proposition 3.6.1. *Given a recursive theory T, the set $\{\ulcorner\mathcal{M}\urcorner \mid (\forall\varphi)(\varphi \in T \Rightarrow \ulcorner\mathcal{M}\urcorner \models \hat{\varphi})\}$ is Δ^1_1.*

For example, $\{\ulcorner\mathcal{M}\urcorner \mid \ulcorner\mathcal{M}\urcorner \models$ KP + "V = L"$\}$ is Δ^1_1 (where for simplicity we make no distinction between KP+"V = L" and the corresponding sentences interpreted in $\ulcorner\mathcal{M}\urcorner$).

3.6.3 Coding ω- and wellfounded models

We now work in the language of set theory and use "E" in place of the symbol $\hat{\in}$. The best models of set theory are of course those that are wellfounded.

Proposition 3.6.2. $\{E \mid \langle \omega, E\rangle$ *codes a wellfounded model*$\}$ *is* Π^1_1.

Proof. $\langle \omega, E\rangle$ codes a wellfounded model if and only if there is no function $f : \omega \to \omega$ such that for any pair (i, j), $i < j$ if and only if $\langle \omega, E\rangle \models f(j)Ef(i)$. □

Coding wellfounded models is a complicated matter, though such models are the "bona fide models" of set theory. The next best thing is to code ω-models. The reason for studying ω-models is that in considering reals or sets of reals in such models, one is assured that the "natural numbers" involved are standard numbers.

Proposition 3.6.3. $\{E \mid \langle \omega, E\rangle$ *codes an* ω*-model*$\}$ *is* Δ^1_1.

Proof. $\langle \omega, E\rangle$ codes an ω-model if and only if there is an $m \in \omega$ and an injection $f : \omega \to \omega$ such that

(i) $\langle \omega, E\rangle \models$ "m is an infinite ordinal";

(ii) f is a bijection between ω and $\{k \mid kEm\}$, and

(iii) for any pair (i, j), $i < j$ if and only if $\langle \omega, E\rangle \models f(i)Ef(j) \wedge f(j)Em$.

Hence the set $\{E \mid \langle \omega, E\rangle$ is an ω-model$\}$ is Σ^1_1.

On the other hand, $\langle \omega, E\rangle$ codes an ω-model if and only if there is an $m \in \omega$ such that

(iv) $\langle \omega, E\rangle \models$ "m is an infinite ordinal";

(v) for every injection f from ω into $\{k \mid kEm\}$, there is a pair (i, j) such that $i < j$ and $\langle \omega, E\rangle \models f(i)Ef(j)$, and

(vi) for every order-preserving function f from ω into $\{k \mid kEm\}$ and for any k with $\langle \omega, E\rangle \models kEm$, there is a j such that $\langle \omega, E\rangle \models kEf(j)$.

Thus $\{E \mid \langle \omega, E\rangle$ is an ω-model$\}$ is also Π^1_1 and hence Δ^1_1. □

Proposition 3.6.4.

(i) *The set* $(\omega, \omega_E, \mathscr{E}) = \{(n, n_E, E) \mid n_E$ *codes* n *in* $\ulcorner\mathcal{M}\urcorner\}$ *is arithmetical;*

(ii) *Suppose that* $\ulcorner\mathcal{M}\urcorner = \langle \omega, E\rangle$ *codes an* ω*-model* $\mathcal{M}$. *Then for any real* $x \in \mathcal{M}$, x *is arithmetical in* E;

(iii) *If* m *codes* ω *in* $\ulcorner\mathcal{M}\urcorner$ *and* $f : \omega \to \{k \mid \ulcorner\mathcal{M}\urcorner \models kEm\}$ *is an order-preserving bijection, then every real* $x \in \mathcal{M}$ *is recursive in* $f \oplus E$.

Proof. (i) $(n, m, E) \in (\omega, \omega_E, \mathscr{E})$ if and only if there is a finite sequence $\{k_j\}_{j\le n}$ such that

- $k_n = m \wedge \ulcorner\mathcal{M}\urcorner \models (\forall k)(\neg kEk_0)$;
- for any $j < n$, $\ulcorner\mathcal{M}\urcorner \models k_jEk_{j+1} \wedge (\forall k)(kEk_{j+1} \to kEk_j \vee k = k_j)$, and
- for all $j < n$, $\ulcorner\mathcal{M}\urcorner \models k_jEm$.

Hence (ω, ω_E, E) is arithmetical.

(ii) Given $x \in \mathcal{M}$, let m code x in $\ulcorner\mathcal{M}\urcorner$. Then for any n, $n \in x$ if and only if $(n, n_E, E) \in (\omega, \omega_E, \mathscr{E})$ and $\ulcorner\mathcal{M}\urcorner \models n_EEm$. Thus x is arithmetical in E.

(iii) For $x \in \mathcal{M}$, let n code x in $\ulcorner\mathcal{M}\urcorner$. Then for any i, $i \in x$ if and only if $\ulcorner\mathcal{M}\urcorner \models f(i)En$, so that $x \le_T E \oplus f$. □

By (iii) of Proposition 3.6.4, if f is recursive, then every real in $\mathcal{M}$ is recursive in E. If such a recursive f exists, we say that $\langle\omega, E\rangle$ is a *nicely coded ω-model.*

Proposition 3.6.5. *If $\langle\omega, E\rangle$ codes an ω-model, then there is a nicely coded ω-model $\langle\omega, E_0\rangle \cong \langle\omega, E\rangle$. Moreover, the set $\{E \mid \langle\omega, E\rangle$ is a nicely coded ω-model$\}$ is Δ^1_1.*

Proof. Suppose that $\langle\omega, E\rangle$ codes an ω-model and m codes ω in $\langle\omega, E\rangle$. Without loss of generality, we may assume that $\{k \mid \langle\omega, E\rangle \models kEm\}$ is infinite. Let $f : \omega \to \{k \mid \langle\omega, E\rangle \models kEm\}$ be an order-preserving bijection. Let $g : \omega \to \omega$ be any bijection such that $g(2n) = f(n)$ for every n. Now define iE_0j if and only if $g(i)Eg(j)$. Then $g : \langle\omega, E\rangle \cong \langle\omega, E_0\rangle$. Obviously $nE_0g^{-1}(m)$ if and only if n is an even number. Let $h(n) = 2n$. Then h is recursive and $h : \omega \to \{i \mid \langle\omega, E_0\rangle \models iE_0g^{-1}(m)\}$ is an order-preserving bijection. Thus $\langle\omega, E_0\rangle$ is a nicely coded ω-model.

We leave the proof of the second part of Proposition 3.6.5 to the reader. □

From the above it is harmless to assume that an ω-model $\langle\omega, E\rangle$ is always nicely coded.

3.6.4 The second least admissible ordinal

An application of the previous results yields

Lemma 3.6.6. *ω_1^{CK} is the second least admissible ordinal.*

Proof. Let α be the second least admissible ordinal. By Proposition 3.3.12, $\alpha \geq \omega_1^{\mathrm{CK}}$.

Let $\mathscr{X} = \{E \mid \langle\omega, E\rangle$ nicely codes an ω-model of KP$\}$. By Proposition 3.6.5, $\mathscr{X}$ is a Δ^1_1-set. By Gandy's Basis Theorem (2.5.3), there is an $E \in \mathscr{X}$ such that $\omega_1^E = \omega_1^{\mathrm{CK}}$. Let $\langle M, \in\rangle$ be the ω-model coded by $\langle\omega, E\rangle$ and M_1 its standard part. By Lemma 3.5.4, $\langle M_1, \in\rangle \models$ KP. We claim that $\beta < \omega_1^{\mathrm{CK}}$ for any $\beta \in M_1$. Assume this is false. Let $m \in \omega$ be a real which codes ω_1^{CK} in $\langle\omega, E\rangle$. Then the set

$$\{(i,j) \mid \langle\omega, E\rangle \models iEj \wedge jEm\}$$

is an E-recursive well-ordering of order type ω_1^{CK}, contradicting $\omega_1^E = \omega_1^{\mathrm{CK}}$. Thus $o(M_1) = \omega_1^{\mathrm{CK}}$ and hence $\alpha = \omega_1^{\mathrm{CK}}$. Hence ω_1^{CK} is the second least admissible ordinal. □

By Theorem 3.3.6 and Lemma 3.6.6, we have

Corollary 3.6.7. *$L_{\omega_1^{\mathrm{CK}}}$ is admissible.*

Theorem 3.6.8. *A real x is hyperarithmetic if and only if $x \in L_{\omega_1^{\mathrm{CK}}}$.*

Proof. To prove that every hyperarithmetic real belongs to $L_{\omega_1^{\mathrm{CK}}}$, it suffices to prove $H_n \in L_{\omega_1^{\mathrm{CK}}}$ for any $n \in \mathcal{O}$. By Proposition 2.1.5, H_n is Turing reducible to some Π^0_1-singleton $f_n \in \omega^\omega$. So it is sufficient to prove that $f_n \in L_{\omega_1^{\mathrm{CK}}}$.

Since f_n is a Π^0_1-singleton, there is a recursive tree $T_n \in \omega^{<\omega}$ such that f_n is the unique infinite path in T. Then for every $m \in \omega$ and $\sigma \in \omega^m \setminus \{f_n \upharpoonright m\}$, $T_\sigma = \{\tau \in T_n \mid$

$\tau \succ \sigma\}$ is wellfounded. Since $L_{\omega_1^{\mathrm{CK}}}$ is admissible, there is an $h_\sigma : T_\sigma \to \omega_1^{\mathrm{CK}}$ in $L_{\omega_1^{\mathrm{CK}}}$ such that for any $\tau_1 \succ \tau_0$ in T_σ, $h_\sigma(\tau_1) < h_\sigma(\tau_0)$. Hence there is a $\beta_\sigma \in \omega_1^{\mathrm{CK}}$ such that $h_\sigma \in L_{\beta_\sigma}$. Since $\omega^m \setminus \{f_n \restriction m\} \in L_{\omega_1^{\mathrm{CK}}}$, by Theorem 3.3.2 and the Σ-Collection Theorem (3.1.6), there is a $\beta_m < \omega_1^{\mathrm{CK}}$ such that $\beta_\sigma < \beta_m$ for all $\sigma \in \omega^m \setminus \{f_n \restriction m\}$. By Theorem 3.1.6 again, there is a $\beta < \omega_1^{\mathrm{CK}}$ which is an upper bound of all the β_m's. Then $\sigma \prec f_n$ if and only if

$$\langle L_\beta, \in\rangle \models (\forall \tau \in \omega^{|\sigma|} \setminus \{\sigma\})(\exists \gamma)(\exists h)(\forall \tau_1, \tau_2 \in T_\tau)(h : T_\tau \to \gamma \wedge \tau_2 \succ \tau_1 \to h(\tau_2) < h(\tau_1)).$$

Thus $f_n \in L_{\beta+1} \subset L_{\omega_1^{\mathrm{CK}}}$.

Suppose that $x \in L_{\omega_1^{\mathrm{CK}}}$. Since $L_{\omega_1^{\mathrm{CK}}}$ is admissible, we have $\omega_1^x = \omega_1^{\mathrm{CK}}$. Assume that x is not hyperarithmetic. Define

$$\mathscr{X} = \{E \mid \langle \omega, E\rangle \text{ nicely codes an } \omega\text{-model of KP}\}.$$

By Proposition 3.6.5, $\mathscr{X}$ is a Δ_1^1-set. By the Kreisel Basis Theorem (2.5.6), there is an $E_0 \in \mathscr{X}$ with $\omega_1^{E_0} = \omega_1^{\mathrm{CK}}$ and $x \not\leq_h E_0$. Let $\langle M, \in\rangle$ be the ω-model coded by $\langle \omega, E_0\rangle$ and let M_1 be the standard part of M. Then every real $z \in M_1$ is recursive in E_0. Hence $x \notin M_1$. By Proposition 3.5.5, $L_{\omega_1^{\mathrm{CK}}} \subseteq M_1$. Thus $x \notin L_{\omega_1^{\mathrm{CK}}}$. □

This gives a model-theoretic characterization of the set of hyperarithmetic reals:

Corollary 3.6.9. *A real x is hyperarithmetic if and only if it belongs to every ω-model of* KP.

Proof. By Theorem 3.6.8, x is hyperarithmetic if and only if $x \in L_{\omega_1^{\mathrm{CK}}}$. By Theorem 3.5.5 and Corollary 3.6.7, $x \in L_{\omega_1^{\mathrm{CK}}}$ if and only if x belongs to every ω-model of KP. □

3.6.5 The Cantor–Bendixon construction revisited

We now return to the recursion-theoretic proof of Theorem 1.1.10 and calculate the least ordinal γ for which $T_{\gamma+1} = T_\gamma$.

Proposition 3.6.10. *Suppose $T \subseteq \omega^{<\omega}$ is a recursive tree with uncountably many infinite paths. Then there is a $\Sigma_1(L_{\omega_1^{\mathrm{CK}}})$-function $f : \omega_1^{\mathrm{CK}} \to L_{\omega_1^{\mathrm{CK}}}$ such that for any $\gamma_1 < \gamma_2 < \omega_1^{\mathrm{CK}}$, $f(\gamma_2) = T_{\gamma_2} \subseteq f(\gamma_1) = T_{\gamma_1} \subseteq T$ and $\bigcap_{\gamma<\omega_1^{\mathrm{CK}}} T_\gamma$ is a perfect subtree of T.*

Proof. Since T is recursive, we have $T \in L_{\omega+1}$. By Theorem 3.1.8, there is a $\Sigma_1(L_{\omega_1^{\mathrm{CK}}})$-function $f : \omega_1^{\mathrm{CK}} \to L_{\omega_1^{\mathrm{CK}}}$ such that

- $f(0) = T$;
- for any $\gamma < \omega_1^{\mathrm{CK}}$, $f(\gamma+1) = T_{\gamma+1} = \{\sigma \in T_\gamma \mid (\exists \tau_0, \tau_1 \succ \sigma)(\tau_0, \tau_1 \in T_\gamma \wedge \tau_0 | \tau_1)\} \subseteq f(\gamma)$;
- if γ is a limit ordina, then $f(\gamma) = T_\gamma = \bigcap_{\beta<\gamma} T_\beta$.

First note that $|[T_\gamma] \setminus [T_{\gamma+1}]| \leq \aleph_0$ for each $\gamma < \omega_1^{\mathrm{CK}}$, and hence $[T_{\omega_1^{\mathrm{CK}}}] \neq \emptyset$. We claim that $T_{\omega_1^{\mathrm{CK}}} = \bigcap_{\gamma<\omega_1^{\mathrm{CK}}} T_\gamma \subseteq T$ is a perfect tree. First note that given $\sigma \in T_{\omega_1^{\mathrm{CK}}}$, there is a $\tau \succ \sigma$ such that $\tau \in T_{\omega_1^{\mathrm{CK}}}$. Otherwise, let $g_\sigma(n)$ be the least ordinal $\gamma < \omega_1^{\mathrm{CK}}$ for which $\sigma^\frown n \notin T_\gamma$. Then g_σ is a total function and Σ_1 in $L_{\omega_1^{\mathrm{CK}}}$. By the admissibility of ω_1^{CK} (Corollary 3.6.7), there is an ordinal $\beta < \omega_1^{\mathrm{CK}}$ so that $g_\sigma(n) < \beta$ for all n. Then $\sigma \notin T_{\beta+1}$, a contradiction.

Now suppose that $\sigma \in T_{\omega_1^{\mathrm{CK}}}$ is an isolated path, i.e. $[\sigma] \cap [T_{\omega_1^{\mathrm{CK}}}]$ contains exactly one path. Let $h_\sigma(n)$ be the least ordinal $\gamma < \omega_1^{\mathrm{CK}}$ such that $|\{\tau \mid \tau \in T_\gamma \wedge \tau \succ \sigma \wedge |\tau| = |\sigma|+n\}| = 1$. Then h_σ is a total function and Σ_1 in $L_{\omega_1^{\mathrm{CK}}}$. By Corollary 3.6.7 again, there is an ordinal $\beta < \omega_1^{\mathrm{CK}}$ such that $h_\sigma(n) < \beta$ for all n. Then for any $\tau_0, \tau_1 \succ \sigma$, if $\tau_0, \tau_1 \in T_\beta$, then τ_0 and τ_1 are compatible. So $\sigma \notin T_{\beta+1}$, a contradiction. □

3.6.6 Exercises

Exercise 3.6.1. *Complete the proof of Proposition 3.6.5.*

Exercise 3.6.2. *Relativize and prove the results in § 3.6.4.*

Exercise 3.6.3. *Prove that there is a recursive tree T with uncountably many infinite paths (as in Proposition 3.6.10) such that for each $\gamma < \omega_1^{\mathrm{CK}}$, T_γ is not a perfect tree.*

Exercise 3.6.4. *Suppose that R_0 and R_1 code well-ordering relations of ω and the order type of R_0 is not greater than that of R_1. Show that there is an order-preserving function $f \leq_h R_0 \oplus R_1$ which embeds R_0 onto an initial segment of R_1.*

Exercise 3.6.5. *Given a hyperarithmetic real x, show that there exist two recursive well-orderings $R_0 \cong R_1$ of ω such that there is no isomorphism between R_0 and R_1 recursive in x.*

Exercise 3.6.6. *Without using Kreisel's Basis Theorem, give a direct proof of Theorem 3.6.8.*

3.7 The Spector–Gandy Theorem

3.7.1 First version

Theorem 3.7.1 (Spector[156], Gandy[34]). *$x \in 2^\omega$ is Π^1_1 if and only if there is a Σ_1-formula $\varphi(u)$ such that $n \in x \Leftrightarrow \langle L_{\omega_1^{\mathrm{CK}}}, \in\rangle \models \varphi(n)$.*

Proof. Suppose that $x \in 2^\omega$ is Π^1_1. By Proposition 1.1.14, there is a recursive tree $T \subseteq \omega \times \omega^{<\omega}$ and a recursive function f such that $n \in x$ if and only if $T_{f(n)} = \{\sigma \mid (f(n), \sigma) \in T\}$ is a wellfounded tree. Since T is recursive, we have $T \in L_{\omega_1^{\mathrm{CK}}}$. Given n, we say that $g : T_{f(n)} \to \gamma$ is order preserving if $\sigma \succ \tau$ implies $g(\sigma) < g(\tau)$. By Corollary 3.6.7,

$L_{\omega_1^{\mathrm{CK}}}$ is admissible. Hence by Theorem 3.1.8, there is an ordinal $\gamma < \omega_1^{\mathrm{CK}}$ and an order-preserving function $g : T_{f(n)} \to \gamma$.

Let $\varphi(n)$ be the statement

$$(\exists\gamma)(\exists g)(\gamma \text{ is an ordinal } \wedge g \text{ is an order-preserving function from } T_{f(n)} \text{ into } \gamma),$$

In other words, $n \in x \Rightarrow \langle L_{\omega_1^{\mathrm{CK}}}, \in\rangle \models \varphi(n)$. On the other hand, if $\langle L_{\omega_1^{\mathrm{CK}}}, \in\rangle \models \varphi(n)$, then $T_{f(n)}$ is wellfounded. Thus $n \in x$.

For the other direction, suppose that there is a Σ_1-formula $\varphi(u)$ such that $n \in x \Leftrightarrow \langle L_{\omega_1^{\mathrm{CK}}}, \in\rangle \models \varphi(n)$. Since $\varphi(u)$ is Σ_1, by Proposition 3.5.5, $\langle L_{\omega_1^{\mathrm{CK}}}, \in\rangle \models \varphi(n)$ if and only if for any E, if $\langle\omega, E\rangle$ codes an ω-model of KP, then $\langle\omega, E\rangle \models \varphi(n_E)$ where n_E denotes the number in $\langle\omega, E\rangle$ which codes n. By Proposition 3.6.3, x is Π_1^1. □

Notwithstanding the remark in [107] that "the Spector–Gandy Theorem does not have many applications but it is undoubtedly one of the jewels of the effective theory", it is a major result and has seen important applications in higher recursion theory. For example, the theorem allows one to develop a robust computation theory for the structure $\langle L_{\omega_1^{\mathrm{CK}}}, \in\rangle$, including implementation of the priority method to study the metadegrees of Π_1^1-sets.

A generalization of Theorem 3.7.1 that lifts the result to sets of reals is the following.

Corollary 3.7.2. *For any real y, a set $A \subseteq 2^\omega$ is $\Pi_1^1(y)$ if and only if there is a Σ_1-formula $\varphi(u, v)$ such that $x \in A \Leftrightarrow \langle L_{\omega_1^{x\oplus y}}[x \oplus y], \in\rangle \models \varphi(x, y)$.*

It should be pointed out that the statement in the corollary remains valid if one replaces 2^ω with ω^ω in Corollary 3.7.2.

3.7.2 Second version

Let $\mathcal{O}^*$ be as defined in § 2.6. Fix a number $i \in \mathcal{O}^* \setminus \mathcal{O}$. Then by the results in the same section, the set $\mathcal{O}_i^* = \{j \mid j <_{\mathcal{O}^*} i\}$ is an r.e. set such that $\mathcal{O}_i^* \cap \mathcal{O}$ is an initial segment of $\mathcal{O}^*$ of order type ω_1^{CK}. By Lemma 2.2.4, there is a recursive function g such that if $n \in \mathcal{O}$, then $\mathcal{O}_n = \Phi_{g(n)}^{H_{2^n}}$.

For any real z and $n \in \omega$, let $(z)_n = \{m \mid 2^n \cdot 3^m \in z\}$. Let $A(n, z)$ be an arithmetical predicate defined as follows: There is an $m <_{\mathcal{O}^*} i$ such that

(1) $(z)_1 = \emptyset$;
(2) $(\forall j)(j \leq_{\mathcal{O}^*} 2^m \to (z)_{2^j} = ((z)_j)')$;
(3) $(\forall e)(3 \cdot 5^e <_{\mathcal{O}^*} 2^m \to (\forall k)((z)_{3\cdot 5^e})_k = (z)_{\Phi_e(k)})$;
(4) $n \in \Phi_{g(m)}^{(z)_{2^m}}$, and
(5) $(\forall j)(j \leq_{\mathcal{O}^*} 2^m \to (z)_j \leq_T (z)_{2^m})$.

Roughly speaking, $A(n, z)$ says that there is an $m <_{\mathcal{O}^*} i$ such that z codes an iteration of Turing jumps along the set $\{j \mid j \leq_{\mathcal{O}^*} m\}$ and $n \in \mathcal{O}_m = \Phi_{g(m)}^{(z)_{2^m}}$. Then for any $n \in \mathcal{O}_i^* \cap \mathcal{O}$, $(z)_n = H_n$.

Theorem 3.7.3 (Spector [156], Gandy [34]). *$x \in 2^\omega$ is Π^1_1 if and only if there is an arithmetical predicate $B(n, z)$ such that $n \in x \Leftrightarrow (\exists z \in \Delta^1_1)(B(n, z))$.*

Proof. By Corollary 2.2.12, if there is an arithmetical predicate $B(n, z)$ such that $n \in x \Leftrightarrow (\exists z \in \Delta^1_1)(B(n, z))$, then x is Π^1_1.

Suppose that x is Π^1_1. Then there is a recursive function f such that $n \in x \Leftrightarrow f(n) \in \mathcal{O}$. Let $A(n, z)$ be as defined earlier. If $n \in x$, then $f(n) \in \mathcal{O}_{2^{f(n)}}$. Let $m <_{\mathcal{O}^*} i$ be in $\mathcal{O}$ so that $|m| > |f(n)|$. Let z be a real such that for any $j \leq_{\mathcal{O}} 2^m$, $(z)_j = H_j$ and $(z)_j = \emptyset$ for the other j's. Then $f(n) \in \mathcal{O}_m = \Phi^{(z)_{2^m}}_{g(m)}$. Clearly z is hyperarithmetic. Thus $n \in x$ implies that there exists a hyperarithmetic real z satisfying $A(f(n), z)$

Now suppose that z is hyperarithmetic and $A(f(n), z)$. Then there is an $m <_{\mathcal{O}^*} i$ with properties (1)–(5). We claim that $m \in \mathcal{O}$. Otherwise, for any $k \in \mathcal{O}$, $H_k \leq_T (z)_{2^m}$, contradicting the fact that z is hyperarithmetic. Thus $f(n) \in \mathcal{O}_m = \Phi^{(z)_{2^m}}_{g(m)}$ and so $n \in x$.

Let $B(n, z)$ be $A(f(n), z)$. Then $n \in x$ if and only if there is a hyperarithmetic real z such that $B(n, z)$. □

We leave it to the reader to prove the following generalization of Theorem 3.7.3 to sets of reals:

Corollary 3.7.4. *$A \subseteq 2^\omega$ is $\Pi^1_1(y)$ if and only if there is an arithmetical predicate B such that $x \in A \Leftrightarrow (\exists z \leq_h x \oplus y)(B(x, z, y))$.*

Theorem 3.7.3 says that if we consider Π^1_1-sets to be r.e. sets, then hyperarithmetic reals are analogs of finite sets. In particular, there are infinite sets of natural numbers which are "finite". This is a departure from the classical notion where infinite sets are never "finite".

As mentioned in § 3.7.1, the Gandy–Spector Theorem presents an effective – hence recursion-theoretic – way to enumerate members of a Π^1_1-set. The enumeration is carried out in $L_{\omega_1^{\mathrm{CK}}}$ through ordinals less than ω_1^{CK}.

The situation is more problematic with Π^1_1-sets of reals. Corollary 3.7.2 says that a set $A \subseteq 2^\omega$ is Π^1_1 if and only if there is a Σ_1 formula $\varphi(u)$ such that $x \in A \Leftrightarrow \langle L_{\omega_1^x}[x], \in\rangle \models \varphi(x)$ (rather than $\langle L_{\omega_1^x}, \in\rangle \models \varphi(x)$). This makes analyzing Π^1_1-sets of reals a more delicate proposition. For each real x, one has to work within $L_{\omega_1^x}[x]$ to search for a solution to the Σ_1-formula $\varphi(u)$. This lack of uniformity is an issue when one wishes to choose the x "enumerated at the least stage" satisfying φ. In the case of the second version of the Gandy–Spector Theorem (3.7.3), the corresponding procedure for "enumerating" a Π^1_1-set of reals is to "enumerate" the H^x-sets in order to compute a witness z. Here we face a problem similar to that for version one, in that there is no uniform way to execute this procedure. We give below an example to illustrate how such a difficulty may be addressed.

Definition 3.7.5. A *norm* on a set $A \subseteq 2^\omega$ is a function $\varphi : A \to \alpha$ for some ordinal α.

Norms are among the most important objects in descriptive set theory. A norm on a set A acts as an "enumerating function" for A.

Proposition 3.7.6. *Let $A \subseteq 2^\omega$ be Π^1_1. There is a norm φ on A, a Π^1_1-relation $P(x,y)$ and a Σ^1_1-relation $Q(x,y)$ such that for any $y \in A$,*

$$(\forall x)(x \in A \wedge \varphi(x) \leq \varphi(y) \leftrightarrow (P(x,y) \leftrightarrow Q(x,y))).$$

Proof. Since A is Π^1_1, by Corollary 3.7.4 there is an arithmetical predicate $B(x,z)$ such that $x \in A \Leftrightarrow (\exists z \leq_h x)(B(x,z))$. If $x \in A$, choose $n \in \mathcal{O}^x$ so that $|n|_x$ is minimal and $B(x, \Phi_e^{H_n^x})$ holds for some e. Let $\varphi(x) = |n|_x$. Define $P(x,y)$ as follows:

(1) $(\exists n_0 \in \mathcal{O}^x)(\exists e)(B(x, \Phi_e^{H^x_{n_0}}) \wedge (\forall m \in \mathcal{O}^x_{n_0})\neg(\exists e)(B(x, \Phi_e^{H^x_m})$;

(2) $(\exists n_1 \in \mathcal{O}^y)(\exists e)(B(y, \Phi_e^{H^y_{n_1}}) \wedge (\forall m \in \mathcal{O}^y_{n_1})\neg(\exists e)(B(y, \Phi_e^{H^y_m})$, and

(3) there exists an isomorphism from $\{m \mid m <_{\mathcal{O}^x} n_0\}$ onto an initial segment of $\{m \mid m <_{\mathcal{O}^y} n_1\}$.

Clause (3) is Π^1_1 since the isomorphism (if there is one) must be hyperarithmetic in $x \oplus y$.

By relativizing the results in § 2.6, there is a Σ^1_1-set $\{(x, \mathcal{O}^{x,*}) \mid x \in 2^\omega\} \subseteq 2^\omega \times \omega$ so that $<_{\mathcal{O}^{x,*}}$ is a binary realign on $\mathcal{O}^{x,*}$ with the following properties:

(i) $\mathcal{O}^x \subseteq \mathcal{O}^{x,*}$;

(ii) $<_{\mathcal{O}^x}$ is the restriction of $<_{\mathcal{O}^{x,*}}$ to $\mathcal{O}^x$;

(iii) for any n, $n \in \mathcal{O}^x$ if and only if $<_{\mathcal{O}^{x,*}}$ restricted to $\{m \mid m <_{\mathcal{O}^{x,*}} n\}$ is a well-ordering.

Let $Q(x,y)$ be the formula

$$(\forall n \in \mathcal{O}^y)(\exists e_0)(B(y, \Phi_{e_0}^{H^y_n}) \rightarrow (\exists m \in \mathcal{O}^{x,*})(\exists f)(\exists e_1)$$
$$(f : \{j \mid j <_{\mathcal{O}^{x,*}} m\} \rightarrow \{j \mid j \in \mathcal{O}^y_n\} \text{ is order preserving } \wedge B(x, \Phi_{e_1}^{H^x_m})).$$

Now it is easy to see that if $y \in A$, then $(\forall x)(x \in A \wedge \varphi(x) \leq \varphi(y))$ if and only if $P(x,y)$, and if and only if $Q(x,y)$. □

Proposition 3.7.6 essentially says that if y belongs to A, then the collection of reals enumerated into A "before y is enumerated" is $\Delta^1_1(y)$.

3.7.3 An application

We apply the Spector–Gandy Theorem to study Σ^1_2-sets of reals. For each ordinal $\alpha < \omega_1$, let r_α be a real coding a well-ordering of ω of order type α. In the next two lemmas, let $R_\alpha = \emptyset$ if α is not a countable admissible ordinal (since ω_1^x is admissible for any $x \subseteq \omega$).

Lemma 3.7.7. *For any $\alpha < \omega_1$, the set $R_\alpha = \{x \mid \omega_1^x = \alpha\}$ is $\Delta^1_1(r_\alpha)$.*

Proof. For any n, let $|n|_{r_\alpha}$ be the order of n under the ordering r_α. Then $x \in R_\alpha$ if and only if

$(\forall n \in \mathcal{O}^x)(\exists f)(f$ is an embedding of $\{m \mid m <_{\mathcal{O}^x} n\}$ into $r_\alpha) \wedge$

$(\exists g)(\exists i \in \mathcal{O}^{*,x})(g$ is an isomorphic map from r_α onto an initial segment of $\mathcal{O}^{*,x}_i)$.

Thus R_α is $\Sigma^1_1(r_\alpha)$. Let $r^+_\alpha \le_T r_\alpha$ be a well-ordering of ω of order type $\alpha + 1$. Then $x \in R_\alpha$ if and only if

$$(\forall f)(\forall n \in \mathcal{O}^{*,x})(f \text{ is not an embedding of } r^+_\alpha \text{ into } \mathcal{O}^{*,x}_n) \wedge$$
$$(\forall n)(\exists m \in \mathcal{O}^x)(\exists g)(g \text{ is an isomorphism of } \{m \mid m <_{r_\alpha} n\} \text{ into}$$
$$\text{an initial segment of } \{i \mid i <_{o^x} m\}).$$

By Exercise 3.6.4, one may take g in the above expression to be hyperarithmetic in $r_\alpha \oplus \{i \mid i <_{o^x} m\}$ which is in turn hyperarithmetic in $r_\alpha \oplus x$. Hence R_α is also $\Pi^1_1(r_\alpha)$ and thus $\Delta^1_1(r_\alpha)$. □

Lemma 3.7.8. *For any $\alpha < \omega_1$, the relation $C_\alpha(x, y) \Leftrightarrow y \le_h x \wedge x \in R_\alpha$ is $\Delta^1_1(r_\alpha)$.*

Proof. By Corollary 2.2.2 and Lemma 3.7.7, C_α is $\Pi^1_1(r_\alpha)$.

We prove that C_α is $\Sigma^1_1(r_\alpha)$. Now $y \le_h x$ if and only if there is an $n \in \mathcal{O}^x$ such that $y \le_T H^x_n$, and by the previous lemma the latter holds if and only if there is an $n \in \mathcal{O}^{x,*}$ and an $f : \{m \mid m \le_{o^{x,*}} n\} \to \omega$ such that for all $i <_{o^{x,*}} j$, $(f(i), f(j)) \in r_\alpha$. Then $y \le_h x$ is $\Sigma^1_1(r_\alpha)$. Thus C_α is $\Sigma^1_1(r_\alpha)$. □

Lemma 3.7.9. (i) *Every Π^1_1-set of reals is the union of $\aleph_1$-many Borel sets.*
(ii) *Every Σ^1_1-set of reals is the union of $\aleph_1$-many Borel sets.*

Proof. (i) Suppose that $A \subseteq 2^\omega$ is Π^1_1. By Corollary 3.7.4, there is an arithmetical predicate $A(x, z)$ such that $x \in A \Leftrightarrow (\exists z \le_h x)(A(x, z))$. For any $\alpha < \omega_1$, let $A_\alpha = A \cap R_\alpha$. Then A_α is $\Pi^1_1(r_\alpha)$. By Lemma 3.7.8, A_α is also $\Sigma^1_1(r_\alpha)$. Thus A_α is $\Delta^1_1(r_\alpha)$ and hence Borel.

(ii) Suppose that $A \subseteq 2^\omega$ is Σ^1_1. Then there is a Π^0_1 set $B \subseteq \omega^\omega \times 2^\omega$ such that $x \in A$ if and only if $(\exists f)B(f, x)$. For $\alpha < \omega_1$, let $A_\alpha = A \cap R_\alpha$. By Lemma 3.7.7, A_α is $\Sigma^1_1(r_\alpha)$.

By the Gandy Basis Theorem 2.5.3, $x \in A$ if and only if $(\exists f <_h \mathcal{O}^x)B(f, x)$. Since $\omega_1^{r_\alpha} > \omega_1^x$, we have $x \not\ge_h r_\alpha$. Relativizing the proof of Corollary 2.4.10 to r_α gives $\mathcal{O}^x \le_h x \oplus r_\alpha$. Thus $x \in A_\alpha$ if and only if $x \in R_\alpha$ and $(\exists f \le_h x \oplus r_\alpha)B(f, x)$. We conclude that A_α is $\Pi^1_1(r_\alpha)$ and hence $\Delta^1_1(r_\alpha)$. □

Corollary 3.7.10 (Lusin [86]). *Every Π^1_1-set of reals is measurable.*

Proof. Following the notations of Lemma 3.7.9, for each $n \in \omega$ and $\alpha < \omega_1$, let

$$D_{n,\alpha} = \{x \mid n \in \mathcal{O}^{x,*} \wedge \langle\{m \mid m <_{o^{x,*}} n\}, <_{o^{x,*}}\rangle \cong r_\alpha\}.$$

Then $D_{n,\alpha}$ is $\Sigma^1_1(r_\alpha)$. By the proof of Lemma 3.7.8 and Exercise 3.6.4,

$$D_{n,\alpha} = \{x \mid n \in \mathcal{O}^x \wedge \langle\{m \mid m <_{o^x} n\}, <_{o^x}\rangle \cong r_\alpha\}$$

and is $\Pi^1_1(r_\alpha)$. Thus $D_{n,\alpha}$ is Borel. Note that $D_{n,\alpha} \cap D_{n,\beta} = \emptyset$ for $\alpha \neq \beta$. Hence for any n, there exists an $\alpha_n < \omega_1$ such that $\mu(D_{n,\beta}) = 0$ for $\beta \ge \alpha_n$. Let $\alpha_\omega < \omega_1$ be an upper bound of these α_n's.

Now suppose that A is Π^1_1. Then for any $\alpha > \alpha_\omega$, $x \in A_\alpha$ implies that there exists an n such that $x \in D_{n,\alpha_\omega}$. In other words, $\bigcup_{\alpha > \alpha_\omega} A_\alpha \subseteq \bigcup_n D_{n,\alpha_\omega}$. By the choice of α_ω,

$\mu(\bigcup_{\alpha>\alpha_\omega} A_\alpha) = 0$. But $\bigcup_{\alpha\leq\alpha_\omega} A_\alpha$ is a Borel set which is measurable. We conclude that $A = \bigcup_{\alpha<\omega} A_\alpha$ is measurable. □

We will compute the value of α_ω in Chapter 14.

Theorem 3.7.11 (Sierpinski [135]). *Every Σ^1_2-set of reals is a union of ω_1-many Borel sets.*

Proof. Suppose that $A \subseteq 2^\omega$ is Σ^1_2. Then there is a Π^1_1-set $B \subseteq \omega^\omega$ such that $x \in A$ if and only if $(\exists f \in \omega^\omega)B(f, x)$. By (i) of Lemma 3.7.9, there is an ω_1-sequence of Borel sets $\{B_\alpha\}_{\alpha<\omega_1}$ such that $\bigcup_{\alpha<\omega_1} B_\alpha = B$. Then $C_\alpha = \{x \mid (\exists f \in \omega^\omega)B_\alpha(f, x)\}$ is $\underset{\sim}{\Sigma}^1_1$ for each $\alpha < \omega_1$. Clearly $A = \bigcup_{\alpha<\omega_1} C_\alpha$. By relativizing (ii) of Lemma 3.7.9, each C_α is a union of ω_1-many Borel sets. Thus A is a union of ω_1-many Borel sets. □

Corollary 3.7.12. *Every Σ^1_2-set of reals of cardinality greater than $\aleph_1$ has a perfect subset.*

Proof. By Theorem 3.7.11, every Σ^1_2-set A of reals is the union of $\aleph_1$-many Borel sets. If A has cardinality greater than $\aleph_1$, then one of these Borel sets is uncountable and so has a perfect subset. □

3.7.4 Exercises

Exercise 3.7.1. *The following is another proof of Theorem 2.4.7.*
(i) *Show that $\mathcal{O}$ is Σ_1-definable over $L_{\omega_1^{\mathrm{CK}}}$;*
(ii) *show that for any reals x and y, $x \leq_h y$ if and only if $x \in L_{\omega_1^y}[y]$ and hence $\mathcal{O} \leq_h y$ if and only if $\omega_1^{\mathrm{CK}} < \omega_1^y$.*

Exercise 3.7.2. *Prove that there exists a sequence of reals $\{x_n\}_{n<\omega}$ such that for each n, $x_n \geq_T x'_{n+1}$.*

Exercise 3.7.3. *Prove that if $\{x_n\}_{n<\omega}$ is a sequence as above, then for each n, x_n is not hyperarithmetic.*

Exercise 3.7.4. *Use the first version of the Spector–Gandy Theorem (3.7.1) to prove Theorem 3.7.11.*

Exercise 3.7.5. *Prove that if x is a real so that every nonempty Π^0_1 set $A \subseteq \omega^\omega$ has a member hyperarithmetical in x, then $x \geq_h \mathcal{O}$.*

Exercise 3.7.6. *Prove that the set $\{(x, y) \mid \omega_1^y > \omega_1^x\}$ is Π^1_1 but not $\underset{\sim}{\Sigma}^1_1$.*

Exercise 3.7.7. *Assume MA + $2^{\aleph_0} > \aleph_1$ (cf. Definition 6.1.5). Prove that every $\underset{\sim}{\Sigma}^1_2$-set is measurable.*

4 The theory of Π_1^1-sets

Π_1^1-sets of reals are endowed with rich structural properties, including those that are best understood from the perspective of the constructible universe L. In this chapter a deeper analysis of such sets is presented to elucidate this point.

4.1 A Π_1^1-basis theorem

4.1.1 Tree presentations for Π_1^1-sets

Recall that given a Π_1^1-set $A \subseteq 2^\omega$, there is a recursive predicate $P \subseteq 2^\omega \times \omega^\omega$ such that $x \in A$ if and only if $(\forall f)(\exists n)\neg P(x \upharpoonright n, f \upharpoonright n)$. In other words, $x \in A$ if and only if the x-recursive tree $T_x = \{\sigma \mid (\forall \tau)(\tau \preceq \sigma \rightarrow P(x \upharpoonright |\tau|, \tau))\}$ is wellfounded under the Kleene–Brouwer ordering. Then there is a function $p_x : T_x \rightarrow \omega_1^x$ with $p_x \in L_{\omega_1^x}[x]$ such that p_x is order preserving under the Kleene–Brouwer ordering.

Now let $T \subseteq 2^{<\omega} \times \omega_1^{\omega^{<\omega}}$ be a set so that $(\sigma, p) \in T$ if and only if p is an order-preserving finite function from $\{\tau \mid \tau \in |\sigma|^{\leq|\sigma|} \wedge (\forall \tau')(\tau' \preceq \tau \rightarrow P(\sigma \upharpoonright |\tau'|, \tau'))\}$ to ω_1. Define $(\sigma', p') \succeq (\sigma, p)$ if and only if $\sigma' \succeq \sigma$ and p' extends p. Then T is a tree under $\succ$. Thus $x \in A$ if and only if there is an $f : T_x \rightarrow \omega_1$ such that (x, f) is an infinite path in T, and this holds if and only if there is an $f : T_x \rightarrow \omega_1$ with $f \in L_{\omega_1^x}[x]$ such that (x, f) is an infinite path in T.

The advantage with a tree presentation is that the companion path f of x exists in $L_{\omega_1^x}[x]$.

Given $(\sigma, p), (\sigma', p') \in 2^{<\omega} \times \omega_1^{\omega^{<\omega}}$, we say that $(\sigma', p') <_l (\sigma, p)$ if

- either $(\sigma', p') \succ (\sigma, p)$, or
- there is an i such that $\sigma'(i) < \sigma(i)$ and for all $j < i$, $\sigma'(j) = \sigma(j)$, or
- there is a $\nu \in \omega^{<\omega}$ such that $p'(\nu) < p(\nu)$ and for all $\nu' <_{\mathrm{KB}} \nu$, $\nu' \in \mathrm{Dom}(\sigma) \Rightarrow p'(\nu') = p(\nu')$.

An infinite path (x, f) in T is the *leftmost* path if there is no infinite path in $\{(\sigma, p) \mid (\sigma, p) <_l (x \upharpoonright |\sigma|, f \upharpoonright \mathrm{Dom}(p))\}$.

Lemma 4.1.1. *Suppose that (x, f) is the leftmost path of T, then $f \in L_{\omega_1^x}[x]$.*

Proof. Suppose that (x, f) is the leftmost path of T. Then f is the unique order-preserving function from the x-recursive tree $T_x = \{\sigma \mid P(x \upharpoonright |\sigma|, \sigma)\}$ onto an initial segment of ω_1^x. Since $L_{\omega_1^x}[x]$ is admissible, by the Σ-induction Theorem (3.1.8), we have $f \in L_{\omega_1^x}[x]$. □

4.1.2 A constructible basis theorem

Theorem 4.1.2 (Guaspari [42]). *If $A \subseteq 2^\omega$ is a nonempty Π_1^1-set, then there is an $x \in A$ such that $x \in L_{\omega_1^x}$.*

Proof. Let T be the tree presentation for A and (x, f) its leftmost path. We claim that $x \in L_{\omega_1^x}$.

By Lemma 4.1.1, $f \in L_{\omega_1^x}[x]$. Hence there is an ordinal $\alpha_0 < \omega_1^x$ such that $\mathrm{Rang}(f) = \alpha_0$. We define a tree T^{α_0} by replacing ω_1 with α_0 in T everywhere. Obviously $T^{\alpha_0} \in L_{\gamma_0}$ for some $\gamma_0 < \omega_1^x$ and (x, f) is an infinite path in T^{α_0}. Moreover, for any $(\sigma, p) <_l (x \upharpoonright |\sigma|, f \upharpoonright \mathrm{Dom}(p))$, there is no infinite path in $T^{\alpha_0}_{(\sigma,p)} = \{(\sigma', p') \in T^{\alpha_0} \mid (\sigma', p') \succ (\sigma, p)\}$. Otherwise, any such infinite path would also be in T and to the left of (x, f), which contradicts the choice of (x, f) in T. Then for any $(\sigma, p) <_l (x \upharpoonright |\sigma|, f \upharpoonright \mathrm{Dom}(p))$, there exist ordinals $\alpha_{(\sigma,p)} < \omega_1^x$, $\beta_{(\sigma,p)} < \alpha_{(\sigma,p)}$ and a function $g_{(\sigma,p)} \in L_{\alpha_{(\sigma,p)}}$ such that $g_{(\sigma,p)}$ is an order-preserving function from $T^{\alpha_0}_{(\sigma,p)}$ to $\beta_{(\sigma,p)}$. Moreover, the function from $(\sigma, p) <_l (x \upharpoonright |\sigma|, f \upharpoonright \mathrm{Dom}(p))$ to $\alpha_{(\sigma,p)} < \omega_1^x$ is Σ_1 in $L_{\omega_1^x}[x]$. Since the set $\{(\sigma, p, T^{\alpha_0}_{(\sigma,p)}) \mid (\sigma, p) <_l (x \upharpoonright |\sigma|, f \upharpoonright \mathrm{Dom}(p))\}$ is an element of $L_{\omega_1^x}[x]$, there is a $\gamma < \omega_1^x$ that bounds any $\alpha_{(\sigma,p)}$ where $(\sigma, p) <_l (x \upharpoonright |\sigma|, f \upharpoonright \mathrm{Dom}(p))$.

We may assume that $\gamma > \alpha_0$. Now working in L_γ, we define an infinite path as follows:

$(\sigma_0, p_0) = \emptyset$.

$(\sigma_{i+1}, p_{i+1}) \succ (\sigma_i, p_i)$ in T^{α_0} is of length $i + 1$ so that

(1) for all $(\sigma, p) <_l (\sigma_{i+1}, p_{i+1})$ not extending (σ_{i+1}, p_{i+1}), there is an order-preserving function $g_{(\sigma,p)} \in L_\gamma$ from $T^{\alpha_0}_{(\sigma,p)}$ into an ordinal $\beta_{(\sigma,p)} < \gamma$, and

(2) there is no order-preserving function $g \in L_\gamma$ from $T^{\alpha_0}_{(\sigma_{i+1},p_{i+1})}$ into any ordinal $\beta < \gamma$.

By the assumption of γ, we have $\bigcup_{i<\omega}(\sigma_i, p_i) = (x, f)$. Clearly the sequence $\{(i, \sigma_i, p_i)\}_{i<\omega}$ is an element of $L_{\gamma+\omega}$. Hence $(x, f) \in L_{\gamma+\omega} \subseteq L_{\omega_1^x}$, and so $x \in L_{\omega_1^x}$. □

We have as an immediate consequence of Theorem 4.1.2,

Corollary 4.1.3. *If x is a Π_1^1-singleton, then $x \in L_{\omega_1^x}$.*

4.1.3 Exercises

Exercise 4.1.1. *Use Theorem 4.1.2 to prove Proposition 4.2.5.*

Exercise 4.1.2. *Prove that there is a nonconstructible real if and only if there is a nonempty Π_2^1-set which does not contain a constructible real.*

4.2 Π^1_1-uniformization

4.2.1 Uniformizing Π^1_1-relations

Given a relation $R \subseteq 2^\omega \times 2^\omega$, a relation R_0 is a *uniformization* of R if for any x and y, $R_0(x, y)$ implies $R(x, y)$ and for any x, if there is a y such that $R(x, y)$, then there is a unique y such that $R_0(x, y)$. Uniformization of R is in effect a choice function for R.

Theorem 4.2.1 (Kondô [74] and Addison [1]). *Suppose that $R \subseteq 2^\omega \times 2^\omega$ is Π^1_1. Then there is a Π^1_1-relation R_0 that uniformizes R.*

Proof. Suppose that $R \subseteq 2^\omega \times 2^\omega$ is Π^1_1. By Corollary 3.7.2, there is a Σ_0-formula $\varphi(u, v, w)$ such that

$$(x, y) \in R \Leftrightarrow \langle L_{\omega_1^{x \oplus y}}[x \oplus y], \in\rangle \models (\exists w)\varphi(x, y, w).$$

Define $R_0 \subseteq 2^\omega \times 2^\omega$ so that $(x, y) \in R_0$ if and only if

(i) $y \in L_{\omega_1^{x \oplus y}}[x]$;

(ii) $(\exists w \in L_{\omega_1^{x \oplus y}}[x])(\langle L_{\omega_1^{x \oplus y}}[x], \in\rangle \models \varphi(x, y, w))$; and

(iii) $\langle L_{\omega_1^{x \oplus y}}[x], \in\rangle \models (\forall z)(\forall w')((z, w') <_{L[x]} (y, w) \to \neg\varphi(x, z, w'))$.

By Theorem 4.1.2, if there is a y such that $(x, y) \in R$, then there is one such that $R_0(x, y)$. By the definition of R_0, R_0 uniformizes R. Note that both (i) and (ii) above are Σ_1-statements. By Proposition 3.3.3, (iii) is also a Σ_1-statement. By Corollary 3.7.2, R_0 is a Π^1_1-relation. □

There are a number of interesting applications of the Π^1_1-Uniformization Theorem (4.2.1) which we now discuss:

Proposition 4.2.2. *Given Π^1_1-sets A_0 and B_0, there exist disjoint Π^1_1-sets $A_1 \subseteq A_0$, $B_1 \subseteq B_0$ such that $A_0 \cup B_0 = A_1 \cup B_1$.*

Proof. Let $R \subseteq 2^\omega \times 2$ be a Π^1_1-relation such that $(x, i) \in R$ if and only if either $i = 0$ and $x \in A_0$ or $i = 1$ and $x \in B_0$. By Theorem 4.2.1, there is a Π^1_1-relation R_0 uniformizing R. Define $x \in A_1$ if and only if $(x, 0) \in R_0$ and $x \in B_1$ if and only if $(x, 1) \in R_1$. Then A_1, B_1 are as required. □

Corollary 4.2.3. *Given disjoint Σ^1_1-sets A_0, B_0, there exist disjoint Δ^1_1-sets A_1, B_1 such that $A_0 \subseteq A_1$, $B_0 \subseteq B_1$ and $A_1 \cup B_1 = 2^\omega$.*

Proof. The hypothesis implies that $2^\omega \setminus A_0$ and $2^\omega \setminus B_0$ are Π^1_1. By Proposition 4.2.2, there exist two Π^1_1-sets $A_1 \supseteq B_0$,$B_1 \supseteq A_0$ such that A_1 and B_1 are disjoint and $A_1 \cup B_1 = 2^\omega$. Hence these sets are Δ^1_1. □

4.2.2 Kreisel compactness

Kreisel's Compactness Theorem is a generalization of the Compactness Theorem in model theory to $L_{\omega_1^{\mathrm{CK}}}$-logic. In this setting, "recursively enumerable" is Π^1_1, and "finite" is Δ^1_1. It is a striking application of the Π^1_1-Uniformization Theorem.

Theorem 4.2.4 (Kreisel). *Let $A \subseteq \omega \times 2^\omega$ be a Π^1_1-set such that for each n, $A_n = \{x \mid (n,x) \in A\}$ is Δ^1_1. Let z be Π^1_1. If $\bigcap_{n\in y} A_n \neq \emptyset$ for every hyperarithmetic $y \subseteq z$, then $\bigcap_{n\in z} A_n \neq \emptyset$.*

Proof. Suppose not. Then z is properly Π^1_1 and $\bigcap_{n\in z} A_n = \emptyset$. Then for each x, there is an n such that $x \notin A_n$. Define a Π^1_1-relation $R \subseteq 2^\omega \times \omega$ such that $R(x,n) \Leftrightarrow n \in z \wedge x \notin A_n$. By the Uniformization Theorem (4.2.1), there is a total Π^1_1-function f uniformizing R. Then the set

$$y = \{n \mid (\exists x)(f(x) = n)\}$$

is contained in z and is Σ^1_1. If $x \in \bigcap_{n\in y} A_n$, then $f(x) \in \omega \setminus y$. But no $m \in \omega \setminus y$ belongs to $f[2^\omega]$. Thus $\bigcap_{n\in y} A_n = \emptyset$. By Corollary 4.2.3 restricted to ω, there is a Δ^1_1-real y_0 with $y \subseteq y_0 \subseteq z$. Then $\bigcap_{n\in y_0} A_n = \emptyset$, which is a contradiction. □

Theorem 4.2.4 was further generalized by Barwise [3] to countable admissible sets as a compactness theorem for infinitary logic.

4.2.3 Π^1_1-singletons

Proposition 4.2.5. *Every nonempty Π^1_1-set of reals contains a Π^1_1-singleton.*

Proof. Suppose that $A \subseteq 2^\omega$ is Π^1_1 and nonempty. Let $(n,x) \in R$ if and only if $x \in A$. Then R is a Π^1_1-relation. By Theorem 4.2.1, there is a Π^1_1-relation R_0 uniformizing R. Then the unique real x such that $(0,x) \in R_0$ is a Π^1_1-singleton. □

By Proposition 4.2.5, we know that every nonempty Π^1_1-set contains a Π^1_1-singleton.

Proposition 4.2.6. *The set $\{x \mid x$ is a Π^1_1-singleton$\}$ is Π^1_1.*

Proof. Let $P \subseteq \omega \times 2^\omega$ be a universal Π^1_1-set. By Proposition 4.2.5, there is Π^1_1 -set $P_0 \subseteq P$ uniformizing P. Let $A = \{x \mid (\exists n)(n,x) \in P_0\}$. Then A is a Π^1_1 set which contains precisely all the Π^1_1-singletons. □

Since there are only countably many Π^1_1-singletons, there is a countable Π^1_1-set which is a basis for Π^1_1-sets of reals. This gives the following basis result for Σ^1_2-sets.

Corollary 4.2.7. *Every nonempty Σ^1_2-set of reals contains a Δ^1_2-member.*

Proof. Suppose A is a nonempty Σ^1_2-set of reals. Then $x \in A$ if and only if $(\exists y)B(x,y)$ for some Π^1_1-relation B. By Proposition 4.2.5, there is a Π^1_1-singleton, and hence a Δ^1_2-real, (x_0, y_0) such that $B(x_0,y_0)$. Then $x_0 \in A$ is Δ^1_2. □

By the proof of Corollary 4.2.7, we have the following characterization of Δ^1_2-reals.

Corollary 4.2.8. *A real x is Δ^1_2 if and only if x is recursive in a Π^1_1-singleton. Hence the set of Δ^1_2-reals is Σ^1_2.*

Corollary 4.2.9 (Shoenfield [129]). (i) *If $A \subseteq 2^\omega$ is nonempty and Σ^1_2, then there is an $x \in A \cap L$ recursive in a Π^1_1-singleton.*
(ii) *Every Δ^1_2-real is constructible.*

Proof. (i) If $A \subseteq 2^\omega$ is a nonempty Σ^1_2-set, then there is a Π^1_1-set $B \subseteq 2^\omega \times 2^\omega$ such that $x \in A$ if and only if there exists a y such that $B(x, y)$. By Proposition 4.2.5, there is a Π^1_1-singleton $B(x_0, y_0)$. By Corollary 4.1.3, $(x_0, y_0) \in L$. Thus $x_0 \in A$ is a constructible real recursive in the Π^1_1-singleton $x_0 \oplus y_0$.

(ii) Immediate from (i), since every Δ^1_2-real is a Σ^1_2-singleton. □

Corollary 4.2.9 (i) is known in the literature as the *Shoenfield Absoluteness Lemma*. There are many applications of this in set theory. For example, given a Σ^1_2-predicate P, one has $\langle L, \in\rangle \models P$ if and only if $\langle V, \in\rangle \models P$.

4.2.4 Exercises

Exercise 4.2.1. *Prove that there is a Σ^1_1-relation $R \subseteq 2^\omega \times 2^\omega$ which is not uniformizable by a Σ^1_1-relation.*

Exercise 4.2.2. *Prove that there is no Π^1_1-set $x \subseteq \mathcal{O}$ where $<_o$ restricted to x is a well-ordering of order type ω_1^{CK} and such that there is a partial recursive function $p : \omega^2 \to \omega$ satisfying properties* (i)–(iii) *of Exercise* 1.4.1.

Exercise 4.2.3. *Prove Theorem* 4.1.2 *from Theorem* 4.2.1.

4.3 Characterizing thin Π^1_1-sets

A set is *thin* if it contains no perfect subset. There is a beautiful characterization of Π^1_1-thin sets in terms of constructibility.

4.3.1 The largest thin Π^1_1-set

Let $\mathcal{C} = \{x \mid x \in L_{\omega_1^x}\}$.

Lemma 4.3.1. *$\mathcal{C}$ is a thin Π^1_1-set.*

Proof. $x \in L_{\omega_1^x}$ if and only if $\langle L_{\omega_1^x}[x], \in\rangle \models (\exists\beta)(x \in L_\beta)$. By Theorem 3.3.2 and Corollary 3.7.2, $\mathcal{C}$ is Π^1_1.

Suppose $\mathcal{C}$ is not thin. Let

$$B = \{T \subseteq 2^{<\omega} \mid T \text{ is a perfect tree in which every infinite path belongs to } \mathcal{C}\}.$$

Then B is a Π^1_1-set. By Theorem 4.1.2, there is a $T \in B$ with $T \in L_{\omega_1^T}$. By relativizing the Gandy Basis Theorem (2.5.3) to T, there is a real $x \in [T] \subseteq \mathcal{C}$ such that $x \not\leq_h T$ and $\omega_1^x \leq \omega_1^{x\oplus T} = \omega_1^T$. Hence $x \notin L_{\omega_1^x}$, a contradiction. □

Lemma 4.3.2 (Mansfield [87], Solovay [149]). *Let $A \subseteq 2^\omega$ be Π^1_1. If there is an $x \in A$ such that $x \notin L_{\omega_1^x}$, then there is a perfect tree $T \in L$ all of whose infinite paths belong to A.*

Proof. Let $\mathcal{C}$ be defined as before. Suppose $A \subseteq 2^\omega$ is Π^1_1 and there is an $x \in A \setminus \mathcal{C}$. Let α_0 be the least ordinal α such that $\alpha = \omega_1^x$ for some $x \in A \setminus \mathcal{C}$.

Let T_0 be a tree presentation of A and T_0^α be as defined in the proof of Theorem 4.1.2. Since $\mathcal{C}$ is a Π^1_1-set, there is a recursive tree $T_1 \subseteq 2^{<\omega} \times \omega^{<\omega}$ such that $x \notin \mathcal{C}$ if and only if $(\exists y)(x, y) \in [T_1]$. Now let $T_2 \subseteq 2^{<\omega} \times \alpha^{\omega^{<\omega}} \times 2^{<\omega}$ be a tree defined as follows: $(\sigma, p, \tau) \in T_2$ if and only if $(\sigma, p) \in T_0^\alpha$ and $(\sigma, \tau) \in T_1$. By assumption, there is an infinite path (x, f, y) in T_2.

We claim that for any $(\sigma, p, \tau) \in T_2$, if there is an infinite path in $T_{2,(\sigma,p,\tau)} = \{(\sigma', p', \tau') \in T_2 \mid (\sigma', p', \tau') \succ (\sigma, p, \tau)\}$, then there are (σ_0, p_0, τ_0) and (σ_1, p_1, τ_1) in $T_{2,(\sigma,p,\tau)}$, with σ_0 and σ_1 incompatible, such that both $T_{2,(\sigma_0,p_0,\tau_0)}$ and $T_{2,(\sigma_1,p_1,\tau_1)}$ have infinite paths. Otherwise, fix a $(\sigma, p, \tau) \in T_2$ for which this is false. Then there is an infinite path (x, f, y) in $T_{2,(\sigma,p,\tau)}$. Note that $\omega_1^x \geq \alpha_0$. By assumption, for any $\sigma' \succ \sigma$ that is not an initial segment of x, the tree

$$T_{2,\sigma'} = \{(\sigma'', p'', \tau'') \in T_{2,(\sigma,p,\tau)} \mid \sigma' \succ \sigma\}$$

is wellfounded. Then by the same proof as that of Theorem 4.1.2, there is a $\gamma < \omega_1^x$ such that the set $\{\sigma' \mid \sigma' \not\prec x \wedge (\exists p')(\exists \tau')(\sigma', p', \tau') \in T_{2,(\sigma,p,\tau)}\}$ belongs to L_γ. Then $x \in L_{\gamma+1} \subseteq L_{\omega_1^x}$, contradicting the fact that $x \notin \mathcal{C}$.

By the claim, one may carry out the first construction of Theorem 1.1.10 to obtain a perfect tree all of whose paths belong to A. Thus the set $\{T \mid T \text{ is perfect} \wedge [T] \subseteq A\}$ is a nonempty Π^1_1-set. By Theorem 4.1.2, there is a perfect tree $T \in L$ such that $[T] \subseteq A$. □

By Lemma 4.3.1 and Lemma 4.3.2, we have the following characterization of the largest thin Π^1_1-set.

Theorem 4.3.3. *$\mathcal{C}$ is the largest thin Π^1_1-set.*

4.3.2 Some properties of $\mathcal{C}$

Since every Π^1_1-set of reals contains an $x \in \mathcal{C}$ that is also a Π^1_1-singleton, one may ask if every real in $\mathcal{C}$ is a Π^1_1-singleton. This is false since $\mathcal{C}$ is uncountable in L.

Proposition 4.3.4. *For any real $x \in L$, there is a $z \in \mathcal{C}$ such that $z \geq_T x$.*

Proof. Let $x \in L$. By Corollary 3.3.8, $x \in J_{\alpha+1} \setminus J_\alpha$ for some $\alpha < (\omega_1)^L$. Then x is $\Sigma_n(J_{\alpha+1})$ for some n. By Theorem 3.4.11, there is a real $z \in J_{\alpha+1} \setminus J_\alpha$ which is a Σ_n-master code. Thus $\rho^n_\alpha = \omega$. By the definition of Σ_n-master code, there is a z-recursive well-ordering of α. In other words, $\omega_1^z > \alpha$. Thus $z \in L_{\omega_1^z}$ and $x \leq_T z$. □

Proposition 4.3.5. *There is an $x \in \mathcal{C}$ in which every Π^1_1-singleton is recursive.*

Proof. By Proposition 4.1.3, every Π^1_1-singleton is in L. Since $\langle L, \in\rangle \models$ ZFC, the statement "there are only countably many Π^1_1-singletons" is true in L. Hence there is a real $y \in L$ in which every Π^1_1-singleton in L is recursive. By Proposition 4.3.4, $y \leq_T z$ for some $z \in \mathcal{C}$. □

Corollary 4.3.6.
(i) *$\mathcal{C}$ is the largest countable Π^1_1-set of reals if and only if $(\omega_1)^L < \omega_1$.*
(ii) *There is an uncountable thin Π^1_1-set if and only if $(\omega_1)^L = \omega_1$.*

Proof. (i) By Proposition 4.3.4, if $\mathcal{C}$ is countable, then $L \cap 2^\omega$ is countable. So $(\omega_1)^L$ is countable.

Suppose that $(\omega_1)^L$ is countable. Then by Proposition 4.3.2, every Π^1_1-countable set of reals is a subset of $\mathcal{C}$. By Corollary 3.3.8, $\mathcal{C}$ is countable.

(ii) follows from (i). □

4.3.3 Exercise

Exercise 4.3.1. *Show that if $\omega_1 = (\omega_1)^L$, then there is no largest countable Π^1_1-set of reals.*

4.4 Σ^1_2-sets

Σ^1_2-sets and Π^1_1-sets of reals are similar in several respects. We list some of them here.

Corollary 4.4.1. *$2^\omega \cap L$ is Σ^1_2.*

Proof. By Proposition 4.3.4, $x \in L$ if and only if there is a $y \in \mathcal{C}$ such that $y \geq_T x$. □

Proposition 4.4.2. *$x \leq_L y$ is a Σ^1_2-relation.*

Proof. $x \leq_L y$ if and only if $x, y \in L$ and there is a binary relation $E \subseteq \omega$ such that
- $\langle \omega, E\rangle$ is wellfounded;
- $\langle \omega, E\rangle \models V = L + \mathrm{KP}$;
- $\langle \omega, E\rangle \models x \leq_L y$.

Thus the relation $x \leq_L y$ is Σ^1_2. □

The following is an analog of the Spector–Gandy Theorem for Σ^1_2-sets.

Proposition 4.4.3. *A set $A \subseteq 2^\omega$ is Σ^1_2 if and only if there is a Σ_1-formula φ such that*

$$x \in A \Leftrightarrow \langle L[x], \in\rangle \models \varphi \Leftrightarrow \langle L_{(\omega_1)^{L[x]}}[x], \in\rangle \models \varphi.$$

Proof. Suppose that $A \subseteq 2^\omega$ is Σ^1_2. Then there is a Π^1_1-relation $B \subseteq 2^\omega \times 2^\omega$ such that

$$x \in A \Leftrightarrow (\exists y)B(x, y).$$

By Corollary 3.7.2, there is a Σ_1-formula ψ such that

$$B(x, y) \Leftrightarrow \langle L_{\omega_1^{x\oplus y}}[x \oplus y], \in\rangle \models \psi.$$

Since for every x, $B_x = \{y \mid B(x, y)\}$ is $\Pi^1_1(x)$, by relativizing Theorem 4.1.2 there is a $y \in B_x$ such that $y \in L_{\omega_1^{x\oplus y}}[x]$.

Hence

$$x \in A \Leftrightarrow (\exists y)(y \in L_{\omega_1^{x\oplus y}}[x] \wedge \langle L_{\omega_1^{x\oplus y}}[x], \in\rangle \models \psi).$$

Now let φ be

$$(\exists y)(\exists \alpha > \omega)(\alpha \text{ is the least admissible ordinal such that } y \in L_\alpha[x] \wedge \langle L_\alpha[x], \in\rangle \models \psi).$$

By Theorem 3.3.2, φ is Σ_1. By the discussion above, $x \in A$ if and only if $\langle L[x], \in\rangle \models \varphi$ if and only if $\langle L_{(\omega_1)^{L[x]}}[x], \in\rangle \models \varphi$.

For the other direction, suppose there is a Σ_1-formula φ such that $x \in A \Leftrightarrow \langle L[x], \in\rangle \models \varphi$. Then for each x, there is a cardinal κ such that $x \in A \Leftrightarrow \langle L_\kappa[x], \in\rangle \models \varphi$. By the Condensation Theorem (3.3.7), there is a countable admissible ordinal α such that $x \in A \Leftrightarrow \langle L_\alpha[x], \in\rangle \models \varphi$. Since φ is Σ_1, we have $x \in A \Leftrightarrow \langle L[x], \in\rangle \models \varphi \Leftrightarrow \langle L_{(\omega_1)^{L[x]}}[x], \in\rangle \models \varphi$.

By Theorem 3.3.4, $\langle L_{(\omega_1)^{L[x]}}[x], \in\rangle \models \varphi$ if and only if there is a countable transitive wellfounded model $\langle M, \in\rangle \models \mathrm{KP} + \text{“}V = L[x]\text{”} + \varphi$. Hence $\langle L_{(\omega_1)^{L[x]}}[x], \in\rangle \models \varphi$ if and only if

$$(\exists E)(\langle \omega, E\rangle \text{ codes a wellfounded transitive model } \wedge \langle \omega, E\rangle \models \mathrm{KP} + \text{“}V = L[x]\text{”} + \varphi).$$

By Proposition 3.6.2, the set $\{x \mid \langle L_{(\omega_1)^{L[x]}}[x], \in\rangle \models \varphi\} = \{x \mid \langle L[x], \in\rangle \models \varphi\}$ is Σ^1_2. □

Thus one may regard L_{ω_1} as the structure where Σ^1_2-sets are based. In a way, Σ^1_2-ness draws the boundary between recursion theory and set theory. Within ZFC, it is not possible to generalize results in the previous sections to a level beyond Σ^1_2. One needs additional set-theoretic assumptions for this. How this is to be implemented is a major subject in descriptive set theory. From this perspective, higher recursion theory serves as a launch pad for the study of descriptive set theory beyond the level of Σ^1_2. Let

$$\delta^1_2 = \text{supremum of the } \Delta^1_2 \text{ well-orderings of } \omega,$$

and

$$\delta = \min\{\alpha \mid L \setminus L_\alpha \text{ contains no } \Pi^1_1\text{-singleton}\}.$$

Proposition 4.4.4. $\delta = \delta^1_2$.

Proof. If $\alpha < \delta$, then there is a Π^1_1-singleton $x \in L_\delta \setminus L_\alpha$. Since $x \in L_{\omega_1^x}$ and ω_1^x is a $\Pi^1_1(x)$ well-ordering, $\alpha < \omega_1^x < \delta^1_2$. Hence $\delta \leq \delta^1_2$.

If $\alpha < \delta^1_2$, then there is a Δ^1_2 well-ordering relation $R \subseteq \omega \times \omega$ of order type α. Then there exist arithmetical relations $S, T \subseteq (\omega^\omega)^2 \times \omega^2$ such that

$$R(n, m) \Leftrightarrow (\exists f)(\forall g)S(f, g, n, m), \text{ and}$$
$$\neg R(n, m) \Leftrightarrow (\exists f)(\forall g)T(f, g, n, m).$$

Define Π^1_1-sets

$$R_0 = \{(h, \langle n, m\rangle) \mid h(0) = 0 \wedge (\exists f)(\forall g)S(f, g, n, m) \wedge (\forall n)(f(n) = h(n + 1)))\}$$

and

$$R_1 = \{(h, \langle n, m\rangle) \mid h(0) = 1 \wedge (\exists f)(\forall g)T(f, g, n, m) \wedge (\forall n)(f(n) = h(n + 1)))\}.$$

By the Π^1_1-Uniformization Theorem (4.2.1), the two relations may be uniformized by partial Π^1_1-functions p_{R_0} and $p_{R_1} : \omega \to \omega^\omega$. Let $p = p_{R_0} \cup p_{R_1}$. Then p is a total Π^1_1-function and may be viewed as a Π^1_1-singleton. Then R is arithmetical in p and hence $\alpha < \omega_1^p < \delta$. Thus $\delta^1_2 = \delta$. □

This gives a finer version of the Spector–Gandy Theorem for Σ^1_2-sets:

Proposition 4.4.5. *A real x is Σ^1_2 if and only if x is Σ_1-definable over $\langle L_{\delta^1_2}, \in\rangle$.*

Proof. Suppose that x is Σ^1_2. Then there is a Σ_0-formula $\varphi(n, y)$ such that

$$n \in x \Leftrightarrow (\exists y)(\langle L_{\omega_1^y}[y], \in\rangle \models (\exists s)\varphi(n, s, y)).$$

By Corollary 4.2.9 and Proposition 4.4.4, we have

$$n \in x \implies \langle L_{\delta^1_2}, \in\rangle \models (\exists s)(\exists y)(s \in L_{\omega_1^y}[y] \wedge \varphi(n, s, y)).$$

Assume $\langle L_{\delta^1_2}, \in\rangle \models (\exists s)(\exists y)(s \in L_{\omega_1^y}[y] \wedge \varphi(n, s, y))$ and fix such a real y. Then $\langle L_{\omega_1^y}[y], \in\rangle \models (\exists s)\varphi(n, s, y)$. Hence $n \in x$. Thus x is Σ_1-definable over $\langle L_{\delta^1_2}, \in\rangle$.

Conversely, suppose that x is Σ_1-definable over $\langle L_{\delta^1_2}, \in\rangle$ via the formula $\varphi(n)$. By Proposition 4.4.4, we have $n \in x$ if and only if there exists a $z \in L_{\omega_1^z}$ such that $\langle L_{\omega_1^z}, \in\rangle \models \varphi(n)$. Thus x is Σ^1_2. □

It is possible to relativize "Δ^1_2-ness" in the obvious way and define the relation "x is Δ^1_2 in y" and thereby introduce the notion of a Δ^1_2-degree. Let $\delta^{1,x}_2$ be the supremum of all well-orderings of ω that are Δ^1_2 in x. The following result says in a strong way that Post's problem has a negative solution for the Δ^1_2-degrees.

Proposition 4.4.6.
(i) *If x is constructible and not Δ^1_2, then $\delta^{1,x}_2 > \delta^1_2$.*
(ii) *Assume $V = L$. Then $\leq_{\Delta^1_2}$ is a well-ordering of the Δ^1_2-degrees and its order type is ω_1. Moreover, if $x <_{\Delta^1_2} y$, then $\delta^{1,x}_2 < \delta^{1,y}_2$.*

Proof. (i) Let

$$A = \{r \mid r \text{ codes a well-ordering of } \omega \wedge x \in L_{|r|}\},$$

where $|r|$ denotes the order type of r as a well-ordering. Then A is $\Delta^1_2(x)$. Let $r \in A$. Since x is not Δ^1_2 we must have $|r| > \delta^1_2$. Hence $\delta^{1,x}_2 > \delta^1_2$.

(ii) If $V = L$, then by Proposition 4.4.2, $\leq_L$ is a Δ^1_2 well-ordering of the reals. Suppose that $x \leq_L y$. Then by the proof of (i), there is a $\Delta^1_2(y)$-real r coding a well-ordering of ω such that $y \in L_{|r|}$ and so $x \in L_{|r|}$. Then $x \in L_{\omega_1^r}$ and hence $x \leq_h r \leq_{\Delta^1_2} y$. Thus $x \leq_{\Delta^1_2} y$. So $\leq_{\Delta^1_2}$ is a well-ordering of Δ^1_2-degrees and its order type is ω_1. If $x <_{\Delta^1_2} y$, then by the same argument as that in (i) but relativized to y, we have $\delta^{1,x}_2 < \delta^{1,y}_2$. □

Proposition 4.4.6 indicates that while L is an appropriate ground model for hyperarithmetic theory, it is less so for a Δ^1_2-theory as the well-ordering of the $\leq_{\Delta^1_2}$ relation makes the structure of the corresponding degrees relatively simple and hence less interesting.

4.4.1 Exercises

Exercise 4.4.1 (Solovay [149] and Mansfield [87]). *Prove that every Σ^1_2-set of reals containing a nonconstructible real has a perfect subset.*

Exercise 4.4.2. *$2^\omega \cap L$ is the largest countable Σ^1_2-set of reals if and only if $(\omega_1)^L < \omega_1$.*

Exercise 4.4.3. *If $\omega_1 = (\omega_1)^L$, then there is no largest countable Σ^1_2-set of reals.*

Exercise 4.4.4. *Prove that there is a real x and a nonempty Π^1_2-set $P \subseteq 2^\omega$ such that $\delta^{1,x}_2 > \delta^1_2$ but there is no real $z \in P$ such that $z \leq_{\Delta^1_2} x$.*

Exercise 4.4.5. *Suppose that $\mathscr{L}$ is a countable language and T is a consistent theory of $\mathscr{L}$. Then T has, up to elementary equivalence, either $\leq \aleph_0$ or $2^{\aleph_0}$-many countable models.*

5 Recursion-theoretic forcing

Although historically Cohen introduced forcing for set-theoretic considerations, the method was adapted to study problems in recursion theory soon after by Feferman [25] (indeed, one detects a hint of Cohen forcing in the theorems of Post and Kleene on the structure of the Turing degrees proved in the 1950s). Here we study three classic examples of forcing (due to Cohen, Sacks and Steel respectively) and their applications in recursion theory. It turns out that the ramified analytical hierarchy provides an ideal platform for this purpose.

5.1 Ramified analytical hierarchy

5.1.1 The structure $\mathfrak{A}(\omega_1^{\mathrm{CK}}, x)$

The ramified analytical hierarchy was introduced by Kleene [72] and forcing constructions over it goes back to Cohen [18] and continued in Feferman [25]. It may be viewed as an analog of the constructible hierarchy defined in the language of second order arithmetic, for the study of hyperarithmetic theory.

The language for the hierarchy, denoted $\mathscr{L}(\omega_1^{\mathrm{CK}}, \dot{x})$, consists of the following:

- number variables: $i, j, k, m, n, \ldots$
- numerals: $\bar{0}, \bar{1}, \bar{2}, \ldots$
- set constants: $\dot{x}, \dot{y}, \ldots$
- ranked set variables: $x^\beta, y^\beta, \ldots$ for $\beta < \omega_1^{\mathrm{CK}}$
- unranked set variables: $x, y, \ldots$
- function and relation symbols: $+$, $\cdot$ (times), $'$ (successor) and $\in$.

Formulas in this language are defined by induction in the usual way. As one might expect, the presence of ranked set variables implies that there is no recursive coding of $\mathscr{L}(\omega_1^{\mathrm{CK}}, \dot{x})$. Instead, coding is carried out in the countable admissible set $L_{\omega_1^{\mathrm{CK}}}$. First choose a Π^1_1-path $\mathcal{O}_1$ through $\mathcal{O}$ which exists according to Theorem 2.6.4. Code the ranked set variable x^β by the number $(2, n)$ where $n \in \mathcal{O}_1$ and $|n| = \beta$. Other symbols are coded recursively. With this, formulas are coded in $L_{\omega_1^{\mathrm{CK}}}$. A formula is *ranked* if all of its set variables are ranked. A ranked formula φ has rank at most β if

- for any γ, if x^γ is a free variable occurring in φ, then $\gamma \leq \beta$;
- for any γ, if x^γ is a bounded variable occurring in φ, then $\gamma < \beta$.

Note that

- the set of Gödel number of formulas is Π^1_1;
- the set of Gödel numbers of ranked formulas of rank at most β is r.e. uniformly in the unique notation for β in $\mathcal{O}_1$.

By Theorem 1.3.4, there is a recursive function h such that for $n \in \mathcal{O}_1$, $W_{h(n)}$ is the set of Gödel numbers of ranked formulas of rank at most $|n|$.

We now define the structure $\mathfrak{A}(\omega_1^{\mathrm{CK}}, x) = \bigcup_{\beta<\omega_1^{\mathrm{CK}}} \mathfrak{A}(\beta, x)$, where x is a real that interprets $\dot{x}$. $\mathfrak{A}(\omega_1^{\mathrm{CK}}, x)$ is a collection of reals defined in an analogous way as Gödel's L. The construction is completed in ω_1^{CK}-many stages. Any ranked set variable x^β is to be interpreted by a real in $\bigcup_{\gamma\leq\beta} \mathfrak{A}(\gamma, x)$. With this as our convention, the satisfaction relation $\bigcup_{\gamma\leq\beta} \mathfrak{A}(\gamma, x) \models \varphi$ is well defined for a ranked formula φ of rank at most β. We define $\mathfrak{A}(\omega_1^{\mathrm{CK}}, x)$ by induction on the recursive ordinals, as follows:

$$\mathfrak{A}(0, x) = \{x\} \cup \omega.$$

Suppose $\beta > 0$ and $\mathfrak{A}(\gamma, x)$ is defined for all $\gamma < \beta$. Given a ranked formula $\varphi(n)$ with rank at most β and whose only free variable is n, let

$$i \in z_\varphi \Leftrightarrow \bigcup_{\gamma<\beta} \mathfrak{A}(\gamma, x) \models \varphi(i).$$

Let B be the collection of reals z_φ for such φ's. Set

$$\mathfrak{A}(\beta, x) = \bigcup_{\gamma<\beta} \mathfrak{A}(\gamma, x) \cup B.$$

Thus the reals constructed at stage β are precisely those arithmetically defined by the reals constructed at earlier stages.

For a formula φ, the satisfaction relation $\mathfrak{A}(\omega_1^{\mathrm{CK}}, x) \models \varphi$ is defined by induction:

- if φ is ranked with rank at most β, then $\mathfrak{A}(\omega_1^{\mathrm{CK}}, x) \models \varphi$ if and only if $\bigcup_{\gamma\leq\beta} \mathfrak{A}(\gamma, x) \models \varphi$;
- if φ is $(\exists z)\psi(z)$, then $\mathfrak{A}(\omega_1^{\mathrm{CK}}, x) \models \varphi$ if and only if there is a $\beta < \omega_1^{\mathrm{CK}}$ and a z^β such that $z^\beta \in \mathfrak{A}(\omega_1^{\mathrm{CK}}, x)$ and $\mathfrak{A}(\omega_1^{\mathrm{CK}}, x) \models \psi(z^\beta)$;
- if φ is $\neg\psi$, then $\mathfrak{A}(\omega_1^{\mathrm{CK}}, x) \models \varphi$ if and only if $\mathfrak{A}(\omega_1^{\mathrm{CK}}, x) \not\models \psi$.

A sentence φ in $\mathscr{L}(\omega_1^{\mathrm{CK}}, \dot{x})$ is Σ_1^1 if it is ranked, or of the form $(\exists x_1)\cdots(\exists x_n)\psi$ for some formula ψ with no unranked set variables bounded by a quantifier. It is not difficult to see that the set of Gödel numbers of Σ_1^1-sentences is Π_1^1. Let $\ulcorner\varphi\urcorner$ be the code of the formula φ.

Theorem 5.1.1. *The set $\{(x, \ulcorner\varphi\urcorner) \mid \varphi$ is $\Sigma_1^1 \wedge \mathfrak{A}(\omega_1^{\mathrm{CK}}, x) \models \varphi\}$ is Π_1^1.*

Proof. By Σ-induction (Theorem 3.1.8) and the definition of $\mathfrak{A}(\omega_1^{\mathrm{CK}}, x)$, there is a Σ_1-formula $\psi(u)$ such that $\langle L_{\omega_1^x}[x], \in\rangle \models \psi(\ulcorner\varphi\urcorner)$ if and only if φ is ranked $\wedge$ $\mathfrak{A}(\omega_1^{\mathrm{CK}}, x) \models \varphi$. Hence by the Spector–Gandy Theorem (3.7.2), the set $\{(x, \ulcorner\varphi\urcorner) \mid \varphi$ is ranked $\wedge$ $\mathfrak{A}(\omega_1^{\mathrm{CK}}, x) \models \varphi\}$ is Π_1^1. Now if φ is of the form $(\exists z)\varphi_0(z)$, then $\mathfrak{A}(\omega_1^{\mathrm{CK}}, x) \models (\exists z)\varphi_0(z)$ if and only if there is an $n \in \mathcal{O}_1$ such that $\mathfrak{A}(\omega_1^{\mathrm{CK}}, x) \models (\exists z^{|n|})\varphi_0(z^{|n|})$. It follows that the set $\{(x, \ulcorner\varphi\urcorner) \mid \varphi$ is $\Sigma_1^1 \wedge \mathfrak{A}(\omega_1^{\mathrm{CK}}, x) \models \varphi\}$ is Π_1^1. □

Corollary 5.1.2. *For $\beta < \omega_1^{\mathrm{CK}}$, the set $\{(x, \ulcorner\varphi\urcorner) \mid rank(\varphi)$ is at most β $\wedge$ $\mathfrak{A}(\omega_1^{\mathrm{CK}}, x) \models \varphi\}$ is Δ_1^1.*

Proof. Note that the set $\{\ulcorner\varphi\urcorner \mid \text{rank}(\varphi) \text{ is at most } \beta\}$ is r.e. and hence Δ^1_1. By Theorem 5.1.1, both the sets $\{(x, \ulcorner\varphi\urcorner) \mid \text{rank}(\varphi) \text{ is at most } \beta \wedge \mathfrak{A}(\omega_1^{\text{CK}}, x) \models \varphi\}$ and $\{(x, \ulcorner\varphi\urcorner) \mid \text{rank}(\varphi) \text{ is at most } \beta \wedge \mathfrak{A}(\omega_1^{\text{CK}}, x) \models \neg\varphi\}$ are Π^1_1 and hence Δ^1_1. □

5.1.2 Characterizing $\mathfrak{A}(\omega_1^{\text{CK}}, x)$

Proposition 5.1.3. *For any real z, $z \in \mathfrak{A}(\omega_1^{\text{CK}}, x)$ if and only if there is an $m \in \mathcal{O}$ such that $z \leq_T \hat{H}^x_m$.*

Proof. Suppose that $z \in \mathfrak{A}(\omega_1^{\text{CK}}, x)$. Then there is a ranked formula $\varphi(n)$ such that

$$n \in z \Leftrightarrow \mathfrak{A}(\omega_1^{\text{CK}}, x) \models \varphi(n).$$

By Corollary 5.1.2, the set $C = \{(x, n) \mid \mathfrak{A}(\omega_1^{\text{CK}}, x) \models \varphi(n)\}$ is Δ^1_1. By the Π^1_1-Uniformization Theorem (4.2.1), there is a total Π^1_1 and hence Δ^1_1-function $f : 2^\omega \to \omega$ such that $f(x) = m$ if and only if $m \in \mathcal{O}^x$ and $(\exists e)(\forall n)((x, n) \in C \Leftrightarrow \Phi_e^{H^x_m}(n) = 1)$. By relativized Σ^1_1-boundedness (Corollary 1.5.2), there is an ordinal $\beta < \omega_1^{\text{CK}}$ such that $|f(x)|_x \leq \beta$ for all x. Thus if $z \in \mathfrak{A}(\omega_1^{\text{CK}}, x)$, then there is an $m \in \mathcal{O}$ such that $z \leq_T \hat{H}^x_m$.

For the other direction, suppose there is an $m \in \mathcal{O}$ such that $z \leq_T \hat{H}^x_m$. It suffices to prove that $\hat{H}^x_m \in \mathfrak{A}(|m| + 2, x)$. We show this by induction on $|m|$.

Suppose that $m = 2^n$. Then $\hat{H}_n \in \mathfrak{A}(|m| + 1, x)$. So $\hat{H}^x_m \in \mathfrak{A}(|m| + 2, x)$.

Suppose that $m = 3 \cdot 5^e$ and $\hat{H}^x_n \in \mathfrak{A}(\omega_1^{\text{CK}}, x)$ for all $n <_o m$. Then $i \in \hat{H}^x_m$ if and only if

$$\mathfrak{A}(|m| + 1, x) \models (\exists z^{|m|})(\exists j)(\exists l)(\hat{H}(\Phi_e(l), z^{|m|}, x) \wedge i = \langle j, l\rangle \wedge j \in z^{|m|}).$$

Thus $\hat{H}^x_m \in \mathfrak{A}(|m| + 2, x)$. □

Theorem 2.3.3 and Proposition 5.1.3 immediately give the following

Corollary 5.1.4. *Suppose that $\omega_1^x = \omega_1^{\text{CK}}$. Then for any z, $z \in \mathfrak{A}(\omega_1^{\text{CK}}, x)$ if and only if $z \leq_h x$. In particular, z is hyperarithmetic if and only if $z \in \mathfrak{A}(\omega_1^{\text{CK}}, \emptyset)$.*

5.1.3 Δ^1_1-comprehension

The *Δ^1_1-comprehension scheme* states:

$$(\forall n)((\exists z)\varphi(n, z) \leftrightarrow (\forall z)\psi(n, z)) \to (\exists y)(\forall n)(n \in y \leftrightarrow (\exists z)\varphi(n, z)),$$

where both $\varphi(n, z)$ and $\psi(n, z)$ are arithmetical. In forcing constructions, satisfying Δ^1_1-comprehension in $\mathfrak{A}(\omega_1^{\text{CK}}, x)$ is used to guarantee $\omega_1^x = \omega_1^{\text{CK}}$, as shown by the following theorem.

Theorem 5.1.5. $\omega_1^x = \omega_1^{\text{CK}}$ *if and only if $\mathfrak{A}(\omega_1^{\text{CK}}, x)$ satisfies the Δ^1_1-comprehension axiom.*

Proof. If $\omega_1^x = \omega_1^{\mathrm{CK}}$, then by Corollary 5.1.4, for any real z, $z \in \mathfrak{A}(\omega_1^{\mathrm{CK}}, x)$ if and only if $z \leq_h x$. Now suppose that $\mathfrak{A}(\omega_1^{\mathrm{CK}}, x) \models (\forall n)((\exists z)\varphi(n, z) \leftrightarrow (\forall z)\psi(n, z))$. Then for each n, let $f(n)$ be the $<_O$-least $m \in \mathcal{O}_1$ such that $\mathfrak{A}(\omega_1^{\mathrm{CK}}, x) \models (\exists z^{|m|})(\varphi(n, z^{|m|}) \vee \neg\psi(n, z^{|m|}))$. f is a total Π_1^1, and hence Δ_1^1, function. It follows that there is an $n_0 \in \mathcal{O}_1$ such that $f(n) <_O n_0$ for all n. Let y be a real such that $n \in y$ if and only if $\mathfrak{A}(\omega_1^{\mathrm{CK}}, x) \models \exists z^{|m_0|}(\varphi(n, z^{|n_0|})$. Then $y \in \mathfrak{A}(\omega_1^{\mathrm{CK}}, x)$ and

$$\mathfrak{A}(\omega_1^{\mathrm{CK}}, x) \models n \in y \leftrightarrow (\exists z)\varphi(n, z).$$

Suppose that $\mathfrak{A}(\omega_1^{\mathrm{CK}}, x)$ satisfies the Δ_1^1-comprehension axiom. For the sake of contradiction, assume that $\omega_1^x > \omega_1^{\mathrm{CK}}$. Let $n \in \mathcal{O}^x$ be such that $n = 3 \cdot 5^e$ and $|n|_x = \omega_1^{\mathrm{CK}}$. Note that $H_m^x \in \mathfrak{A}(\omega_1^{\mathrm{CK}}, x)$ for $m <_{O^x} n$. Let $\varphi(n, z)$ be

$$(\exists m)(\exists i)(\hat{H}(\Phi_e^x(m), z, x) \wedge n = \langle m, i\rangle \wedge i \in z).$$

If $m <_{O^x} n$, then $\hat{H}^x_{\Phi_e^x(m)}$ is the unique real z such that $\hat{H}(\Phi_e^x(m), z, x)$. Thus

$$\mathfrak{A}(\omega_1^{\mathrm{CK}}, x) \models (\forall n)((\exists z)\varphi(n, z) \leftrightarrow (\forall z)((\exists i)(\langle m, i\rangle = n \wedge i \notin z) \to \neg\varphi(n, z)).$$

Then by Δ_1^1-comprehension, there is a $y \in \mathfrak{A}(\omega_1^{\mathrm{CK}}, x)$ such that

$$\mathfrak{A}(\omega_1^{\mathrm{CK}}, x) \models (\forall n)(n \in y \leftrightarrow (\exists z)\varphi(n, z)).$$

Since $\hat{H}^x_{\Phi_e^x(m)} \in \mathfrak{A}(\omega_1^{\mathrm{CK}}, x)$ for $m <_{O^x} n$, we have $y = \hat{H}_n^x$, contradicting Proposition 5.1.3. □

Remark 5.1.6. The results above may be generalized. For example, one may introduce the language $\mathcal{L}(\omega_1^z, \dot{x}, \dot{z})$, define the structure $\mathfrak{A}(\omega_1^z, x, z)$, and show that all the results above hold. Then $\{(x, \ulcorner\varphi\urcorner, z) \mid \varphi \text{ is } \Sigma_1^1 \wedge \mathfrak{A}(\omega_1^z, x, z) \models \varphi\}$ is Π_1^1. Moreover, $\mathfrak{A}(\omega_1^z, x, z)$ satisfies Δ_1^1-comprehension if and only if $\mathfrak{A}(\omega_1^z, x, z)$ is precisely the collection of reals hyperarithmetic in $x \oplus z$, and if and only if $\omega_1^{z \oplus x} = \omega_1^z$.

5.1.4 Exercise

Exercise 5.1.1. *Prove that $\mathfrak{A}(\omega_1^{\mathrm{CK}}, \emptyset)$ is the smallest ω-model satisfying the Δ_1^1-comprehension axiom.*

5.2 Cohen forcing

Cohen's original notion of forcing was defined over a structure similar to the ramified analytical hierarchy, although subsequent developments in set theory have since followed a different approach. For hyperarithmetic theory, this hierarchy is the most natural setting in which to study forcing.

The collection **P** of forcing conditions we consider here are relatively simple, such as finite binary strings or hyperarithmetically encoded perfect sets. A forcing relation $\Vdash$ is a binary relation that is a subset of $\mathbf{P}\times\{\ulcorner\varphi\urcorner \mid \varphi$ is a formula in $\mathscr{L}(\omega_1^{\mathrm{CK}}, \dot{x})\}$. The objective is to make $\Vdash$ as simply defined as possible. For example, to be hyperarithmetic when the formula φ being forced is ranked, or Π_1^1 when φ is Σ_1^1. The generic set x will be a real that is obtained from a collection of pairwise compatible conditions through a union or intersection operation. The model $\mathfrak{A}(\omega_1^{\mathrm{CK}}, x)$ is then a generic extension of $\mathfrak{A}(\omega_1^{\mathrm{CK}}, \emptyset)$. We always make $\mathfrak{A}(\omega_1^{\mathrm{CK}}, x)$ satisfy Δ_1^1-comprehension so that $\omega_1^x = \omega_1^{\mathrm{CK}}$. With this, one can construct generic x's with a prescribed property.

5.2.1 Cohen forcing

A Cohen condition p is a finite string $\sigma \in 2^{<\omega}$ which may be identified with the basic open set $[\sigma]$. Define a partial ordering $\leq$ such that $p \leq q$ if and only if $p \succeq q$.

Given formulas $\varphi(x)$ and $\psi(n)$, denote by $\varphi(x \mid \psi)$ the formula obtained upon replacing every occurrence of $n \in x$ in φ with $\psi(n)$ and every occurrence of $n \notin x$ in φ with $\neg\psi(n)$.

For a sentence $\varphi \in \mathscr{L}(\omega_1^{\mathrm{CK}}, \dot{x})$, define $p \Vdash \varphi$ by induction on the complexity of φ:
- $p \Vdash \bar{n} \in \dot{x}$ if and only if $p(n) = 1$;
- $p \Vdash t_1 = t_2$ if and only if $\mathfrak{A}(\omega_1^{\mathrm{CK}}, \emptyset) \models t_1 = t_2$, where t_1, t_2 are terms involving only numerals and arithmetic operations;
- $p \Vdash t \in \dot{x}$ if and only if $\mathfrak{A}(\omega_1^{\mathrm{CK}}, \emptyset) \models t = \bar{n}$ for some n and $p \Vdash \bar{n} \in \dot{x}$, where t is a term involving only numerals and arithmetic operations;
- $p \Vdash (\exists n)\varphi(n)$ if and only if $p \Vdash \varphi(\bar{n})$ for some $n \in \omega$;
- $p \Vdash (\exists z^{\beta})\varphi(z^{\beta})$ if and only if $p \Vdash \varphi(z^{\beta} \mid \psi)$ for some ranked formula $\psi(\bar{n})$ with rank at most β;
- $p \Vdash (\exists z)\varphi(z)$ if and only if $p \Vdash (\exists z^{\beta})\varphi(z^{\beta})$ for some $\beta < \omega_1^{\mathrm{CK}}$;
- $p \Vdash \varphi \wedge \psi$ if and only if $p \Vdash \varphi$ and $p \Vdash \psi$;
- $p \Vdash \neg\varphi$ if and only if $\forall q \leq p(q \nVdash \varphi)$.

The following basic facts are easy to verify.

Proposition 5.2.1. *For any formula φ,*
(i) $(\forall p)(p \Vdash \varphi \vee p \Vdash \neg\varphi)$;
(ii) $(\forall p)(\exists q \leq p)(q \Vdash \varphi \vee q \Vdash \neg\varphi)$;
(iii) $(\forall p)(\forall q \leq p)(p \Vdash \varphi \rightarrow q \Vdash \varphi)$.

Theorem 5.2.2. *The set $\{(p, \ulcorner\varphi\urcorner) \mid p \Vdash \varphi \wedge \varphi$ is $\Sigma_1^1\}$ is Π_1^1.*

Proof. Note that the set of Σ_1^1-sentences is Π_1^1, and if $(\exists x)\theta(x)$ is Σ_1^1 where $\theta(x)$ is arithmetical, then $p \Vdash (\exists x)\theta(x)$ if and only if there exists an $n \in \mathcal{O}_1$ such that $p \Vdash \theta(x^{|n|})$. By the Σ-Induction Theorem (3.1.8) and the definition of forcing, there is a Σ_1-

formula $\psi(u, v)$ in the language of set theory such that for a Σ_1^1-sentence φ, $\langle L_{\omega_1^{\mathrm{CK}}}, \in\rangle \models \psi(p, \ulcorner\varphi\urcorner)$ if and only if $p \Vdash \varphi$. By the Spector–Gandy Theorem (3.7.1), the set $\{(p, \ulcorner\varphi\urcorner) \mid p \Vdash \varphi \wedge \varphi \text{ is } \Sigma_1^1\}$ is Π_1^1. □

A set $D \subseteq 2^{<\omega}$ is *dense* if for any $p \in 2^{<\omega}$, there is a $q \in D$ such that $q \leq p$. Given a sentence φ, let $D_\varphi = \{p \mid p \Vdash \varphi \vee p \Vdash \neg\varphi\}$. The next lemma follows from (ii) of Proposition 5.2.1.

Lemma 5.2.3. *Given a formula φ, D_φ is dense.*

A real x is *Cohen generic* if for any sentence φ there is an n such that $x \upharpoonright n \Vdash \varphi$ or $x \upharpoonright n \Vdash \neg\varphi$. By Lemma 5.2.3 and induction on the set of Gödel numbers of formulas, it is an easy exercise to show that a Cohen generic real exists. The following theorem states that for a generic real, forcing equals truth.

Theorem 5.2.4. *Suppose that x is Cohen generic. Then for any sentence φ, $\mathfrak{A}(\omega_1^{\mathrm{CK}}, x) \models \varphi$ if and only if there is an n such that $x \upharpoonright n \Vdash \varphi$.*

Proof. Either direction of the theorem may be proved by induction on the complexity of formulas. We show the nontrivial case when φ is of the form $\neg\psi$.

Suppose that there is an n such that $x \upharpoonright n \Vdash \neg\psi$. We claim that $\mathfrak{A}(\omega_1^{\mathrm{CK}}, x) \not\models \psi$. Otherwise, by induction, there is an m such that $x \upharpoonright m \Vdash \psi$. By Lemma 5.2.1, $x \upharpoonright (m + n) \Vdash \psi \wedge \neg\psi$, contradicting Lemma 5.2.1. Hence $\mathfrak{A}(\omega_1^{\mathrm{CK}}, x) \models \neg\psi$, i.e. $\mathfrak{A}(\omega_1^{\mathrm{CK}}, x) \models \varphi$.

Assume that $\mathfrak{A}(\omega_1^{\mathrm{CK}}, x) \models \neg\psi$. By induction, $x \upharpoonright n \nVdash \psi$ for any n. Since x is Cohen generic, there exists an m such that $x \upharpoonright m \Vdash \psi$ or $x \upharpoonright m \Vdash \neg\psi$. Hence $x \upharpoonright m \Vdash \neg\psi$. □

Theorem 5.2.5. *If x is Cohen generic, then $\mathfrak{A}(\omega_1^{\mathrm{CK}}, x)$ satisfies Δ_1^1-comprehension.*

Proof. Suppose that x is Cohen generic and

$$\mathfrak{A}(\omega_1^{\mathrm{CK}}, x) \models (\forall n)((\exists z)\varphi(n, z) \leftrightarrow (\forall z)\psi(n, z)),$$

where both $\varphi(n, z)$ and $\psi(n, x)$ are arithmetical. Then

$$\mathfrak{A}(\omega_1^{\mathrm{CK}}, x) \models (\forall n)(\exists z)(\varphi(n, z) \vee \neg\psi(n, z)).$$

By Theorem 5.2.4, there exists a $p \prec x$ such that

$$p \Vdash (\forall n)(\exists z)(\varphi(n, z) \vee \neg\psi(n, z)).$$

By the definition of forcing,

$$(\forall q \leq p)(q \nVdash (\exists n)\neg(\exists z)(\varphi(n, z) \vee \neg\psi(n, z))).$$

Hence

$$(\forall q \leq p)(\forall n)(q \nVdash \neg(\exists z)(\varphi(n, z) \vee \neg\psi(n, z))).$$

Therefore

$$(\forall q \leq p)(\forall n)(\exists r \leq q)(r \Vdash (\exists z)(\varphi(n, z) \vee \neg\psi(n, z))).$$

By the Π^1_1-Uniformization Theorem (4.2.1), there is a total Π^1_1 and so Δ^1_1-function $f : \{q \mid q \leq p\} \times \omega \to \omega$ such that $f(q, n) = m$ if and only if $m \in \mathcal{O}_1$ and there is an $r \leq q$ with $r \Vdash (\exists z^{|m|})(\varphi(n, z^{|m|}) \vee \neg\psi(n, z^{|m|}))$. Let S be the range of f.

By Theorem 5.2.2, f is a well-defined Π^1_1-function. Since f is total on its domain, it is Δ^1_1. By Corollary 1.5.2, there is an $n_0 \in O_1$ such that for any $m \in S$, $m <_o n_0$. Then

$$(\forall q \leq p)(\forall n)(\exists r \leq q)(r \Vdash (\exists z^{|n_0|})(\varphi(n, z^{|n_0|}) \vee \neg\psi(n, z^{|n_0|}))).$$

Hence

$$(\forall n)(\forall q \leq p)(q \nVdash \neg(\exists z^{|n_0|})(\varphi(n, z^{|n_0|}) \vee \neg\psi(n, z^{|n_0|}))),$$

and so

$$(\forall n)p \Vdash \neg\neg(\exists z^{|n_0|})(\varphi(n, z^{|n_0|}) \vee \neg\psi(n, z^{|n_0|})).$$

Since $x \succ p$, by Theorem 5.2.4,

$$\mathfrak{A}(\omega_1^{\mathrm{CK}}, x) \models (\forall n)(\exists z)^{|n_0|}(\varphi(n, z^{|n_0|}) \vee \neg\psi(n, z^{|n_0|})).$$

Let y be such that $n \in y$ if and only if $\mathfrak{A}(\omega_1^{\mathrm{CK}}, x) \models (\exists z^{|n_0|})\varphi(n, z^{|n_0|})$. Then $y \in \mathfrak{A}(\omega_1^{\mathrm{CK}}, x)$. Now if $n \in y$, then $\mathfrak{A}(\omega_1^{\mathrm{CK}}, x) \models (\exists z)\varphi(n, z)$. If $n \notin y$, then $\mathfrak{A}(\omega_1^{\mathrm{CK}}, x) \models (\exists z^{|n_0|})\neg\psi(n, z^{|n_0|})$. Then by assumption, $\mathfrak{A}(\omega_1^{\mathrm{CK}}, x) \models (\forall z)\neg\varphi(n, z)$. Thus $\mathfrak{A}(\omega_1^{\mathrm{CK}}, x)$ satisfies Δ^1_1-comprehension. □

Corollary 5.2.6. *If x is Cohen generic, then*
(i) $\omega_1^x = \omega_1^{\mathrm{CK}}$.
(ii) *For any real $y \leq_h x$, there is an $n \in \mathcal{O}$ such that $y \leq_T x \oplus H_n$.*

Proof. Theorem 5.2.5 and Theorem 5.1.5 give (i) as an immediate consequence.

For (ii), suppose $y \leq_h x$. Then by Corollary 5.1.4 and (i), $y \in \mathfrak{A}(\omega_1^{\mathrm{CK}}, x)$. By Theorem 5.2.4, there is a ranked formula φ such that for any k, $y(k) = 1 \Leftrightarrow (\exists m)(x \upharpoonright m \Vdash \varphi(k))$ and $y(k) = 0 \Leftrightarrow (\exists m)(x \upharpoonright m \Vdash \neg\varphi(k))$. By Theorem 5.2.2,

$$R(p, k, i) \Leftrightarrow (i = 1 \wedge p \Vdash \varphi(k)) \vee (i = 0 \wedge p \Vdash \neg\varphi(k))$$

is Π^1_1. Since $\mathcal{O}$ is Π^1_1-complete (Theorem 1.5.1), there is a recursive function $f : 2^{<\omega} \times \omega \times 2 \to \omega$ such that $f(p, k, i) \in \mathcal{O}$ if and only if $R(p, k, i)$ holds.

Define $R^x \subseteq \omega \times 2 \times \mathcal{O}$ so that $R^x(k, i, n)$ if and only if $n \in \mathcal{O}$ and there is an m with $f(x \upharpoonright m, k, i) = n$. Then R^x is $\Pi^1_1(x)$. By Theorem 4.2.1, there is a total $\Pi^1_1(x)$, hence $\Delta^1_1(x)$, function $h : \omega \to 2 \times \omega$ uniformizing R^x. Thus Range(h) is a $\Delta^1_1(x)$-set. By (i) and Corollary 1.5.2, there is an $n_0 \in \mathcal{O}$ such that Range$(h) \subseteq 2 \times \mathcal{O}_{n_0}$. Hence for any k, $y(k) = 1 \Leftrightarrow (\exists m)(f(x \upharpoonright m, k, 1) \in \mathcal{O}_{n_0})$, and $y(k) = 0 \Leftrightarrow (\exists m)(f(x \upharpoonright m, k, 0) \in \mathcal{O}_{n_0})$. It follows that $y \leq_T x \oplus \mathcal{O}_{n_0}$ and hence there is an $n \in \mathcal{O}$ such that $y \leq_T x \oplus H_n$. □

From the topological point of view, a Cohen generic real is one which meets all definable dense open sets. Thus the set of such reals is co-meager. In Chapter 14, we will develop a parallel notion of forcing based on the co-null property.

5.2.2 Applications to hyperdegrees

We now use Cohen forcing to study the structure of hyperdegrees.

Proposition 5.2.7. *If z is not hyperarithmetic, then there exists an x such that $x \not\leq_h z$ and $x \not\geq_h z$.*

Proof. Assume that z is not hyperarithmetic. Fix an enumeration $\{z_i\}_{i<\omega}$ of reals hyperarithmetic in z and an enumeration $\{\varphi_i\}_{i<\omega}$ of all formulas.

We define a sequence $\emptyset = p_0 \geq p_1 \geq \cdots$ of conditions as follows:

Suppose that p_i is defined. At stage $i+1$, first let $q_i^0 \leq p_i$ so that there exists a j with $q_i^0(j) \neq z_i(j)$. Then let $q_i^1 \leq q_i^0$ so that $q_i^1 \in D_{\varphi_i}$. This is possible according to Lemma 5.2.3.

See if $\varphi_i(n,k)$ is ranked and whether for all n, there exists an $r \leq q_i^1$ and k such that $r \Vdash \varphi_i(n,k)$. If not, let $p_{i+1} = q_i^1$ and go to next stage. Otherwise, see if for all n and $r \leq q_i^1$, if $r \Vdash \varphi_i(n,k)$ then k is unique. If this is so, let $p_{i+1} = q_i^1$ and go to next stage. If not, let $p_{i+1} \leq q_i^1$ so that there exists an n with $p_{i+1} \Vdash \varphi(n,k)$ for some $k \neq z(n)$.

Let $x = \bigcup_{i<\omega} p_i$. Then x is Cohen generic and $x \not\leq_h z$. By Corollary 5.2.6, $\omega_1^x = \omega_1^{\mathrm{CK}}$. If $x \geq_h z$, then by Corollary 5.1.4, there is a ranked formula $\varphi_i(n,k)$ in the enumeration list such that $z(n) = k$ if and only if $\mathfrak{A}(\omega_1^{\mathrm{CK}}, x) \models \varphi_i(n,k)$. By Theorem 5.2.4, for each n, there exist k and m such that $x \upharpoonright m \Vdash \varphi_i(n,k)$. Then by the construction at stage $i+1$, it must be the case that for every n and $r \leq q_i^1$, there is a unique k satisfying $r \Vdash \varphi_i(n,k)$. Define $\hat{z}(n) = k$ if and only if there is an $r \leq q_i^1$ with $r \Vdash \varphi_i(n,k)$. By Theorem 5.2.2, $\hat{z}$ is Π_1^1 and hence hyperarithmetic. But $z = \hat{z}$, a contradiction.

Thus $x \not\leq_h z$ and $x \not\geq_h z$. □

5.2.3 Exercises

Exercise 5.2.1. *Prove that there exists a Cohen generic $x \leq_T \mathcal{O}^{(\omega)}$.*

Exercise 5.2.2. *Prove that there is a sequence $\{x_i\}_{i<\omega}$ of reals such that for each j, $x_j \not\leq_h \{\langle i,n\rangle \mid n \in x_i \wedge i \neq j\}$.*

Exercise 5.2.3. *Prove that any countable partial ordering is embeddable into the set of hyperdegrees.*

Exercise 5.2.4. *Prove that the Σ_1-theory of the hyperdegrees as a partial ordering is decidable.*

Exercise 5.2.5. *Prove that for any nonhyperarithmetic x, the set $\{y \mid y \not\geq_h x \wedge y \not\leq_h x\}$ is co-meager.*

5.3 Sacks forcing

The notion of forcing with perfect sets over the ramified analytical hierarchy was introduced by Gandy and Sacks [36] in their construction of a set with minimal hyperdegree. The origin of perfect set forcing, however, traces back to Spector's proof of the existence of a minimal Turing degree [154]. As will be seen, the construction of a minimal hyperdegree is much more intricate as it requires a detailed analysis of the complexity of the forcing relation $\Vdash$.

5.3.1 Perfect set forcing over the ramified analytical hierarchy

Let $\mathcal{S}$ be the collection of hyperarithmetic perfect sets. Each $S \in \mathcal{S}$ is a condition for *Sacks forcing*. Given two conditions $S, T \in \mathcal{S}$, we say that $S \leq T$ if $S \subseteq T$.

Let φ be a sentence in the language of ramified analytical hierarchy $\mathcal{L}(\omega_1^{\mathrm{CK}}, \dot{x})$. Define $T \Vdash \varphi$ as follows.

- If φ is ranked, then $T \Vdash \varphi$ if $(\forall x \in T)(\mathfrak{A}(\omega_1^{\mathrm{CK}}, x) \models \varphi)$;
- if $\varphi(y)$ is unranked and $T \Vdash \varphi(y \mid \psi)$ for some ψ of rank at most β, then $T \Vdash (\exists y^\beta)\varphi(y^\beta)$;
- $T \Vdash (\exists y)\varphi(y)$ if and only if $T \Vdash (\exists y^\beta)\varphi(y^\beta)$ for some $\beta < \omega_1^{\mathrm{CK}}$;
- if $\varphi(n)$ is unranked and $T \Vdash \varphi(\bar{m})$ for some number m, then $T \Vdash (\exists n)\varphi(n)$;
- if φ and ψ are unranked, $T \Vdash \varphi$ and $T \Vdash \psi$, then $T \Vdash \varphi \wedge \psi$;
- if φ is unranked and $(\forall S)(S \subseteq T \to \neg S \Vdash \varphi)$, then $T \Vdash \neg\varphi$.

Note that while the definition of Cohen forcing for ranked sentences is by induction on the complexity of the sentence, that of Sacks forcing is given directly. This immediately implies the following.

Lemma 5.3.1. *If $T \Vdash \varphi$ and $S < T$, then $S \Vdash \varphi$.*

By Theorem 5.1.1 and the definition of forcing, we have

Lemma 5.3.2. *The set*

$$\{(T, \ulcorner\varphi\urcorner) \mid \varphi \text{ is } \Sigma_1^1 \wedge T \Vdash \varphi\}$$

is Π_1^1.

Lemma 5.3.3. *(Fusion Lemma) Let T be a condition and $\{(i, \varphi_i)\}_{i<\omega}$ be a hyperarithmetic set of Σ_1^1-formulas. Assume that*

$$(\forall i)(\forall P \leq T)(\exists S \leq P)(S \Vdash \varphi_i).$$

Then $(\exists S \leq T)(\forall i)(S \Vdash \varphi_i)$.

Proof. Using the notation $P \upharpoonright n = \{\tau \in 2^{\leq n} \mid \tau \in P\}$, define $\mathcal{R}$ by

$$\mathcal{R}(i, \sigma, R, P) \Leftrightarrow \sigma \in R \wedge P \subseteq R \wedge P \upharpoonright |\sigma| = \{\tau \mid \tau \prec \sigma\} \wedge P \Vdash \varphi_i\}.$$

Note that $\mathcal{R}$ is a Π^1_1-relation. Then $\mathcal{R}$ may be uniformized by a partial Π^1_1-function $F : \omega \times 2^{<\omega} \times \mathcal{S} \to \mathcal{S}$. By assumption, for any $R \leq T$, $\sigma \in R$ and $i \in \omega$, $F(R, i, \sigma)$ is defined. Using F, a hyperarithmetic family $\{P_\sigma \mid \sigma \in 2^{<\omega}\}$ is defined by recursion on σ:

$$P_\emptyset = T.$$

If $\sigma \notin P_\sigma$, let $P_{\sigma^\frown 0} = P_{\sigma^\frown 1} = \emptyset$. Otherwise let $P_{\sigma^\frown 0}$ and $P_{\sigma^\frown 1}$ be disjoint conditions stronger than (i.e. contained in) P_σ and $P_{\sigma^\frown 0} \upharpoonright |\sigma|, P_{\sigma^\frown 1} \upharpoonright |\sigma| = \{\tau \mid \tau \prec \sigma\}$ so that $P_{\sigma^\frown 0}, P_{\sigma^\frown 1} \Vdash \bigwedge_{j \leq |\sigma|} \varphi_j$.

Let $Q = \bigcap_n \bigcup_{|\sigma|=n} P_\sigma$. Then $Q \in \mathcal{S}$.

For each i and $\sigma \in 2^{i+1}$, we have either $P_\sigma = \emptyset$ or $P_\sigma \Vdash \varphi_i$. Suppose that φ_i is of the form $(\exists z)\psi_i(z)$ for some ranked $\psi_i(z)$. Then $P_\sigma \Vdash (\exists z^{\beta_\sigma})\psi_i(z^{\beta_\sigma})$ for some β_σ whenever $P_\sigma \neq \emptyset$. Let $\beta = \bigcup_{\sigma \in 2^{i+1}} \beta_\sigma$. By the definition of the forcing relation, we have $Q \Vdash (\exists z^\beta)\psi_i(z^\beta)$.

Hence for every i, $Q \Vdash \varphi_i$. □

Lemma 5.3.4. *For any sentence φ and condition T, there is an $S \leq T$ such that either $S \Vdash \varphi$ or $S \Vdash \neg\varphi$.*

Proof. If φ is unranked, then the lemma follows immediately from the definition. For φ of rank β, we prove the lemma by induction on β. Without loss of generality, assume that φ is of the form $(\exists x^\beta)\psi(x^\beta)$ for some ψ with rank less than β. Let $\{\hat{\phi}_i(n)\}_{i<\omega}$ be a hyperarithmetic enumeration of all the formulas of rank at most β whose only free variable is n. If there is an i and $S \leq T$ such that $S \Vdash \psi(x^\beta \mid \hat{\phi}_i)$, then we are done. Otherwise, by Lemma 5.3.3 and the inductive hypothesis, there is an $S \leq T$ such that $S \Vdash \neg\psi(x^\beta \mid \hat{\phi}_i)$ for every i. Then $S \Vdash \neg\varphi$. □

5.3.2 Sacks generic reals

A set $D \subseteq \mathcal{S}$ is *dense* if for any condition T, there is an $S \in D$ such that $S \leq T$. A real x is *Sacks generic* if for any dense set $D \subseteq \mathcal{S}$ arithmetical in $\mathcal{O}$, there is a condition $T \in \mathcal{D}$ such that $x \in [T]$. Note that by Lemma 5.3.2 and Lemma 5.3.4, for any sentence φ the set $D_\varphi = \{T \mid T \Vdash \varphi \vee T \Vdash \neg\varphi\}$ is dense and arithmetical in $\mathcal{O}$.

Lemma 5.3.5. *Let x be Sacks generic and $x \in [T_0] \cap [T_1]$. Then there is a $T \subseteq T_0 \cap T_1$ such that $x \in [T]$.*

Proof. Let $i \in \{0, 1\}$. Since T_i is hyperarithmetic, there is a ranked formula φ_i such that

$$(\forall\sigma)(\sigma \in T_i \Leftrightarrow \mathfrak{A}(\omega_1^{\mathrm{CK}}, \emptyset) \models \varphi_i(\sigma)).$$

Hence for $i < 2$,

$$(\forall z)(z \in [T_i] \Leftrightarrow \mathfrak{A}(\omega_1^{\mathrm{CK}}, z) \models (\forall n)\varphi_i(z \upharpoonright n)).$$

Since $x \in [T_0] \cap [T_1]$,

$$\mathfrak{A}(\omega_1^{CK}, x) \models (\forall n)\varphi_0(x \restriction n) \wedge (\forall n)\varphi_1(x \restriction n).$$

By the genericity of x and the definition of forcing for ranked formulas, there is a condition T such that $x \in [T]$ and

$$T \Vdash (\forall n)\varphi_0(\dot{x} \restriction n) \wedge (\forall n)\varphi_1(\dot{x} \restriction n).$$

This means that

$$(\forall z)(z \in [T] \Leftrightarrow \mathfrak{A}(\omega_1^{CK}, z) \models (\forall n)\varphi_0(z \restriction n) \wedge (\forall n)\varphi_1(z \restriction n))$$

and so $T \leq T_0, T_1$. □

Lemma 5.3.6. *Suppose that x is Sacks generic. Then for any sentence φ,*

$$\mathfrak{A}(\omega_1^{CK}, x) \models \varphi \Leftrightarrow (\exists T)(x \in [T] \wedge T \Vdash \varphi).$$

Proof. If φ is ranked, then by the definition of forcing, $(\exists T)(x \in [T] \wedge T \Vdash \varphi)$ implies $\mathfrak{A}(\omega_1^{CK}, x) \models \varphi$. Conversely suppose that $\mathfrak{A}(\omega_1^{CK}, x) \models \varphi$. The genericity of x implies that there is a condition T such that $x \in [T]$ and either $T \Vdash \varphi$ or $T \Vdash \neg\varphi$. Then by the definition of forcing again, $T \Vdash \varphi$.

If φ is unranked, then one may prove the lemma by induction on the complexity of φ as was done in Theorem 5.2.4. The only difference with Cohen forcing is when φ is the conjunction of two sentences, but this is resolved by applying Lemma 5.3.5. □

Lemma 5.3.7. *If x is Sacks generic, then $\mathfrak{A}(\omega_1^{CK}, x)$ satisfies Δ_1^1-comprehension and hence $\omega_1^x = \omega_1^{CK}$.*

Proof. Suppose that

$$\mathfrak{A}(\omega_1^{CK}, x) \models (\forall n)((\exists z)\varphi(n, z) \leftrightarrow (\forall z)\psi(n, z)),$$

where both $\varphi(n, z)$ and $\psi(n, x)$ are arithmetical.

Thus

$$\mathfrak{A}(\omega_1^{CK}, x) \models (\forall n)(\exists z)(\varphi(n, z) \vee \neg\psi(n, z)).$$

Let T be a condition such that

$$T \Vdash (\forall n)(\exists z)(\varphi(n, z) \vee \psi(n, z)).$$

By the definition of forcing,

$$(\forall S \leq T)(\neg S \Vdash (\exists n)\neg(\exists z)(\varphi(n, z) \vee \neg\psi(n, z))).$$

Hence

$$(\forall S \leq T)(\forall n)(\neg S \Vdash \neg\exists z(\varphi(n, z) \vee \neg\psi(n, z))),$$

implying that

$$(\forall P \leq T)(\forall n)(\exists S \leq P)(S \Vdash \exists z(\varphi(n, z) \vee \neg\psi(n, z))).$$

By Lemma 5.3.3,

$$(\exists S \leq T)(\forall n)(S \Vdash \exists z(\varphi(n, z) \vee \neg\psi(n, z))).$$

By Lemma 5.3.2, define a Π^1_1-relation $R \subseteq \omega \times \omega$ such that $R(n, m)$ if and only if

$$m \in \mathcal{O}_1 \wedge S \Vdash (\exists z^{|m|})(\varphi(n, z^{|m|}) \vee \neg\psi(n, z^{|m|})).$$

By Theorem 4.2.1, there is a total Π^1_1 and hence Δ^1_1-function f uniformizing R. By Corollary 1.5.2, there is an $m_0 \in O_1$ such that for any n, $f(n) <_o m_0$, so that

$$(\forall n)(S \Vdash (\exists z^{|m_0|})(\varphi(n, z^{|m_0|}) \vee \neg\psi(n, z^{|m_0|}))).$$

Hence for any condition T with $T \Vdash (\forall n)(\exists z)(\varphi(n, z) \vee \neg\psi(n, z))$, there is an $S \leq T$ and $\beta < \omega_1^{\mathrm{CK}}$ such that

$$(\forall n)(S \Vdash (\exists z^{\beta})(\varphi(n, z^{\beta}) \vee \neg\psi(n, z^{\beta}))).$$

Let

$$\begin{aligned}\mathcal{D} = \{S \mid S \Vdash \neg(\forall n)(\exists z)(\varphi(n, z) \vee \neg\psi(n, z)) \vee \\ (\exists \beta < \omega_1^{\mathrm{CK}}(S \Vdash (\forall n)(\exists z^{\beta})(\varphi(n, z^{\beta}) \vee \neg\psi(n, z^{\beta})))\}.\end{aligned}$$

Then $\mathcal{D}$ is a dense set of conditions arithmetical in $\mathcal{O}$. By the genericity of x there is a $T \in \mathcal{D}$ such that $x \in [T]$. By Lemma 5.3.6 and the assumption, $T \nVdash \neg(\forall n)(\exists z)(\varphi(n, z) \vee \neg\psi(n, z))$. Thus

$$T \Vdash (\forall n)(\exists z^{\beta})(\varphi(n, z^{\beta}) \vee \neg\psi(n, z^{\beta})))$$

for some $\beta < \omega_1^{\mathrm{CK}}$. By the definition of forcing,

$$\mathfrak{A}(\omega_1^{\mathrm{CK}}, x) \models (\forall n)(\exists z^{\beta})(\varphi(n, z^{\beta}) \vee \neg\psi(n, z^{\beta}))).$$

Thus as in the proof of Theorem 5.2.5, $\mathfrak{A}(\omega_1^{\mathrm{CK}}, x)$ satisfies Δ^1_1-comprehension. □

Theorem 5.3.8 (Gandy and Sacks [36]). *Suppose that x is Sacks generic and $y \leq_h x$ is nonhyperarithmetic. Then $x \leq_h y$, i.e. x is a set of minimal hyperdegree.*

Proof. Suppose that $y \leq_h x$ is nonhyperarithmetic. By Corollary 5.1.4 and Lemma 5.3.7, there is a ranked formula φ such that

$$(\forall n)(y(n) = i \Leftrightarrow \mathfrak{A}(\omega_1^{\mathrm{CK}}, x) \models \varphi(n, i)).$$

By Lemma 5.3.6, let T be a condition such that $x \in [T]$ and

$$T \Vdash (\forall n)(\exists! i)(i < 2 \wedge \varphi(n, i)).$$

Claim. *For any $S \leq T$, there exist $Q_i \leq S$, $i < 2$, and an n such that $Q_i \Vdash \varphi(\bar{n}, i)$.*

Otherwise, fix S such that for all Q_i and all n, it is not the case that $Q_0 \Vdash \varphi(\bar{n}, 0)$ and $Q_1 \Vdash \varphi(\bar{n}, 1)$. Then

$$n \in y \Leftrightarrow (\exists Q \leq S) Q \Vdash \varphi(n, 0)$$

and

$$n \notin y \Leftrightarrow (\exists Q \leq S) Q \Vdash \varphi(\bar{n}, 1),$$

implying by Lemma 5.3.2 that y is hyperarithmetic, which is a contradiction.

By the Claim and Lemma 5.3.3, one may define a condition $S \leq T$ such that

$$(\forall \sigma, \sigma' \in S)(\exists m)(\exists j < 2)(\sigma|\sigma' \to S \cap [\sigma] \Vdash \varphi(m, j) \land S \cap [\sigma'] \Vdash \varphi(m, 1-j)).$$

Furthermore, there is a hyperarithmetic function f mapping (σ, σ') to an appropriate $(m, j, 1-j)$. (*)

Let

$$D_\varphi = \{S \mid S \Vdash \neg(\forall n)(\exists! i < 2)\varphi(n, i) \text{ or } S \text{ satisfies } (*)\}.$$

Then D_φ is a dense set arithmetic in $\mathcal{O}$. By the assumption on x, there is an $S \in D_\varphi$ such that $x \in [S]$ and S satisfies (*). The function f defined on S effectively separates the reals in $[S]$, and hence x is recursive in $y \oplus f$. Since f is hyperarithmetic, we conclude that $x \leq_h y$. □

5.3.3 A Σ^1_1-set of minimal hyperdegrees

Let φ be ranked and

$$\hat{D}_\varphi = \{S \mid S \Vdash \neg(\forall n)(\exists! i < 2)\varphi(n, i) \text{ or } (\forall n)(\exists! i < 2) S \Vdash \varphi(n, i) \text{ or } S \text{ satisfies } (*) \text{ defined in the proof of Theorem 5.3.8}\}.$$

Define x to be *weakly Sacks generic* if for any ranked $\varphi(n, i)$, there is a condition $S \in \hat{D}_\varphi$ such that $x \in [S]$. Note that the set $\{(S, \ulcorner\varphi(n, i)\urcorner) \mid S \in \hat{D}_\varphi\}$ is Π^1_1. By the proof of Theorem 5.3.8, $\hat{D}_\varphi$ is dense for every ranked $\varphi(n, i)$. Moreover, by going through the proof of Theorem 5.3.8, we have the following lemma.

Lemma 5.3.9. *Let x be weakly Sacks generic and $n \in \mathcal{O}$. If $\emptyset <_h y \leq_T \hat{H}^x_n$, then $x \leq_h y$.*

Let $\{\varphi_\beta\}_{\beta<\omega_1^{\mathrm{CK}}}$ be a $\Sigma_1(L_{\omega_1^{\mathrm{CK}}})$-enumeration of all ranked sentences. By the Fusion Lemma (5.3.3) and Theorem 3.1.8, we may define by induction a $\Sigma_1(L_{\omega_1^{\mathrm{CK}}})$-enumeration of sets of conditions $\{\mathcal{I}_\beta\}_{\beta<\omega_1^{\mathrm{CK}}}$ such that for any $\gamma < \beta < \omega_1^{\mathrm{CK}}$,

(i) $(\forall S \in \mathcal{I}_\gamma)(\exists S_0, S_1 \in \mathcal{I}_\beta)(S_0 \leq S \cap [t_S(0)] \land S_1 \leq S \cap [t_S(1)])$, where t_S is the natural homeomorphic map from $2^{<\omega}$ to S;

(ii) $(\forall S \in \mathcal{I}_\beta)(\exists n)(\exists \{S_i\}_{i\leq n} \subseteq \mathcal{I}_\gamma)([S] \subseteq \bigcup_{i\leq n}[S_i])$, and

(iii) $(\forall S \in \mathcal{I}_\beta)(\exists n)(\forall \sigma \in 2^n)(S \cap [t_S(\sigma)] \in \hat{D}_{\varphi_\beta})$.

Let $\mathcal{I}_0 = \{2^{<\omega}\}$. At stage $\beta + 1$, for any $S \in \mathcal{I}_\beta$, by the density of $\hat{D}_{\varphi_\beta}$, let $S_0, S_1 \in \hat{D}_{\varphi_\beta}$ be the $<_L$-least pair of conditions such that $S_0 \leq S \cap [t_S(0)]$ and $S_1 \leq S \cap [t_S(1)]$. Let $S_0, S_1 \in \mathcal{I}_{\beta+1}$.

At limit stage λ, let $\{\beta_i\}_{i<\omega}$ be the $<_L$-least increasing ω-sequence of ordinals with limit λ. For $\beta < \lambda$, let i_β be the least i such that $\beta \leq \beta_i$. By the inductive hypothesis, it is not difficult to see that there is a $<_L$-least fusion sequence $\{S_\sigma\}_{\sigma\in 2^{<\omega}}$ such that

- $S_\emptyset = S$;
- $(\forall n)(\forall \sigma \in 2^{n+1})(S_\sigma \in \mathcal{I}_{\beta_{i_\beta+n}})$.

Let $T = \bigcap_{n<\omega} \bigcup_{\sigma\in 2^n} S_\sigma$. Put $T \cap [t_T(0)]$ and $T \cap [t_T(1)]$ into $\mathscr{I}_\lambda$. We leave it to the reader to verify that $\{\mathscr{I}_\beta\}_{\beta<\omega_1^{\mathrm{CK}}}$ satisfies (i)–(iii). The following is an immediate consequence of (iii).

Lemma 5.3.10. *If for any $\beta < \omega_1^{\mathrm{CK}}$ there is a condition $S_\beta \in \mathscr{I}_\beta$ such that $x \in [S_\beta]$, then x is weakly Sacks generic.*

Now let

$$A = \{(n, x) \mid (\exists S)(x \in [S] \wedge S \in \mathscr{I}_{|n|}) \wedge n \in \mathcal{O}_1\}.$$

Since every $S \in \mathscr{I}_\beta$ is hyperarithmetic, by Corollary 2.2.12, A is Π^1_1. Moreover, for any n, $A_n = \{x \mid (n, x) \in A\}$ is Δ^1_1.

Lemma 5.3.11. *$\bigcap_{n\in\mathcal{O}_1} A_n$ is an uncountable Σ^1_1-set.*

Proof. Obviously $\bigcap_{n\in\mathcal{O}_1} A_n$ is Σ^1_1. By the definition of $\{\mathscr{I}_\beta\}_{\beta<\omega_1^{\mathrm{CK}}}$, for any $n \in \mathcal{O}_1$, $\bigcap_{m<_o n} A_m \neq \emptyset$. Hence for any Δ^1_1 $y \subseteq \mathcal{O}_1$, $\bigcap_{n\in y} A_n \neq \emptyset$. By the Kreisel Compactness Theorem (4.2.4), $\bigcap_{n\in\mathcal{O}_1} A_n$ is nonempty. Since no member of this set is hyperarithmetic, by Lemma 2.5.4 it is uncountable. □

Theorem 5.3.12 (Simpson [138]). *There is an uncountable Σ^1_1-set of minimal hyperdegrees.*

Proof. Let $M = \{x \mid x \notin \mathsf{HYP} \wedge x \in \bigcap_{n\in\mathcal{O}_1} A_n \wedge \omega_1^x = \omega_1^{\mathrm{CK}}\}$. Then M is Σ^1_1 and by Lemma 5.3.11 and the Gandy Basis Theorem (2.5.3), M is nonempty. By Lemma 2.5.4, M is uncountable. By Lemma 5.3.9, every $x \in M$ is of minimal hyperdegree. □

By relativizing the proof of Theorem 5.3.12 (see [138]), one can show that there is a Σ^1_1-set $M \subseteq 2^\omega \times 2^\omega$ such that for any x, there is a y with $(x, y) \in M$, and for any $(x, y) \in M$, y's hyperdegree is a minimal cover of the hyperdegree of x.

5.3.4 Exercises

Exercise 5.3.1 (Sacks [123]). *Prove that x is Sacks generic if and only if for every sentence φ, there is an $S \in \hat{D}_\varphi$ such that $x \in [S]$.*

Exercise 5.3.2 (Simpson [138]). *Prove that there are two reals x, y of minimal hyperdegree such that $x \oplus y \geq_h \mathcal{O}$.*

Exercise 5.3.3. *Prove that no weakly Sacks generic real is hyperarithmetic.*

Exercise 5.3.4. *Let X be a Π^1_1-set of ranked sentences. Assume that if $A \subseteq X$ is Δ^1_1, then there is an x such that $\mathfrak{A}(\omega_1^{\mathrm{CK}}, x)$ is a model of A. Show that there is an x such that $\mathfrak{A}(\omega_1^{\mathrm{CK}}, x)$ is a model of X.*

Exercise 5.3.5 (Simpson [138]). *If there is an x such that $2^\omega = L[x] \cap 2^\omega$, then for any z, there is a $y \geq_h z$ whose hyperdegree is not a minimal cover of any hyperdegree.*

5.4 Characterizing countable admissible ordinals

By relativizing the proof of Corollary 3.6.7, one can show that ω_1^x is a countable admissible ordinal for any x. The converse was proved by Sacks [122]: every countable admissible ordinal $\alpha > \omega$ is ω_1^x for some x. His proof produces in addition an x such that no $y <_h x$ satisfies $\omega_1^y = \alpha$. We prove a weaker result, that $\alpha = \omega_1^x$ for some x, using a notion of forcing introduced by Steel [158]. Let α be a countable admissible ordinal greater than ω for the rest of the section.

5.4.1 Generalized ramified analytical hierarchy

Let $\mathscr{L}(\alpha, \dot{x})$ be a language consisting of the following:
- number variables: $i, j, k, m, n, \cdots$
- numerals: $\bar{0}, \bar{1}, \bar{2}, \ldots$
- constants: $\dot{x}, \dot{y}, \ldots$
- ranked set variables: $x^\gamma, y^\gamma, \ldots$ where $\gamma < \alpha$
- unranked set variables: $x, y, \ldots$
- operations and relation: $+, \cdot$ (times), $'$ (successor) and $\in$.

Formulas are defined by induction. The set of number-theoretic terms is the set of terms closed under $+, \cdot, '$ over numerals and number variables. Atomic formulas are formulas of the form $t = s$ and $t \in x$, where t and s are number-theoretic terms. The other formulas are generated in the usual way. A formula is *ranked* if all of its set variables are ranked. A ranked formula φ has rank at most γ if
- for any β, if x^β is a free variable occurring in φ, then $\beta \leq \gamma$;
- for any β, if x^β is a bound variable occurring in φ, then $\beta < \gamma$.

By Σ-induction on L_α, we have
- the set of formulas is Σ_1 in L_α;
- the set $\{(\gamma, \varphi) \mid \varphi$ is a ranked formulas of rank at most $\gamma\}$ is Δ_1 in L_α.

Given x, one now defines the structure $\mathfrak{A}(\alpha, x) = \bigcup_{\gamma<\alpha} \mathfrak{A}(\gamma, x)$, where x interprets the symbol $\dot{x}$ as a countable union of collations of reals. During the construction, a ranked variable x^γ is to be interpreted as a real in $\bigcup_{\beta\leq\gamma} \mathfrak{A}(\beta, x)$. The satisfaction relation $\bigcup_{\beta\leq\gamma} \mathfrak{A}(\beta, x) \models \varphi$, for a ranked formula φ with rank at most γ, is well defined. $\mathfrak{A}(\alpha, x)$ is constructed by induction on ordinals less than α:

$\gamma = 0$: $\mathfrak{A}(\gamma, x) = \{x\} \cup \omega$.

$\gamma > 0$: Suppose $\mathfrak{A}(\beta, x)$ is defined for all $\beta < \gamma$. For a ranked formula $\varphi(n)$ with rank at most γ and n as the only free variable, let

$$i \in z_\varphi \Leftrightarrow \bigcup_{\beta<\gamma} \mathfrak{A}(\beta, x) \models \varphi(i).$$

Let B be the collection of such z_φ's. Set

$$\mathfrak{A}(\gamma, x) = \bigcup_{\beta<\gamma} \mathfrak{A}(\beta, x) \cup B.$$

The satisfaction relation $\mathfrak{A}(\alpha, x) \models \varphi$ is defined inductively:

- if φ is ranked with rank at most γ, then $\mathfrak{A}(\alpha, x) \models \varphi$ if and only if $\bigcup_{\beta\leq\gamma} \mathfrak{A}(\beta, x) \models \varphi$;
- if φ is $(\exists z)\psi(z)$, then $\mathfrak{A}(\alpha, x) \models \varphi$ if and only if there is a $\gamma < \alpha$ and a z^γ such that $z^\gamma \in \mathfrak{A}(\alpha, x)$ and $\mathfrak{A}(\alpha, x) \models \psi(z^\gamma)$;
- if φ is $\neg\psi$, then $\mathfrak{A}(\alpha, x) \models \varphi$ if and only if $\mathfrak{A}(\alpha, x) \not\models \psi$.

We leave the proof of the following results to the reader.

Theorem 5.4.1. *If $\omega_1^x \geq \alpha$, then the set $\{(x, \varphi) \mid \varphi$ is ranked $\wedge\, \mathfrak{A}(\alpha, x) \models \varphi\}$ is Δ_1 in $L_{\omega_1^x}[x]$.*

Theorem 5.4.2. *Assume that $\omega_1^x \geq \alpha$. Then $\omega_1^x = \alpha$ if and only if $\mathfrak{A}(\alpha, x)$ satisfies the Δ_1^1-comprehension axiom.*

5.4.2 Steel forcing

Steel forcing is a notion of forcing whose conditions are finite tagged trees p with the following properties:

- $p = (\hat{t}, h)$ where $\hat{t} \subseteq \omega^{<\omega}$ and $h : \hat{t} \to \alpha \cup \{\infty\}$ is a function such that $h(\emptyset) = \infty$ and $(\forall \sigma, \tau \in \mathrm{Dom}(h))(\sigma \succ \tau \to h(\sigma) < h(\tau))$. We adopt the convention that $\infty > \beta$ for all $\beta < \alpha$;
- $p = (\hat{t}_p, h_p) < q = (\hat{t}_q, h_q)$ if and only if $\hat{t}_q$ is a subtree of $\hat{t}_p$ and $h_p \restriction \hat{t}_q = h_q$.

Fix a recursive coding of $\omega^{<\omega}$ so that a string in $\omega^{<\omega}$ is identified with a natural number. Since the generic real is intended to be a tree, we also use $\dot{T}$ in place of $\dot{x}$.

Given a sentence $\varphi \in \mathscr{L}(\alpha, \dot{T})$, define $p = (t_p, h_p) \Vdash \varphi$ by induction on the complexity of φ:

- $p \Vdash \bar{\sigma} \in \dot{T}$ if and only if $\sigma = \emptyset \vee (\sigma \restriction (|\sigma| - 1) \in \hat{t}_p \wedge h_p(\sigma \restriction (|\sigma| - 1)) > 0))$;
- $p \Vdash t_1 = t_2$ if and only if $\mathfrak{A}(\alpha, 0) \models t_1 = t_2$, where t_1, t_2 are number-theoretic terms;
- $p \Vdash t \in \dot{T}$ if and only if $\mathfrak{A}(\alpha, 0) \models t = \sigma$ for some σ and $p \Vdash \bar{\sigma} \in \dot{T}$, where t is a number-theoretic term;
- $p \Vdash (\exists n)\varphi(n)$ if and only if $p \Vdash \varphi(\bar{n})$ for some $n \in \omega$;
- $p \Vdash (\exists z^\gamma)\varphi(z^\gamma)$ if and only if $p \Vdash \varphi(z \mid \psi)$ for some ranked formula $\psi(\bar{n})$ of rank at most γ;
- $p \Vdash (\exists z)\varphi(z)$ if and only if $p \Vdash (\exists z^\gamma)\varphi(z^\gamma)$ for some $\gamma < \alpha$;
- $p \Vdash \varphi \wedge \psi$ if and only if $p \Vdash \varphi$ and $p \Vdash \psi$;
- $p \Vdash \neg\varphi$ if and only if $(\forall q \leq p)(\neg q \Vdash \varphi)$.

The following basic facts follow immediately from the definition of forcing.

Proposition 5.4.3. *For any sentence φ,*
(i) $(\forall p)(p \nVdash \varphi \vee p \nVdash \neg\varphi)$;
(ii) $(\forall p)(\exists q \leq p)(q \Vdash \varphi \vee q \Vdash \neg\phi)$;
(iii) $(\forall p)(\forall q \leq p)(p \Vdash \varphi \implies q \Vdash \varphi)$.

Analyzing the complexity of $\Vdash$ defined above is not a straightforward matter. The difficulty is that Steel forcing, unlike Cohen forcing, is essentially a form of "class forcing". Because of tagging, the forcing conditions range over the entire L_α. This makes the complexity of the statement "$p \Vdash \neg\varphi$" harder to determine. Our major effort is then to handle this problem.

Definition 5.4.4. Suppose that p, q are conditions and $\gamma < \alpha$. Then $\mathrm{Ret}(\gamma, p, q)$, a retagging of q by p, holds if and only if
- $\hat{t}_p = \hat{t}_q$;
- $(\forall\sigma)((h_p(\sigma) < \gamma \vee h_p(\sigma) = 0 \vee h_q(\sigma) = 0) \to h_p(\sigma) = h_q(\sigma))$, and
- $(\forall\sigma)(h_p(\sigma) \geq \gamma \to h_q(\sigma) \geq \gamma)$.

We will use retagging to reduce a complex condition with a simpler condition while preserving the forcing relation.

Lemma 5.4.5. *If $\gamma_0 < \gamma_1 < \alpha$ and $\mathrm{Ret}(\omega\gamma_1, p, q)$, then*

$$(\forall p_1 \leq p)(\exists q_1 \leq q)\mathrm{Ret}(\omega\gamma_0, p_1, q_1).$$

Proof. The reason for using $\omega\gamma_0$ and $\omega\gamma_1$ is to set aside some space to do the retagging. Assume that $\gamma_0 < \gamma_1 < \alpha$ and $\mathrm{Ret}(\omega\gamma_1, p, q)$. Let $p_1 \leq p$. Define q_1 as follows:

Set $\hat{t}_{q_1} = \hat{t}_{p_1}$ and
Case (1) $\sigma \in \hat{t}_q$ and $h_q(\sigma) < \omega\gamma_0$. Let $h_{q_1}(\sigma) = h_q(\sigma)$;
Case (2) otherwise. Let $h_{q_1}(\sigma) = \omega\gamma_0 + |\hat{t}_{q_1,\sigma}|$ where $\hat{t}_{q_1,\sigma} = \{\tau \geq \sigma \mid \tau \in \hat{t}_{q_1}\}$.
We leave it to the reader to verify that q_1 is as required. □

It is not difficult to see that there is a one-one $\Delta_1(L_\alpha)$-function $f : \{\varphi \mid \varphi \text{ is ranked}\} \to \alpha$ such that if the rank of φ is less than that of ψ, then $f(\varphi) < f(\psi)$.

Lemma 5.4.6. *If $f(\varphi) = \gamma$ and $\mathrm{Ret}(\omega\gamma, p, q)$, then $p \Vdash \varphi \Leftrightarrow q \Vdash \varphi$.*

Proof. We prove the lemma by induction on ordinals. The only nontrivial case is when φ is of the form $\neg\psi$ and $f(\psi) < f(\varphi)$. Suppose $\mathrm{Ret}(\omega\gamma, p, q)$ where $f(\varphi) = \gamma$.

If $p \Vdash \neg\varphi$ but $q \nVdash \neg\varphi$, then $(\forall p_1 \leq p)(p_1 \nVdash \psi)$ but $(\exists q_1 \leq q)(q \Vdash \psi)$. By Lemma 5.4.5, there is a $p_2 \leq p$ such that $\mathrm{Ret}(\omega\gamma_1, p_2, q_1)$ where $f(\psi) = \gamma_1 < \gamma$. By induction, $p_2 \Vdash \varphi$, a contradiction. Hence $p \Vdash \varphi \Rightarrow q \Vdash \varphi$. A similar argument shows that $q \Vdash \varphi \Rightarrow p \Vdash \varphi$. □

Theorem 5.4.7. *The set $\{(p, \ulcorner\varphi\urcorner) \mid p \Vdash \varphi \wedge \varphi \text{ is ranked}\}$ is $\Delta_1(L_\alpha)$.*

Proof. Suppose that $f(\varphi) = \gamma$. For any condition p, $p \Vdash \neg\varphi$ if and only if $(\forall q \leq p)$ $(q \nVdash \varphi)$. Fix a condition $q \leq p$ and define q' so that $\hat{t}_{q'} = \hat{t}_q$ and

Case (1) $\sigma \in \hat{t}_p$ or $h_q(\sigma) < \omega\gamma$. Let $h_{q'}(\sigma) = h_q(\sigma)$;

Case (2) otherwise. Let $h_{q'}(\sigma) = \omega\gamma + |\hat{t}_{q,\sigma}|$ where $\hat{t}_{q,\sigma} = \{\tau \succeq \sigma \mid \tau \in \hat{t}_q\}$.

Hence $\mathrm{Ret}(\omega\gamma, q, q')$ holds. By Lemma 5.4.6, $q \Vdash \varphi \Leftrightarrow q' \Vdash \varphi$. Thus

$$(\forall q \leq p)(\mathrm{Range}(h_q) \subseteq \{\infty\} \cup \sum_{\sigma \in \hat{t}_p} h_p(\sigma) + \omega \cdot (\gamma + 1)) \to q \nVdash \varphi).$$

The theorem now follows from this by applying the Σ-Induction Theorem (3.1.8). □

Let $\{D_i\}_{i<\omega}$ be an enumeration of all dense sets of conditions definable over $\langle L_\alpha, \in\rangle$. Then there is a sequence $\{p_i\}_{i<\omega}$ such that $p_0 \geq p_1 \geq \cdots$ and $p_i \in D_i$ for each i. Let $T = \bigcup_{i<\omega} \hat{t}_{p_i}$ and $h = \bigcup_{i<\omega} h_{p_i}$. The following lemma is immediate.

Lemma 5.4.8. *For any $\gamma < \alpha$,*
- *The set $D^0_\gamma = \{p \mid (\exists\sigma)(\sigma \in \hat{t}_p \wedge h_p(\sigma) = \gamma)\}$ is dense.*
- *The set $D^1_\gamma = \{p \mid (\forall\sigma)(\sigma \in \hat{t}_p \wedge h_p(\sigma) > \gamma \to (\exists\tau \succ \sigma)(\tau \in \hat{t}_p \wedge h_p(\tau) = \gamma))\}$ is dense.*

Lemma 5.4.9. $\omega_1^T \geq \alpha$.

Proof. By Lemma 5.4.8, for any $\gamma < \alpha$, there is a $\sigma \in T$ such that $h(\sigma) = \gamma$. By induction on γ, it is straightforward to verify that the subtree $T_\sigma = \{\tau \in T \mid \tau \succeq \sigma\}$ is well ordered under the Kleene–Brouwer ordering with order type at least γ. Hence $\omega_1^T \geq \alpha$. □

The following forcing lemma is proved by a routine argument and is left to the reader.

Lemma 5.4.10. *For any sentence φ, $\mathfrak{A}(\alpha, T) \models \varphi$ if and only if there is a p with $\hat{t}_p \subseteq T$ such that $p \Vdash \varphi$.*

Lemma 5.4.11. *$\mathfrak{A}(\alpha, T)$ satisfies the Δ^1_1-comprehension axiom.*

Proof. Let φ and ψ be arithmetical such that

$$\mathfrak{A}(\alpha, T) \models (\forall n)((\exists z)\varphi(n, z) \leftrightarrow (\forall z)\psi(n, z)).$$

Then

$$\mathfrak{A}(\alpha, T) \models (\forall n)(\exists z)(\varphi(n, z) \vee \neg\psi(n, z)).$$

As in the proof of Theorem 5.2.5, there is a p with $\hat{t}_p \subseteq T$ such that

$$(\forall q \leq p)(\forall n)(\exists r \leq q)(r \Vdash (\exists z)(\varphi(n, z) \vee \neg\psi(n, z))).$$

Hence

$$(\forall q \leq p)(\forall n)(\exists r \leq q)(\exists\gamma)(r \Vdash (\exists z^\gamma)(\varphi(n, z^\gamma) \vee \neg\psi(n, z^\gamma))).$$

By Lemma 5.4.6 and Lemma 5.4.5, we have an increasing sequence $\gamma_0 < \gamma_1 < \cdots$ in L_α such that

(i) $\text{Range}(h_p) \subseteq \gamma_0 \cup \{\infty\}$;

(ii) for any m, $f((\exists z^{\gamma_m})(\varphi(n, z^{\gamma_m}) \vee \neg\psi(n, z^{\gamma_m})) < \gamma_{m+1}$ and for any n and $p_n \leq p$ in L_{γ_m}, there exists a $q_n \leq p_n$ with $q_n \in L_{\gamma_{m+1}}$ such that $q_n \Vdash (\exists z^{\gamma_{m+1}})(\varphi(n, z^{\gamma_{m+1}}) \vee \neg\psi(n, z^{\gamma_{m+1}}))$, and

(iii) for each m, $\gamma_{m+1} = \omega\gamma_{m+1}$.

Let $\beta = \bigcup_{m<\omega} \gamma_m = \omega\beta$. Then $\beta < \alpha$. We claim that

$$p \Vdash (\forall n)(\exists z^{\beta})(\varphi(n, z^{\beta}) \vee \neg\psi(n, z^{\beta})).$$

The forcing relation is equivalent to

$$(\forall n)(\forall q \leq p)(\exists r \leq q)(q \Vdash (\exists z^{\beta})(\varphi(n, z^{\beta}) \vee \neg\psi(n, z^{\beta}))).$$

Given n and $q \leq p$, let q' be a condition such that $\hat{t}_{q'} = \hat{t}_q$ and $h_{q'}(\sigma) = h_q(\sigma)$ for $h_q(\sigma) < \beta$ and $h_{q'}(\sigma) = \infty$ otherwise. By (i), $q' \leq p$ and $q' \in L_{\gamma_m}$ for some m. By (ii), there is a condition $r' \in L_\beta$ such that $r' \leq q'$ and

$$r' \Vdash (\exists z^{\gamma_m})(\varphi(n, z^{\gamma_m}) \vee \neg\psi(n, z^{\gamma_m})).$$

By (iii), $\text{Ret}(\omega\beta, q, q')$ holds and

$$\omega \cdot f((\exists z^{\gamma_m})(\varphi(n, z^{\gamma_m}) \vee \neg\psi(n, z^{\gamma_m}))) < \beta.$$

By Lemma 5.4.5 and Lemma 5.4.6, there is a condition $r \leq q$ such that

$$r \Vdash (\exists z^{\gamma_m})(\varphi(n, z^{\gamma_m}) \vee \neg\psi(n, z^{\gamma_m})).$$

Hence

$$r \Vdash (\exists z^{\beta})(\varphi(n, z^{\beta}) \vee \neg\psi(n, z^{\beta})).$$

By Lemma 5.4.10,

$$\mathfrak{A}(\alpha, T) \models (\forall n)(\exists z^{\beta})(\varphi(n, z^{\beta}) \vee \neg\psi(n, z^{\beta})).$$

Let y be defined such that $n \in y$ if and only if $\mathfrak{A}(\alpha, T) \models (\exists z^{\beta})\varphi(n, z^{\beta})$. Then $y \in \mathfrak{A}(\alpha, T)$ and thus $\mathfrak{A}(\omega_1^{\mathrm{CK}}, T)$ satisfies Δ_1^1-comprehension. □

In summary, we have the following:

Theorem 5.4.12 (Sacks [122]). *For any countable admissible ordinal $\alpha > \omega$, there is a real x such that $\omega_1^x = \alpha$.*

Remark 5.4.13. In the literature, there are several proofs of Theorem 5.4.12. Sacks's original proof was by the method of perfect set forcing and provided a stronger conclusion. The second proof, due to Friedman and Jensen [31], was by an application of the Barwise Compactness Theorem [3]. The third proof, which remains unpublished, was due to Jensen [54] and uses a class of Levy collapse operations which may be viewed as a variant of Steel forcing.

5.4.3 Exercises

Exercise 5.4.1. *Prove that for any x with $x \notin L_{\omega_1^x}$, there is a y such that $y \not\geq_h x$, $y \not\leq_h x$ and $\omega_1^y = \omega_1^x$.*

Exercise 5.4.2. *Prove that $x \in L_{\omega_1^x}$ if and only if for all y, $\omega_1^y \geq \omega_1^x$ implies $y \geq_h x$.*

Exercise 5.4.3. *Let α_n be the n-th admissible ordinal. Then $\bigcup_{n<\omega} \alpha_n$ is not admissible.*

Exercise 5.4.4 (Chong [11]). *Suppose that $\gamma > \omega$ is an ordinal so that $J_{\gamma+1} \setminus J_\gamma \cap 2^\omega \neq \emptyset$. Then for any admissible ordinal $\omega < \alpha < \gamma^+$, where γ^+ is the least admissible ordinal greater than γ, there is an $x \in J_{\gamma^+} \setminus J_\gamma$ such that $\omega_1^x = \alpha$.*

Exercise 5.4.5 (Jensen [54]). *Suppose that α is a countable limit of admissible ordinals. Then there is a set $A \subseteq \alpha$ so that for any admissible $\beta < \alpha$,*

$$\langle L_\beta[A], \in, A \cap L_\beta[A] \rangle$$

is admissible and

$$\langle L_\alpha[A], \in, A \cap L_\beta[A] \rangle \models \text{“}\beta \text{ is countable”}.$$

Exercise 5.4.6 (Jensen [54]). *Suppose that $\omega < \alpha_0 < \cdots < \alpha_n$ is a finite sequence of countable admissible ordinals. Prove that there is an x such that for any $i \leq n$, α_i is the $i+1$-th countable admissible ordinal in $L[x]$.*

6 Set theory

This chapter reviews results in set theory that have recursion-theoretic applications. Some of these will be stated without proof for future reference.

6.1 Set-theoretic forcing

The technique of forcing was discussed in the context of recursion theory in Chapter 5. The original version of the method as presented in Cohen [18] was very similar to that developed in § 5.2. Here we follow the approach of unramified forcing introduced by Shoenfield [130] (see Kunen [78] for an excellent exposition of the subject).

6.1.1 Forcing and genericity

Let $\mathcal{M} = \langle M, \in\rangle$ be a countable transitive model of ZFC and let $\mathbb{P} = \langle \mathbf{P}, \leq\rangle$ be a partial ordering in $\mathcal{M}$. An element $p \in \mathbf{P}$ is called a *condition*. Given two conditions $p, q \in \mathbf{P}$, we say that q is stronger than p if $q \leq p$. $D \subseteq \mathbf{P}$ is *dense* if for any $p \in \mathbf{P}$, there is a $q \in D$ such that $q \leq p$.

A *filter* $F \subseteq \mathbf{P}$ is a nonempty set such that

- $(\forall p)(\forall q)(q \leq p \wedge q \in F \rightarrow p \in F)$;
- $(\forall p)(\forall q)(\exists r)(p \in F \wedge q \in F \rightarrow r \in F \wedge r \leq p \wedge r \leq q)$.

A *generic* set $G \subseteq \mathbf{P}$ is a filter that meets every dense set. In general, a generic set does not belong to M except for very special partial orderings. If $G \notin M$, then there is a model $\mathcal{M}[G] = \langle M[G], \in\rangle$ of ZFC such that $M \cup \{G\} \subseteq M[G]$ and $o(M) = o(M[G])$. This is defined as follows.

First, a $\mathbb{P}$-*name* in $\mathcal{M}$ is defined by induction:

- $\emptyset$ is a $\mathbb{P}$-name;
- if $\tau \subseteq$ the set of $\mathbb{P}$-names $\times\, \mathbf{P}$, then τ is a $\mathbb{P}$-name.

We use $M^{\mathbb{P}}$ to denote the class of $\mathbb{P}$-names. Note that $M^{\mathbb{P}} \subset M$ and is not a member of M. Suppose G is generic. Define $\mathrm{val}(\tau, G)$ by induction:

- if τ is not a $\mathbb{P}$-name, then $\mathrm{val}(\tau, G) = \tau$;
- otherwise, $\mathrm{val}(\tau, G) = \{\mathrm{val}(\sigma, G) \mid (\sigma, p) \in \tau \wedge p \in G\}$.

Let

$$M[G] = \{\mathrm{val}(\tau, G) \mid \tau \in M\}.$$

Thus every element $x \in M[G]$ has a corresponding $\mathbb{P}$-name. We use $\dot{x}$ to denote the $\mathbb{P}$-name of x. We use $\check{a}$ to denote $\{(\check{b}, p) \mid b \in a \wedge p \in \mathbf{P}\}$ for any $a \in M$ by induction on the rank of a. Since for any $a \in M$, $\mathrm{val}(\check{a}, G) = a$, we have $M \subseteq M[G]$. Moreover $\dot{G} = \{(\check{p}, p) \mid p \in \mathbf{P}\}$ is a $\mathbb{P}$-name and $\mathrm{val}(\dot{G}, G) = G$. Hence $G \in M[G]$. $\langle M[G], \in\rangle$ is a transitive model of ZFC and $o(M) = o(M[G])$.

Definition 6.1.1. Given a condition $p \in \mathbf{P}$ and a formula $\varphi(\tau_0, \dots, \tau_n)$ with $\tau_0, \dots, \tau_n \in M^{\mathbb{P}}$, we say that p *forces* $\varphi(\tau_0, \dots, \tau_n)$, or $p \Vdash \varphi(\tau_0, \dots, \tau_n)$, if and only if for any generic set G with $p \in G$,

$$\langle M[G], \in\rangle \models \varphi(\mathrm{val}(\tau_0, G), \dots, \mathrm{val}(\tau_n, G)).$$

Theorem 6.1.2.
(i) *The class* $\{(p, \varphi, (\tau_0, \dots, \tau_n)) \mid p \Vdash \varphi(\tau_0, \dots, \tau_n)\}$ *is definable in* $\mathcal{M}$;
(ii) *for every generic G and formula* $\varphi(\tau_0, \dots, \tau_n)$ *with* $\tau_0, \dots, \tau_n \in M^{\mathbb{P}}$,

$$\langle M[G], \in\rangle \models \varphi(\mathrm{val}(\tau_0, G), \dots, \mathrm{val}(\tau_n, G)) \Leftrightarrow (\exists p \in G)(p \Vdash \varphi(\tau_0, \dots, \tau_n)).$$

Theorem 6.1.2 is not surprising when taken together with the definition of $\Vdash$ in § 5.2 (see [78]). The following facts about the forcing relation are standard.

Proposition 6.1.3. *For any formula* φ, *condition p and* $\mathbb{P}$-*names* $\tau_0, \dots, \tau_n$,
(i) $q \le p \wedge p \Vdash \varphi \implies q \Vdash \varphi$;
(ii) $p \Vdash \varphi \wedge \psi$ *if and only if* $p \Vdash \varphi$ *and* $p \Vdash \psi$;
(iii) $p \Vdash \neg\varphi$ *if and only if* $(\forall q \le p)(q \nVdash \varphi)$;
(iv) *the set* $\{p \mid p \Vdash \neg\varphi \vee p \Vdash \varphi\}$ *is dense;*
(v) $p \Vdash (\exists x)\varphi(x)$ *if and only if the set* $\{q \mid q \nleq p \vee (\exists \tau \in M^{\mathbb{P}})(q \Vdash \varphi(\tau))\}$ *is dense;*
(vi) $p \Vdash (\exists x)\varphi(x)$ *if and only if there is a* $\mathbb{P}$-*name* $\tau \in M^{\mathbb{P}}$ *such that* $p \Vdash \varphi(\tau)$.

6.1.2 κ-chain condition and Martin's Axiom

Given a partial ordering $\mathbb{P} = (\mathbf{P}, \le)$, a set $A \subseteq \mathbf{P}$ is an *antichain* if for $p \neq q$ in A, there is no r satisfying $r \le p$ and $r \le q$. $\mathbb{P}$ has the *κ-chain condition*, or κ-c.c., if in $\mathcal{M}$ every antichain of $\mathbb{P}$ has cardinality less than κ. In particular, ω_1-c.c. is known as the countable chain condition, or c.c.c. A chain condition restricts the possible values of a function in a generic extension. In fact, chain conditions are used as basic tools for the preservation of cardinals in a generic extension. For example, if $\mathbb{P} = \langle \mathbf{P}, \le\rangle$ satisfies c.c.c. and $f : \omega \to (\omega_1)^{\mathcal{M}}$ is a function in $\mathcal{M}[G]$ for some generic G, then for every $n < \omega$, the set $A_n = \{\alpha < (\omega_1)^{\mathcal{M}} \mid (\exists p)(p \Vdash \dot{f}(\check{n}) = \check{\alpha})\} \in M$ is countable in $\mathcal{M}$ so that $\bigcup_n A_n \in M$ is also countable in $\mathcal{M}$. Thus $\mathbb{P}$ does not collapse $(\omega_1)^{\mathcal{M}}$.

Lemma 6.1.4. *If* $\mathbb{P}$ *has the* κ-*chain condition, then for any generic set G*, $(\kappa)^{\mathcal{M}} = (\kappa)^{\mathcal{M}[G]}$.

Definition 6.1.5 (Martin's Axiom [90]). Let κ be a cardinal. MA_κ states that for any c.c.c. partial ordering $\mathbb{P} = \langle \mathbf{P}, \le\rangle$ and any κ-sequence of dense sets $\{D_\alpha\}_{\alpha<\kappa}$, there is a filter $G \subseteq \mathbf{P}$ that meets every D_α. We use MA to denote $(\forall \kappa < 2^{\aleph_0})\mathrm{MA}_\kappa$.

Obviously MA_ω is a theorem of ZFC. It is known that if ZFC is consistent then so is ZFC + MA. The κ-chain condition is also used to compute the cardinality of a power set.

Lemma 6.1.6. *Suppose that $\kappa > \aleph_0$ is a regular cardinal and $\mathbb{P} = \langle \mathbf{P}, \leq \rangle$ satisfies κ-c.c. in $\mathcal{M}$ with $|\mathbf{P}| = \lambda$. Then for any regular cardinal $\eta \in M$ and generic set G, $\mathcal{M}[G] \models 2^\eta \leq ((\lambda^{<\kappa})^\eta)^M$.*

Proof (Sketch). Every $x \in M[G] \cap 2^\eta$ corresponds to a $\mathbb{P}$-name $\dot{x} \subseteq \eta \times \mathscr{P}(\mathbf{P})$. Moreover, we may assume that for any $\gamma < \eta$, $\{p \mid (\check{\gamma}, p) \in \dot{x}\}$ is an antichain. There are $\lambda^{<\kappa}$-many antichains in $\mathcal{M}$. Hence there are at most $(\lambda^{<\kappa})^\eta$-many such $\mathbb{P}$-names. In other words, $\mathcal{M}[G] \models 2^\eta \leq ((\lambda^{<\kappa})^\eta)^M$ (see [78] for a detailed proof). □

6.1.3 κ-closedness

Two conditions p and q in $\mathbb{P}$ are compatible if $p \leq q$ or $q \leq p$. Given a cardinal κ, $\mathbb{P}$ has the *κ-closed* property (in $\mathcal{M}$) if for any sequence $\{p_\alpha\}_{\alpha<\nu}$ of compatible conditions of length $\nu < \kappa$, there is a q such that $q \leq p_\alpha$ for each $\alpha < \nu$. Satisfying the κ-closed property is another basic tool for preserving cardinals.

Lemma 6.1.7. *If $\mathbb{P}$ has the κ-closed property and $A \in M$ with $|A| < \kappa$, then any function f in a generic extension $\mathcal{M}[G]$ with domain A belongs to $\mathcal{M}$.*

It follows that κ-closed property preserves all cardinals not greater than κ.

6.1.4 Product forcing

Definition 6.1.8. Let I be an index set and let $\mathbb{P}_i = \langle \mathbf{P}_i, \leq_i \rangle$ be a notion of forcing for $i \in I$. Let $C \subseteq \mathscr{P}(I)$. The *$C$-product forcing* $\mathbb{P} = \langle \mathbf{P}, \leq \rangle$ is a partial ordering satisfying the following for each $p \in \mathbf{P}$:
- there is a $J \in C$ such that $p(i)$ is defined if and only if $i \in J$ and $p(i) \in \mathbf{P}_i$;
- $q \leq p$ in $\mathbf{P}$ if and only if for any $i \in I$, $p(i)$ is defined implies $q(i)$ is defined and $q(i) \leq_i p(i)$.

If C is the collection of all finite subsets of I, then $\mathbb{P}$ is known as product forcing with *finite support*. If C is the collection of all countable subsets of I, then $\mathbb{P}$ is product forcing with *countable support*. I is often taken to be an ordinal.

Let I be an ordinal γ. If $C = I$, we call $\mathbb{P}$ simply the product of $\{\mathbb{P}_\alpha\}_{\alpha<\gamma}$. Then the C-product forcing $\mathbb{P} = \Pi_{\alpha<\gamma}\mathbb{P}_\alpha$ may be decomposed into $\Pi_{\alpha<\beta}\mathbb{P}_\alpha \times \Pi_{\beta\leq\alpha<\gamma}\mathbb{P}_\alpha$. In other words, $\mathbb{P} \cong \Pi_{\alpha<\beta}\mathbb{P}_\alpha \times \Pi_{\beta\leq\alpha<\gamma}\mathbb{P}_\alpha$.

Theorem 6.1.9. *Suppose that $\mathbb{P} = \langle \mathbf{P}, \leq_P \rangle$ and $\mathbb{Q} = \langle \mathbf{Q}, \leq_Q \rangle$ are two notions of forcing. Let $\mathbb{P} \times \mathbb{Q}$ be the product of $\mathbb{P}$ and $\mathbb{Q}$. Then for any set $G \subseteq \mathbf{P} \times \mathbf{Q}$, G is generic over $\mathcal{M}$ if* *and only if $G = G_1 \times G_2$ where $G_1 \subseteq \mathbf{P}$ is generic over $\mathcal{M}$ and G_2 is generic over $\mathcal{M}[G_1]$. Moreover, $\mathcal{M}[G] = \mathcal{M}[G_1][G_2]$.*

In general, the c.c.c. property is not preserved under product forcing. This limitation is addressed by the following stronger property.

Definition 6.1.10. A forcing notion has *property* (K) if every uncountable set of conditions has an uncountable subset of mutually compatible elements.

The next theorem is in Jech [53].

Theorem 6.1.11.
(i) *Let $\gamma \geq \omega$ be an ordinal and assume that for each $\alpha < \gamma$, $\mathbb{P}_\alpha$ has property* (K). *Then the product forcing $\mathbb{P} = \Pi_{\alpha<\gamma}\mathbb{P}_\alpha$ with finite support has property* (K);
(ii) *Suppose that $\kappa \geq \aleph_1$, $\kappa^{\aleph_0} = \kappa$ and $|\mathbf{P}_i| \leq \kappa$ for each $i \in I$. Then the product forcing $\Pi_{i\in I}\mathbb{P}_i$ with countable support satisfies κ^+-c.c.*

6.1.5 Iterated forcing

Product forcing is less amenable to fine tuning the construction of a model. The notion of iterated forcing, introduced by Solovay and Tennenbaum [147], offers more "flexibility" in this respect. Given an ordinal α, let $I \subseteq \mathscr{P}(\alpha)$ be an ideal such that $\{0\} \in I$. We define iterated forcing $\mathbb{P}_\beta = \langle \mathbf{P}_\beta, \dot{Q}_\gamma, \gamma \leq \beta, I\rangle$ on ordinals $\beta \leq \alpha$ with support I by induction on β.

Definition 6.1.12. For $\beta = 0$, $\mathbb{P}_0 = \langle \mathbf{P}_0, \dot{Q}_0, I\rangle$ satisfies
- Q_0 is a notion of forcing;
- $p \in \mathbf{P}_0$ if and only if p is a function such that $\mathrm{Dom}(p) = \{0\}$ and $p(0) \in Q_0$.

For $\beta = \xi + 1$, a forcing condition $p \in \mathbf{P}_\beta$ is a partial function whose domain belongs to I so that
- $p \restriction \xi = \{(\eta, r) \mid \eta < \xi \wedge p(\eta) = r\} \in \mathbf{P}_\xi$;
- $\dot{Q}_\beta$ is a $\mathbf{P}_\xi$-name;
- if $\beta \in \mathrm{Dom}(p)$, then $p \restriction \xi \Vdash p(\beta) \in \dot{Q}_\beta$.

If $q \in \mathbf{P}_\beta$, define $p \leq_\beta q$ if $p \restriction \xi \leq q \restriction \xi$ and $p \restriction \xi \Vdash p(\beta) \leq q(\beta)$.

If β is a limit ordinal, then a forcing condition $p \in \mathbf{P}_\beta$ is a partial function whose domain belongs to I_α so that
- for every $\xi < \beta$, $p \restriction \xi = \{(\eta, r) \mid \eta < \xi \wedge p(\eta) = r\} \in \mathbf{P}_\xi$;
- $\{\xi < \beta \mid \xi \in \mathrm{Dom}(p)\} \in I$.

If $q \in \mathbf{P}_\beta$, define $p \leq_\beta q$ if and only if $p \restriction \xi \leq q \restriction \xi$ for every $\xi < \beta$.

The two most useful notions of iterated forcing are those defined in terms of finite and countable support.

Proposition 6.1.13. *Let α be an infinite ordinal. Assume that $\mathcal{M} \models \mathrm{ZFC} \wedge$ "$\kappa > \omega$ is regular", $I = \{s \mid s \subseteq \alpha \wedge |s| < \aleph_0\}$ and let $\mathbb{P}_\alpha = \langle \mathbf{P}_\alpha, \dot{Q}_\gamma, \gamma \leq \alpha, I\rangle$. If for each p and $\beta \leq \alpha$, $p \restriction \beta \Vdash$ "$\dot{Q}_\beta$ satisfies κ-c.c.", then for each $\beta \leq \alpha$, $\mathbb{P}_\beta$ satisfies κ-c.c. in $\mathcal{M}$.*

Remark 6.1.14. A similar statement holds for proper forcing in the case when I is countably supported. The proofs can be found in [53].

6.1.6 Embeddings and isomorphisms

Definition 6.1.15. A function f from $\langle \mathbf{P}, \leq\rangle$ into $\langle \mathbf{Q}, \leq\rangle$ is a *dense embedding* if
- $(\forall p_0)(\forall p_1)(p_0 \leq p_1 \to f(p_0) \leq f(p_1))$;
- $(\forall p_0)(\forall p_1)(p_0 | p_1 \to f(p_0) | f(p_1))$;
- Range(f) is a dense subset of $\mathbf{Q}$.

A function f from $\langle \mathbf{P}, \leq\rangle$ into $\langle \mathbf{Q}, \leq\rangle$ is an *isomorphism* if f is a bijection such that $(\forall p_0)(\forall p_1)(p_0 \leq p_1 \leftrightarrow f(p_0) \leq f(p_1))$. Clearly every isomorphism is a dense embedding.

Definition 6.1.16. Suppose that $f : \mathbb{P} \to \mathbb{Q}$. For any $\tau \in M^{\mathbb{P}}$, recursively define the corresponding $\mathbb{Q}$-name of τ induced by f as

$$f_*(\tau) = \{(f_*(\sigma), f(p)) \mid (\sigma, p) \in \tau\}.$$

This allows one to prove the following proposition.

Proposition 6.1.17. *Let $\mathcal{M} \models \mathrm{ZFC}$ and let $f : \mathbb{P} = \langle \mathbf{P}, \leq\rangle \to \mathbb{Q} = \langle \mathbf{Q}, \leq\rangle$ be a dense embedding.*
(i) *If H is $\mathbb{Q}$-generic over $\mathcal{M}$, then $f^{-1}(H)$ is $\mathbb{P}$-generic over $\mathcal{M}$ and $M[f^{-1}(H)] \subseteq M[H]$.*
(ii) *For any formula $\varphi(v_0, \ldots, v_n)$,*

$$p \Vdash \varphi(\tau_0, \ldots, \tau_n) \Leftrightarrow f(p) \Vdash \varphi(f_*(\tau_0), \ldots, f_*(\tau_n)).$$

6.1.7 Exercise

Exercise 6.1.1. *Prove Proposition* 6.1.17.

6.2 Some examples of forcing

6.2.1 Cohen forcing

Definition 6.2.1. Cohen forcing $\mathbb{C} = \langle 2^{<\omega}, \leq\rangle$ is the notion of forcing such that for any $p, q \in 2^{<\omega}$, $q \leq p$ if and only if $q \supseteq p$.

Since $2^{<\omega}$ is countable, the following implies that cardinals are preserved under a Cohen extension:

Proposition 6.2.2. *Cohen forcing satisfies property* (K).

Given a Cohen generic set G, the set $g = \bigcup_{\hat{\sigma} \in G} \hat{\sigma}$ is called a Cohen real. The next two results are proved in [78].

Proposition 6.2.3. *Suppose that $g = g_1 \oplus g_2$. Then g is a Cohen real over $\mathcal{M}$ if and only if g_1 is a Cohen real over $\mathcal{M}$ and g_2 is a Cohen real over $\mathcal{M}[g_1]$.*

Proposition 6.2.4. *Suppose $\kappa \geq \omega$ and $S \in \mathcal{M}$. Let $\mathbb{P}$ be a notion of product Cohen forcing of length κ and let G be generic. Then for any $A \subseteq S$ in $M[G]$, there is an $\alpha \leq \min\{\kappa, |S|^{\mathcal{M}}\}$ such that $A \in \mathcal{M}[G \cap \mathbf{P}_{<\alpha}]$ where $\mathbf{P}_{<\alpha} = \Pi_{\gamma<\alpha} 2^{<\omega}$.*

6.2.2 Random forcing

Random forcing was introduced by Solovay [150] to construct a model of ZF + DC in which every set of reals is Lebesgue measurable.

Definition 6.2.5. Let μ denote the Lebesgue measure. Random forcing $\mathbb{R} = \langle \mathbf{T}, \leq \rangle$ is a notion of forcing where
- every $T \in \mathbf{T}$ is a subtree of $2^{<\omega}$ such that $\mu([T]) > 0$;
- $T_1 \leq T_2$ if and only if $T_1 \subseteq T_2$.

Proposition 6.2.6. *Random forcing satisfies c.c.c.*

Proof. Suppose that there is an uncountable antichain $\{T_\alpha\}_{\alpha<\omega_1}$. Then for $\alpha \neq \beta < \omega_1$, $\mu([T_\alpha] \cap [T_\beta]) = 0$. Let $S_\alpha = [T_\alpha] \setminus \bigcup_{\beta<\alpha} [T_\beta]$. Then $\{S_\alpha\}_{\alpha<\omega_1}$ is an uncountable sequence of pairwise disjoint Borel sets with $\mu(S_\alpha) > 0$ for each $\alpha < \omega_1$. Let n be such that the set $R_n = \{S_\alpha \mid \mu(S_\alpha) > \frac{1}{n}\}$ is uncountable. Then $1 = \mu(2^\omega) \geq \mu(\bigcup_{S_\alpha \in R_n} S_\alpha) = \infty$, a contradiction. □

A real x is *random* if it is $\mathbb{R}$-generic.

6.2.3 Collapsing cardinals

Let κ be a cardinal. There is a notion of forcing that collapses all cardinals below κ while preserving κ: Let $\mathbb{P}(\kappa) = \langle \kappa^{<\omega}, \leq \rangle$ and for $p, q \in \kappa^{<\omega}$, define $q \leq p$ if and only if $q \supseteq p$. Then $\mathbb{P}(\kappa)$ satisfies κ^+-c.c. Moreover, for any generic filter G, the function $g = \bigcup_{p \in G} p$ maps ω onto κ. Thus we have

Proposition 6.2.7. *Suppose that $\kappa \geq \omega_1$ is a cardinal. Then*
(i) *$\mathbb{P}(\kappa)$ satisfies κ^+-c.c.;*
(ii) *if G is generic, then $\kappa^+ = (\omega_1)^{\mathcal{M}[G]}$.*

Note that $\mathbb{P}(\kappa)$ only preserves the successor cardinal κ^+. The next notion of forcing allows one to preserve limit cardinals which are regular.

Definition 6.2.8. Let $\kappa \geq \omega$ be a cardinal. Define *Lévy forcing for* κ, $\mathbb{L}(\kappa) = \langle \mathbf{L}_\kappa, \leq \rangle$, as follows:

- $p \in \mathbf{L}_\kappa$ if and only if p is a finite function mapping a subset of $\kappa \times \omega$ into κ such that for any $\alpha < \kappa$ and $n < \omega$, if $p(\alpha, n)$ is defined, then $p(\alpha, n) < \alpha$.
- For any $p, q \in \mathbf{L}_\kappa$, $q \leq p$ if and only if $q \supseteq p$.

$\mathbb{L}(\kappa)$ may be considered as product forcing with finite support of length κ, where at each $\alpha < \kappa$ the forcing action collapses α onto ω.

Proposition 6.2.9. *Let* $\kappa \geq \omega$ *be a regular cardinal and let* G *be* $\mathbb{L}(\kappa)$*-generic. Then* $\kappa = (\omega_1)^{\mathcal{M}[G]}$.

Proof. It is clear that for any $\alpha < \kappa$, $G(\alpha) = \bigcup_{n<\omega, p \in G} p(\alpha, n)$ maps ω onto α. Hence $\mathcal{M}[G] \models \kappa \leq \omega_1$. We prove that $\mathbb{L}(\kappa)$ satisfies κ-c.c.

Suppose that $A \subseteq \mathbf{L}_\kappa$ is an antichain of cardinality κ. Without loss of generality, we may assume that there is an n^* such that $|\mathrm{Dom}(p)| = n^*$ for every $p \in A$. We claim that there is an $\alpha < \kappa$ such that $A_\alpha = \{p \in A \mid (\exists n)(\alpha, n) \in \mathrm{Dom}(p)\}$ has cardinality κ. Otherwise, there exist $p, q \in A$ with disjoint domains such that p and q are compatible, which is a contradiction. Hence let α_0 be the least α such that A_α has cardinality κ. Let $B_{\alpha_0} = \{p \in A_{\alpha_0} \mid (\forall \beta < \alpha_0)(\forall n)(\beta, n) \notin \mathrm{Dom}(p)\}$. By the assumption on α_0, $|B_{\alpha_0}| = \kappa$.

Since $|B_{\alpha_0}| = \kappa$ and $|\alpha_0| < \kappa$, we may assume that for any $p, q \in B_{\alpha_0}$, $\{(\alpha_0, n) \mid (\alpha_0, n) \in \mathrm{Dom}(p)\} = \{(\alpha_0, n) \mid (\alpha_0, n) \in \mathrm{Dom}(q)\}$. Furthermore since $p(\alpha, n) < \alpha$ in general, we may even assume that for any $p, q \in B_{\alpha_0}$, $(\alpha_0, n) \in \mathrm{Dom}(p)$ implies $p(\alpha_0, n) = q(\alpha_0, n)$.

Then by the same argument as that for α_0, there is an $\alpha_1 > \alpha_0$ which is the least ordinal $> \alpha_0$ such that $B_{\alpha_0} \supseteq A_{\alpha_1}$ and the latter has size κ. Let $B_{\alpha_1} = A_{\alpha_1} \cap \{p \mid (\forall \beta)(\forall n)(\alpha_0 < \beta < \alpha_1 \to (\beta, n) \notin \mathrm{Dom}(p))\}$. Then $|B_{\alpha_1}| = \kappa$.

Thus as before we may define a sequence of ordinals $\alpha_0 < \alpha_1 < \cdots < \alpha_{n^*}$ and a sequence of sets $B_{\alpha_0} \supseteq B_{\alpha_1} \supseteq \cdots \supseteq B_{\alpha_{n^*}}$ such that $|B_{\alpha_{n^*}}| = \kappa$ and for all $p \in B_{\alpha_{n^*}}$, $\mathrm{Dom}(p) = \{\alpha_0, \ldots, \alpha_{n^*}\}$. Then $|B_{\alpha_n}| \leq |\alpha_{n^*}{}^{\leq n^*}| < \kappa$, a contradiction. □

6.2.4 Sacks forcing

Forcing with perfect sets was introduced by Spector [154] in his construction of a set of minimal Turing degree. This method was generalized by Sacks and applied to the hyperarithmetic and set-theoretic setting to produce sets of minimal hyperdegree and minimal degree of constructibility [121] respectively.

Definition 6.2.10. *Sacks forcing* is a notion of forcing $\mathbb{S} = (\mathbf{T}, \leq)$, such that

- $\mathbf{T}$ is the collection of perfect trees $\subseteq 2^{<\omega}$;
- for any $T_1, T_2 \in \mathbf{T}$, $T_1 \leq T_2$ if and only if $T_1 \subseteq T_2$.

A Sacks generic real is one which meets every dense collection of perfect sets. The following theorem says that such a real has the "minimality property".

Theorem 6.2.11 (Sacks [121]). *Suppose that g is Sacks generic over a model $\mathcal{M}$ of* ZFC *and $x \in M[g] \setminus M$ is a real. Then there are reals $z_0, z_1 \in M$ such that $x \oplus z_0 \geq_T g$ and $x \leq_T z_1 \oplus g$.*

Proof. Suppose that g is Sacks generic produced by a generic filter G and let $x \in M[g] \setminus M$. By Theorem 6.1.2, there is a condition T^0 such that $T^0 \in G$ (i.e. $g \in [T^0]$) and

$$T^0 \Vdash \dot{x} \notin \check{M} \wedge \dot{x} \in \dot{2^\omega}.$$

Let $T^1 \leq T^0$. Define a function $T : 2^{<\omega} \to T^1$ and a sequence of conditions $\{T_\sigma\}_{\sigma \in 2^{<\omega}}$ and numbers n_σ by induction as follows:

Let $T(\emptyset) = \emptyset$ and $T_\emptyset = T^1$.

Suppose $\sigma \in 2^{<\omega}$, $T(\sigma)$ is defined and $T_\sigma \subseteq T^1$. Then $T_\sigma \Vdash \dot{x} \notin \check{M} \wedge \dot{x} \in \dot{2^\omega}$, and there exist an $n_\sigma > |\sigma|$ as well as conditions $T_0, T_1 \subseteq T_\sigma \cap [T(\sigma)]$ such that $T_0 \Vdash \dot{x}(\check{n}_\sigma) = 0$ and $T_1 \Vdash \dot{x}(\check{n}_\sigma) = 1$, where $[T(\sigma)]$ is the set of strings extending $T(\sigma)$.

Then there is a $\tau \succ T(\sigma)$ and an $i \in \{0, 1\}$ such that $\tau \in T_\sigma$, $[\tau^\frown i] \cap T_0 \neq \emptyset$ and $[\tau^\frown(1-i)] \cap T_1 \neq \emptyset$. Let $T(\sigma^\frown i) = \tau^\frown i$, $T_{\sigma^\frown i} = [\tau^\frown i] \cap T_i$ for $i \in \{0, 1\}$. Then $T_{\sigma^\frown i} \Vdash \dot{x}(n_\sigma) = 0$ and $T_{\sigma^\frown(1-i)} \Vdash \dot{x}(n_\sigma) = 1$.

Let $T^2 = \{\tau \mid (\exists\sigma)(T(\sigma) \succ \tau)\} \subseteq T^1$. Then for any σ, $T^2 \cap [T(\sigma)] \leq T_\sigma$. Thus

$$(\forall\sigma)(\forall i < 2)(T^2 \cap [T(\sigma^\frown i)] \Vdash \dot{x}(\check{n}_\sigma) = \check{i}).$$

Let

$$\begin{aligned} D = \{T^2 \mid (\forall T)(T \nleq T^2 \vee T \nleq T^0) \vee \\ & (\exists\{n_\sigma\}_{\sigma \in 2^{<\omega}})(\exists T)(T : 2^{<\omega} \to 2^{<\omega}(\forall\sigma)(\forall\tau)(\forall i < 2)(\exists j < 2) \\ & (T^2 = \{\tau \mid (\exists\nu)(\tau \prec T(\nu))\}) \wedge \\ & (\sigma \prec \tau \leftrightarrow T_\tau \leq T_\sigma) \wedge (n_\sigma > |\sigma|) \wedge \\ & (T^2 \cap [T(\sigma^\frown i)] \Vdash \dot{x}(\check{n}_\sigma) = \check{j} \wedge T^2 \cap [T(\sigma^\frown(1-i))] \Vdash \dot{x}(\check{n}_\sigma) = \check{1} - \check{j}))\}. \end{aligned}$$

It is clear that D is a dense set. Hence there is a $T^2 \in D \cap G$. Furthermore, there exist a sequence $\{n_\sigma\}_{\sigma \in 2^{<\omega}}$ and an order-preserving map $T : 2^{<\omega} \to 2^{<\omega}$ such that

- $T^2 = \{\tau \mid (\exists\sigma)(\tau \prec T(\sigma))\}$;
- $(\forall\sigma)(n_\sigma > |\sigma|)$;
- $(\forall\sigma)(\forall i < 2)(\exists j < 2)(T^2 \cap [T(\sigma^\frown i)] \Vdash \dot{x}(\check{n}_\sigma) = \check{j} \wedge T^2 \cap [T(\sigma^\frown(1-i))] \Vdash \dot{x}(\check{n}_\sigma) = \check{1} - \check{j}))$.

Let

$$z_0 = \{(\sigma, T(\sigma^\frown i), n_\sigma, j) \mid \sigma \in 2^{<\omega} \wedge i \in 2 \wedge j \in 2 \wedge T^2 \cap [T(\sigma^\frown i)] \Vdash \dot{x}(\check{n}_\sigma) = \check{j}\}.$$

Then $z_0 \in M$ and may be coded as a real.

Note that $g \in [T^2]$. Since $x \in M[g]$, to compute $g \restriction n$ for a given n, search for a $|\tau| \geq n$ and a σ such that for any $k \leq |\sigma|$, there are unique $n_{\sigma \restriction k}$ and $\tau_k \prec \tau$ such that $(\sigma \restriction k, \tau_k, n_{\sigma \restriction k}, j) \in z_0$ if and only if $x(n_{\sigma \restriction k}) = j$. Then $\tau \restriction n = g \restriction n$. Hence $g \leq_T x \oplus z_0$.

To see that there is a real $z_1 \in M$ such that $g \oplus z_1 \geq_T x$, assume that

$$T^2 \cap [T(\sigma)] \Vdash x \restriction |\check{\sigma}| = \check{\tau}$$

for some $\tau \in 2^\sigma$. Let $z_1 = \{(\sigma, \tau) \mid T^2 \cap [T(\sigma)] \Vdash x \restriction |\check{\sigma}| = \check{\tau}\}$. Then $x \leq z_1 \oplus g$. □

The method in the proof of Theorem 6.2.11 is typical of what is known as a *fusion construction*. This line of construction allows one to find, in $\mathcal{M}$, a fusion sequence $\{T_\sigma\}_{\sigma \in 2^{<\omega}}$ for which there is a condition T^2 in M that is an intersection of $\{T_\sigma\}_{\sigma \in 2^{<\omega}}$, such that $(\forall \sigma)(\forall i < 2)(T^2 \cap [T(\sigma^\frown i)] \Vdash \dot{x}(\check{n_\sigma}) = \check{i})$. This "splitting property" facilitates the decoding of G from x and the real z.

Corollary 6.2.12. *If g is Sacks generic over $\mathcal{M} = \langle M, \in \rangle$, then every countable set of ordinals in $M[g]$ is a subset of a countable set in M and $(2^{\aleph_0})^{\mathcal{M}[g]} = (2^{\aleph_0})^{\mathcal{M}}$.*

Hence Sacks forcing preserves $\aleph_1$. It is known, however, that Sacks forcing may collapse cardinals (see [6]).

6.2.5 Destroying the Axiom of Choice

In general there is no direct way of destroying the Axiom of Choice (AC) by a forcing extension of a transitive model of ZFC. One introduces an intermediate model to achieve this.

Definition 6.2.13.

- A set X is *ordinal definable* in $\mathcal{M}$, written $X \in \mathrm{OD}^{\mathcal{M}}$, if there is a formula $\varphi(x, v_0, \ldots, v_n)$ together with ordinals $\alpha_0, \ldots, \alpha_n, \beta$ in M such that $a \in X \Leftrightarrow V_\beta^{\mathcal{M}} \models \varphi(a, \alpha_0, \ldots, \alpha_n)$, where $V_\beta^{\mathcal{M}}$ is the collection of sets in $\mathcal{M}$ of rank less than β.
- X is *hereditarily ordinal definable* in $\mathcal{M}$, written $X \in \mathrm{HOD}^{\mathcal{M}}$, if $\mathrm{TC}(\{X\}) \subset \mathrm{OD}^{\mathcal{M}}$.

If $\mathcal{M}$ is a transitive model of ZFC, then $\langle \mathrm{HOD}^{\mathcal{M}}, \in \rangle$ is a transitive model of ZFC (see [53]). The notion may be generalized to a wider class by defining $\mathrm{OD}(A)^{\mathcal{M}}$ and $\mathrm{HOD}(A)^{\mathcal{M}}$. If $\mathcal{M}$ is a transitive model of ZFC and $A \in M$, then $\langle \mathrm{HOD}(A)^{\mathcal{M}}, \in \rangle$ is a transitive model of ZF (see [53]). We use $\mathrm{HOD}(A)^{\mathcal{M}}$ to obtain a model that destroys AC.

Definition 6.2.14. Let $A \in M$,

- A set X is ordinal definable relative to A in $\mathcal{M}$, written $X \in \mathrm{OD}(A)^{\mathcal{M}}$, if there exist ordinals $\alpha_0, \ldots, \alpha_n$, elements $a_0, \ldots, a_m \in A$ and a formula $\varphi(x, \alpha_0, \ldots, \alpha_n, a_0, \ldots, a_m, A)$ in the expanded language defining X in the structure $\langle \mathcal{M}, A, a_0, \ldots, a_m, \in \rangle$.
- A set X is in $\mathrm{HOD}(A)^{\mathcal{M}}$ if $\mathrm{TC}(\{X\}) \subset \mathrm{OD}(A)^{\mathcal{M}}$.

Theorem 6.2.15. *Suppose that $\mathcal{M} \models V = L$. Let $\mathbb{P} = \Pi_{i\in\omega}\mathbb{C}$ be a notion of product Cohen forcing with finite support and let G be a $\mathbb{P}$-generic filter. If $A = \{g \mid (\exists i)(\forall j)g(j) = \bigcup_{p\in G} p(i)(j)\}$, then there is no well-ordering of the reals in $\mathcal{N} = \langle \mathrm{HOD}(\mathrm{TC}(\{A\}))^{\mathcal{M}[G]}, \in\rangle$.*

Proof. Suppose that there is a bijection $F : 2^\omega \to \alpha$ in $\mathcal{N}$ for some $\alpha \in o(N)$. We may identify G with the ω-sequence of generic reals $\{(i, g) \mid g(j) = \bigcup_{p\in G} p(i)(j)\}$.

Let

$$\dot{A} = \{(\dot{G}_i, p) \mid i < \omega \wedge p \in \mathbf{P}\},$$

where $G_i = (\forall j)(\exists p)(p \in G \wedge G_i(j) = (p(i))(j))$. Obviously $\mathrm{val}(\dot{A}, G) = A$. Then there is a formula $\varphi_0(x, \nu, \vec{\alpha}, \vec{a}, A)$ defining $F(x) = \nu$ via A, ordinals $\vec{\alpha}$ and $\vec{a} \in A^{<\omega}$. Let $\varphi_1(x, \nu, \vec{\alpha}, \vec{a}, A)$ be

$$x \text{ is the unique real such that } \varphi_0(x, \nu, \vec{\alpha}, \vec{a}, A).$$

Then for any real $x \in M[G]$, there is a $\nu < \omega_1$ such that

$$\mathcal{M}[G] \models \varphi_1(x, \nu, \vec{\alpha}, \vec{a}, A).$$

Let n be the least number such that $\vec{a} \in \mathcal{M}[G_{<n}]$ where $G_{<n} = \{(i, j) \mid i < n \wedge (\forall j)(\exists p)(p \in G \wedge (p(i))(j) = 1)\}$. Then $G_n \in \mathcal{N}$, and hence

$$\mathcal{M}[G] \models \varphi_1(G_n, \nu, \vec{\alpha}, \vec{a}, A)$$

for some $\nu < \alpha$.

Thus there is a condition $p \in G$ such that

$$p \Vdash \varphi_1(\dot{G}_n, \check{\nu}, \check{\vec{\alpha}}, \check{\vec{a}}, \dot{A}).$$

Let $m > n$ be the least number not in $\mathrm{Dom}(p)$. Set $\pi : \omega \to \omega$ to be a permutation of ω that only swaps m with n. Then π induces an automorphism $\bar{\pi}$ of $\mathbb{P}$: For every $p \in \mathbf{P}$,

$$\mathrm{Dom}(\bar{\pi}p) = \{\pi(i) \mid i \in \mathrm{Dom}(p))\})$$

and

$$(\bar{\pi}p)(\pi(i)) = p(i).$$

Obviously $\bar{\pi}(\check{\nu}) = \check{\nu}$, $\bar{\pi}(\check{\vec{\alpha}}) = \check{\vec{\alpha}}$, $\bar{\pi}(\check{\vec{a}}) = \check{\vec{a}}$, $\bar{\pi}(\dot{G}_n) = \dot{G}_m$ and $\bar{\pi}(\dot{A}) = \dot{A}$. Moreover, p is compatible with $\bar{\pi}(p)$. Hence there is a $q \in \mathbb{P}$ such that $q \le p, \bar{\pi}(p)$. Then

$$q \Vdash \varphi_1(\dot{G}_n, \check{\nu}, \check{\vec{\alpha}}, \check{\vec{a}}, \dot{A})$$

and

$$q \Vdash \varphi_1(\dot{G}_m, \check{\nu}, \check{\vec{\alpha}}, \check{\vec{a}}, \dot{A}).$$

This implies that $q \Vdash G_n \neq G_m$ which is a contradiction. □

6.2.6 Exercises

Exercise 6.2.1. *Prove that if* ZF *is consistent, then so is* ZF+ *"There is an infinite set A of reals all of whose subsets are either finite or cofinite in A".*

Exercise 6.2.2. *Prove that Sacks forcing does not satisfy c.c.c.*

Exercise 6.2.3. *Prove that if g is a real Cohen generic over* $\mathcal{M} \models$ ZFC, *then for any real* $x \in M[g] \setminus M$, *there is a* $z \in M$ *such that* $x \leq_T g \oplus z$.

Exercise 6.2.4. *Prove that if g is a real Cohen generic over* $\mathcal{M} \models$ ZFC, *then every analytical real in* $M[g]$ *is a member of* M.

Exercise 6.2.5 (Levy [84]). *Prove that if g is a real Cohen generic over L, then*

$$(\mathrm{HOD})^{L[g]} = L.$$

6.3 A cardinality characterization of Π_1^1-sets

In this section, we present an application of Martin's Axiom to the characterization of $\underset{\sim}{\Pi}_1^1$-sets due to Martin and Solovay [90].

6.3.1 Almost disjoint forcing

Definition 6.3.1. A set $A \subseteq 2^\omega$ of reals is *almost disjoint* if for any $x \in A$, $|x| = \infty$ and for any $x \neq y \in A$, $x \cap y$ is finite.

The following notion of forcing was introduced by Silver.

Definition 6.3.2. Given an index set I, an almost disjoint set $A = \{x_i\}_{i \in I}$ and $X \subset I$, $\mathbb{P} = \langle \mathbf{P}, \leq \rangle$ is an *almost disjoint forcing* corresponding to A and X if every $p \in \mathbf{P}$ is a partial function from ω to 2 such that
- $(\forall i \in X)(|\mathrm{Dom}(p) \cap x_i| < \infty)$; and
- $|\{n \mid p(n) = 1\}| < \infty$.

We write $p \leq_P q$ if $p \supseteq q$.

Lemma 6.3.3. *Given an almost disjoint set* $A = \{x_i\}_{i \in I}$ *on an index set I and* $X \subset I$, *the almost disjoint forcing corresponding to A and X is c.c.c.*

Proof. Given two conditions p and q, p is incompatible with q if and only if $\{n \mid p(n) = 1\} \neq \{n \mid q(n) = 1\}$. Since there are only countably many finite subset of ω, the lemma is immediate. □

Lemma 6.3.4.
(i) *For each* $i \in I \setminus X$, $D_{0,i} = \{p \in \mathbf{P} \mid x_i \subseteq \mathrm{Dom}(p)\}$ *is dense.*
(ii) *For each* $i \in X$ *and* $i \in \omega$, $D_{1,i} = \{p \in \mathbf{P} \mid |\{n \in x_i \mid p(n) = 1\}| > i\}$ *is dense.*

Proof. (i) is obvious. To see (ii), given any condition p, we have $|\mathrm{Dom}(p) \cap x_i| < \infty$. Then simply extend p to a condition q such that $|\{n \in x_i \mid q(n) = 1\}| > i$. □

Theorem 6.3.5. *If* G *is a generic set meeting all* $D_{0,i}$*'s and* $D_{1,i}$*'s for all* $i \in I$*, then the generic set* $g = \bigcup_{p \in G} p$ *is a real such that for any* $i \in I$, $i \in X$ *if and only if* $g \cap x_i$ *is infinite.*

Proof. Let G be a generic set meeting all $D_{0,i}$'s and $D_{1,i}$'s for all $i \in I$. If $i \in X$, then by (ii) of Lemma 6.3.4, $x_i \cap g$ is infinite. If $i \notin X$, then by (i) of Lemma 6.3.4, $x_i \subseteq \mathrm{Dom}(p)$ for some $p \in G$. But $|\{n \mid p(n) = 1\}| < \infty$, and so $x_i \cap g$ is finite. □

Hence almost disjoint forcing produces a real g which codes a large set X via an almost disjoint set A.

6.3.2 Martin–Solovay's characterization

Lemma 6.3.6. *Assume* $\mathrm{MA}_{\aleph_1}$ *and suppose that* $(\omega_1)^{L[x]} = \omega_1$ *for some* $x \in 2^\omega$*. If* $B \subseteq 2^\omega$ *has cardinality* $\aleph_1$*, then* B *is* $\underset{\sim}{\Pi}^1_1$*.*

Proof. We assume that $x(0) = 1$ for any $x \in B$.

Since $(\omega_1)^{L[x]} = \omega_1$, let $C = \{z \mid z(0) = 0 \wedge z \in L_{\omega_1^{z \oplus x}}[z \oplus x]\}$. Then C is a $\underset{\sim}{\Pi}^1_1$-set of cardinality $\aleph_1$.

Let $f : \omega \to 2^{<\omega}$ be a recursive bijection. Define

$$T = \{\sigma \in 2^{<\omega} \mid (\exists \tau \in 2^{<\omega})(\forall n < |\sigma|)(\sigma(n) = 1 \leftrightarrow f(n) \preceq \tau))\}.$$

Note that $[T]$ is an almost disjoint set. Moreover T is a recursive perfect tree and there is a recursive homeomorphic $h : 2^\omega \to [T]$. Also note that any $x \in [T]$ is recursive in each of its infinite subsets. Let $C_0 = \{y \mid (\exists x \in C)(h(x) = y)\}$ and $B_0 = \{y \mid (\exists x \in B)(h(x) = y)\}$. Suppose $p_0, p_1, \ldots$ is a recursive enumeration of the prime numbers. Let

$$B_0 = \{x_\alpha\}_{\alpha<\aleph_1},$$

$$C_0 = \{y_\alpha\}_{\alpha<\aleph_1},$$

$$B_1 = \{x_{\alpha,n} \mid (\forall i \in \omega)(x_{\alpha,n}(i) = 1 \leftrightarrow (\exists j)(i \text{ is the } p_n^j\text{-th element of } x_\alpha))\}$$

and

$$C_1 = \{y_{\alpha,n} \mid (\forall i \in \omega)(y_{\alpha,n}(i) = 1 \leftrightarrow (\exists j)(i \text{ is the } p_n^j\text{-th element of } y_\alpha))\}.$$

Set $A = B_0 \cup C_0$. Then A is almost disjoint, and $x \in C_1$ if and only if
(i) there is a $z \leq_T x$ and an $n \in \omega$ such that $z \in C_0$, and
(ii) $(\forall i \in \omega)(x(i) = 1 \leftrightarrow (\exists j)(i \text{ is the } p_n^j\text{-th element of } z))$.

Then C_1 is $\underset{\sim}{\Pi}_1^1$. By $\mathrm{MA}_{\aleph_1}$, Lemma 6.3.3 and Theorem 6.3.5, there is a real g such that for any $\alpha < \aleph_1$ and $n \in \omega$,

(1) $g \cap x_{\alpha,2n+1}$ is infinite if and only if $n \in y_\alpha$; and

(2) $g \cap y_{\alpha,2n+2}$ is infinite if and only if $n \in x_\alpha$.

Let h_0 be a function so that $h_0(x, n) = \{i \mid (\exists j)(i \text{ is the } p_n^j\text{-th element of } x)\}$. Note that $h_0(x_\alpha, n) = x_{\alpha,n}$ and $h_0(y_\alpha, n) = y_{\alpha,n}$.

Set $h_1(x) = \{n \mid |h_0(x, 2n+1) \cap g| = \infty\}$ and $h_2(x) = \{n \mid |h_0(x, 2n+2) \cap g| = \infty\}$. Then for any $x \in 2^\omega$,

$$x \in B_0 \leftrightarrow h_1(x) \in C_0 \wedge h_2(h_1(x)) = x,$$

showing that B is $\underset{\sim}{\Pi}_1^1$. □

Theorem 6.3.7 (Martin and Solovay [90]). *Assume* $\mathrm{MA}_{\aleph_1}$. *Then* $(\omega_1)^{L[x]} = \omega_1$ *for some* $x \in 2^\omega$ *if and only if every subset of* 2^ω *with cardinality* $\aleph_1$ *is* $\underset{\sim}{\Pi}_1^1$.

Proof. Suppose that $(\omega_1)^{L[x]} < \omega_1$ for all $x \in 2^\omega$. Let $A = \{x_\alpha\}_{\alpha<\omega_1}$ be a maximal chain of Turing degrees. Then A has cardinality $\aleph_1$ and contains no perfect subset. It follows that for any real x, A must contain a real not constructible in $L[x]$. If A is $\Pi_1^1(x_0)$ for some x_0, then by Lemma 4.3.2, A contains a perfect subset which is a contradiction.

Conversely suppose that $(\omega_1)^{L[x]} = \omega_1$ for some $x \in 2^\omega$. By Lemma 6.3.6, every subset of 2^ω of cardinality $\aleph_1$ is $\underset{\sim}{\Pi}_1^1$. □

Remark 6.3.8. Theorem 6.3.7 may be viewed as a result that goes "beyond the limits" of recursion theory, in which typically a lightface version of a theorem is proved and then relativized to a boldface version. However, in this case there is no lightface version for Theorem 6.3.7.

An immediate consequence of Theorem 6.3.7 (see Exercise 6.3.3) is that if $\mathrm{MA}_{\aleph_1}$ holds, then $(\omega_1)^{L[x]} = \omega_1$ for some $x \in 2^\omega$ if and only if every union of $\aleph_1$-many Borel sets is $\underset{\sim}{\Sigma}_2^1$. In contrast to this, Harrington [44] proved that for any cardinal $\kappa < 2^{\aleph_0}$, there is a model of ZFC in which any union of κ-many Borel sets is $\underset{\sim}{\Pi}_2^1$.

6.3.3 Exercises

Exercise 6.3.1 (Martin and Solovay [90]). *Assume* MA. *Prove that for any* $X \subset 2^{\aleph_0}$ *with* $|X| < 2^{\aleph_0}$, *there is a real* g *such that* $L[g] = L[X]$.

Exercise 6.3.2 (Martin and Solovay [90]). *Assume* MA. *Prove that* $2^\kappa = 2^{\aleph_0}$ *for any* $\kappa < 2^{\aleph_0}$.

Exercise 6.3.3 (Martin and Solovay [90]). *Assume* $\mathrm{MA}_{\aleph_1}$. *Prove that* $(\omega_1)^{L[x]} = \omega_1$ *for some* $x \in 2^\omega$ *if and only if every union of* $\aleph_1$*-many Borel sets is* $\underset{\sim}{\Sigma}_2^1$.

Exercise 6.3.4 (Harrington [44]). *If* $\mathcal{M} \models \mathrm{ZFC} + \omega_1 = \omega_1^L$ *and* $A \subseteq 2^\omega$ *in* M, *then there is a c.c.c. extension* $\mathcal{N}$ *of* $\mathcal{M}$ *in which* A *is* $\underset{\sim}{\Pi}_2^1$.

6.4 Large cardinals

Despite a fairly comprehensive set of axioms for a theory of sets, a considerable number of important mathematical problems are known to be independent of ZF. Additional assumptions have been introduced to study such problems, and among the assumptions are those involving the notion of a large cardinal. We discuss some of them here.

6.4.1 Inaccessible cardinals

Definition 6.4.1.
- A cardinal κ is *inaccessible* if κ is regular and for every cardinal $\nu < \kappa$, $2^\nu < \kappa$;
- A cardinal κ is *weakly inaccessible* if κ is a regular limit cardinal.

We only consider $\kappa > \omega$ here. The assumption that there is an inaccessible cardinal is the weakest large cardinal axiom. If κ is inaccessible, then ZFC $\vdash$ "$\langle V_\kappa, \in\rangle \models$ ZFC". In other words, ZFC + $(\exists\kappa)(\kappa$ is inaccessible) implies that ZFC is consistent. By Gödel's second Incompleteness Theorem, ZFC + "ZFC is consistent" does not imply the consistency of "ZFC + $(\exists\kappa)(\kappa$ is inaccessible)". This means that one is not able to construct a model of "ZFC + $(\exists\kappa)(\kappa$ is inaccessible)" by assuming that there is a model of ZFC.

At first sight, the notion of an inaccessible cardinal appears to be disconnected from the structure of the set of reals. Rather surprisingly, the two are closely linked, as illustrated by the following example.

Proposition 6.4.2 (Solovay [149]). *If there is no uncountable $\underset{\sim}{\Pi}^1_1$-thin set, then ω_1 is inaccessible in L.*

Proof. Clearly ω_1 is regular in L. Since $\langle L, \in\rangle \models$ GCH, it is sufficient to show that ω_1 is a limit cardinal in L.

Suppose not. Then there is a countable ordinal $\gamma < \omega_1$ such that $\langle L, \in\rangle \models \omega_1 = \gamma^+ = |2^\gamma|$. Since γ is countable, there is an x which computes a well-ordering of ω of order type γ. Then $\langle L[x], \in\rangle \models 2^\omega = \omega_1$. Relativizing the proof of Corollary 4.3.6 to x, we have $\{z \mid z \in L_{\omega_1^{x\oplus z}}[x]\}$ to be an uncountable $\Pi^1_1(x)$-thin set, a contradiction. □

Proposition 6.4.2 is concerned with the classical topic in descriptive set theory of the existence of perfect subsets of an uncountable definable set of reals. It says that to require all $\underset{\sim}{\Pi}^1_1$-sets to be "well-behaved", one would have to appeal at least to the existence of an inaccessible cardinal. This situation may be viewed from the perspective of reverse mathematics, a program which investigates set existence axioms that are necessary and sufficient to prove theorems in "ordinary mathematics". In the present context, one is interested in set-theoretic axioms that are necessary and sufficient to establish the consistency of a statement in descriptive set theory, relative to the consistency of ZF or ZFC. More precisely, given such a statement φ in descriptive set theory, is there a large cardinal axiom ψ such that ZFC + φ (or ZF + φ) is consistent if and only if ZFC + ψ is consistent? When this holds, we say that φ is *equiconsistent* with ψ. The next result due to Solovay exemplifies this point.

Proposition 6.4.3 (Solovay [149]). *The statement that every uncountable $\underset{\sim}{\Pi}^1_1$-set has a perfect subset is equiconsistent with the existence of an inaccessible cardinal.*

Proof. By Proposition 6.4.2, the statement that every uncountable $\underset{\sim}{\Pi}^1_1$-set has a perfect subset implies ω_1 is inaccessible in L. Hence over ZFC, the consistency of the statement implies the consistency of the existence of an inaccessible cardinal.

Now suppose that there exists an inaccessible cardinal κ in a transitive model $\mathcal{M}$ of ZFC. Fix an $\mathbb{L}(\kappa)$-generic set G over $\mathcal{M}$. By Proposition 6.2.9, $\kappa = (\omega_1)^{\mathcal{M}[G]}$. Let $x \in \mathcal{M}[G]$. Then there is a sequence of conditions $\cdots \leq p_n \leq \cdots \leq p_0$ such that for any n, $p_n \in G$ forces the value of $x(n)$. Since $\kappa = (\omega_1)^{\mathcal{M}[G]}$, there is a regular cardinal $\eta < \kappa$ such that $p_n \in \mathbf{L}(\eta)$ for every n and hence $x \in \mathcal{M}[G \cap \mathbf{L}(\eta)]$. Thus $(\omega_1)^{L[x]} \leq (\omega_1)^{\mathcal{M}[G\cap\mathbf{L}(\eta)]} < \eta^+ < \kappa$. In other words,

$$\mathcal{M}[G] \models \text{“}(\omega_1)^{L[x]} < \omega_1\text{”}.$$

Relativizing the proof of Corollary 4.3.6 to x, we conclude that the largest $\Pi^1_1(x)$-thin set $\{z \mid z \in L_{\omega_1^{x\oplus z}}[x]\}$ is countable. Then

$$\mathcal{M}[G] \models \text{“Every uncountable } \underset{\sim}{\Pi}^1_1\text{-set has a perfect subset”}. \qquad \square$$

6.4.2 Measurable cardinals

For a given infinite cardinal κ, consider $\mathscr{P}(\kappa)$ as a partially ordered set under $\subseteq$.

Definition 6.4.4.

- A set $A \subseteq \mathscr{P}(\kappa)$ is an *ultrafilter* over κ if it is a filter and for any $a \subseteq \kappa$, either $a \in A$ or $\kappa \setminus a \in A$;
- a filter A is *nonprincipal* if for any $a \in A$, there exists a $b \in A$ such that $a \not\subseteq b$;
- let $\eta \leq \kappa$ be a cardinal. A filter A is *η-complete* if for any $\beta < \eta$ and any sequence $\{a_\alpha\}_{\alpha<\beta}$ in A, $\bigcap_{\alpha<\beta} a_\alpha \in A$.

A κ-complete nonprincipal ultrafilter A over κ induces a κ-additive zero-one measure μ on $\mathscr{P}(\kappa)$ defined so that for any $a \subseteq \kappa$, $\mu(a) = 1$ if and only if $a \in A$. It is not difficult to construct under ZFC an ω-complete nonprincipal ultrafilter A over ω. For $\kappa > \omega$, however, no κ-complete nonprincipal ultrafilter A over κ can be proved to exist within the same axiom system.

Definition 6.4.5. We say that $\kappa > \omega$ is *measurable* if there exists a κ-complete nonprincipal ultrafilter A over κ.

It can be proved that every measurable cardinal is inaccessible (see [53]). Existence of a measurable cardinal is a very strong large cardinal assumption. Although it is a major notion in set theory, none of the results proved in this book requires this assumption.

6.4.3 $0^\sharp$

Compared to a measurable cardinal, $0^\sharp$ is a relatively small large cardinal. Nevertheless, it plays a significant role in the study of definability. According to Slaman (Gödel Lecture, 2001), "$0^\sharp$ is the recursion theorist's passport to set theory". In a rather striking way, the definition of $0^\sharp$ is more involved than those of the large cardinals considered thus far.

Definition 6.4.6. Assume that
(i) for any pair of uncountable cardinals $\eta < \kappa$, $\langle L_\eta, \in\rangle \prec \langle L_\kappa, \in\rangle$;
(ii) there is a unique closed and unbounded class I of ordinals containing all uncountable cardinals such that for any cardinal $\kappa > \omega$,
  (a) $|I \cap \kappa| = \kappa$;
  (b) for any $\psi(v_0, \dots, v_k)$, any two increasing sequences $\gamma_0^0 <, \cdots, < \gamma_k^0$ and $\gamma_1^1 <, \cdots, < \gamma_k^1$ in $I \cap \kappa$, we have $\langle L_\kappa, \in\rangle \models \psi(\gamma_0^0, \cdots, \gamma_k^0) \Leftrightarrow \langle L_\kappa, \in\rangle \models \psi(\gamma_1^1, \cdots, \gamma_k^1)$;
  (c) every element $a \in L_\kappa$ is definable in $\langle L_\kappa, \in\rangle$ with parameters in $I \cap \kappa$.

Then $n \in 0^\#$ if and only if n is the Gödel number of a formula φ such that $\langle L_{\aleph_\omega}, \in\rangle \models \varphi(\aleph_1, \dots, \aleph_{m+1})$.

The term $0^\#$ was introduced by Solovay. Gaifman [32] proved that the existence of $0^\sharp$ follows from the existence of a measurable cardinal. Silver [137] derived the same conclusion assuming the existence of a Ramsey cardinal, known to be weaker than that of a measurable cardinal. Solovay [148] proved that $0^\#$ is a nonconstructible Δ_3^1-set of integers. Since every Δ_2^1-real belongs to L (Corollary 4.2.9 (ii)), this gives Δ_2^1 as the optimal bound for constructibility of reals under a large cardinal assumption.

Though the definition of $0^\sharp$ appears to be undefinable in the language of set theory, it can be reformulated model-theoretically. This enabled Silver to prove that $0^\sharp$ is a Π_2^1-singleton and is hence Δ_3^1. $0^\sharp$ may well be the weakest large cardinal axiom that is incompatible with $V = L$.

Proposition 6.4.7. *If $0^\sharp$ exists, then every constructible real is recursive in $0^\sharp$. Hence there are only countably many constructible reals.*

Proof. Suppose that $0^\sharp$ exists and x is constructible. Since $x \in L_{\omega_1}$, by Definition 6.4.6 (iic) there is a formula $\varphi_x(v, u_1, \dots, u_m)$ and an increasing sequence $i_1, \dots, i_m$ in $I \cap L_{\omega_1}$ such that for any $n < \omega$, $n \in x$ if and only if $\langle L_{\omega_1}, \in\rangle \models \varphi_x(n, i_1, \dots, i_m)$.

For each n, let $\varphi_n(v)$ be a formula defining n so that

$$(\exists v_0)\cdots(\exists v_n)(\forall v)(v \notin v_0 \wedge v_1 = v_0 \cup \{v_0\} \wedge v_2 = v_1 \cup \{v_1\} \wedge \cdots \wedge v_n = v_{n-1} \cup \{v_{n-1}\} \wedge v = v_n).$$

Then

$$n \in x \Leftrightarrow \langle L_{\omega_1}, \in\rangle \models (\exists v)(\varphi_n(v) \wedge \varphi_x(v, i_1, \ldots, i_m)).$$

By Definition 6.4.6 (i) and (ii),

$$n \in x \Leftrightarrow \langle L_{\aleph_\omega}, \in\rangle \models (\exists v)(\varphi_n(v) \wedge \varphi_x(v, \aleph_1, \ldots, \aleph_m)).$$

Clearly the set $\{\ulcorner(\exists v)(\varphi_n(v) \wedge \varphi_x(v, u_1, \ldots, u_m))\urcorner \mid n < \omega\}$ is recursive and

$$n \in x \Leftrightarrow \ulcorner(\exists v)(\varphi_n(v) \wedge \varphi_x(v, u_1, \ldots, u_m))\urcorner \in 0^\sharp.$$

Thus $x \leq_T 0^\sharp$. □

Proposition 6.4.7 implies that $0^\sharp$ has extremely high computational power. Recursion-theoretically, one may view $0^\sharp$ as the "jump" of $\emptyset$ under the #-operator.

While $0^\sharp$ collapses cardinals in L, its absence presents quite a different picture provided by Jensen's remarkable Covering Lemma.

Theorem 6.4.8 (Jensen [21]). *If $0^\sharp$ does not exist then every uncountable set of ordinals is contained in a constructible set of the same cardinality.*

6.5 Axiom of determinacy

In this book our focus is mostly on low level definable sets of reals such as those that are Π^1_1 or Σ^1_1. To go "higher up", one resorts to the use of games and strategies which lie at the heart of the Axiom of Determinacy. It turns out that this axiom is at its core a large cardinal axiom.

6.5.1 Games and strategies

Let $A \subseteq \omega^\omega$. An infinite game G_A with perfect information has two players labeled **I** and **II**. The game is played by choosing a natural number alternately between the two players and ends in ω-many steps. Each game generates a real $x = (n_0, m_0, \ldots, n_i, m_i, \ldots) \in \omega^\omega$ where n_i and m_i are the numbers played by **I** and **II** respectively at step i. If $x \in A$, then **I** wins the game. Otherwise, **II** wins.

Definition 6.5.1. Given a set $A \subseteq \omega^\omega$, **I** has a *winning strategy* $\hat{\sigma} : \omega^{<\omega} \to \omega$ for the game G_A if for every $g \in \omega^\omega$, the real $\hat{\sigma} * g = (n_0, g(0), \ldots, n_i, g(n_i), \ldots) \in A$ where $n_0 = \hat{\sigma}(\emptyset)$ and $n_i = \hat{\sigma}((n_0, g(0), \ldots, n_{i-1}, g(i-1)))$ for $i > 0$. The string

$(n_0, g(0), \dots, n_i)$ is called a play at step i by **I** *following* $\hat{\sigma}$. **II** has a *winning strategy* $\hat{\tau} : \omega^{<\omega} \to \omega$ for the game G_A if for every $f \in \omega^\omega$, the real $f * \hat{\tau} = (f(0), m_0, \dots, f(i), m_i, \dots) \notin A$ where $m_0 = \hat{\tau}(f(0))$ and $m_i = \hat{\tau}((f(0), m_0, , \cdots, f(i)))$ for $i > 0$. The string $(f(0), m_0, \dots, m_i)$ is called a play by **II** at step i following $\hat{\tau}$. G_A is *determined* if either of the two players has a winning strategy.

Definition 6.5.2. The *Axiom of Determinacy* (AD) states that for any $A \subseteq \omega^\omega$, either **I** or **II** has a winning strategy for G_A.

It is not difficult to see that AC contradicts AD. However, G_A is determined whenever A is simply defined.

Proposition 6.5.3. *If A is a closed set, then G_A is determined.*

Proof. Since A is closed, there is a tree $T \subseteq \omega^{<\omega}$ such that $A = [T]$. Suppose that **II** has no winning strategy. We define a **I**-winning strategy $\hat{\sigma}$ as follows:

Let $\hat{\sigma}(\emptyset)$ be an n_0 such that **II** has no winning strategy on $G_{[T_{(n_0)}]}$ where $T_{(n_0)} = \{\tau \in T \mid n_0 \preceq \tau\}$, where n_0 is the string of length 1 whose value at 0 is n_0. Note that n_0 exists by the assumption that **II** has no winning strategy.

By induction assume that $\hat{\sigma}$ is defined for sequences of length up to $i - 1$, and **II** has no winning strategy on $G_{[T_{(n_0, m_0, \cdots, n_{i-1})}]}$ where

$$T_{(n_0, m_0, \cdots, n_{i-1})} = \{\tau \in T \mid (n_0, m_0, \dots, n_{i-1}) \preceq \tau\},$$

for the sequence $(n_0, m_0, \dots, m_i)$ played by **I** following $\hat{\sigma}$. Then there is an n_i such that **II** has no winning strategy on $G_{[T_{(n_0, m_0, \dots, m_i, n_i)}]}$ where

$$T_{(n_0, m_0, \dots, m_i, n_i)} = \{\tau \in T \mid (n_0, m_0, \dots, m_{i-1}, n_i) \preceq \tau\}.$$

Let

$$\hat{\sigma}((n_0, m_0, \dots, m_{i-1})) = n_i.$$

Then for any $g \in \omega^\omega$ and number k, $\hat{\sigma} * g \upharpoonright k \in T$. Hence $\hat{\sigma} * g \in A$ since A is closed. □

It is not always true that if G_A is determined, then so is $G_{\omega^\omega \setminus A}$ (see Exercise 6.5.1). However, the next proposition implies that this complementation property holds when A is closed.

Proposition 6.5.4. *Suppose that $\Gamma \subseteq \mathscr{P}(\omega^\omega)$ so that for any $A \in \Gamma$ and string $\sigma \in \omega^{<\omega}$, the set $A_\sigma = \{x \mid \sigma^\frown x \in A\} \in \Gamma$. If G_A is determined for every $A \in \Gamma$, then so is $G_{\omega^\omega \setminus A}$ for every $A \in \Gamma$.*

Proof. Suppose that $A \in \Gamma$. Note that for any n, $A_n = \{x \mid n^\frown x \in A\} \in \Gamma$ and so G_{A_n} is determined.

Case (1) There exists an n_0 such that player **II** has a winning strategy $\hat{\tau}$ for $G_{A_{n_0}}$. Let $\hat{\sigma}(0) = n_0$. For any string $(n_0, m_0, \dots, m_i)$, let $\hat{\sigma}((n_0, m_0, \dots, m_i))$ be $\hat{\tau}(m_0, \dots, m_i)$. Then $\hat{\sigma}$ is a winning strategy for player **I** for $G_{\omega^\omega \setminus A}$.

Case (2) Otherwise. Then for every n, player **I** has a winning strategy $\hat{\sigma}_n$ for G_{A_n}. Define $\hat{\tau}$ so that for each n, $\hat{\tau}(n) = \hat{\sigma}_n(\emptyset)$ and for any string $(n_0, m_0, \dots, n_i)$, $\hat{\tau}(n_0, m_0 \dots, n_i) = \hat{\sigma}_{n_0}((m_0, \dots, n_i))$. Then $\hat{\tau}$ is a winning strategy for **II** for the game $G_{\omega^\omega \setminus A}$. □

The optimal result on determinacy provable in ZF is due to Martin. Optimality here means that any stronger version is essentially a large cardinal axiom. Let DC denote the Axiom of Dependent Choice:

Theorem 6.5.5 (Martin [93]). ZF + DC ⊢ *"Every Borel set is determined"*.

Theorem 6.5.6 (Martin [92] and Harrington [45]). *Π^1_1-determinacy is equivalent to the existence of* $0^\sharp$.

The status concerning AD is more complex. A detailed discussion would involve the notion of a Woodin cardinal which we will not define. Nonetheless, Woodin, Martin and Steel have shown that ZF + AD is equiconsistent with ZFC+ "There exist infinitely many Woodin cardinals". Woodin also showed the following:

Theorem 6.5.7 (Woodin). *The following are equivalent:*
(i) ZF + DC + *Δ^1_2-Determinacy is consistent;*
(ii) ZF + DC + *Analytical Determinacy is consistent;*
(iii) ZFC + *"There exists a Woodin cardinal" is consistent.*

Woodin as well as Martin and Steel [95] have also provided a level-by-level analysis of determinacy in the projective hierarchy (known as Projective Determinacy).

6.5.2 Exercise

Exercise 6.5.1. *Assume* ZFC. *Prove that there is a set A such that G_A is determined but $G_{\omega^\omega \setminus A}$ is not.*

6.6 Recursion-theoretic aspects of determinacy

The investigation of determinacy from the perspective of recursion theory was initiated by Martin. A natural starting point is to analyze the complexity of winning strategies.

6.6.1 Basis theorems for winning strategies

If A is Δ^1_1, then both the sets $\{\hat{\sigma} \mid (\forall g)(\hat{\sigma} * g \in A)\}$ and $\{\hat{\tau} \mid (\forall f)(f * \hat{\tau} \notin A)\}$ are Π^1_1. By Corollary 4.2.7, there is a Δ^1_2-winning strategy for either **I** or **II**. However, even for Π^0_1-sets, it is not possible to improve the upper bound significantly.

Theorem 6.6.1 (Blass [7]). *Suppose that A is Π^0_1. If **I** has a winning strategy for G_A, then **I** has a winning strategy recursive in $\mathcal{O}$; if **II** has a winning strategy for G_A, the **II** has a hyperarithmetic winning strategy.*

Proof. Since A is Π^0_1, there is a recursive tree $T \subseteq \omega^{<\omega}$ such that $A = [T]$. Suppose that **I** has a winning strategy. Then the set

$$B = \{\hat{\sigma} \mid (\forall i)(\forall \sigma \in \omega^i)((\sigma \text{ is a play at step } i \text{ by } \mathbf{I} \text{ following } \hat{\sigma}) \to (\exists x)(x \succ \sigma \wedge x \in A))\}$$

is a nonempty Σ^1_1-set. By Gandy's Basis Theorem (2.5.3), there is a $\hat{\sigma}_0 \leq_T \mathcal{O}$ in B. Obviously $\hat{\sigma}_0$ is a winning strategy for player **I**.

Now suppose that **I** has no winning strategy. Let $n \in \mathcal{O}^* \setminus \mathcal{O}$. Then $\mathcal{O}^*_n = \{m \mid m \leq_{\mathcal{O}^*} n\}$ is an r.e. set with an initial segment of order type ω_1^{CK} under the r.e. linear ordering relation $<_{\mathcal{O}^*}$. Define an arithmetical set C of functions from $\omega^{<\omega}$ into $\mathcal{O}^*_n$ as follows: $f \in C$ if and only if for any σ,

(i) if $[\sigma] \subseteq [T]$, then $f(\sigma)$ is undefined;

(ii) if $[\sigma] \cap [T] = \emptyset$, then $f(\sigma) = 1$;

(iii) suppose σ satisfies neither (i) nor (ii). (a) If $|\sigma|$ is even, then $f(\sigma) = 2^i$ if $i = \sup\{k \mid k = f(\sigma^\frown j)\}$ exists where sup is in the sense of $<_{\mathcal{O}^*}$; otherwise, $f(\sigma)$ is undefined. (b) If $|\sigma|$ is odd, then $f(\sigma) = 2^i$ if $i = \inf\{k \mid k = f(\sigma^\frown m)\}$ exists where inf is in the sense of $<_{\mathcal{O}^*}$; otherwise, $f(\sigma)$ is undefined.

Since $\mathcal{O} \cap \mathcal{O}^*_n$ is the largest well-ordered segment of $\mathcal{O}^*_n$, for any $f \in C$ and σ, if $f(\sigma)$ is defined then $f(\sigma) \in \mathcal{O}$.

Claim 1. *Suppose σ satisfies (iiia). If for any m, there is a hyperarithmetic partial function $f_m \in C$ such that $f_m(\sigma^\frown m)$ is defined, then there is a hyperarithmetic partial function $f \in C$ such that $f(\sigma)$ is defined.*

Proof of Claim 1. Define a Π^1_1-relation $R \subseteq \omega \times (\omega \times \mathcal{O})$ as follows: $(m, (e, i)) \in R$ if and only if $i \in \mathcal{O}$ and $\Phi_e^{H_i} \in C$ so that $\Phi_e^{H_i}(\sigma^\frown m)$ is defined. By Theorem 4.2.1, there is a Π^1_1-function $g : \omega \to \omega \times \mathcal{O}$ uniformizing R. By the assumption, g is total and hence hyperarithmetic.

Define f to be a hyperarithmetic function such that for any m and $\tau \succeq \sigma^\frown m$, $f(\tau) = \Phi_{m_0}^{H_{m_1}}(\tau)$ where $g(m) = (m_0, m_1)$. Obviously $f \in C$. Moreover, since f is hyperarithmetic, $\sup\{k \mid k = f(\sigma^\frown m)\}$ exists and so $f(\sigma)$ is defined.

Claim 2. *There exists a hyperarithmetic $g \in C$ such that $g(\emptyset)$ is defined.*

Proof of Claim 2. Suppose not. We obtain a winning strategy $\hat{\sigma}$ for **I** playing G_A as follows:

By Claim 1, there is an m such that no hyperarithmetic function $f \in C$ exists for which $f(m)$ is defined. Let $\hat{\sigma}(\emptyset) = m$. Note that by (iiib), for any hyperarithmetic $f \in C$ and number j, $f(m^\frown j)$ is undefined.

By induction on k, for any play $(i_0, j_0, \ldots, i_k)$ at step k by **I** following $\hat{\sigma}$, there is no hyperarithmetic $f \in C$ such that $f((i_0, j_0, \cdots, i_k))$ is defined. Then for any j_k and any hyperarithmetic $f \in C$, $f((i_0, j_0, \ldots, i_k, j_k))$ is undefined. By Claim 1, there is a i_{k+1} for which there is no hyperarithmetic function $f_{i_{k+1}} \in C$ so that $f_{i_{k+1}}((i_0, j_0, \ldots, i_k, j_k, i_{k+1}))$ is defined. Let $\hat{\sigma}((i_0, j_0, \ldots, i_k, j_k) = i_{k+1}$. Then $\hat{\sigma}$ is a winning strategy for **I**, contradicting the assumption.

Thus there is a hyperarithmetic function $g \in C$ such that $g(\emptyset) \in \mathcal{O}$. We define a hyperarithmetic strategy $\hat{\tau}_g$ for **II** as follows: For each n, let $\hat{\tau}_g(n) = m$ where $g(m) <_{o^*} g(\emptyset)$. For each k and $\sigma \in \omega^{<\omega}$ with even length, let $\hat{\tau}_g(\sigma^\frown k) = m$ where $g(\sigma^\frown k^\frown m) <_{o^*} g(\sigma)$ (if it exists).

We claim that $\hat{\tau}_g$ is **II**'s winning strategy. To see this, for any $h \in \omega^\omega$, $g(h * \hat{\tau}_g \upharpoonright 2k+2) <_{o*} g(h * \hat{\tau}_g \upharpoonright 2k)$ for any k with $g(h * \hat{\tau}_g \upharpoonright 2k) \neq 1$. Since $g(\emptyset) \in \mathcal{O}$, there is a k such that $g(h * \hat{\tau}_g \upharpoonright 2k) = 1$. Thus $h * \hat{\tau}_g \in \omega^\omega \setminus A$. □

A general basis theorem for the analytical hierarchy is the following Third Periodicity Theorem due to Moschovakis.

Theorem 6.6.2 (Moschovakis [107]).
(i) *Suppose that every $\underset{\sim}{\Sigma}^1_{2n}$-set is determined. If **I** has a winning strategy for G_A where A is Σ^1_{2n}, then **I** has a Δ^1_{2n+1}-winning strategy.*
(ii) *Suppose that every $\underset{\sim}{\Delta}^1_{2n}$-set is determined. If **I** has a winning strategy for G_A where A is Π^1_{2n+1}, then **I** has a Δ^1_{2n+2}-winning strategy.*

6.6.2 Cones of Turing degrees

A set $\mathbf{A}$ of Turing degrees is a *cone* if there exists a $\mathbf{z}$ such that $\mathbf{A} = \{\mathbf{x} \mid \mathbf{x} \geq \mathbf{z}\}$. We say that $\mathbf{z}$ is the *base* of $\mathbf{A}$.

Proposition 6.6.3. *Assume* ZF + AD. *For any set* $\mathbf{A}$ *of Turing degrees, either* $\mathbf{A}$ *or its complement contains a cone of Turing degrees.*

Proof. Let $A = \{x \mid \deg(x) \in \mathbf{A}\}$. Suppose that **I** has a winning strategy $\hat{\sigma}$ for G_A. Then for any $g \geq_T \hat{\sigma}$, $\hat{\sigma} * g \equiv_T g$. Thus $g \in A$. Hence $\mathbf{A}$ contains a cone of Turing degrees.

If **II** has a winning strategy $\hat{\tau}$ for G_A, then by the same argument, the complement of $\mathbf{A}$ contains a cone of Turing degrees. □

Although the proof of Proposition 6.6.3 looks straightforward, it reveals the remarkable fact that the global structure of the Turing degrees may be sensitive to set-theoretic assumptions.

Definition 6.6.4.
- $A \subseteq \omega^\omega$ is *cofinal* if for every x there is a $y \in A$ such that $y \geq_T x$.
- A tree $T \subseteq \omega^{<\omega}$ is *pointed* if T is perfect and for any $x \in [T]$, $x \geq_T T$.
- A pointed tree T is *uniformly pointed* if there exists an e such that for any $x \in [T]$, $T = \Phi_e^x$.

Proposition 6.6.5 (Martin [91]). ZF + DC + AD *implies that every cofinal set of reals has a uniformly pointed subtree.*

Proof. Suppose that $A \subseteq \omega^\omega$ is cofinal. Let $B \subseteq \omega^\omega$ be such that $z = x \oplus y \in B$ if and only if either
- $x \not\leq_T y$; or
- $x \leq_T y$ and $y \notin A$.

By AD, G_B is determined. Suppose that **I** has a winning strategy $\hat{\sigma}$. Let $y \in A$ so that $y \geq_T \hat{\sigma}$. Then $\hat{\sigma} * y = y \oplus x$ for some $x \leq_T y$. Hence $\hat{\sigma} * y \notin B$.

Thus **II** has a winning strategy $\hat{\tau}$. Let $T_0 \subseteq 2^{<\omega}$ be a uniformly pointed tree such that $T_0 \equiv_T \hat{\tau}$ (see Exercise 6.6.1). Let T_1 be defined such that $(n_0, \dots, n_i) \in T_1$ if and only if there exists a finite string $(m_0, \dots, m_i) \in T_0$ such that $(m_0, n_0, \dots, m_i, n_i)$ is a play by **II** at step i following $\hat{\tau}$. Clearly $T_1 \leq_T \hat{\tau}$. Since $\hat{\tau}$ is **II**'s winning strategy, for any $y \in [T_1]$, there is an $x \in [T_0]$ such that $y = x * \hat{\tau}$ and $\hat{\tau} \leq_T x \leq_T y$. Moreover T_1 is a perfect tree: Every $\tau = (n_0, \dots, n_i) \in T_1$ that is the image of a $\sigma = (m_0, \dots, m_i) \in T_0$ (i.e. a play by **II** of $(m_0, n_0, \dots, m_i, n_i)$ at step i following $\hat{\tau}$) has a pair of (immediate) incompatible extensions τ_1 and τ_2 in T_1 so that there exist $x_1, x_2 \succ \sigma$ with $x_1 * \hat{\tau} \succ \tau_1$ and $x_2 * \hat{\tau} \succ \tau_2$, since otherwise there is a unique y such that for any $x \succ \sigma$ in $[T_0]$, $x \leq_T x * \hat{\tau} = y$ and this would imply that there are uncountably many reals recursive in y, which is a contradiction. We conclude that T_1 is a pointed subtree of A.

We follow the idea in [58] to construct a subtree T_2 of T_1 so that T_2 is uniformly pointed. Define a sequence of trees $T_1 = T_1^{-1} \supseteq T_1^0 \supseteq T_1^1 \dots$ which are recursive in T_1 as follows:

Stage e:

Case (1) There is an n such that $\Phi_e^\sigma(n) \neq T_1(n)$ for some $\sigma \in T_1^{e-1}$. Let $T_1^e = \{\tau \mid \tau \in T_1^{e-1} \wedge (\tau \succeq \sigma \vee \tau \preceq \sigma)\}$ and proceed to the next stage.

Case (2) Otherwise.

Subcase (i) $(\forall n)(\forall \sigma \in T_1^{e-1})(\exists \tau \succ \sigma)(\tau \in T_1^{e-1} \wedge \Phi_e^\tau(n)[|\tau|] \downarrow)$. Then it is not difficult to define a T_1-recursive perfect tree $T_1^e \subseteq T_1^{e-1}$ such that for every $x \in [T_1^e]$, we have $\Phi_e^x = T_1$. In this case, terminate the construction since $T_1^e \leq_T T_1$ and T_1^e is a uniformly pointed tree.

Subcase (ii) Otherwise. Then there exist n_0 and $\sigma_0 \in T_1^{e-1}$ such that for every $\tau \succeq \sigma_0$ in T_1^{e-1}, $\Phi_e^\tau(n)[|\tau|] \uparrow$. Let $T_1^e = \{\tau \mid \tau \in T_1^{e-1} \wedge (\tau \succeq \sigma_0 \vee \tau \preceq \sigma)\}$. Proceed to the next stage.

If the construction does not terminate at any stage e, then there is a real $x \in \bigcap_e [T_1^e]$. By the definition of the sequence $\{T_1^e\}$, we have $\Phi_e^x \neq T_1$, which contradicts the fact that T_1 is a pointed tree. Thus the construction must terminate at some stage e where Subcase (i) holds. Let $T_2 = T_1^e$. □

6.6.3 Exercise

Exercise 6.6.1. *Prove that for any real z, there is a uniformly pointed tree* $T \subseteq 2^{<\omega}$ *such that* $T \equiv_T z$.

6.6.4 Turing determinacy

Definition 6.6.6.

- *Turing determinacy* (TD) states that any set of Turing degrees either contains or is disjoint from a cone.
- If Γ is a class of sets of Turing degrees, then Γ*-Turing determinacy* states that every set in Γ either contains or is disjoint from a cone.

Turing determinacy plays a significant role in descriptive set theory, as shown by the following theorems which are stated without proof:

Theorem 6.6.7 (Harrington [45]). ZFC+$\utilde{\Sigma}_1^1$*-determinacy is consistent if and only if* ZFC+ $\utilde{\Sigma}_1^1$*-Turing determinacy is consistent.*

Theorem 6.6.8 (Woodin).

(i) *For any* $n \geq 2$, ZFC + $\utilde{\Sigma}_n^1$*-determinacy is consistent if and only if* ZFC + $\utilde{\Sigma}_n^1$*-Turing determinacy is consistent.*

(ii) ZF + DC + AD *is consistent if and only if* ZF + DC + TD *is consistent.*

6.6.5 Martin's conjecture

A function $F : 2^\omega \to 2^\omega$ is *degree invariant* on a cone (of reals) $C = \{x \mid x \geq_T z\}$ for some z if any $x, y \geq_T z$ in C of the same Turing degree satisfy $F(x) \equiv_T F(y)$. F is *increasing* on a cone C if $F(x) \geq_T x$ for $x \in C$. If F and G are degree invariant on a cone, write $F \leq_M G$ if $F(x) \leq_T G(x)$ for all $x \in C$. Martin made the following conjecture.

Conjecture 6.6.9. *Assume* ZF + AD + DC.

1. *Every degree invariant function that is not increasing on a cone is a constant on a cone;*
2. *$\le_M$ prewellorders the degree invariant functions which are increasing on a cone. Furthermore, if the $\le_M$-rank of F is α, then F' has $\le_M$-rank $\alpha + 1$, where $F'(x) = (F(x))'$, the Turing jump of $F(x)$.*

Martin's conjecture is one of the deepest conjectures in recursion theory. Essentially it says that the only "natural unsolvable problem" is the halting problem. We will take up this topic in detail in Chapter 7.

6.6.6 Exercises

Exercise 6.6.2 (Blass [7]). *Prove that for any hyperarithmetic real x, there is a Π^0_1-set A such that **II** has a winning strategy for G_A but has no winning strategy recursive in x.*

Exercise 6.6.3. *Suppose that A is a Π^1_1-set such that for any Turing degree $\mathbf{x}$, $\mathbf{x} \cap A \neq \emptyset$. Then there is a pointed tree T so that $[T] \subseteq A$.*

Exercise 6.6.4 (Blass [7]). *Prove that there is a Π^0_1-set A such that **I** has a winning strategy but has no hyperarithmetic winning strategy for G_A.*

Exercise 6.6.5. *Prove that there exists a Turing degree $\mathbf{a}$ such that*

$$\langle \mathscr{D}(\le \mathbf{a}), \le \rangle \equiv \langle \mathscr{D}(\le \mathbf{b}), \le \rangle$$

for any Turing degree $\mathbf{b} \ge \mathbf{a}$, where $\mathscr{D}(\le \mathbf{a}) = \{\mathbf{c} \mid \mathbf{c} \le \mathbf{a}\}$.

Exercise 6.6.6. *Assume ZF + PD. Prove that there exists a Turing degree $\mathbf{a}$ such that $\langle \mathscr{D}(\ge \mathbf{a}), \le \rangle \equiv \langle \mathscr{D}(\ge \mathbf{b}), \le \rangle$ for any Turing degree $\mathbf{b} \ge \mathbf{a}$, where $\mathscr{D}(\ge \mathbf{a}) = \{\mathbf{c} \mid \mathbf{c} \ge \mathbf{a}\}$.*

Exercise 6.6.7 (Jockusch and Soare [59]). *Prove that $\langle \mathscr{A}, \le \rangle \not\equiv \langle \mathscr{D}, \le \rangle$ where $\mathscr{A}$ is the set of Turing degrees of arithmetical sets.*

Exercise 6.6.8. *Assume ZFC. Prove that there is a sequence $\{f_n\}_{n<\omega}$ of functions such that $f_{n+1} <_M f_n$ for all n.*

Exercise 6.6.9 (Simpson [138]). *Assume ZF + Π^1_1-determinacy. Prove that there is an x such that for any $y \ge_h x$, the hyperdegree of y is a minimal cover.*

***Exercise 6.6.10** (Harrington and Kechris [46]). *Let $P \subseteq 2^\omega$ be a Π^0_1-set that is cofinal with respect to the Turing degrees of hyperarithmetic sets, i.e. for any $x \in$ HYP there is a $y \in P$ such that $y \ge_T x$. Prove then any $x \ge_T \mathcal{O}$ is Turing equivalent to a $y \in P$.*

Part II: **The story of Turing degrees**

7 Classification of jump operators

The Turing jump operator $x \mapsto x'$ is degree invariant, i.e. if $x \equiv_T y$ then $x' \equiv_T y'$. The property of degree invariance is a fundamental property of the operator that motivated the introduction of Martin's conjecture (§ 6.6.5). In this chapter we discuss a stronger one called *uniform degree invariance* and give a classification of functions that satisfy this property.

7.1 Uniformly degree invariant functions

Definition 7.1.1.

- If $e = \langle e_0, e_1 \rangle$, we say that $x \equiv_T y$ *via* e if $x = \Phi^y_{e_0}$ and $y = \Phi^x_{e_1}$;
- a function $F : 2^\omega \to 2^\omega$ is *uniformly degree invariant* on a cone $C = \{x \mid x \geq_T z\}$ with base z if there is a function $t : \omega \to \omega$ such that for any $x, y \geq_T z$ with $x \equiv_T y$ via e, we have $F(x) \equiv_T F(y)$ via $t(e)$.

Note that in the definition above, the value of $t(e)$ is independent of the reals x, y.

In 1967, Sacks made a conjecture on the existence of a degree invariant function $x \mapsto W^x$ for a general solution of Post's problem. This conjecture contradicts Martin's conjecture since the latter asserts that any such function must be at least as strong as the jump operator:

Conjecture 7.1.2 (Sacks[119]). *There is a Σ^0_1-predicate $W \subseteq 2^\omega \times \omega$ such that, if $W^x = \{n \mid W(x, n)\}$, then $x <_T W^x < x'$ and for any $y \equiv_T x$, $W^x \equiv_T W^y$.*

In 1975, Lachlan [80] proved that this conjecture failed for uniformly degree invariant W.

7.1.1 Nonincreasing uniformly degree invariant functions

Uniformly degree invariant functions are well-behaved under ZF + DC + AD. The following theorem says that a uniformly degree invariant function is either a constant or increasing on a cone.

Theorem 7.1.3 (Slaman and Steel [142]). *Assume* ZF + DC + AD. *If F is uniformly degree invariant and not increasing on a cone, then F is a constant on a cone.*

We decompose the proof of Theorem 7.1.3 into a series of lemmas. Let F be uniformly degree invariant via t and not increasing on a cone.

Lemma 7.1.4. *Assume* ZF+DC+AD. *If F is uniformly degree invariant and nonincreasing on a cone, then $F(x) <_T x$ on a cone.*

Proof. Suppose that F is uniformly degree invariant and not increasing on a cone. Then the set $\{x \mid F(x) \not\geq_T x\}$ is cofinal in the Turing degrees. By Proposition 6.6.5, there is a pointed tree T such that F is uniformly degree invariant on T via a function $t : \omega \to \omega$ and $F(x) \not\geq_T x$ for any $x \in [T]$.

Given x, let $x^+(i) = x(i+1)$ for $i < \omega$. Let $A \subseteq \omega^\omega$ be such that $x = x_0 \oplus x_1 \in A$ if and only if either $x_1^+ \notin 2^\omega$ or

- $x_0^+ \in 2^\omega$;
- $x_0(0) = \langle e, n\rangle$ for some $e, n < \omega$;
- either $x_0 \not\leq_T x_1$ or $F(x_0^+)(n) = 0 \Leftrightarrow x_1(x_1(0)) \neq 0$ and
- $x_0^+ \in [T] \wedge \Phi_e^{x_0^+} = x_1$.

Consider a two person game $\mathcal{G}_A$ played by I and II, choosing numbers alternately as usual. By AD, either I or II has a winning strategy for $\mathcal{G}_A$. Suppose that I has a winning strategy $\hat{\sigma}$. We show that $x \leq_T F(x)$ on the cone above $T \oplus \hat{\sigma}$, which will give a contradiction.

Let $z_0 \geq_t T \oplus \hat{\sigma}$ and $x_1^m = m^\frown z_0$ for $m < \omega$. Let x_0^m be the real in $[T]$ played by I in response to II playing x_1^m, i.e. $\hat{\sigma}(x_1^m) = \langle e_0, n_0\rangle^\frown x_0^m$, where $\langle e_0, n_0\rangle$ codes the pair of numbers e_0, n_0 and $\hat{\sigma}(\emptyset) = \langle e_0, n_0\rangle$ is the first move dictated by $\hat{\sigma}$. Since I wins, $x^m = x_0^m \oplus x_1^m \in A$ for every m. As $x_1^m \geq_T \hat{\sigma}$, we have $x_1^m \geq_T x_0^m$. Then for every m,

$$F((x_0^m)^+)(n_0) = 0 \Leftrightarrow x_1^m(m) = 1.$$

Since $\Phi_{e_0}^{(x_0^m)^+} = x_1^m$ and $x_1^m = m^\frown z_0$, there is an i such that $\Phi_i^{(x_0^m)^+} = (x_0^{m+1})^+$ for every m. Similarly, there is a j such that $\Phi_j^{(x_0^{m+1})^+} = (x_0^m)^+$ for every m. Hence for each m, $(x_0^m)^+ \equiv_T (x_0^{m+1})^+$ via $\langle i, j\rangle$, and so $F((x_0^m)^+) \equiv_T F((x_0^{m+1})^+)$ via $t(i,j) = \langle i', j'\rangle$. Since F is degree invariant and $z_0 \equiv_T x_1^0 \equiv_T (x_0^0)^+$, we have $\Phi_k^{F(z_0)} = F((x_0^0)^+)$ for some k. Let $l_m = F((x_0^m)^+)(n_0)$. Note that $F((x_0^0)^+) = \Phi_k^{F(z_0)}$ and $F((x_0^{m+1})^+) = \Phi_{i'}^{F((x_0^m)^+)}$. So $\{l_m\}_{m\in\omega}$ is $F(z_0)$-recursive. Note that $z_0(m) = x_1^{m+1}(m+1) = 1 - l_{m+1}$. Thus $z_0 \leq_T F(z_0)$, a contradiction.

Hence II has a winning strategy $\hat{\tau}$. We show that $F(x) <_T x$ on the cone above $T \oplus \hat{\tau}$. Fix a real $x_0 \in [T]$ so that $x_0 \geq_T T \oplus \hat{\tau}$. By the Recursion Theorem relativized to x_0, there is an e_0 such that $\hat{\tau}(\langle e_0, n\rangle^\frown x_0)(m) = \Phi_{e_0}^{x_0}(\langle n, m\rangle)$ for every n, m. For each n, let $x_1^n = \hat{\tau}(\langle e_0, n\rangle^\frown x_0)$. Since II wins, it must be the case that $x_0 \leq_T x_1^n$ and $F(x_0^+)(n) = x_1^n(x_1^n(0))$. Then

$$F(x_0^+)(n) = x_1^n(x_1^n(0)) = \Phi_{e_0}^{x_0}(\langle n, \Phi_{e_0}^{x_0}(\langle n, 0\rangle)\rangle)$$

and so $F(x_0) \leq_T x_0$. Since F is degree invariant, we conclude that $F(x) <_T x$ on a cone. □

By Lemma 7.1.4, fix x_0 such that $F(x) <_T x$ on the cone $\{x \mid x \geq_T x_0\}$. Then there is an e_0 such that the set $C = \{x \mid \Phi_{e_0}^x = F(x) \wedge x \geq_T x_0\}$ is cofinal. By Proposition 6.6.5, there is a perfect tree T_0 and a number i such that $[T_0] \subseteq C$ and $T_0 = \Phi_i^x$ for every $x \in [T_0]$.

Lemma 7.1.5. *There exists a perfect tree $T_1 \subseteq T_0$ such that $T_1 \leq_T T_0$ and F is either injective or a constant on $[T_1]$.*

Proof. If there exists a $\sigma \in T_0$ such that $\Phi_{e_0}^x = \Phi_{e_0}^y$ for all $x, y \in [T_0] \cap [\sigma]$, then let $T_1 = T_0 \cap \{\tau \in 2^{<\omega} \mid \tau \succ \sigma\}$. Obviously F is a constant on $[T_1]$. Otherwise, T_0-recursively finds a perfect subtree $T_1 \subseteq T_0$ such that F is injective on $[T_1]$. □

By Lemma 7.1.5, we may assume that F is injective on T_0.

Let

$$S_0 = \{\tau \in 2^{<\omega} \mid (\exists \sigma \in T_0)(\forall n \leq |\tau|)(\tau(n) = \Phi_{e_0}^{\sigma}(n))\}.$$

Then S_0 is a T_0-recursive perfect tree so that Φ_{e_0} is a homeomorphism from $[T_0]$ to $[S_0]$.

Since T_0 is perfect, let $p : 2^{<\omega} \to T_0$ be an order-preserving, T_0-recursive bijection. Then p may be viewed as a homeomorphism from 2^ω to $[T_0]$. Let $\pi : 2^{<\omega} \to 2^{<\omega}$ be a recursive function such that for any σ and $i < 2$, $\pi(\sigma^\frown i) = \pi(\sigma)^\frown i$ if $\sigma \neq \emptyset$ and $\sigma(j) = 1$ for some $j < |\sigma|$; else $\pi(\sigma^\frown i) = \sigma^\frown(1-i)$. Then π can be viewed as a homeomorphism from 2^ω to 2^ω by letting $\pi(x) = \bigcup_n \pi(x \upharpoonright n)$. Let

$$p_0 = p \circ \pi \circ p^{-1}.$$

Then p_0 is a homeomorphism from T_0 to T_0. We leave it to the reader to verify the following lemma.

Lemma 7.1.6. *For any σ, $\{\pi^{(n)}(\sigma)\}_{n<2^{|\sigma|}}$ is a permutation of $2^{|\sigma|}$.*

Let $p_1 = F \circ p_0 \circ F^{-1}$ be a homeomorphism from $[S_0]$ to $[S_0]$ and fix an $x_0 \in [T_0]$. Since T_0 is uniformly pointed, there is a pair (i_0, j_0) such that for any n, $p_0^{(n)}(x_0) \equiv_T p_0^{(n+1)}(x_0)$ via (i_0, j_0). Then $F(p_0^{(n)}(x_0)) \equiv_T F(p_0^{(n+1)}(x_0))$ via $(i_1, j_1) = t(i_0, j_0)$. Note that for any n, $F(p_0^{(n)}(x_0)) = p_1^{(n)}(F(x_0))$. So the real $\oplus_{n<\omega} p_1^{(n)}(F(x_0))$ is recursive in $p_1^{(0)}(x_0) = F(x_0)$.

Given a perfect tree T and a number n, we say that a finite tree $S \subseteq T$ *codes a 2^n copy of T* if

- S is isomorphic to 2^n;
- for any $\tau \in T \setminus S$, there exists an end node $\sigma \in S$ such that $\tau \succ \sigma$.

Lemma 7.1.7. *S_0 is recursive in $F(x_0)$.*

Proof. Note that we may view $F : [T_0] \to [S_0]$ as a homeomorphism. Hence for any σ, there is an m such that $p^{-1}(F^{-1}((F(x_0) \upharpoonright m)))$ is of length $2^{|\sigma|}$. By Lemma 7.1.6, $\{\pi^{(n)}(p^{-1}(F^{-1}((F(x_0) \upharpoonright m))))\}_{n<2^{|\sigma|}}$ is a permutation of $2^{|\sigma|}$. Since both F and p are homeomorphisms, the finite tree

$$S_0^{|\sigma|} = \{\tau \mid (\exists n < 2^{|\sigma|})(\tau \preceq p_1^{(n)}(F(x_0) \upharpoonright m)\}$$

codes a $2^{|\sigma|}$ copy of S_0.

Thus $\sigma \in S_0$ if and only if $\sigma \in S_0^{|\sigma|}$. Obviously $S_0^{|\sigma|}$ can be effectively computed from $\{p_1^{(n)}(F(x_0))\}_{n<2^{|\sigma|}}$. Hence $S_0 \leq_T F(x_0)$. □

This shows that S_0 is a pointed tree. Then there exists an $x_1 \in [S_0]$ such that $T_0 \oplus S_0 \leq_T x_1$. Since Φ_{e_0} is a homeomorphism from T_0 to S_0, it is not difficult to see that the unique real $y \in [T_0]$ such that $\Phi^y_{e_0} = x_1$ is recursive in x_1. In other words, $F(y) = x_1 \geq_T y$, a contradiction.

7.1.2 Exercise

Exercise 7.1.1. *Prove Lemma 7.1.6.*

7.2 Martin's conjecture for uniformly degree invariant functions

The results discussed in this section are due to Steel [159].

Lemma 7.2.1. *Assume* ZF + AD + DC. *Then* $\leq_M$ *is a linear ordering on the collection of functions which are increasing and uniformly degree invariant.*

Proof. Suppose that F and G are increasing uniformly degree invariant functions via t on a cone $\{x \mid x \geq_T z_0\}$. Define a set $A \subseteq \omega^\omega$ so that $x_0 \oplus x_1 \in A$ if and only if either $x_1^+ \notin 2^\omega$ or

- $x_0^+ \in 2^\omega$;
- $\Phi^{x_0^+}_{e_0} = x_1^+$ where $x_0(0) = \langle e_0, n_0\rangle$, and
- either $\Phi^{x_1^+}_{e_1} \neq x_0^+$ or $n_0 \in F(x_0^+) \Leftrightarrow n_1 \notin G(x_1^+)$ where $x_1(0) = \langle e_1, n_1\rangle$.

Consider a two person game $\mathcal{G}_A$ played by I and II on A, with I playing $\langle e_0, n_0\rangle$ as the first move. Suppose that I has a winning strategy $\hat{\sigma}$ for $\mathcal{G}_A$. Fix a real $y \geq_T \hat{\sigma} \oplus z_0 \oplus t$. By the Recursion Theorem relativized to y, for any number m, there is an e_1^m such that $\Phi^y_{e_1^m} = (\hat{\sigma}(x_1^m))^+$ where $x_1^m = (\langle e_1^m, m\rangle)^\frown y$. Note that the function $m \mapsto e_1^m$ is recursive. Let $x_0^m = \hat{\sigma}(x_1^m)$ and $x_0^m(0) = \langle e_0^m, n_0^m\rangle$. Note that the function $m \mapsto \langle e_0^m, n_0^m\rangle$ is y-recursive (actually a constant function). Since $\hat{\sigma}$ is a winning strategy for I, $\Phi^{(x_0^m)^+}_{e_0^m} = y$ and

$$n_0^m \in F((x_0^m)^+) \Leftrightarrow m \notin G(x_1^-) = G(y).$$

Note that $(x_0^m)^+ \equiv_T y$ via (e_0^m, e_1^m), and so we have $F((x_0^m)^+) \equiv_T F(y)$ via $t(e_0^m, e_1^m) = (i_0^m, i_1^m)$. Since $F(y) \geq_T y$, $t \leq_T y$ and $m \mapsto (e_0^m, e_1^m)$ is y-recursive, the function $m \mapsto (i_0^m, i_1^m)$ is $F(y)$-recursive. Then for any m, we may $F(y)$-recursively find n_0^m and i_1^m such that $\Phi^{F(y)}_{i_1^m}(n_0^m) = F((x_0^m)^+)(n_0^m) = 1 - G(y)(m)$. Hence $G(y) \leq_T F(y)$ and so $G \leq_M F$.

If II has a winning strategy, then by a similar argument one shows that $F \leq_M G$. □

Given x and y, we say that $x \equiv_m y$ via (e_0, e_1) if there are two recursive functions Φ_{e_0} and Φ_{e_1} such that for any n,

- $n \in y \Leftrightarrow \Phi_{e_0}(n) \in x$; and
- $n \in x \Leftrightarrow \Phi_{e_1}(n) \in y$.

Lemma 7.2.2. *Assume* ZF+AD+DC. *If F and G are increasing uniformly degree invariant functions via t such that $F <_M G$, then $F' \leq_M G$.*

Proof. Suppose that $F <_M G$ on the cone $\{x \mid x \geq_T z_0\}$ but $F' \not\leq_M G$. Let $p : 2^{<\omega} \to \omega$ be a recursive bijection and for any x, let $G_1(x) = y$ where

$$y = \{n \mid (\forall m)(p(G(x) \upharpoonright m) \neq n)\}.$$

Then for each x, $G_1(x) \equiv_T G(x)$. Moreover, since G is uniformly degree invariant, so is G_1.

We apply the same method as that in Lemma 7.2.1. Define $A \subseteq \omega^\omega$ such that $x_0 \oplus x_1 \in A$ if and only if either $x_1^+ \notin 2^\omega$ or

- $x_0^+ \in 2^\omega$;
- $\Phi_{e_0}^{x_0^+} = x_1^+$ where $x_0(0) = \langle e_0, n_0\rangle$, and
- either $\Phi_{e_1}^{x_1^+} \neq x_0^+$ or $n_0 \in F'(x_0^+) \Leftrightarrow n_1 \notin G_1(x_1^+)$, where $x_1(0) = \langle e_1, n_1\rangle$.

A game $\mathcal{G}_A$ is played by the two players I and II. As in Lemma 7.2.1, if II has a winning strategy for $\mathcal{G}_A$, then $F' \leq_M G_1$ which is a contradiction.

Hence I has a winning strategy $\hat{\sigma}$ for $\mathcal{G}_A$. Fix $y \geq_T \hat{\sigma} \oplus z_0 \oplus t$. By the Recursion Theorem relativized to y, for each m there is an e_1^m such that $\Phi_{e_1^m}^y = (\hat{\sigma}(x_1^m))^+$ where $x_1^m = \langle e_1^m, m\rangle^\frown y$. Note that the function $m \mapsto e_1^m$ is recursive. Let $x_0^m = \hat{\sigma}(x_1^m)$ and $x_0^m(0) = \langle e_0^m, n_0^m\rangle$. The function $m \mapsto \langle e_0^m, n_0^m\rangle$ is y-recursive. Since $\hat{\sigma}$ is a winning strategy for I, $\Phi_{e_0^m}^{(x_0^m)^+} = y$ and

$$n_0^m \in F'((x_0^m)^+) \Leftrightarrow m \notin G_1(x_1^-) = G_1(y).$$

Note that $(x_0^m)^+ \equiv_T y$ via (e_0^m, e_1^m). We then have $F((x_0^m)^+) \equiv_T F(y)$ via $t(e_0^m, e_1^m) = (i_0^m, i_1^m)$. Since $t \leq_T y$ and $m \mapsto (e_0^m, e_1^m)$ is y-recursive, the function $m \mapsto (i_0^m, i_1^m)$ is y-recursive as well. Hence there is a y-recursive function $m \mapsto (j_0^m, j_1^m)$ such that $F'((x_0^m)^+) \equiv_m F'(y)$ via (j_0^m, j_1^m). Thus for any m, one may y-recursively find n_0^m and an j_1^m such that

$$F'(y)(\Phi_{j_1^m}(n_0^m)) = F'((x_0^m)^+)(n_0^m) = 1 - G_1(y)(m).$$

Since $F(y) \geq_T y$, the set $E_1 = \{m \mid F'(y)(\Phi_{j_1^m}(n_0^m)) = 1\}$ is infinite and $F(y)$-r.e. Let $E_2 \subseteq E_1$ be infinite $F(y)$-recursive. Then for any $m \in E_2$, there is an n_m such that $p^{-1}(m) = G_1(y) \upharpoonright n_m$. Thus $G(y) \leq_T F(y)$ and so $G \leq_M F$, contradicting the hypothesis. We conclude that $F' \leq_M G$. □

Now suppose that there is a $\leq_M$-sequence $\{F_n\}_{n<\omega}$ such that for each n,
(a) F_n is uniformly degree invariant via t_n, and
(b) $F_{n+1} <_M F_n$ on the cone $\{x \mid x \geq_T z_n\}$.

Let $z = \oplus_{n<\omega} z_n$ and $t(n, e_0, e_1) = t_n(e_0, e_1)$. For each n, we define sets A_n and B_n for the pair (F_{n+1}, F_n) to be the corresponding set A in Lemma 7.2.1 and Lemma 7.2.2 respectively. Then for every n, I has a winning strategy $\hat{\sigma}_n$ for $\mathcal{G}_{A_n}$ and II has a winning strategy $\hat{\tau}_n$ for $\mathcal{G}_{B_n}$. Let $y = z \oplus t \oplus (\oplus_{n<\omega}\hat{\sigma}_n) \oplus (\oplus_{n\in\omega}\hat{\tau}_n)$. Then as in the proof of Lemma 7.2.1, there is a y-arithmetical relation $P(x_0, x_1)$ such that for any n, $F_{n+1}(y)$ is the unique real x for which $P(x, F_n(y))$ holds. Note that by Lemma 7.2.2, $F'_{n+1}(y) \leq_T F_n(y)$ for every n.

Proposition 7.2.3 (Steel [157]). *Let $R \subseteq 2^\omega \times 2^\omega$ be an arithmetical relation. If $\{x_n\}_{n<\omega}$ is a sequence of reals such that x_{n+1} is the unique y satisfying $R(y, x_n)$, then for some n, $x'_{n+1} \not\leq_T x_n$.*

Proof. Suppose for the sake of contradiction that $x'_{n+1} \leq_T x_n$ for each n. Define

$$P(x, i, j) \Leftrightarrow (\exists k > j+1)(\exists \sigma \in \omega^{<\omega})(|\sigma| = j+1 \wedge \Phi^x_{\sigma(0)} = x \wedge (\forall k' < k) R(\Phi^x_{\sigma(k'+1)}, \Phi^x_{\sigma(k')}) \wedge (\exists e)(\Phi^x_e(i) = 0 \wedge \Phi^x_e = (\Phi^x_{\sigma(k)})^{(j+1)})).$$

By assumption on the sequence $\{x_n\}_{n<\omega}$, for any n we have

$$(\forall i)(\forall j)(P(x_n, i, j) \Leftrightarrow (\exists k > j)(i \notin (x_{n+k})^{(j+1)})).$$

Since P is arithmetical, there exist m and e such that

$$P(x, i, j) \Leftrightarrow \Phi^{x^{(m)}}_e(i, j) = 1.$$

Then for any n, i, j,

$$\Phi^{(x_n)^{(m)}}_e(i, j) = 1 \Leftrightarrow (\exists k > j+1)(i \notin (x_{n+k})^{(j+1)}).$$

Setting $j = m$, there is an e_0 such that for every i,

$$\Phi^{(x_n)^{(m)}}_{e_0}(i) = 1 \Leftrightarrow \Phi^{(x_n)^{(m)}}_e(i, m) = 1 \Leftrightarrow (\exists k > m+1)(i \notin (x_{n+k})^{(m+1)}).$$

When $i = e_0$, we have for any n,

$$e_0 \in (x_n)^{(m+1)} \Leftrightarrow (\exists k > m+1)(e_0 \notin (x_{n+k})^{(m+1)}).$$

Thus $e_0 \in (x_0)^{(m+1)}$ if and only if $(\exists k_0 > m+1)(e_0 \notin (x_{k_0})^{(m+1)})$, and $e_0 \in (x_{k_0})^{(m+1)}$ if and only if $(\exists k_1 > m+1)(e_0 \notin (x_{k_0+k_1})^{(m+1)})$. Then $e_0 \in (x_0)^{(m+1)}$ if and only if $e_0 \in (x_{k_0+k_1})^{(m+1)}$. By the equivalence above again, $e_0 \in (x_0)^{(m+1)}$ if and only if $e_0 \notin (x_0)^{(m+1)}$, a contradiction. □

Relativizing Proposition 7.2.3 to y, there is an n such that $F_{n+1}(y)' \not\leq_T F_n(y)$, which contradicts our assumption on the sequence $\{F_n\}_{n<\omega}$ and Lemma 7.2.2. We therefore have:

Lemma 7.2.4. *Assume ZF+AD+DC, the partial order $\leq_M$ on increasing uniformly degree invariant functions is well-ordering.*

In conclusion, we have the following theorem.

Theorem 7.2.5 (Steel [159], Slaman and Steel [142]). *Assume* ZF + AD + DC, *Martin's conjecture (6.6.9) holds for uniformly degree invariant functions.*

Remark 7.2.6. H. Friedman [30] proved that there is a hyperarithmetic predicate R and a sequence $\{x_n\}_{n<\omega}$ such that $x'_{n+1} \leq_T x_n$ and x_{n+1} is the unique y satisfying $R(y, x_n)$.

7.2.1 Exercises

Exercise 7.2.1. *Prove that there exists an arithmetical predicate P such that for any $n > 0$, there exists a sequence $\{x_j\}_{i\leq n}$ satisfying:* (i) *for any $i < n$, $x'_{i+1} \leq_T x_i$,* (ii) *$P(x_{i+1}, x_i)$ and* (iii) *if $P(y, x_i)$ then $y = x_{i+1}$.*

Exercise 7.2.2. *Prove that there is an arithmetical predicate R and a $<_h$-descending sequence $\{x_n\}_{n<\omega}$ such that for any n, x_{n+1} is the unique real y satisfying $R(y, x_n)$.*

7.3 The Posner–Robinson Theorem

7.3.1 The original Posner–Robinson Theorem

Theorem 7.3.1 (Posner and Robinson [114]). *For any nonrecursive $x \leq_T \emptyset'$, there is a g such that $x \oplus g \equiv_T g' \equiv_T \emptyset'$.*

Thus given a nonrecursive Δ^0_2-real x, Theorem 7.3.1 says that there is a g relative to which x is its Turing jump.

Proof. To prove this, we use a method due to Jockusch and Shore [57]. Fix $x >_T \emptyset$ to be Δ^0_2. We may assume that x is not r.e. For every e, consider the following statement:

$$(\forall \sigma \in 2^{<\omega})(\exists n)[(n \in x \wedge (\forall \tau \succeq \sigma^\frown 0^n 1)(\Phi^\tau_e(e)[|\tau|] \uparrow)) \vee (n \notin x \wedge (\exists \tau \succeq \sigma^\frown 0^n 1)(\Phi^\tau_e(e)[|\tau|] \downarrow)]. \qquad (\sharp_e)$$

Since x is not r.e., $(\sharp_e)$ is true for every e. We construct g by induction.

Step 0: Define $g_0 = \emptyset$.

Step $s + 1$:

Substep 1. Define $g(s, 1) = g_s^\frown \emptyset'(s)$.

Substep 2. Find the least n such that either

$$(n \in x \wedge \forall \tau \succeq g(s, 1)^\frown 0^n 1(\Phi^\tau_s(s)[|\tau|] \uparrow))$$

or

$$(n \notin x \wedge (\exists \tau \succeq g(s, 1)^\frown 0^n 1)(\Phi^\tau_s(s)[|\tau|] \downarrow)).$$

By ($\sharp_s$), such an n exists. In the first case, define $g_{s+1} = g(s,1)^\frown 0^n 1$. In the second case, select the lexicographically least τ such that $\tau \succeq g(s,1)^\frown 0^n 1(\Phi_s^\tau(s)[|\tau|]\downarrow)$. Define $g_{s+1} = g(s,1)^\frown \tau$.

Let $g = \bigcup_{s<\omega} g_s$. Note that the construction is uniformly $x \oplus \emptyset'$-recursive. Hence $\emptyset' \geq_T g \oplus x$. We show that $\emptyset'(s)$ and g_s are uniformly $g \oplus x$-recursive. Suppose we have computed g_s and $\emptyset'(s)$. Then $\emptyset'(s+1) = g(|g_s|+1)$. To compute g_{s+1}, search for the unique n such that $(g \upharpoonright |g_s|+1)^\frown 0^n 1 \prec g$. By the construction, either

(1) $n \in x$ or

(2) $n \notin x$ and there is a $\tau \succeq (g \upharpoonright |g_s|+1)^\frown 0^n 1$ such that $\tau \prec g$ and $\Phi_s^\tau(s)[|\tau|]\downarrow$.

If (1) holds, then $g(s+1) = (g \upharpoonright |g_s|+1)^\frown 0^n 1$. If (2) holds, then there is a unique $\tau \succeq (g \upharpoonright |g_s|+1)^\frown 0^n 1$ such that $\tau \prec g$ and $\Phi_s^\tau(s)[|\tau|]\downarrow$. It follows that $g_{s+1} = \tau$.

Thus $\emptyset' \equiv_T g \oplus x$.

To prove $g' \leq_T g \oplus \emptyset'$, observe that for every e there is an n such that either $\Phi_e^{g\upharpoonright n}(e)\downarrow$ or $\Phi_e^\tau(e)\uparrow$ for any $\tau \succ g \upharpoonright n$. Using this, one verifies immediately that $g' \leq_T g \oplus \emptyset'$. □

7.3.2 A transfinite Posner–Robinson Theorem

Theorem 7.3.1 may be iterated into the transfinite in several different ways. Here is the first version.

Theorem 7.3.2 (Slaman and Steel [142]). *Let $0 < \beta < \omega_1^{\mathrm{CK}}$ and let z be a real such that $\emptyset^{(\gamma)} <_T z$ for each $\gamma < \beta$. Then there is a g such that $g \oplus z \equiv_T g^{(\beta)}$.*

The rest of the section is devoted to a proof of Theorem 7.3.2.

Let $a \in \mathcal{O}$. The first step is to introduce an enumeration of arithmetical predicates indexed by $|a|$. By Theorem 1.3.4, the set $\{b \mid b <_o a\}$ is r.e. Applying the Recursion Theorem, we may assume that there is a partial recursive function $\Phi : \omega^2 \to \omega$ such that for each $b <_o a$, $\{\Phi(b,i)\}_{i<\omega}$ is an enumeration of the set of Gödel numbers of "Σ_b^0-predicates". To simplify notations, we identify a predicate with its Gödel number.

We introduce an enumeration of Σ_b^0-predicates by induction. The enumeration of Σ_1^0-predicates is standard. If $b = 2^c \in \mathcal{O}$ and $b <_o a$, then a predicate $P(x,n)$ is the i-th Σ_b^0-predicate, or $P(x,n) = P_{\Phi(b,i)}(x,n)$, if $P(x,n) \Leftrightarrow (\exists m)\neg Q(x,\langle n,m\rangle)$ where $Q(x,\langle n,m\rangle)$ is the i-th Σ_c^0-predicate.

If $b = 3 \cdot 5^e \in \mathcal{O}$ and $b <_o a$, then a predicate $P(x,n)$ is the i-th Σ_b^0-predicate, or $P(x,n) = P_{\Phi(b,i)}(x,n)$, if $P(x,n) \Leftrightarrow P_{\Phi(p(\Phi_i(n)))}(x,n)$, where $p(\Phi_i(n)) = (i_0(n), i_1(n))$ so that $i_0(n) <_o b$ for every n. The next lemma is proved by induction on the notation of b and is left to the reader.

Lemma 7.3.3. *There exist recursive functions $f : \omega \to \omega$ and $g : \omega \to \omega^2$ such that for any $b <_o a$,*

(i) *if $b = 1$ or 2^c, then $P_{\Phi(b,f(b))} \equiv_T \hat{H}_{2^b}^x$ via $g(b)$;*

(ii) *if $b = 3 \cdot 5^e <_o a$, then $P_{\Phi(b,f(b))} \equiv_T \hat{H}_b^x$ via $g(b)$.*

We now define a notion of Cohen forcing over Σ^0_b-predicates for $b <_O a$. If $P(x, n)$ is a Σ^0_1-predicate, then there is a partial recursive function $\Phi^x(n)$ such that $P(x, n) \Leftrightarrow \Phi^x(n)\downarrow$. Define $\sigma \Vdash P(x, n)$ if and only if $\Phi^\sigma(n)\downarrow$.

If $b = 2^c \in \mathcal{O}$, $b <_O a$, and $P(x, n)$ is Σ^0_b, then $P(x, n) \Leftrightarrow (\exists m)\neg Q(x, \langle n, m\rangle)$. Define $\sigma \Vdash P(x, n) \Leftrightarrow (\exists m)(\sigma \Vdash Q(x, \langle n, m\rangle))$.

If $b = 3 \cdot 5^e \in \mathcal{O}$, $b <_O a$, and $P(x, n)$ is a Σ^0_b-predicate, then $P(x, n) \Leftrightarrow P_{\Phi(p(\Phi_i(n)))}(x, n)$ for some i. Define $\sigma \Vdash P(x, n) \Leftrightarrow \sigma \Vdash P_{\Phi(p(\Phi_i(n)))}(x, n)$.

Given $b <_O a$, if $P(x, n)$ is a Σ^0_b-predicate, then $\sigma \Vdash \neg P(x, n) \Leftrightarrow (\forall \tau \succ \sigma)(\tau \nVdash P(x, n))$.

The verification of the following lemma is also left to the reader.

Lemma 7.3.4. *There is a partial recursive function $h : \omega^2 \times 2 \to \omega$ such that for any $b <_O a$,*

(i) *if $b = 1$ or 2^c, then $(\forall n)(\forall m)((\sigma \Vdash P_{\Phi(b,n)}(x, m) \leftrightarrow \Phi^{H_{2^b}}_{h(b,n)}(\sigma, m) = 1) \wedge (\sigma \Vdash \neg P_{\Phi(b,n)}(x, m) \leftrightarrow \Phi^{H_{2^b}}_{h(b,n)}(\sigma, m) = 0))$;*

(ii) *if $b = 3 \cdot 5^e$, then $(\forall n)(\forall m)((\sigma \Vdash P_{\Phi(b,n)}(x, m) \leftrightarrow \Phi^{H_b}_{h(b,n)}(\sigma, m) = 1) \wedge (\sigma \Vdash \neg P_{\Phi(b,n)}(x, m) \leftrightarrow \Phi^{H_b}_{h(b,n)}(\sigma, m) = 0))$.*

A real g is *a-generic* if for any $b <_O a$ and Σ^0_b-predicate $P(x, n)$ there is an m such that either $g \upharpoonright m \Vdash P(x, n)$ or $g \upharpoonright m \Vdash \neg P(x, n)$. This yields the following truth lemma:

Lemma 7.3.5. *Suppose that g is a-generic. For any m, n and $b <_O a$, $P_{\Phi(b,n)}(g, m)$ if and only if there is a k such that $g \upharpoonright k \Vdash P_{\Phi(b,n)}(x, m)$, and $\neg P_{\Phi(b,n)}(g, m)$ if and only if there is a k such that $g \upharpoonright k \Vdash \neg P_{\Phi(b,n)}(x, m)$.*

Lemma 7.3.6. *If g is a-generic, then there is a partial recursive function r such that for any $b <_O a$*

- *if $b = 1$ or 2^c, then $\hat{H}^g_b = \Phi^{g \oplus H_{2^b}}_{r(b)}$;*
- *if $b = 3 \cdot 5^e$, then $\hat{H}^g_b = \Phi^{g \oplus H_b}_{r(b)}$.*

Proof. We give the proof for the case when the notation of b is a successor, and leave the limit case to the reader.

If $b = 1$, then by Lemma 7.3.3, $n \in \hat{H}^g_b$ if and only if $P_{\Phi(1,n)}(g, n)$. By Lemma 7.3.5, there exists a k such that either $g \upharpoonright k \Vdash P_{\Phi(1,n)}(g, n)$ or $g \upharpoonright k \Vdash \neg P_{\Phi(1,n)}(g, n)$. By Lemma 7.3.4, this is decidable by $\Phi^{H_2}_{h(1,n)}(g \upharpoonright k, n)$. Hence $\hat{H}^g_1 = \Phi^{g \oplus H_2}_i$ for some i. Let $r(1) = i$.

If $b = 2^c$, then by Lemma 7.3.3, $n \in \hat{H}^g_b$ if and only if $P_{\Phi(b,n)}(g, n)$. By Lemma 7.3.5, there exists a k such that either $g \upharpoonright k \Vdash P_{\Phi(b,n)}(g, n)$ or $g \upharpoonright k \Vdash \neg P_{\Phi(b,n)}(g, n)$. Since $P_{\Phi(b,n)}(g, n) \Leftrightarrow (\exists m)P_{\Phi(c,n)}(g, n, m)$, we have

$$(\exists m)(g \upharpoonright k \Vdash P_{\Phi(c,n)}(g, n, m))) \vee (\forall \tau \succ g \upharpoonright k)(\forall m)(\tau \nVdash \neg P_{\Phi(c,n)}(g, n, m))).$$

By Lemma 7.3.4, this is decidable by either $g \oplus H'_{2^c} = g \oplus H_{2^b}$ ($c = 2^d$ for some d) or $g \oplus H'_c \leq_T g \oplus H_{2^b}$ ($c = 3 \cdot 5^e$ for some e). Hence $\hat{H}^g_b = \Phi^{g \oplus H_{2^b}}_i$ for some i, a number which may be effectively computed from b. Let $r(b) = i$. □

We are now ready to prove Theorem 7.3.2. We establish the case where β is a successor ordinal and leave the limit case to the reader.

Let $a \in \mathcal{O}$ so that $|a| = \beta$. Fix $z >_T H_b$ for all $b <_o a$. We may assume that z is not r.e. in H_b for any $b <_o a$. Given $b <_o a$ and $i, j < \omega$, consider the following statement:

$$(\forall \sigma \in 2^{<\omega})(\exists n)[(n \in z \wedge \sigma^\frown 0^n 1 \Vdash \neg P_{\Phi(b,i)}(x, j)) \vee \\ (n \notin z \wedge (\exists \tau \succeq \sigma^\frown 0^n 1)(\tau \Vdash P_{\Phi(b,i)}(x, j))]. \qquad (\sharp_{b,\langle i,j\rangle})$$

Since z is not r.e. in H_b for $b <_o a$, by Lemma 7.3.4, $(\sharp_{b,n})$ is true for every $b <_o a$ and $n < \omega$. We construct g in stages as the union of the sequence $\{g_s\}_{s<\omega} \subseteq 2^{<\omega}$ with $g_0 \prec g_1 \prec g_2 \prec \cdots$.

Let $\{(b_s, \langle i_s, j_s\rangle)\}_{s<\omega}$ be an enumeration of $\{b \mid b <_o a\} \times \omega^2$.

Stage 0: Let $g_0 = \emptyset$.

Stage $s + 1$:

Substage 1. Define $g(s, 1) = g_s{}^\frown z(s)^\frown H_a(s)$.

Substage 2. Define $g(s, 2) = g(s, 1)^\frown 0^{\langle b_s, \langle i_s, j_s\rangle\rangle} 1$.

Substage 3.

Case (i): $2^{b_s} \neq a$. Define $g(s, 3) = g(s, 2)^\frown 0^{e_s} 1$ where e_s is an index so that $\Phi^z_{e_s} = \Phi^{H_{2^{b_s}}}_{h(b_s, i_s)}$ ($b_s = 2^c$ for some c) or $\Phi^z_{e_s} = \Phi^{H_{b_s}}_{h(b_s, i_s)}$ ($b_s = 3 \cdot 5^e$ for some e).

Case (ii): $2^{b_s} = a$. Let $g(s, 3) = g(s, 2)^\frown 0^{e_s} 1$ where e_s is an index such that $\Phi^z_{e_s} = \Phi^{H_{b_s}}_{h(c, i_s)}$ ($b_s = 2^c$ for some c) or $\Phi^z_{e_s} = \Phi^{H_{b_s}}_{h(b_s, i_s)}$ ($b_s = 3 \cdot 5^e$ for some e).

Substage 4. Find the least n such that either

$$n \in z \wedge g(s, 3)^\frown 0^n 1 \Vdash \neg P_{\Phi(b_s, i_s)}(x, j_s)$$

or

$$n \notin z \wedge (\exists \tau \succeq g(s, 3)^\frown 0^n 1)(\tau \Vdash P_{\Phi(b_s, i_s)}(x, j_s)).$$

By $(\sharp_{b_s, \langle i_s, j_s\rangle})$, such a number n exists. If it satisfies the first half of the disjunction, define $g_{s+1} = g(s, 3)^\frown 0^n 1$. Otherwise, select the lexicographically least τ such that $\tau \succeq g(s, 3)^\frown 0^n 1$ and $\tau \Vdash P_{\Phi(b_s, i_s)}(x, j_s))$. Define $g_{s+1} = g(s, 3)^\frown \tau$.

Let $g = \bigcup_{s<\omega} g_s$.

It is clear that g is a-generic. By Lemma 7.3.6, $\hat{H}^g_a \equiv_T g \oplus H_a$. By Lemma 7.3.4, the construction can be decoded by $g \oplus H_a$. So $z \oplus g \leq_T g \oplus H_a \equiv_T \hat{H}^g_a$.

To prove that $z \oplus g \geq_T \hat{H}^g_a$, it is sufficient to decode the construction by $z \oplus g$ and compute H_a. To do this, we go through the same steps as in the proof of Theorem 7.3.1. Suppose we know g_s. Then $z(s + 1) = g(|g_s| + 1)$ and $H_a(s) = g(|g_s|_2)$. Then one obtains (b_s, i_s, j_s) by calculating $g(s, 2)$. By calculating $g(s, 3)$, we find an e_s such that $\Phi^z_{e_s}$ computes the set $\{(\sigma, i_s, j_s) \mid \sigma \Vdash P_{\Phi(b_s, i_s)}(x, j_s)\}$. To compute g_{s+1}, search for the unique number n such that $(g \upharpoonright |g(s, 3)|)^\frown 0^n 1 \prec g$. By the construction, either

(1) $n \in z$ or

(2) $n \notin z$ and there is a $\tau \succeq g(s, 3)^\frown 0^n 1(\tau \Vdash P_{\Phi(b_s, i_s)}(x, j_s))$.

If (1) holds, then $g_{s+1} = (g \upharpoonright |g(s,3)|)^\frown 0^n 1$. If (2) holds, then there is a unique $\tau \geq g(s,3)^\frown 0^n 1$ such that $\tau \Vdash P_{\Phi(b_s,i_s)}(x, j_s)$. We may use $\Phi^z_{e_s}$ to find the τ. Then $g_{s+1} = \tau$ and we conclude that $z \oplus g \geq_T H_a$.

7.3.3 Further generalizations

Another generalization of the Posner–Robinson Theorem (7.3.1) into the transfinite is as follows:

Theorem 7.3.7 (Shore and Slaman [134]). *For any recursive ordinal β, if $x \not\leq_T \emptyset^{(\gamma)}$ for $\gamma < \beta$, then there is a g such that $x \oplus g \equiv_T g^{(\beta)}$.*

Theorem 7.3.7 is proved by a forcing argument (Kumabe–Slaman forcing) different from Cohen forcing, and is used to show the definability of the complete Turing degree $\mathbf{0}'$ (see Exercise 7.3.3). On the other hand, it is possible to extend Theorem 7.3.1 beyond recursive ordinals. This was shown by Woodin using a set-theoretic argument:

Theorem 7.3.8 (Woodin [169]). *If z is not hyperarithmetic, then there is a g such that $z \oplus g \equiv_T \mathcal{O}^g$.*

7.3.4 Exercises

Exercise 7.3.1. *Prove Lemma 7.3.3 and Lemma 7.3.4.*

Exercise 7.3.2. *Prove, without using Theorem 7.3.8, that if $x >_T \emptyset^{(\beta)}$ for $\beta < \omega_1^{\mathrm{CK}}$, then there is a g such that $x \oplus g \geq_T g^{(\beta)}$ for all $\beta < \omega_1^{\mathrm{CK}}$.*

Exercise 7.3.3 (Shore and Slaman [134]). *Using Theorem 7.3.7, prove that if for some $n > 1$, the function $\mathbf{x} \mapsto \mathbf{x}^{(\mathbf{n})}$ is definable in $\langle \mathcal{D}, \leq \rangle$ where $\mathcal{D}$ is the collection of Turing degrees, then the function $\mathbf{x} \mapsto \mathbf{x}'$ is also definable in $\langle \mathcal{D}, \leq \rangle$.*

7.4 Classifying order-preserving functions on 2^ω

We say that $f : 2^\omega \to \omega_1$ is degree invariant if $x \equiv_T y$ implies $f(x) = f(y)$.

Lemma 7.4.1. *Suppose that $f : 2^\omega \to \omega_1$ is degree invariant and $f(x) < \omega_1^x$ on a cone. Then f is a constant on a cone.*

Proof. Suppose that $f : 2^\omega \to \omega_1$ is Turing invariant and $f(x) < \omega_1^x$ on $\{x \mid x \geq_T z_0\}$. For any e, let $A_e = \{x \mid x \geq_T z_0 \wedge \Phi^x_e$ is a well-ordering of order type $f(x)\}$. Then there is an e_0 such that A_{e_0} is cofinal. By Proposition 6.6.5, there is a pointed tree T_0 such that $x \in A_{e_0}$ for $x \in [T_0]$. We claim that there is a countable ordinal α_0 such that $f(x) < \alpha_0$ for $x \in [T_0]$. Otherwise for any x, $x \in \mathsf{WO}_1$ if and only if there exists a $y \in [T_0]$

and an order-preserving function f from x into Φ_e^y. In other words, WO_1 is $\Delta_1^1(T_0)$, a contradiction.

By Proposition 6.6.5 again, there is a pointed tree $T_1 \subseteq T_0$ and a countable ordinal $\alpha_1 < \alpha_0$ such that $f(x) = \alpha_1$ for any $x \in [T_1]$. □

If $F : 2^\omega \to 2^\omega$ is degree invariant, then it is *order preserving* if for any $x \leq_T y$, $F(x) \leq_T F(y)$.

Theorem 7.4.2 (Slaman and Steel [142]). *Assume* ZF + AD + DC. *Suppose that* $F : 2^\omega \to 2^\omega$ *is order preserving and increasing on a cone. Then either*
(i) *there is an* $\alpha < \omega_1$ *such that on a cone,* $F(x) \equiv_T x^{(\alpha)}$, *or*
(ii) *on a cone,* $F(x) >_T x^{(\alpha)}$ *for any* $\alpha < \omega_1^x$.

Proof. Assume that F is increasing on the cone C. If (ii) is not true for F, then by Lemma 7.4.1 and Proposition 6.6.5, there exists an ordinal $\alpha < \omega_1$ such that $x^{(\alpha)} \not<_T F(x)$ on a cone $C_1 \subseteq C$. Let α_0 be the least such ordinal. Then $F(x) >_T x^{(\beta)}$ for any $\beta < \alpha_0$. We may assume that $\omega_1^{z_1} > \alpha_0$.

Let $x_0 \in C_1$. By relativizing Theorem 7.3.2 to x_0, there is a $y \geq_T x_0$ such that $y \oplus F(x_0) \equiv_T y^{(\alpha_0)}$. Since F is order preserving and increasing, $F(y) \geq_T y \oplus F(x_0) \equiv_T y^{(\alpha_0)}$. Thus $F(y) \equiv_T y^{(\alpha_0)}$. It follows that the set $\{x \mid F(x) \equiv_T x^{(\alpha_0)}\}$ is cofinal. By Proposition 6.6.5 again, $F(x) \equiv_T x^{(\alpha_0)}$ on a cone. □

7.4.1 Exercise

Exercise 7.4.1. *Assume* ZF + AD + DC. *Suppose that F is order preserving and increasing on a cone. Then either*
(i) $(\exists \alpha < \omega_1)(F(x) \equiv_T x^{(\alpha)})$ *on a cone, or*
(ii) $F(x) \geq_T \mathcal{O}^x$ *on a cone.*

7.5 Pressdown functions

A function $F : 2^\omega \to 2^\omega$ is *pressdown* on a cone C if $F(x) <_T x$ for $x \in C$. By Theorem 7.1.3, every uniformly degree invariant function that is pressdown on a cone is a constant on a cone. We prove in this section a stronger result.

Theorem 7.5.1 (Slaman and Steel [142]). *Assume* ZF+AD+DC. *If F is a degree invariant pressdown function on a cone, then F is constant on a cone.*

Proof of Theorem 7.5.1. Suppose that F is a degree invariant pressdown function on a cone $C = \{x \mid x \geq_T z\}$ for some z. By Proposition 6.6.5, there is an e_0 and a pointed tree T such that $F(x) = \Phi_{e_0}^x$ for all $x \in [T]$. By Lemma 7.1.5, we may assume that Φ_{e_0} is a homeomorphism from $[T]$ to a perfect set $[S]$.

Without loss of generality, we may assume that for any splitting node $\sigma \in T$, $\Phi_{e_0}^{\sigma^\frown 0}$ is incompatible with $\Phi_{e_0}^{\sigma^\frown 1}$. The plan is to find two reals $x_0, x_1 \leq_T T$ in $[T]$ such that $\Phi_{e_0}^{x_0} \oplus \Phi_{e_0}^{x_1} \geq_T T$. Since F is degree invariant and T is pointed, $F(x_0 \oplus x_1) \equiv_T F(x_0) \equiv_T F(x_0) \oplus F(x_1) \geq_T T \equiv_T x_0 \oplus x_0$. Then we have a contradiction to the assumption that $F(x) <_T x$ on C. The following lemma says that for $x \in [T]$, the growth rate of any function $h \leq_T x$ is controlled by $F(x)$.

Lemma 7.5.2. *Suppose that $x \in [T]$ and $h \leq_T x$. For each e, let l^e be a function recursive in $\Phi_{e_0}^x$ so that*

$$l^e(0) = 0$$

and

$$l^e(n+1) = \sum_{i \leq l(n)} \min\{m \mid \Phi_e^{\Phi_{e_0}^x \upharpoonright m}(i)[m] \downarrow\}.$$

Then there is an e such that l^e is total and $(\exists m)(\forall n \geq m)(h(l^e(n)) < l^e(n+1))$.

Proof. Without loss of generality, we may assume that h is increasing. Suppose that the lemma is false. We construct a real $y \leq_T x$ in $[T]$ such that $\Phi_{e_0}^x \not\geq_T \Phi_{e_0}^y$. By shrinking T if necessary, we may assume that $x \leq_T T$. Since T is pointed, we have $y \equiv_T x$ but $\Phi_{e_0}^x \not\geq_T \Phi_{e_0}^y$, a contradiction.

The construction is by full approximation recursive in x. We shall construct an x-recursive sequence $\sigma_0 \prec \sigma_1 \cdots$ and let $y = \bigcup_{s<\omega} \sigma_s$. At any stage s, we may declare a number e to be *discarded*.

Let $\sigma_0 = \emptyset$.

At stage $s+1$, let $\tau \succ \sigma_s$ be the immediate splitting node in T so that $\Phi_{e_0}^{\tau^\frown 0}(k_s) \downarrow \neq \Phi_{e_0}^{\tau^\frown 1}(k_s) \downarrow$ for some k_s. For $i \leq 1$, let $\tau_i \succ \tau^\frown i$ be the next immediate splitting node in T so that $\Phi_{e_0}^{\tau_i^\frown 0}(k_s^i) \downarrow \neq \Phi_{e_0}^{\tau_i^\frown 1}(k_s^i) \downarrow$ for some $k_s^i > k_s$.

Search for the least $e \leq s$ which has not been discarded and

$$\Phi_e^{\Phi_{e_0}^x \upharpoonright (h(|\tau_0|+|\tau_1|))}(k_s)[h(|\tau_0| + |\tau_1|)] \downarrow .$$

If there exists such an e, then let $\sigma_{s+1} = \tau^\frown i$ where i is the least number such that

$$\Phi_e^{\Phi_{e_0}^x \upharpoonright h(|\tau_0|+|\tau_1|)}(k_s)[h(|\tau_0| + |\tau_1|)] \downarrow \neq \Phi_{e_0}^{\tau^\frown i}(k_s)$$

and discard e. Otherwise, let $\sigma_{s+1} = \sigma_s$.

Clearly $y \leq_T x$. Suppose that $\Phi_{e_0}^y = \Phi_e^{\Phi_{e_0}^x}$ for some e. By the assumption, for any m there exists an $n \geq m$ such that $h(l^e(n)) \geq l^e(n+1)$. Let s_0 be a stage large enough so that no $e' \leq e$ is discarded after s_0. Then there is a least $n > |\sigma_{s_0}| + s_0$ such that $h(l^e(n)) \geq l^e(n+1)$. Let s_1 be the largest stage $\geq s_0$ so that $k_{s_1} \leq l^e(n)$. Then at stage s_1+1, either $k_{s_1}^0$ or $k_{s_1}^1$ is greater than $l^e(n)$. Hence $|\tau_0| + |\tau_1| > l^e(n)$. Since $k_{s+1} \leq l^e(n+1)$, we have $\Phi_e^{\Phi_{e_0}^x \upharpoonright (h(|\tau_0|+|\tau_1|))}(k_s)[h(|\tau_0| + |\tau_1|)] \downarrow$. In other words, e is discarded at stage s_1+1. □

Let t be a T-recursive increasing function such that for any k and $\sigma \in 2^{t(k)} \cap T$, $|\Phi_{e_0}^{\sigma}| = k$. Let $\pi : \omega^{<\omega} \to T$ be a function defined by induction as follows:

- $\pi(\emptyset) = \sigma$ where σ is the splitting node in T such that every $\tau \in T$ is compatible with σ;
- $(\forall \sigma)(\pi(\sigma^\frown i) = \pi(\sigma)^\frown \tau)$, where $\pi(\sigma)^\frown \tau$ is a splitting node in T for which there is a finite sequence $\pi(\sigma) = \sigma_0 \prec \sigma_1 \cdots \sigma_{t(i)} \prec \tau$ such that for each $j \le t(i)$, σ_j is the immediate splitting node extending $\sigma_{j-1}{}^\frown 0$ in T and τ is the next immediate splitting node in T extending $\sigma_{t(i)}{}^\frown 1$.

Hence π induces a T-recursive continuous injection from ω^ω to $[T]$ so that for each $h \in \omega^\omega$, $\pi(h) \equiv_T T \oplus h$.

By Lemma 7.5.2, for each h, there is a least e_h for which one chooses the least m_h such that

$$(\forall n \ge m_h)(h(l^{e_h}(n)) < l^{e_h}(n+1)).$$

Note that the function $h \mapsto (e_h, m_h)$ is $\Delta_1^1(T)$. Then for each pair (e, m), the set

$$H_{e,m} = \{h \mid (e_h, m_h) = (e, m)\}$$

is a Borel set and therefore has the Baire property. Then there is a pair (e_1, m_1) such that H_{e_1,m_1} is not meager. In other words, there is a $u \in \omega^{<\omega}$ such that for any $\tau \succ u$, there is an $h \in H_{e_1,m_1}$ with $h \succ \tau$. Now for any $\sigma \in \omega^{<\omega}$, we T-recursively define a finite function l_σ so that

$$l_\sigma(\emptyset) = 0$$

and

$$l_\sigma(n+1) = \sum_{i \le l_\sigma(n)} \min\{m \mid \Phi_{e_1}^{\Phi_{e_0}^{\pi(\sigma)} \upharpoonright m}(i)[m] \downarrow\}.$$

For $h \in \omega^\omega$, let $l_h(n) = k$ if there exists an m such that $l_{h \upharpoonright m}(n) = k$. The proof of the following lemma is straightforward.

Lemma 7.5.3.

(i) $(\forall \sigma)(\forall i)(\forall k)(|\sigma| > k \wedge k = l_\sigma(i) \Rightarrow l_{\sigma \upharpoonright k}(i) = k)$;

(ii) $(\forall \tau \succeq u)(\forall n)(n \ge m_1 \wedge l_\tau(n) \ge |\tau| \Rightarrow n + 1 \notin \mathrm{Dom}(l_\tau))$.

Proof. (i) If $|\sigma| > k$ and $k = l_\sigma(i)$, then $|\pi(\sigma)| \ge t(|\sigma|) \ge t(k)$. Then $|\Phi_{e_0}^{\pi(\sigma \upharpoonright k)}| = k$. Thus $l_{\sigma \upharpoonright k}(i) = k$.

If (ii) is not true, then there exists a $\tau \succ u$ and $n \ge m_1$ such that $l_\tau(n) \ge |\tau|$ and $l_\tau(n+1)$ is defined. Then $\tau(l_\tau(n))$ is undefined. By the assumption on u, there is an $h \succ \tau$ in H_{e_1,m_1} such that $h(l_h(n)) > l_\tau(n+1) = l_h(n+1)$, a contradiction. □

Let $x \equiv_T T$, $m_2 \ge m_1$ and $h \succ u$ so that $l_h(m_2)$ is defined. Let $\tau_0 = \tau_1 = h \upharpoonright l_h(m_2)$. By Lemma 7.5.3, $l_{\tau_0}(m_2)$ is defined and $|\tau_0| = l_{\tau_0}(m_2)$ (so $l_{\tau_0}(m_2 + 1)$ is undefined). We T-recursively construct two sequences $\tau_0 = \tau_0^0 \prec \tau_1^0 \cdots$ and $\tau_1 = \tau_0^1 \prec \tau_1^1 \cdots$ as follows.

At stage $s+1$, suppose that $l_{\tau_s^i}(m_2+s)$ is defined and $|\tau_s^i| = l_{\tau_s^i}(m_2+s)$ (hence $l_{\tau_s^i}(m_2+s+1)$ is undefined) for $i = 0, 1$.

Case 1. $x(s) = 0$. Then by the assumption on u and Lemma 7.5.3, let $\tau_{s+1}^0 \succ \tau_s^0$ so that $l_{\tau_{s+1}^0}(m_2+s+1)$ is defined and $|\tau_s^0| = l_{\tau_s^0}(m_2+s)$. Let $\tau = (\tau_{s+1}^0)^\frown(l_{\tau_{s+1}^0}(m_2+s+1)+1)$. By the assumption on u and Lemma 7.5.3 again, let $\tau_{s+1}^1 \succ \tau$ so that $l_{\tau_{s+1}^1}(m_2+s+1)$ is defined and $|\tau_s^1| = l_{\tau_s^1}(m_2+s+1)$. Note that by the assumption on u, there is an $h \succ \tau_{s+1}^1$ in H_{e_1,m_1}. Hence by Lemma 7.5.3,

$$l_{\tau_{s+1}^1}(m_2+s+1) = l_h(m_2+s+1) \geq h(m_2+s+1) = l_{\tau_{s+1}^0}(m_2+s+1)+1 > l_{\tau_{s+1}^0}(m_2+s+1).$$

Case 2. $x(s) = 1$. Do the same construction as in Case 1 but replace 0 by 1. We have

$$l_{\tau_{s+1}^1}(m_2+s+1) < l_{\tau_{s+1}^0}(m_2+s+1).$$

Let $h_i = \bigcup_s \tau_s^i$ and $x_i = \pi(h_i) \in [T]$ for $i \leq 1$. Then $x_0 \equiv_T x_1 \equiv_T T$. Let l_i be l^{e_1} for x_i as defined in Lemma 7.5.2. Then $l_i \leq_T \Phi_{e_0}^{x_i}$ for $i \leq 1$. Moreover,

$$(\forall i)((l_1(m+i+1) > l_0(m+i+1) \Leftrightarrow x(i) = 0) \wedge (l_1(m+i+1) < l_0(m+i+1) \Leftrightarrow x(i) = 1)).$$

Hence $T \equiv_T x \leq_T \Phi_{e_0}^{x_0} \oplus \Phi_{e_0}^{x_1}$.

The completes the proof of Theorem 7.5.1. □

7.5.1 Exercises

Exercise 7.5.1. *Prove that in Lemma 7.5.2 there is a pointed tree $T_1 \subseteq T$ such that for any $x \in [T_1]$, $F(x) \geq_T \emptyset^{(\alpha)}$ if α is recursive.*

Exercise 7.5.2 (Slaman and Steel). *Suppose that $F : 2^\omega \to 2^\omega$ is a pressdown degree invariant function. Then there exists a z_0 such that for any z_1, there is an $x \geq_T z_1$ such that $F(x) \not\geq_T z_0$.*

Exercise 7.5.3 (Slaman). *There is a continuous injection $F : 2^\omega \to 2^\omega$ such that for every $x \geq_T \emptyset'$, $F(x)' \equiv_T x$ and for any $x =^* y$, $F(x) \equiv_T F(y)$, where $=^*$ denotes the finite difference relation.*

8 The construction of Π^1_1-sets

As seen in the previous chapters, inductive definition plays a key role in hyperarithmetic theory. In fact Gandy (see [106]) held the view that inductive definition rather than computation constituted the core characteristic of any generalized recursion theory. Admissible set theory (Chapter 3) reinforces the line of thought. Here we discuss inductive definition in greater detail and present some applications of this notion to the construction of Π^1_1-sets. The reader is referred to [47] and [106] for more on this subject.

8.1 An introduction to inductive definition

8.1.1 Inductive definability

Definition 8.1.1. Let $\Gamma : 2^\omega \to 2^\omega$ be a function.

- Γ is *progressive* if $x \subseteq \Gamma(x)$ for every $x \in 2^\omega$;
- Γ is *monotonic* if $\Gamma(y) \subseteq \Gamma(x)$ if $y \subseteq x$.

Given a function $\Gamma : 2^\omega \to 2^\omega$, let

$$\begin{aligned}
\Gamma_0 &= \emptyset;\\
\Gamma_{\alpha+1} &= \Gamma(\Gamma_\alpha) \cup \Gamma_\alpha;\\
\Gamma_\lambda &= \bigcup_{\beta<\lambda} \Gamma_\beta \text{ if } \lambda \text{ is a limit ordinal;}\\
\Gamma_\infty &= \bigcup_\alpha \Gamma_\alpha .
\end{aligned}$$

A real $x \subset 2^\omega$ is *inductively defined* by Γ if $x = \Gamma_\infty$. We use $|\Gamma|$ to denote the least ordinal α such that $\Gamma_\alpha = \Gamma_\infty$. Note that $|\Gamma| < (2^{\aleph_0})^+$. It is not difficulty to see that for any real x, there is a function $\Gamma : 2^\omega \to 2^\omega$ such that x is inductively defined by Γ (Exercise 8.1.1).

Theorem 8.1.2 (Gandy). *Suppose that Γ is a progressive function such that the set $\{(n, x) \in \omega \times 2^\omega \mid n \in \Gamma(x)\}$ is Π^0_1. Then $|\Gamma| \leq \omega_1^{\mathrm{CK}}$ and Γ_∞ is Π^1_1.*

Proof. By Theorem 3.1.8, there is a $\Sigma_1(L_{\omega_1^{\mathrm{CK}}})$-definable function $f : \omega_1^{\mathrm{CK}} \to L_{\omega_1^{\mathrm{CK}}}$ such that $f(\beta) = \Gamma_\beta$ for $\beta < \omega_1^{\mathrm{CK}}$. Let x be a real such that

$$n \in x \Leftrightarrow (\exists \beta < \omega_1^{\mathrm{CK}})(n \in f(\beta)).$$

Note that $x = \Gamma_{\omega_1^{\mathrm{CK}}}$. Moreover x is Σ_1-definable in $\langle L_{\omega_1^{\mathrm{CK}}}, \in\rangle$ and hence Π^1_1. Let R be a recursive predicate such that for any i,

$$i \in \Gamma(x) \Leftrightarrow (\forall j) R(i, x \restriction j) .$$

Fix an $i \in \Gamma(x)$. Then $(\forall j)R(i, x \upharpoonright j)$. Now for each $j \in \omega$, there is an ordinal $\beta < \omega_1^{\mathrm{CK}}$ such that $\Gamma_\beta \upharpoonright j = x \upharpoonright j$ and hence $R(i, \Gamma_\beta \upharpoonright k)$ for every $k \leq j$. Since R is recursive, we have

$$\langle L_{\omega_1^{\mathrm{CK}}}, \in \rangle \models (\forall j)(\exists \beta)R(i, \Gamma_\beta \upharpoonright j).$$

By induction, let β_j be the least $\beta > \beta_{j-1}$ such that $\langle L_{\omega_1^{\mathrm{CK}}}, \in \rangle \models R(i, \Gamma_\beta \upharpoonright k)$ for every $k \leq j$. Then the function $p : j \mapsto \beta_j$ is $\Sigma_1(L_{\omega_1^{\mathrm{CK}}})$-definable and hence belongs to $L_{\omega_1^{\mathrm{CK}}}$ by the admissibility of $L_{\omega_1^{\mathrm{CK}}}$. It follows that the ordinal $\beta(i) = \bigcup_{j<\omega} \beta_j$ is a limit ordinal less than ω_1^{CK}. Note that $\Gamma_{\beta_j} \subseteq \Gamma_{\beta_{j+1}}$ for $j < \omega$. Then for each k, there is a $j \geq k$ such that $\Gamma_{\beta(i)} \upharpoonright k = \Gamma_{\beta_j} \upharpoonright k$. Thus

$$\langle L_{\omega_1^{\mathrm{CK}}}, \in \rangle \models R(i, \Gamma_{\beta(i)} \upharpoonright k).$$

In other words,

$$\langle L_{\omega_1^{\mathrm{CK}}}, \in \rangle \models (\forall k)R(i, \Gamma_{\beta(i)} \upharpoonright k).$$

Hence if $i \in \Gamma(x)$, then $i \in \Gamma_{\beta(i)}$ and so $i \in x$. Thus $\Gamma(x) \subseteq x$, i.e. $\Gamma(x) = x$. We conclude that $x = \Gamma_\infty$. □

8.1.2 Monotonic inductive definability of Π^1_1-reals

Proposition 8.1.3. *Let $k > 0$. Suppose that Γ is progressive and monotonic, and the set $\{(n, x) \mid n \in \Gamma(x)\}$ is Π^1_k. Then Γ_∞ is Π^1_k.*

Proof. Since Γ is progressive and monotonic, Γ_∞ is the least real (in the sense of inclusion) x such that $\Gamma(x) \subseteq x$. Then $n \in \Gamma_\infty$ if and only if

$$(\forall x)(\Gamma(x) \subseteq x \to n \in \Gamma(x)).$$

If the set $\{(n, x) \mid n \in \Gamma(x)\}$ is Π^1_k, then Γ_∞ is Π^1_k. □

The following strengthens Theorem 8.1.2 for monotonic functions.

Theorem 8.1.4 (Spector [153]). *Suppose that Γ is progressive and monotonic such that $\{(n, x) \in \omega \times 2^\omega \mid n \in \Gamma(x)\}$ is Π^1_1. Then $|\Gamma| \leq \omega_1^{\mathrm{CK}}$.*

Proof. We prove that $\Gamma(\Gamma_{\omega_1^{\mathrm{CK}}}) = \Gamma_{\omega_1^{\mathrm{CK}}}$ and $|\Gamma| \leq \omega_1^{\mathrm{CK}}$.

Let $C \subseteq \mathscr{P}(\omega \times \omega)$ be a Σ^1_1-set such that $x \in C$ if and only if

- $\{n \mid (1, n) \in x\} = \emptyset$
- $(\forall j)(\Gamma(\{n \mid (j, n) \in x\}) \subseteq \{n \mid (2^j, n) \in x\})$, and
- $(\forall e)(e \in \mathcal{O} \to (\bigcup_{j\in\omega}\{n \mid (\Phi_e(j), n) \in x\} = \{n \mid (3 \cdot 5^e, n) \in x\}))$.

Let

$$z = \bigcap_{x \in C} x.$$

Then z is Π^1_1. We leave it to the reader to verify that

$$(\forall j \in \mathcal{O})(\{n \mid (j, n) \in z\} = \Gamma_{|j|}).$$

Hence $\Gamma_{\omega_1^{CK}} = \{n \mid (\exists j \in \mathcal{O})((j, n) \in z)\}$ is a Π^1_1-real.

Let $i \in \mathcal{O}^* \setminus \mathcal{O}$. Define $D \subseteq \mathcal{O}^*_i \times 2^\omega$ to be a Σ^1_1-set such that $x \in D$ if and only if

$$(\forall n)(\forall j)(j \in \mathcal{O} \cap \mathcal{O}^*_i \to ((j, n) \in z \leftrightarrow (j, n) \in x)).$$

By Gandy's Basis Theorem (2.5.3), there is a $z^* \in D$ such that $\omega_1^{z^*} = \omega_1^{CK}$.

For $j \in \mathcal{O}^*_i$, let

$$z^*_j = \{n \mid (\exists k)(k <_{o^*} j \wedge (k, n) \in z^*)\}.$$

Clearly $\Gamma_{\omega_1^{CK}} = \bigcup_{j \in \mathcal{O} \cap \mathcal{O}^*_i} z^*_j$. Moreover, for any $j_0 <_{o^*} j_1 <_{o^*} i$, $z^*_{j_0} \subseteq z^*_{j_1}$.

Suppose that $n \in \Gamma(\Gamma_{\omega_1^{CK}})$. We claim that there is a $j \in \mathcal{O} \cap \mathcal{O}^*_i$ such that $n \in \Gamma(z^*_j) = \Gamma(\Gamma_{|j|})$. Otherwise, since Γ is monotonic, it must be the case that $n \in \Gamma(z^*_j)$ for any $j \in \mathcal{O}^*_i \setminus \mathcal{O}$. Let

$$j \in y_n \Leftrightarrow n \in \Gamma(z^*_j).$$

Note that $z^*_j \leq_h z^*$ for every $j \in \mathcal{O}^*$. By Corollary 2.2.12, y_n is $\Pi^1_1(z^*)$. By the assumption, $y_n = \mathcal{O}^*_i \setminus \mathcal{O}$. Thus y_n is also $\Sigma^1_1(z^*)$ and hence $\Delta^1_1(z^*)$. Obviously $\omega_1^{\mathcal{O}^*_i \setminus \mathcal{O}} > \omega_1^{CK}$ and hence $\omega_1^{z^*} > \omega_1^{CK}$, contradicting the choice of z^*. We conclude that $\Gamma(\Gamma_{\omega_1^{CK}}) \subseteq \bigcup_{j \in \mathcal{O} \cap \mathcal{O}^*_i} \Gamma(\Gamma_{|j|}) = \bigcup_{j \in \mathcal{O} \cap \mathcal{O}^*_i} \Gamma_{|j|} = \Gamma_{\omega_1^{CK}}$. □

8.1.3 Inductively defining $\mathcal{O}$

As an illustration, we show how $\mathcal{O}$ may be defined inductively. Let Γ be a function such that for any x,

$$\Gamma(x)(n) = \begin{cases} 1 & n \in x \\ 1 & n = 1 \\ 1 & (\exists m \in x)(n = 2^m), \\ 1 & (\exists e)(\forall i)(n = 3 \cdot 5^e \wedge \Phi_e(i) \in x), \\ 0 & \text{Otherwise.} \end{cases}$$

We leave the proof of the following proposition to the reader.

Proposition 8.1.5. *Let Γ be as defined above. Then $\Gamma_\infty = \Gamma_{\omega_1^{CK}} = \mathcal{O}$.*

Thus $\mathcal{O}$ is inductively defined by the arithmetical function Γ. Indeed, every Π^1_1-real is "generated" by a simple monotonic function Γ (Exercise 8.1.3).

8.1.4 Exercises

Exercise 8.1.1. *Prove that for any real x, there is a function $\Gamma : 2^\omega \to 2^\omega$ such that x is inductively defined by Γ.*

Exercise 8.1.2. *Prove Proposition* 8.1.5.

Exercise 8.1.3. *For any Π^1_1-real z, there is a monotonic function Γ such that the set $\{(n,x) \mid n \in \Gamma(x)\}$ is Π^0_1 and z is many-one reducible to Γ_∞.*

Exercise 8.1.4. *Suppose that Γ is a progressive function such that $\{(n,x) \in \omega \times 2^\omega \mid n \in \Gamma(x)\}$ is Π^0_1. Show that if x is Π^1_1, then every n in $\Gamma(x)$ belongs to $\Gamma(y)$ for some hyperarithmetic $y \subseteq x$.*

8.2 Inductively defining Π^1_1-sets of reals

There is a parallel way of defining Π^1_1-sets of reals inductively. In [164], Engelen, Miller and Steel constructed a number of Π^1_1-sets of reals not based on the method of inductive definition.[1] We prove some of the results here via inductive definition (see Chong and Yu [17]).

8.2.1 A Π^1_1-induction principle for reals

Definition 8.2.1. A binary relation $\mathrm{P}(x,y) \subseteq 2^\omega \times 2^\omega$ is *cofinally progressive* if for every real x, the set $\{y \mid P(x,y)\}$ is cofinal in the hyperdegrees, i.e. for each z, there is a $y \geq_h z$ such that $\mathrm{P}(x,y)$ holds.

Let $\mathcal{C} = \{y \mid y \in L_{\omega_1^y}\}$.

Lemma 8.2.2. *Assume $(2^\omega)^L = 2^\omega$. If P is Π^1_1 and cofinally progressive, then for every x and z there is a $y \geq_h z$ such that $y \in \mathcal{C}$ and $\mathrm{P}(x,y)$.*

Proof. Suppose $(2^\omega)^L = 2^\omega$ and P is cofinally progressive. Then for any reals x, z, the set $A = \{y \mid x \oplus z \in L_{\omega_1^y} \wedge \mathrm{P}(x,y)\}$ is a nonempty $\Pi^1_1(x \oplus z)$-set. By Theorem 4.1.2 there is a $y \in A$ such that $y \in L_{\omega_1^{x\oplus y\oplus z}}[x \oplus z]$. Since $x \oplus z \in L_{\omega_1^y}$, we have $x \oplus z \leq_h y$ and so $\omega_1^{x\oplus y\oplus z} = \omega_1^y$. Then $y \in L_{\omega_1^{x\oplus y\oplus z}}[x \oplus z] = L_{\omega_1^y}$ and $\mathrm{P}(x,y)$ holds. □

Given a countable set A of reals and a real x, we say that x *codes* A if $\{(x)_n \mid n < \omega\} = A$ where $(x)_n = \{m \mid m < \omega \wedge \langle n, m\rangle \in x\}$.

A cofinally progressive relation generates a progressive relation as follows: Let $P(x,y)$ be cofinally progressive. For each $X \subseteq 2^\omega$, let $Q(X) = X \cup \{y \mid (\exists x \in X)P(x,y)\}$. Then it is immediate that $Q(X) \supseteq X$ for all X. However, an operation such as Q may be too general to be useful and this motivates a refinement of Q to one that is similar to a uniformization function for well-defined sets. Now if $P(x,y)$ is Π^1_1, then Theorem 4.2.1

1 Attempts were made by several authors to inductively define these sets (see, for example, [100]). In [17] a solution in this direction was given via a recursion-theoretic approach.

guarantees the existence of a uniformization function for P. However, our interest is in *strong* uniformization, i.e. a function that uniformizes “initial segments of reals” that does roughly the following: If x codes a set of reals in $\mathcal{C}$ such that any two elements $w <_L z$ in x satisfy $P(w, z)$, then there is a “least” $y \in \mathcal{C}$ such that $x \leq y$ and $P(x, y)$ holds. This strong uniformization property enables one to implement an inductive construction which is made precise by the Π^1_1-inductive principle $\mathfrak{I}$ defined as follows:

Π^1_1-inductive principle $\mathfrak{I}$. *If $P(x, y)$ is Π^1_1 and cofinally progressive, then there is a Π^1_1-set $A \subseteq \mathcal{C}$ such that:*
(i) *$|\leq_L\restriction A|$, the height of $\leq_L$ restricted to A, is ω_1;*
(ii) *$(\forall y)(y \in A \to (\exists x)(x \in L_{\omega_1^y} \wedge x$ codes the set $\{z \mid z \in A \wedge z <_L y\} \wedge P(x, y)))$.*

Theorem 8.2.3 (Chong and Yu [17]). *$V = L$ implies $\mathfrak{I}$.*

Proof. Suppose $V = L$, and P is Π^1_1 and cofinally progressive. Recall that for each constructibly countable β, there is an $\alpha > \beta$ such that $(L_{\alpha+1} \setminus L_\alpha) \cap 2^\omega \neq \emptyset$. Given ordinals $\beta < \alpha$ and $X \subseteq \alpha \times \omega$, denote by $X[\beta]$ the real $\{n \mid (\beta, n) \in X\}$. We may regard X as a sequence of reals of length α.

Since P is Π^1_1, by the Spector–Gandy Theorem (3.7.1) there is a Σ_0-formula $\psi(x, y, s)$ such that

$$P(x, y) \Leftrightarrow \langle L_{\omega_1^{x\oplus y}}[x, y], \in\rangle \models (\exists s)\psi(x, y, s).$$

Assume $V = L$. We define a partial function F which we will show to be total on $\omega_1 \times \bigcup_{\alpha<\omega_1} 2^{\alpha\times\omega}$. For $\alpha < \omega_1$, $z \in 2^\omega$ and $X \subseteq \alpha \times \omega$, let $\omega_1^{X,z}$ be the least ordinal γ such that $L_\gamma[X, z]$ is admissible. We define $F(\alpha, X)$ to be the real z such that $L_{\omega_1^{X,z}}[X, z]$ satisfies the following properties:

(1) there is a $\beta \geq \alpha$ such that $z \in L_{\beta+1} \setminus L_\beta$;
(2) there is an $x \in L_{\omega_1^{X,z}}$ which codes X;
(3) there is a limit ordinal γ and an element $s \in L_\gamma$ such that $L_\gamma[X, z] \models \psi(x, z, s)$;
(4) if $(t, y, \mu) <_L (s, z, \gamma)$, then $\langle L_\mu[X, y], \in\rangle \models \psi(x, y, t)$ implies $\omega_1^{(X,y)} < \beta$;
(5) if $(t, y, \mu) <_L (s, z, \gamma)$ and $\langle L_\mu[X, y], \in\rangle \models \psi(x, y, t)$, then either no real $r \in L_{\omega_1^{X,y}}$ codes X or $y \notin L_{\omega_1^{(X,y)}}$.

Since $L_{\omega_1^{X,z}}[X, z] = L_{\omega_1^{X,z}}$, $<_L$ is uniformly Δ_1 in $L_{\omega_1^{X,z}}[X, z]$. Moreover, (4) implies that (5) is Σ_1 in $L_{\omega_1^{X,z}}[X, z]$. Hence (1)–(5) are uniformly Σ_1 in $L_{\omega_1^{X,z}}[X, z]$. We show that $F(\alpha, X)$ is defined for $\alpha < \omega_1$ if X is countable.

Fix (α, X). Since $V = L$, there is a countable $\gamma' > \alpha$ with a real $x \in L_{\gamma'}$ coding X. Since P is cofinally progressive, by Lemma 8.2.2 there is a $\beta > \gamma'$ such that $(L_{\beta+1} \setminus L_\beta) \cap 2^\omega \neq \emptyset$, and a real $z \in L_{\beta+1} \setminus L_\beta$ such that $x \oplus z \in L_{\omega_1^z}$, $L_\gamma \models \psi(x, z, s)$ for some limit ordinal $\gamma < \omega_1^z$ and $s \in L_\gamma$. Note that $L_{\omega_1^z} = L_{\omega_1^{X,z}}[X, z]$ and $L_\gamma = L_\gamma[X, z]$ for all sufficiently large $\gamma < \omega_1^z$. Hence (1)–(3) are satisfied. Clearly we can assume that (s, z, γ) is the $<_L$-least satisfying these properties. Thus (4) and (5) are satisfied as well. By the absoluteness of $<_L$, one concludes that F is a well-defined function on each (α, X), where $\alpha < \omega_1$ and X is countable.

Observe that (1)–(4) are Σ_1-statements. By (4), one verifies that (5) is a Δ_1-statement. Hence there is a Σ_0-formula $\varphi(\alpha, X, z, s)$ such that

$$F(\alpha, X) = z \Leftrightarrow L_{\omega_1^{X,z}}[X, z] \models (\exists s)\varphi(\alpha, X, z, s).$$

Thus we may perform transfinite induction on α to define a sequence $\{z_\alpha \mid \alpha < \omega_1\}$ using F. However, the resulting sequence yields in general a set of reals that is Σ_1 over L_{ω_1}, i.e. Σ^1_2 but not necessarily Π^1_1 over second order arithmetic. To ensure that Π^1_1-ness is achieved, we refine the approach as follows.

Define $G(\alpha) = z$ if and only if $\alpha < \omega_1^z$ and there is a function $f : \alpha + 1 \to 2^\omega$ with $f \in L_{\omega_1^z}[z]$ so that for all $\beta \le \alpha$, $f(\beta) = F(\beta, \{(\gamma, n) | n \in f(\gamma) \wedge \gamma < \beta\})$ and $f(\alpha) = z$. Since $L_{\omega_1^z}[z]$ is admissible, $\{f(\gamma) \mid \gamma \le \alpha\} \in L_{\omega_1^z}[z]$. Hence $G(\alpha) = z$ if and only if there is a function $f : \alpha + 1 \to 2^\omega$ with $f \in L_{\omega_1^z}[z]$ such that

$$\langle L_{\omega_1^z}[z], \in\rangle \models ((\forall\beta \le \alpha)(\exists s)\varphi(\beta, \{(\gamma, n) \mid n \in f(\gamma) \wedge \gamma < \beta\}, s, f(\beta))) \wedge f(\alpha) = z.$$

Since $L_{\omega_1^z}[z]$ is admissible, G is Σ_1-definable. In other words, $G(\alpha) = z$ if and only if there is a function $f : \alpha + 1 \to 2^\omega$ with $f \in L_{\omega_1^z}[z]$ such that

$$\langle L_{\omega_1^z}[z], \in\rangle \models ((\exists t)(\forall\beta \le \alpha)(\exists s \in t)\varphi(\beta, \{(\gamma, n) \mid n \in f(\gamma) \wedge \gamma < \beta\}, s, f(\beta))) \wedge f(\alpha) = z.$$

Let A be the range of G. Then $z \in A$ if and only if there exists an ordinal $\alpha < \omega_1^y$ and a function $f : \alpha + 1 \to 2^\omega$ with $f \in L_{\omega_1^z}[z]$ such that

$$\langle L_{\omega_1^z}[z], \in\rangle \models (\exists t)(\forall\beta \le \alpha)(\exists s \in t)\varphi(\beta, \{(\gamma, n) \mid n \in f(\gamma) \wedge \gamma < \beta\}, s, f(\beta)) \wedge f(\alpha) = z.$$

Hence A is Π^1_1.

All that remains is to use an argument as that for Theorem 3.1.8 to show that G is a well-defined total function on ω_1. The only nontrivial part is to prove that the function f defined above exists. This can be done by induction. Let $\alpha < \omega_1$ be given and suppose that $\{f_\beta\}_{\beta<\alpha}$ is a uniformly Σ_1 and increasing (i.e. $f_\beta \supset f_\gamma$ for $\beta > \gamma$) sequence such that each f_β is an f at stage $\beta < \alpha$ as required, Let f_α be such that $f_\alpha \in L_{\omega_1^z}[z]$ and for all $\beta \le \alpha$, $f_\alpha(\beta) = F(\beta, \{(\gamma, n) \mid n \in f_\beta(\gamma) \wedge \gamma < \beta\})$ and $f_\alpha(\alpha) = z$, where $F(\alpha, X) = z$ and $X = \{(\gamma, n) \mid (\exists\beta)(\gamma \le \beta < \alpha \wedge n \in f_\beta(\gamma))\}$. Note that $X = \{(\gamma, n) \mid \gamma \le \alpha \wedge n \in f_\gamma(\gamma))\}$ by absoluteness. Then $f_\alpha \supset f_\beta$ for all $\beta < \alpha$ and satisfies the requirement of such an f at stage α.

By (1) and (2) above, $(x, z) \in L_\gamma$ and $L_\gamma \in L_{\omega_1^z}$. So $A \subset \mathfrak{C}$. Since P is cofinally progressive, (i) of the definition of $\mathfrak{I}$ is satisfied.

For (ii): If $y \in A$ then $y = G(\alpha)$ for some $\alpha < \omega_1$. By the definition of G, there is a real $x \in L_{\omega_1^{G(\alpha)}}$ such that x codes the set $\{G(\beta) \mid \beta < \alpha\}$ and $\mathrm{P}(x, G(\alpha))$ holds. Clearly $\beta < \alpha$ if and only if $G(\beta) <_L G(\alpha)$. Hence (ii) holds. □

It turns out that the Π^1_1-inductive principle $\mathfrak{I}$ is equivalent to the set-theoretic assumption $(\omega_1)^L = \omega_1$, as we now show.

Theorem 8.2.4. $(\omega_1)^L = \omega_1$ *if and only if the* Π_1^1*-inductive principle* $\mathfrak{J}$ *holds.*

Proof. The "if" direction is immediate: Choose $P(x, y)$ to express $x \leq_T y$. Then P is cofinally progressive. Let A be a Π_1^1-set satisfying $\mathfrak{J}$ for P, of $<_L$-height ω_1. Then $(\omega_1)^L = \omega_1$.

For the other direction, suppose P is Π_1^1 and cofinally progressive. Then for any pair of reals $x, z \in L$, the set $U_{x,z} = \{y \mid y \geq_h z \wedge P(x, y)\}$ is a nonempty $\Pi_1^1(x, z)$-set. By Theorem 4.1.2, there is a $y \in U_{x,z} \cap L$. By Theorem 4.1.2 again,

$$\langle L, \in\rangle \models \text{P is cofinally progressive.}$$

By Theorem 8.2.3, there is a Π_1^1-set A such that $(A)^L$ witnesses the correctness of the inductive principle $\mathfrak{J}$ in L.

Since $(\omega_1)^L = \omega_1$, $\mathfrak{J}$ (i) is true in V. Since the statement "$x \in L_{\omega_1^x}$" is Π_1^1 and

$$\langle L, \in\rangle \models (\forall x)(x \in A \to x \in L_{\omega_1^x}),$$

by the Shoenfield Absoluteness Lemma (Corollary 4.2.9), $\langle V, \in\rangle \models (\forall x)(x \in A \to x \in L_{\omega_1^x})$. Hence $A \subset \mathfrak{C}$.

Let $y \in A$. Since $A \subseteq L$, $y \in L$. Then there exists an $x \in L_{\omega_1^y}$ such that

$$\langle L, \in\rangle \models x \text{ codes the set } \{z \mid z \in A \wedge z <_L y\} \wedge P(x, y).$$

Since $A \subset L$, x codes the set $\{z \mid z \in A \wedge z <_L y\}$. Since the relation P is Π_1^1 and $x, y \in L$, by the Shoenfield Absoluteness Lemma (Corollary 4.2.9) again, $P(x, y)$ holds. Hence (ii) follows. □

One may relativize the Π_1^1-inductive principle $\mathfrak{J}$ to admit real parameters and obtain its boldface version $\underset{\sim}{\mathfrak{J}}$. Then Theorem 8.2.4 may be generalized to state: Boldface $\underset{\sim}{\Pi}_1^1$-*inductive principle* $\underset{\sim}{\mathfrak{J}}$ *fails if and only if there is no real x such that* $(\omega_1)^{L[x]} = \omega_1$. This leads to the following conclusion:

Corollary 8.2.5. *The statement "ZFC +* $\underset{\sim}{\Pi}_1^1$*-inductive principle* $\underset{\sim}{\mathfrak{J}}$ *is false" is consistent if and only if "ZFC+ There exists an inaccessible cardinal" is consistent.*

Proof. If $\underset{\sim}{\mathfrak{J}}$ fails, then $(\omega_1)^{L[x]} < \omega_1$ for all x, so that the latter is inaccessible in L. Conversely, suppose ZFC+ "There is an inaccessible cardinal" is consistent. Then by Proposition 6.2.9 (the Levy collapsing map), ZFC+ "ω_1 is inaccessible in L" is consistent. Now in a model $\mathcal{M}$ with such a property, $\omega_1 > (\omega_1)^{L[x]}$ for all x. Hence $\underset{\sim}{\mathfrak{J}}$ fails in $\mathcal{M}$. □

8.2.2 Exercise

Exercise 8.2.1 (Miller [100]). *A subset of* 2^ω *is almost disjoint if any two of its distinct members have finite intersection. Assume* $2^\omega = (2^\omega)^L$. *Prove that there is a* Π_1^1*-maximal almost disjoint set.*

8.3 Π^1_1-maximal chains and antichains of Turing degrees

We now apply the Π^1_1-inductive principle $\mathfrak{I}$ to construct some special Π^1_1-sets of Turing degrees.

8.3.1 Chains in the Turing degrees

Given a partial ordering $(P, \leq_P)$, $p \in P$ is a *minimal cover* of $A \subseteq P$ if $q <_P p$ for all $q \in A$ and there is no $r \in P$ such that $r <_P p$ having the same property. A set $A \subseteq P$ is a *chain* if for any $x, y \in A$, either $x \leq_P y$ or $y \leq_P x$. A chain A is *maximal* if there is no chain $B \supset A$. Define $\mathrm{P}(x, y)$ if and only if

$$y \text{ is a minimal cover of } \{(x)_n \mid n \in \omega\} \text{ under } \leq_T .$$

Lemma 8.3.1. $\mathrm{P}(x, y)$ *is a cofinally progressive relation.*

Proof. We prove that for any countable set $A = \{x_i\}_{i<\omega}$ and any real x, there is a minimal cover z of A such that $z'' \geq_T x$.

Fix an effective enumeration of partial recursive oracle functions $\{\Phi_e\}_{e<\omega}$.

Given a perfect tree T, define the n-th level $Lev_n(T)$ of T to be

$$\{\sigma \notin Lev_{n-1}(T) \mid (\exists\tau)(\tau \in T \wedge \sigma \preceq \tau \wedge \tau \text{ is an } n\text{-th splitting node on } T)\}.$$

Let $\{p_n\}_{n<\omega}$ be a recursive one-one enumeration of prime numbers. Allowing ambiguity, we will also use p_σ to denote the prime number which codes the string σ. Given a recursive oracle function Φ, we use $\Phi^y = T$ to express the following:

(1) Φ^y is total;

(2) $(\forall n)(\Phi^y(n) = \prod_{\sigma \in Lev_n(T)} p_\sigma)$.

Given a tree T and a string $\nu \in T$, let T^ν be the subtree of T consisting of all $\nu' \in T$ extending ν. Define $L_n(T)$ to be the leftmost $n+1$-th splitting node and $R_n(T)$ to be the rightmost $n+1$-th splitting node on T.

For each finite string σ^*, real x and perfect tree $T \subseteq 2^{<\omega}$ with $\sigma^* \in T$, define $\mathfrak{T}(\sigma^*, x, T)$ to be the perfect tree that is the intersection of a sequence of perfect trees T_n given as follows:

$$T_0 = T^{\sigma^*}.$$

$$T_{n+1} = \{\tau \mid (\exists\sigma \in Lev_n(T_n))(\tau \preceq \sigma)\} \cup S_{n+1} \subseteq T_n$$

where S_{n+1} is defined as:

Case 1. $x(n) = 0$: $\sigma \in S_{n+1}$ if and only if there exists $\tau \in Lev_{n+1}(T_n)$ so that $\sigma \succeq \tau$ and $\tau = L_1(T_n^\nu)$ for some ν which is an n-th splitting node of T_n.

Case 2. $x(n) = 1$: $\sigma \in S_{n+1}$ if and only if there exists $\tau \in Lev_{n+1}(T_n)$ so that $\sigma \succeq \tau$ and $\tau = R_1(T_n^\nu)$ for some ν which is an n-th splitting node of T_n.

Define

$$\mathcal{T}(\sigma^*, x, T) = \bigcap_n T_n.$$

In other words, $\mathcal{T}(\sigma^*, x, T)$ is a subtree which, roughly speaking, codes $x(n)$ at a $2n$-th splitting node of T^{σ^*}. Note that $\mathcal{T}(\sigma^*, x, T) \oplus T \geq_T x$. Moreover, suppose for some recursive oracle function Φ, we have $\Phi^y = T$ for all $y \in [T]$. Then there is a recursive oracle function Ψ such that $\Psi^y = x$ for all $y \in \mathcal{T}(\sigma^*, x, T)$. Furthermore, given an index of the oracle function Φ, an index of the oracle function Ψ may be effectively obtained from σ^*. In other words, there is a recursive function f such that $\Phi^y_{f(\sigma^*)} = x$ for all $y \in [\mathcal{T}(\sigma^*, x, T)]$. Since $\Phi^y = T$ for all $y \in [\mathcal{T}(\sigma^*, x, T)] \subseteq [T]$ and $x \oplus T \geq_T \mathcal{T}(\sigma^*, x, T)$, there is a recursive function g so that $\Phi^y_{g(\sigma^*)} = \mathcal{T}(\sigma^*, x, T)$ for all $y \in [\mathcal{T}(\sigma^*, x, T)]$.

We give a sketch of the idea behind the construction of z. To obtain a minimal cover of a countable set $A = \{x_i\}_{i<\omega}$, one makes appropriate modifications of the construction of a minimal degree (see [82]). To make the minimal degree relatively high (i.e. to make it compute a given x through jumps), one needs to code the indices of the perfect trees in the course of the construction. Were the construction uniform, one could use the Recursion Theorem to code the index of the next perfect tree being defined during the step by step construction. This technique could be applied to code x and x_i into z for each $i < \omega$. However, the construction to achieve minimality is nonuniform (one needs to decide whether the next tree will be an "e-splitting tree" or a "full tree", which in general is a "double jump" question). Although it is highly nonuniform, the construction does become uniform once it is decided which situation one is in (see Substep 3 of the construction below). This is the reason for using z'' to "get up" to x.

We now turn to the construction.

Fix a real x and an enumeration $\{x_i\}_{i\in\omega}$ of A. Suppose the recursive oracle functional Φ_0 satisfies $\Phi_0^y = 2^{<\omega}$ for all reals y. We construct a sequence of perfect trees step by step. At step n, we construct a perfect tree $T_n \leq_T \oplus_{i<n} x_i$ and a finite string σ_n so that $T_n \upharpoonright |\sigma_n| = \{\tau \mid \tau \preceq \sigma_n\}$, $|\sigma_n| > n$ and there is a recursive oracle functional Φ_{e_n} such that for each $y \in [T_n]$, $\Phi^y_{e_n} = T_n$.

Construction

Step 0. Define $T_0 = 2^{<\omega}$ and $\sigma_0 = \emptyset$.

Step $n+1$. There are three substeps:

Substep 1 (Coding x). For each k, if $x(n) = 0$ then define $\sigma^k_{n+1,0} = R_1(T_n^{L_k(T_n)})$. Otherwise, let $\sigma^k_{n+1,0} = L_1(T_n^{R_k(T_n)})$. Define $T_{n+1,0,k} = T_n^{\sigma^k_{n+1,0}}$. Hence there is a recursive oracle functional Φ_{i_k} such that for each $y \in T_{n+1,0,k}$, $\Phi^y_{i_k} = T_{n+1,0,k}$. Note that the function $k \mapsto i_k$ is recursive. It follows from the Recursion Theorem that there is a number k_0 such that $\Phi^y_{i_{k_0}} = \Phi^y_{k_0}$ for all $y \in [T_n^{\sigma^{k_0}_{n+1,0}}]$. Fix this k_0 and define $\sigma^0_{n+1} = \sigma^{k_0}_{n+1,0}$, and $T_{n+1,0} = T_{n+1,0,k_0}$.

Substep 2 (Coding x_n). For each k, define

$$T_{n+1,1,k} = \mathcal{T}(R_1(T_{n+1,0}^{L_k(T_{n+1,0})}), x_n, T_{n+1,0}).$$

By the discussion above, there are recursive functions f, g so that $\Phi^y_{f(k)} = x_n$ and $\Phi^y_{g(k)} = T_{n+1,1,k}$ for all $y \in [T_{n+1,1,k}]$. By the Recursion Theorem, there is a k_1 such that $\Phi^y_{k_1} = \Phi^y_{g(k_1)} = T_{n+1,1,k_1}$ for all $y \in [T_{n+1,1,k_1}]$. Define $\sigma_{n+1,1} = R_1(T_{n+1,0}^{L_{k_1}(T_{n+1,0})})$ and $T_{n+1,1} = T_{n+1,1,k_1}$.

Substep 3 (Forcing a minimal cover). This is the only place where z'' is used.

Case 1. There exists a $\sigma \in T_{n+1,1}$ and a number i_σ such that for all $m > i_\sigma, \tau_1, \tau_2 \succeq \sigma, \{\tau_1, \tau_2\} \subset T_{n+1,1} \wedge \Phi^{\tau_1}_{n+1}(m) \downarrow \wedge \Phi^{\tau_2}_{n+1}(m) \downarrow$ implies $\Phi^{\tau_1}_{n+1}(m) = \Phi^{\tau_2}_{n+1}(m)$. Choose the least such $\sigma \in T_{n+1,1}$ in the sense of the coding of strings. Define $S = T^\sigma_{n+1,1}$. For each k, define $S_k = S^{R_1(S^{L_k(S)})}$. By the Recursion Theorem again, there is a k_2 so that $\Phi^y_{k_2} = S^{R_1(S^{L_{k_2}(S)})}$ for all $y \in [S^{R_1(S^{L_{k_2}(S)})}]$. Define $\sigma_{n+1} = R_1(S^{L_{k_2}(S)})$ and $T_{n+1} = S^{\sigma_{n+1}}$.

Case 2. Otherwise, for each k define $S_k = T_{n+1,1}^{R_1(T_{n+1,1}^{L_k(T_{n+1,1})})}$. Then we can S_k-recursively find a subperfect tree P_k of S_k such that for all $y \in [P_k]$, Φ^y_n is total and for all $\tau_0, \tau_1 \in P_k$, if $\tau_0 | \tau_1$ then there is an i such that for all reals $z_0 \succ \tau_0$ and $z_1 \succ \tau_1$, $\Phi^{z_0}_n(i) \downarrow \neq \Phi^{z_1}_n(i) \downarrow$. By the Recursion Theorem, there is a k_2 such that $\Phi^y_{k_2} = P_{k_2}$ for all $y \in [P_{k_2}]$. Let $T_{n+1} = P_{k_2}$ and $\sigma_{n+1} = R_1(T_{n+1,1}^{L_{k_2}(T_{n+1,1})}) \in T_{n+1}$.

Let $e_{n+1} = k_3$. Then $\Phi^y_{e_{n+1}} = T_{n+1}$ for all $y \in [T_{n+1}]$. This completes the construction at step $n + 1$.

Note that by induction on n, $T_{n+1} \leq_T \oplus_{i<n+1} x_i$.

Finally, let $z = \bigcup_n \sigma_n$.

By the construction, $T_{n+1} \leq_T \oplus_{i<n+1} x_i$ is a pointed tree for each n. Thus $z \geq_T x_n$ for every $n \in \omega$. If Φ^z_n is total and $\Phi^z_{n+1} \not\leq_T \oplus_{i<n+1} x_i$, then we have Case (2) at Substep 3 of Step $n + 1$. Then it is obvious that $z \leq_T \Phi^z_{n+1}$. Hence z is a minimal cover of A.

We show that $z'' \geq_T x$. This is proved by induction on n that $x(n)$ may be z''-uniformly computed. To do this, we argue that the index e_n is uniformly computed from z''.

Suppose we are at step $n + 1$. By inductive hypothesis, assume that z'' recursively computes the index e_n. By the construction, we have $\Phi^y_{e_n} = T_n$ for all $y \in [T_n]$. Use z to recursively find an initial segment of z which is $R_1(T_n^{L_{k_0}(T_n)})$ or $L_1(T_n^{R_{k_0}(T_n)})$ for some k_0. In the first case, $x(n) = 0$. In the second case, $x(n) = 1$. Note that $\Phi^z_{k_0}$ is the perfect tree in Substep 1 of the construction. Now z-recursively find k_1 such that $R_1((\Phi^z_{k_0})^{L_{k_1}(\Phi^z_{k_0})}) \preceq z$. Then $\Phi^z_{k_1} = \mathcal{T}(R_1((\Phi^z_{k_0})^{L_{k_1}(\Phi^z_{k_0})}), x_n, \Phi^z_{k_0}) = T_{n+1,1}$. Use z'' to decide which case $\Phi^z_{k_1}$ is in. In Case 1, we z''-recursively find the least $\sigma \in T_{n+1,1}$ with the required property. Let $S = T^\sigma_{n+1,1}$. Then use z to compute the k_2 such that $\Phi^z_{k_2} = S^{R_1(S^{L_{k_2}(S)})}$. Then $e_{n+1} = k_2$ and $T_{n+1} = S^{R_1(S^{L_{k_2}(S)})} = \Phi^z_{k_2}$. In Case 2, use z to compute the k_2 such that $R_1(T_{n+1,1}^{L_{k_2}(T_{n+1,1})}) \preceq z$. Then by the construction, $e_{n+1} = k_2$ and $T_{n+1} = \Phi^z_{k_2}$.

Thus $z'' \geq_T x$. □

Theorem 8.3.2 (Chong and Yu [15]). *Assume $(\omega_1)^L = \omega_1$. There is a Π^1_1-maximal chain in the Turing degrees.*

Proof. By Lemma 8.3.1, P is a Δ^1_1-cofinally progressive relation. Hence by Theorem 8.2.4, there is a Π^1_1-set A as prescribed. $\mathfrak{I}$(ii) implies that A is a maximal chain. □

8.3.2 Antichains in the Turing degrees

Given a partial ordering $(P, <_P)$, a set $A \subseteq P$ is an *antichain* if any distinct $x, y \in A$ are $<_P$-incomparable. A is *maximal* if it is not a proper subset of an antichain.

Define a binary relation $\mathrm{P}(x, y)$ to hold if either (1) or (2) below is satisfied:

(1) There exist $n \neq m$ such that $(x)_n \neq (x)_m$ but $(x)_n$ is Turing comparable with $(x)_m$;

(2) $\{(x)_n \mid n \in \omega\} \cup \{y\}$ is an antichain such that

  (2a) $x \oplus (y)_0 \in L_{\omega_1^y}$,

  (2b) $\{(x)_n \mid n \in \omega\} \cup \{(y)_0\}$ is an antichain, and

  (2c) for every $r <_L (y)_0$, $\{(x)_n \mid n \in \omega\} \cup \{r\}$ is not an antichain.

Lemma 8.3.3. *If $2^\omega = (2^\omega)^L$, then* P *is a cofinally progressive relation.*

Proof. Fix y_0. Given x, let $X = \{(x)_n \mid n \in \omega\}$. Assume that the set $X \cup \{y_0\}$ is an antichain of Turing degrees. We show that for any $y_1 \geq (x \oplus y_0)''$, there is a z such that

  (1) $z'' \geq_T y_1$,

  (2) $y_0 \leq_T z \leq_T y_1$,

  (3) $X \cup \{z\}$ is an antichain.

Then given a constructible y_{-1} and a countable antichain $X \cup \{y_0\}$ such that for every $r <_L y_0$, $X \cup \{r\}$ is not an antichain, by Proposition 4.3.4, there is a $y_1 \geq_T (y_{-1} \oplus X \oplus y_0)''$ such that $y_1 \in L_{\omega_1^{y_1}}$. Then by (1)–(3) above, there is a z such that $z \equiv_h y_1$ (and hence $y_0 \oplus z \in L_{\omega_1^{y_0 \oplus z}}$) and $X \cup \{z\}$ is an antichain. Thus $\{z \mid \mathrm{P}(x, y_0 \oplus z)\}$ is cofinally progressive.

We construct a real z such that $(z \oplus x_0)'' \geq_T y_1$ and the following requirements are satisfied:

$$N_{e,i} : \Phi_e^{z \oplus y_0} \text{ is total } \Rightarrow \Phi_e^{z \oplus y_0} \neq (x)_i.$$

Then $X \cup \{z \oplus y_0\}$ is an antichain. The real z is obtained as the union of a sequence of finite strings $\sigma_0 \prec \sigma_1 \prec \cdots$.

Construction

Step 0. Let $\sigma_0 = \emptyset$.

Step $n + 1 = \langle e, i\rangle$.

Substep 1 (Satisfying $N_{e,i}$). Consider the following statement:

$$(\exists\tau \succeq \sigma_n)(\forall\tau_0, \tau_1 \succeq \tau)(\forall m)(\Phi^{\tau_0 \oplus y_0}(m)\downarrow \wedge \Phi^{\tau_1\oplus y_0}(m)\downarrow \to \Phi^{\tau_0\oplus y_0}(m) = \Phi^{\tau_1\oplus y_0}(m)).$$

If the statement is true, then find the (lexicographically) least such τ and define $\sigma^0_{n+1} = \tau$. Then for every real $z \succ \sigma^0_{n+1}$, $\Phi^{z\oplus y_0}_e$ is total implies $\Phi^{z\oplus y_0}_e \leq_T y_0$. Thus $\Phi^{z\oplus y_0}_e \neq (x)_i$ since $X \cup \{y_0\}$ is an antichain. If the statement is false, then find the least $\tau_0 \succeq \sigma_n$ for which there exists a (least) $\tau_1 \succeq \sigma_n$ such that $\Phi^{\tau_0\oplus y_0}(m)\downarrow \neq \Phi^{\tau_1\oplus y_0}(m)\downarrow$ for some m. Define $\sigma^0_{n+1} = \tau_k$ for the $k < 2$ where $\Phi^{\tau_k\oplus y_0}(m) \neq (x)_i(m)$.

Substep 2 (Coding y_1). Define $\sigma_{n+1} = (\sigma^0_{n+1})^\frown(y_1(n))$.

Finally, let $z = \bigcup_n \sigma_n$.

Since $y_0 \not\geq_T (x)_i$ for all i, the same conclusion holds for $z\oplus y_0$. By the construction, $z \oplus y_0 \not\leq_T (x)_i$ for all i, and so $X \cup \{z \oplus y_0\}$ is an antichain.

To see that $(z \oplus y_0)'' \geq_T y_1$, consider the statement in Substep 1. The statement is decidable by y_0''. If it is true, then we can y_0''-recursively find the least such τ. Then $\tau = \sigma^0_{n+1}$. Otherwise, we can y_0''-recursively find both τ_0 and τ_1. In this case we use z to decide which string is the σ^0_{n+1}. Thus $y_1(n) = 0$ if and only if $z(|\sigma^0_{n+1}| + 1) = 0$. Moreover, $\sigma_{n+1} = z \upharpoonright (|\sigma^0_{n+1}| + 1)$. Hence the sequence $\{\sigma_n\}_n$ can be computed from $z \oplus y_0''$. Therefore $(z \oplus y_0)'' \geq_T z \oplus y_0'' \geq_T y_1$.

Clearly the entire construction can be decoded by y_1. Thus $y_1 \geq_T z$. □

Theorem 8.3.4 (Chong and Yu [16]). *Assume $(2^\omega)^L = 2^\omega$. There exists a Π^1_1-thin maximal antichain in the Turing degrees.*

Proof. By Lemma 8.3.3, P is a cofinally progressive relation. Note that (2c) is equivalent to

$$L_{\omega_1^{(x,y)}}[x, y] \models (\exists\beta)((y)_0 \in L_\beta \wedge (\forall z \in L_\beta)(z <_L (y)_0 \to (\exists n)((x)_n \geq_T y \vee y \geq_T (x)_n)).$$

Thus P is Π^1_1. Hence by Theorem 8.2.4, there is a Π^1_1 set A with the prescribed properties. Using $\mathfrak{I}$(ii), we show by induction on $<_L$ that A is an antichain.

Suppose $x \in A$ and $\{y \mid y \in A \wedge y <_L x\}$ is an antichain. By $\mathfrak{I}$(ii), there is a z coding $\{y \mid y \in A \wedge y <_L x\}$ so that P(z, y) holds. Then by (2), $\{(z)_n \mid n \in \omega\} \cup \{y\}$ is an antichain. So A is an antichain. Note that $A \subset \mathfrak{C}$. This together with Lemma 4.3.1 implies that A is a thin set. By (2b) above, A is maximal. □

8.3.3 Exercises

Exercise 8.3.1. *Prove that there is no $\underset{\sim}{\Sigma}^1_1$-maximal chain in the Turing degrees.*

Exercise 8.3.2. *Prove that* ZFC+ *"There is no $\underset{\sim}{\Pi}^1_1$-maximal chain in the Turing degrees" is consistent if and only if* ZFC+ *"There is an inaccessible cardinal" is consistent.*

Exercise 8.3.3. *Prove that there exists a thin maximal antichain in the Turing degrees.*

Exercise 8.3.4. *Prove that every maximal antichain in the Turing degrees is of size* $2^{\aleph_0}$.

Exercise 8.3.5. *Prove that if there exists a* Π^1_1*-thin maximal antichain in the Turing degrees, then every real is constructible.*

Exercise 8.3.6. *Prove that there exists no* Π^1_1*-maximal chain of hyperdegrees.*

***Exercise 8.3.7** (Harrington and Kechris [46]). *Prove that every Turing degree* $\geq \deg(\mathcal{O})$ *is a minimal cover.*

8.4 Martin's conjecture for Π^1_1-functions

8.4.1 Π^1_1-Martin's conjecture in L

Since Borel determinacy is a theorem of ZF + DC [93], the results in Chapter 7 hold under ZF + DC for $\underset{\sim}{\Delta}^1_1$-functions. Since Chapter 7 is concerned with total functions, by Proposition 1.1.8 (iv) the results apply to total $\underset{\sim}{\Pi}^1_1$-functions as well. However, the situation changes when it comes to partial Π^1_1-functions. Throughout this section, a function F is said to be Π^1_1 if its graph G^F is Π^1_1. We prove that Martin's conjecture fails for certain Π^1_1-functions under the assumption $V = L$. For a start, Lemma 7.2.1 says (under ZF $+AD+DC$) that $\leq_M$ is a linear ordering for increasing uniformly degree invariant functions. The conclusion fails if one assumes $V = L$.

Theorem 8.4.1. *Assume* $V = L$. *There is a* Π^1_1*-uniformly degree invariant function* F *that is increasing and order preserving on a cone such that for all* y, *there exist reals* $x_0, x_1 \geq_T y$ *satisfying* $F(x_0) \equiv_T x_0$ *and* $F(x_1) \geq_T \mathcal{O}^{x_1}$.

Proof. Assume $V = L$. Define P(x, y) as

$$y \in L_{\omega_1^y} \wedge \mathcal{O}^{\mathcal{O}^x} = (y)_0 \wedge (y)_1 = x \wedge (y)_2 \text{ is the } <_L\text{-least real such that } (y)_2 \not\leq_T (y)_0.$$

Since "$\mathcal{O}^{\mathcal{O}^x} = (y)_0$" is Π^1_1, P is a Π^1_1-cofinally progressive relation. By Theorem 8.2.4, there is a Π^1_1-set A satisfying the prescribed properties in $\mathfrak{I}$.

By the definition of P, every real is Turing below some real in A. Moreover, P ensures that $<_T$ is a well ordering and consistent with $<_L$ on A.

Define $F(x) = y$ if

(1) $y \in A$;

(2) $x \leq_T y$;

(3) $x >_T \emptyset \Rightarrow (\forall n)(x \not\leq_T ((y)_1)_n)$.

Roughly speaking, y is the $<_T$-least real in A such that $x \leq_T y$.

By (3) and the property of A, F is well defined. Furthermore, it is Π^1_1, increasing, order persevering and uniformly degree invariant ($x \equiv_T y$ implies $F(x) = F(y)$). Moreover, since $F(x) = x$ for every $x \in A$, we have $F(x) \equiv_T x$ cofinally.

For every real s, choose $x_s \in A$ such that $x_s \geq_T s$. Then for the $<_T$-least real $y \in A$ above x_s, there is a z coding $\{x \mid x <_L y \wedge x \in A\} = \{x \mid x <_T y \wedge x \in A\}$ so that $\mathrm{P}(z, y)$ holds. Obviously $x_s' \leq_T z' \leq_T \mathcal{O}^z$. By the definition of P,

$$\mathcal{O}^{x_s'} \leq_T \mathcal{O}^{\mathcal{O}^z} \leq_T y,$$

and $(y)_1 = z$. Thus

$$\{((y)_1)_n \mid n \in \omega\} = \{r \mid r \leq_T x \wedge r \in A\}.$$

Since $x_s' \not\leq_T r$ for all $r \in A$ with $r \leq_T x_s$, we have $x_s' \not\leq_T ((y)_1)_n$ for all $n \in \omega$. Hence $F(x_s') = y$. In other words, $F(x_s') \geq_T \mathcal{O}^{x_s'}$. □

Thus the function F is M-incomparable with the function G where $G(x) \equiv_T x'$ for every x.

Assume $\mathrm{ZF} + DC + AD$. By Lemma 7.1.4, if F is uniformly degree invariant and nonincreasing on a cone, then $F(x) <_T x$ on a cone $C = \{x \mid x \geq_T z_0\}$ for some z_0. We show that this also fails for Π^1_1-functions under $V = L$.

Theorem 8.4.2. *Assume $V = L$. There is a uniform degree invariant, nonincreasing Π^1_1-function F such that $F(x)|_T x$ on a cone. In particular, F is not constant on a cone.*

Proof. Let A be the Π^1_1-set as constructed in Theorem 8.4.1. Define $F(x) = y$ if and only if there is a real $z \equiv_T y'$ such that

(1) $z \in A$;

(2) the degree of y is minimal;

(3) $y = \Phi^z_e$ for some e such that for all $i < e$, if the degree of Φ^z_i is minimal then $(\Phi^z_i)' \not\equiv_T z$, and

(4) $x >_T \emptyset \Rightarrow (\forall n)(x \not\leq_T ((z)_1)_n)$.

We leave it to the reader to verify that F is the desired counterexample. □

By Lemma 7.2.4, $\mathrm{ZF}+AD+DC$ implies that $\leq_M$ is a well-ordering on the set of increasing uniformly degree invariant functions. This again fails under the assumption $V = L$.

Theorem 8.4.3. *Assume $V = L$. There is a sequence of uniform degree invariant, increasing, Π^1_1-functions $\{f_n\}_{n<\omega}$ such that $F_{n+1} <_M F_n$ for all n. Thus $<_M$ is not a prewellordering.*

Proof. Define $\mathrm{P}(x, y)$ by

$$y \in L_{\omega_1^y} \wedge (y)_0 = x \wedge (y)_1 \text{ is the } <_L\text{-least such that}$$
$$(y)_1 \not\leq_T (y)_0 \wedge y \equiv_T ((y)_0 \oplus (y)_1 \oplus (y)_2)'.$$

Clearly $\mathrm{P}(x, y)$ is a Π^1_1-cofinally progressive relation. By Theorem 8.2.4, there is a Π^1_1-set A with the property prescribed by P.

By the definition of P, every real is Turing below some element of A and $<_T$ is a well-ordering consistent with $<_L$ on A.

It is not difficult to see that there is an arithmetical set $W \subseteq \{(m, n, z) \mid m, n < \omega \wedge z \in 2^\omega\}$ such that $\{W^z_n\}_{n<\omega}$, where $W^z_n = \{m \mid (m, n, z) \in W\}$, is a sequence of z-r.e. sets satisfying $z <_T W^z_{n+1} <_T W^z_n$ for each n.

Let $F_n(x) = y$ if there exists a $z <_T y$ such that

(i) $z \in A$;

(ii) $z \geq_T x$;

(iii) $y = W^z_{n+1}$;

(iv) $x >_T \emptyset \Rightarrow (\forall m)(x \not\leq_T ((z)_0)_m)$.

Roughly speaking, y is W^z_{n+1} for the $<_T$-least z in A with $z \geq_T x$.

Clearly $\{f_n\}_{n<\omega}$ is a sequence of Π^1_1-functions. Note that for any $x <_T y$ in A, $x' \leq_T y$. Hence for every n, F_n is degree invariant and increasing.

By (iv) and the property of A, F is well defined. By the choice of $\{W^z_n\}_{n<\omega}$, $\{f_n\}_{n<\omega}$ is a $<_M$-descending sequence. □

8.4.2 The consistency strength of Π^1_1-Martin's conjecture

Given that the results in Chapter 7 are established under AD which is essentially a large cardinal assumption, one would expect a direct connection between Martin's conjecture and large cardinals. The first level for such an investigation is clearly Π^1_1. We prove that if Π^1_1-Martin's conjecture is true, then $0^\sharp$ exists.[2]

Define the *Friedman set* F^* as

$$F^* = \{x \mid (\forall \alpha < \omega^x_1)(\forall a \subseteq \alpha)(a \in L_{\omega^x_1} \rightarrow a \in L_{\alpha+1}[x])\}$$

It is not difficult to see that F^* is Σ^1_1.

Lemma 8.4.4. *F^* is cofinal in the Turing degrees.*

Proof. For any real z, let

$$F(z) = \{x \oplus z \mid (\forall \alpha < \omega^z_1)(\forall a \subseteq \alpha)(a \in L_{\omega^z_1} \rightarrow a \in L_{\alpha+1}[x \oplus z])\}.$$

Then $F(z)$ is $\Sigma^1_1(z)$ and nonempty. By Gandy's Basis Theorem (2.5.3) relativized to z, there is a x such that $\omega^{x\oplus z}_1 = \omega^z_1$ and $x \oplus z \in F(z)$. Then $x \oplus z \in F$. □

We will use the following result due to Harrington. A sketch of the proof is given in Exercises 8.4.3 to 8.4.8.

Lemma 8.4.5 (Harrington [45]). *If $0^\sharp$ does not exist, then $\bar{F} = 2^\omega \setminus F$ is cofinal in the Turing degrees.*

Theorem 8.4.6 (Chong, Wang and Yu [13]). *If there is no uniformly order-preserving Π^1_1-function G such that $\{x \mid G(x) = \mathcal{O}^{\mathcal{O}^x}\}$ and $\{x \mid G(x) = \mathcal{O}^{\mathcal{O}^{\mathcal{O}^x}}\}$ are cofinal in the Turing degrees, then $0^\sharp$ exists.*

2 See [13] for a higher level analysis of Martin's conjecture.

Proof. If $0^\sharp$ does not exist, then by Lemmas 8.4.4 and 8.4.5, both F^* and $\bar{F}^*$ are cofinal. Let $P(x, y)$ be an arithmetical predicate defined as

$$x \in \bar{F}^* \Leftrightarrow (\forall y)P(x, y).$$

Claim 1. *If $x \leq_T y$ are such that $x \in F^*$ and $y \in \bar{F}^*$, then $\mathcal{O}^x \leq_h y$.*

Proof. As $y \notin F^*$, there are $\alpha < \omega_1^y$ and $a \subseteq \alpha$ with $a \in L_{\omega_1^y} \setminus L_{\alpha+1}[y]$. Then $a \notin L_{\alpha+1}[x]$ and $\omega_1^x \leq \alpha < \omega_1^y$. As $x \leq_T y$, $\mathcal{O}^x \leq_h y$. □

By Claim 1, if $x \leq_T y$ are such that $x \in F^*$ and $y \in \bar{F}^*$, then $\mathcal{O}^{\mathcal{O}^x} \leq_T \mathcal{O}^y$.

Now define $G(x) = y$ as follows:

(1) $(y)_0 = 0^\frown\mathcal{O}^{\mathcal{O}^x} \wedge x \in \bar{F}^*$ or $(y)_0 = 1^\frown\mathcal{O}^{\mathcal{O}^{\mathcal{O}^x}} \wedge (\exists v \leq_T (y)_0)\neg P(x, v)$,

(2) $(y)_1 = \mathcal{O}^x$,

(3) Φ_e^x is partial $\Rightarrow (y)_{e+2} = \emptyset$,

(4) $u = \Phi_e^x$ is total $\Rightarrow$

(i) $(y)_0(0) = 0 \wedge (\exists v \leq_T (y)_1)\neg P(u, v) \wedge (y)_{e+2} = 1^\frown\mathcal{O}^{\Phi_i^{(y)_1}}$ where i is the least index so that $\mathcal{O}^{\mathcal{O}^u} = \mathcal{O}^{\Phi_i^{(y)_1}}$, or

(ii) $(y)_0(0) = 1 \wedge (\exists v \leq_T (y)_1)\neg P(u, v) \wedge (y)_{e+2} = 1^\frown\mathcal{O}^{\mathcal{O}^{\mathcal{O}^u}}$, or

(iii) $(\forall v \leq_T (y)_1)P(u, v) \wedge (y)_{e+2} = 0^\frown\mathcal{O}^{\mathcal{O}^u}$.

It is straightforward to verify that G is Π^1_1.

Claim 2. *If $x \in F^*$ then $G(x) \equiv_T \mathcal{O}^{\mathcal{O}^{\mathcal{O}^x}}$.*

Proof. Clearly $x \in F^* \Rightarrow \mathcal{O}^{\mathcal{O}^{\mathcal{O}^x}} \leq_T G(x)$ and $(G(x))_0 \oplus (G(x))_1 \leq_T \mathcal{O}^{\mathcal{O}^{\mathcal{O}^x}}$.

Given $e < \omega$, $\mathcal{O}^{\mathcal{O}^{\mathcal{O}^x}}$ can uniformly decide whether Φ_e^x is total. Suppose that Φ_e^x is total. To calculate $(G(x))_{e+2}(n)$ we need to check clauses (4 ii, 4 iii) above. Now the predicate $(\forall v \leq_T (y)_1)P(u, v)$ is $\Delta^1_1(\mathcal{O}^x)$, and hence recursive in $\mathcal{O}^{\mathcal{O}^{\mathcal{O}^x}}$. Once this predicate is decided, $\mathcal{O}^{\mathcal{O}^{\mathcal{O}^x}}$ will be able to complete the calculation with the use of two recursive functions f, g such that $u = \Phi_e^w \to \mathcal{O}^{\mathcal{O}^u} = \Phi_{f(e)}^{\mathcal{O}^{\mathcal{O}^{\mathcal{O}^w}}}$ and $u = \Phi_e^w \to \mathcal{O}^{\mathcal{O}^{\mathcal{O}^u}} = \Phi_{g(e)}^{\mathcal{O}^{\mathcal{O}^{\mathcal{O}^w}}}$. □

Claim 3. *If $x \in \bar{F}^*$ then $G(x) \equiv_T \mathcal{O}^{\mathcal{O}^x}$.*

Proof. This is similar to the proof of Claim 2, except for he final step in the calculation of $(G(x))_{e+2}(n)$. Now $\mathcal{O}^{\mathcal{O}^x}$ can decide whether (4 i) or (4 iii) holds, as in the above claim. If (4 iii) holds, then the calculation is again similar to the above. Suppose that (4 i) holds. Then $u \in F^*$. By Claim 1, $\mathcal{O}^u \leq_h x$ and thus $\mathcal{O}^{\mathcal{O}^u} \leq_T \mathcal{O}^x = (G(x))_1$. Now the index i prescribed in (4 i) may be found uniformly via a $\Pi^1_1(\mathcal{O}^x)$ procedure. Hence $\mathcal{O}^{\mathcal{O}^x}$ computes $(G(x))_{e+2}(n)$ uniformly. □

It follows from the last two claims that G is degree invariant. Moreover, G preserves $\leq_T$ by Claim 1.

To show that G is uniformly order preserving, let h be a recursive function such that $(\forall x, y, e)(x = \Phi_e^y \to \mathcal{O}^x = \Phi_{h(e)}^{\mathcal{O}^y})$. In addition, let s be another recursive function with

$$(\forall x, y, e, i)(x = \Phi^y_e \rightarrow \Phi^x_i = \Phi^y_{s(i)}).$$

Suppose that $x = \Phi^y_e$. Then

(1) $(G(x))_0 = (G(y))_{e+2}$,

(2) $(G(x))_1 = \Phi^{(G(y))_1}_{h(e)}$, and

(3) $(G(x))_{i+2} = (G(y))_{s(i)+2}$.

Hence G is as desired. □

8.4.3 Exercises

Exercise 8.4.1. *Use Theorem 6.5.6 to prove that if $0^\sharp$ exists, then Martin's conjecture for Π^1_1-uniformly degree invariant functions is true.*

Exercise 8.4.2. *Prove that there is a sentence φ in the language of partial ordering of the Turing degrees such that $\{\mathbf{x} \mid \langle \mathscr{D}(\geq \mathbf{x}), \leq \rangle \models \varphi\}$ is cofinal and if it contains a cone, then $0^\sharp$ exists.*

Exercise 8.4.3. *Suppose that $\alpha > \beta > \omega$ are countable admissible ordinals and g is a generic real produced by Steel forcing over L_α. Then for any ordinal $\gamma < \beta$ and any set $A \subseteq \gamma$, if $A \in L_\alpha \cap L_\beta[g]$, then $A \in L_\beta$.*

Exercise 8.4.4. *Suppose that the Friedman set F^* contains a cone of Turing degrees. Then there is a real x such that for any $y \geq_T x$, any $\alpha < \omega^y_1$ and any $A \subseteq \alpha$, if $A \in L$ then $A \in L_{\omega^y_1}[y]$.*

Exercise 8.4.5. *Assume that the Friedman set F^* contains a cone of Turing degrees. Show that there is a real x such that for any $y \geq_T x$ and countable ordinal α, if $L_\alpha[y]$ is admissible then α is a cardinal in L.*

Exercise 8.4.6. *Assume that the Friedman set F^* contains a cone of Turing degrees. Show that there is a real x such that for any $y \geq_T x$ and ordinal α, if $L_\alpha[y]$ is admissible then α is a cardinal in L.*

Exercise 8.4.7. *Suppose that x is a real such that for any ordinal α, if $L_\alpha[x]$ is admissible then α is a cardinal in L. Furthermore, assume that $V = L[x]$. Then there is an ordinal $\alpha < \aleph_2$ and an elementary embedding $\pi : L_\alpha[x] \rightarrow L_{\aleph_3}[x]$ such that*

- $\aleph_2 \in \text{Range}(\pi)$,
- *there is an ordinal $\beta < \alpha$ such that $\pi(\beta) > \beta$, and*
- $(\text{Range}(\pi))^\omega \subseteq \text{Range}(\pi)$.

Exercise 8.4.8. *As in Exercise 8.4.7, let β be the least ordinal such that $\pi(\beta) > \beta$. Then the set $\{X \in L \mid X \subseteq \beta \wedge \beta \in \pi(X)\}$ is an ω-complete nonprincipal ultrafilter over $L \cap \mathscr{P}(\beta)$. Thus there is a nontrivial elementary embedding of L into L.*

9 Independence results in recursion theory

9.1 Independent sets of Turing degrees

9.1.1 Locally countable partial ordering

Definition 9.1.1. A partially ordered set $\langle \mathbf{P}, \leq_P \rangle$ is *locally countable* if for any $p \in \mathbf{P}$, $|\{q \mid q \leq_P p\}| \leq \aleph_0$.

Definition 9.1.2. A function $f : \langle \mathbf{P}, \leq_P \rangle \to \langle \mathbf{Q}, \leq_Q \rangle$ is an *embedding* if $(\forall p_0, p_1 \in \mathbf{P})$ $(p_0 \leq p_1 \Leftrightarrow f(p_0) \leq_Q f(p_1))$.

Obviously $\langle \mathscr{D}, \leq_T \rangle$, the partial ordering of the Turing degrees, is locally countable. Sacks made the following conjecture.

Conjecture 9.1.3 (Sacks [117]). *Every locally countable partial ordering on 2^ω is embeddable into $\langle \mathscr{D}, \leq_T \rangle$.*

A positive solution is known assuming the Continuum Hypothesis:

Theorem 9.1.4 (Sacks [117]). *Assuming* ZFC + CH. *Every locally countable partially ordered set $\langle 2^\omega, \leq_P \rangle$ is embeddable into $\langle \mathscr{D}, \leq_T \rangle$.*

9.1.2 Independent sets of Turing degrees

As a "test question" for Conjecture 9.1.3, Sacks introduced the notion of independent sets and conjectured that every maximal independent set of Turing degrees has size $2^{\aleph_0}$.

Definition 9.1.5. Let $\langle \mathbf{P}, \leq_P \rangle$ be a partial ordering. $X \subseteq \mathbf{P}$ is independent if for any finite subset Y of X and any $p \in X \setminus Y$, there is a $q \in \mathbf{P}$ such that

$$p \not\leq_P q \wedge (\forall r \in Y)(r \leq_P q).$$

Theorem 9.1.6 (Sacks[117]). *Assume* ZFC + CH. *Every maximal independent set of Turing degrees has size $2^{\aleph_0}$.*

However, the conclusion of Theorem 9.1.6 is independent of ZFC:

Theorem 9.1.7 (Groszek and Slaman [40]). *There is a model $\mathcal{N} \models$ ZFC + ¬CH with a maximal independent set of Turing degrees of size $\aleph_1$.*

The rest of this section is devoted to a proof of Theorem 9.1.7. We begin with a ground model $\mathcal{M} \models$ ZFC + GCH. Let $\mathbb{P} = \langle P, \leq \rangle = \Pi_{\alpha<\omega_1} \mathbb{S}_\alpha$ be product Sacks forcing with countable support, where $\mathbb{S}_\alpha$ is Sacks forcing for $\alpha < \omega_1$. Every $\mathbb{P}$-generic set G may be viewed as an $(\omega_1)^M$-sequence of reals $\{G_\alpha\}_{\alpha<(\omega_1)^M}$ where $G_\alpha = \{i \mid (\alpha, i) \in G\}$. It is not difficult to see that the Turing degrees of $\{G_\alpha\}_{\alpha<(\omega_1)^M}$ form an independent set.

Lemma 9.1.8 (Baumgartner [5]). *If G is $\mathbb{P}$-generic over $\mathcal{M} = \langle M, \in\rangle$, then every countable set of ordinals in $M[G]$ is countable in M.*

Proof. The proof is a fusion argument similar to that of Theorem 6.2.11 and is left to the reader. □

Since $\mathcal{M} \models GCH$ and $\mathbb{P} = \langle \mathbf{P}, \leq\rangle$ is a notion of product forcing with countable support, we have $|\mathbb{P}| \leq (\aleph_1)^{\mathcal{M}}$. Thus $\mathbb{P}$ satisfies $(\aleph_2)^{\mathcal{M}}$-c.c. Together with Lemma 9.1.8, we have:

Lemma 9.1.9. *$\mathbb{P}$ preserves cardinals.*

Lemma 9.1.10. *For any real $x \in M$ and countable ordinal β, there is a countable $\alpha > \beta$ such that $x \leq_T G_\alpha$.*

Proof. We show that for any $x \in M$ and $\beta < (\omega_1)^M$, the set $D_{x,\beta} = \{q \in \mathbf{P} \mid q \Vdash (\exists \check{\alpha} > \check{\beta})(\check{x} \leq_T \dot{G}_{\check{\alpha}})\}$ is dense. Given $p \in \mathbf{P}$, there is a countable $\alpha > \beta$ such that $p(\alpha)$ is undefined. Choose any such ordinal α and let $q \leq p$ such that $q(\gamma) = p(\gamma)$ for $\gamma \neq \alpha$ and $q(\alpha) = \{\sigma \in 2^{<\omega} \mid (\forall n < |\sigma|)(n \text{ is even } \to \sigma(n) = x(\frac{n}{2}))\}$. Then $q(\alpha)$ is a pointed tree in which every infinite path Turing computes x. Hence $q \in D_{x,\beta}$. □

For any countable ordinal $\alpha \in M$, let $R_\alpha \subseteq \omega \times \omega$ be a well-ordering of $\omega \times \omega$ in M of order type α. Let $|n|_{R_\alpha}$ denote the order type of R_α restricted to the set of numbers less than n under R_α. Define $\oplus_{R_\alpha}\{G_\beta\}_{\beta<\alpha} = \{(n, m) \mid G_\beta(m) = 1 \wedge \beta < \alpha \wedge |n|_{R_\alpha} = \beta\}$.

Lemma 9.1.11 (Baumgartner [5]). *For every real $x \in M[G] \setminus M$, there is a countable ordinal α such that $x \leq_T \oplus_{R_\alpha}\{G_\beta\}_{\beta<\alpha}$. Hence $\mathcal{M}[G] \models$ CH.*

Proof. The proof is again a fusion argument similar to that of Theorem 6.2.11. □

Now let $X \supseteq G$ be a set of reals whose Turing degrees are maximally independent in $\mathcal{M}[G]$. We call such an X a maximally independent set.

The next step is to destroy GCH by adding reals to $M[G]$ but keeping X as a maximally independent set. Let $\dot{\mathbb{Q}} = \langle \dot{\mathbf{Q}}, \dot{\leq}\rangle \in M$ be a $\mathbb{P}$-name such that for any $p \in \mathbf{P}$,

$$p \Vdash \text{“}\dot{\mathbb{Q}} \text{ is a product Sacks forcing of length } \check{\omega}_2 \text{ with countable support”}$$

and

$$p \Vdash (\forall \dot{q})(\dot{q} \in \dot{\mathbf{Q}} \wedge \dot{q}(\check{\alpha}) \text{ is defined} \to \check{\omega}_1 \leq \check{\alpha} < \check{\omega}_2).$$

Hence $\mathbb{P} * \dot{\mathbb{Q}}$ is a two step iterated forcing. By Theorem 6.1.11 and the assumption that $\mathcal{M}[G] \models$ CH for any $\mathbb{P}$-generic set G, we have the following lemma.

Lemma 9.1.12. *For any $p \in \mathbf{P}$, $p \Vdash \dot{\mathbb{Q}}$ is $\check{\aleph}_2$-c.c.*

Lemma 9.1.13. *$\mathbb{P} * \dot{\mathbb{Q}}$ preserves cardinals.*

Proof. By Lemma 9.1.9, it is sufficient to prove that for any $\mathbb{P}$-generic set G over $\mathcal{M}$,

$$\mathcal{M}[G] \models \mathbb{Q} \text{ preserves cardinals.}$$

As in Lemma 9.1.8, $\mathcal{M}[G] \models \mathbb{Q}$ preserves $\aleph_1$. By Lemma 9.1.12, $\mathcal{M}[G] \models \mathbb{Q}$ preserves κ for all $\kappa \geq \aleph_2$. Hence $\mathcal{M}[G] \models \mathbb{Q}$ preserves cardinals. □

Let H be a $\mathbb{Q}$-generic set over $\mathcal{M}[G]$. Then H may be viewed as an ω_2-sequence of reals $\{H_\alpha\}_{\omega_1 \leq \alpha < \omega_2}$. Thus $\mathbb{Q}$ adds at least $\aleph_2$-many reals to $\mathcal{M}[G]$ and we have the following:

Lemma 9.1.14. *If $(p, \dot{q}) \in (\mathbf{P}, \dot{\mathbf{Q}})$, then $(p, \dot{q}) \Vdash \neg\mathrm{CH}$.*

Lemma 9.1.15. *If $(p_0, \dot{q_0}) \in (\mathbf{P}, \dot{\mathbf{Q}})$, $\alpha < \omega_1$ and $(p_0, \dot{q_0}) \Vdash$ "$\dot{x}$ is not in $\mathcal{M}[\dot{G}]$", then there is a $z \in M$ and a condition $(p_1, \dot{q_1}) \leq (p_0, \dot{q_0})$ such that $(p_1, \dot{q_1}) \Vdash \dot{G_\alpha} \leq_T \dot{x} \oplus \check{z}$.*

By Lemma 9.1.10 and Lemma 9.1.15, $\{G_\alpha\}_{\alpha<\omega_1}$ is still a maximally independent set in $\mathcal{M}[G][H]$. The rest of this section is devoted to a proof of Lemma 9.1.15.

Given a perfect tree T, we associate with it the natural homeomorphism $t_T : 2^{<\omega} \to T$ such that $t_T(\emptyset)$ is a splitting node in T and for every $\sigma \in T$, $\sigma \preceq t_T(\emptyset)$ or $\sigma \succeq t_T(\emptyset)$. Suppose that H is $\mathbb{Q}$-generic over $\mathcal{M}[G]$. Let $(G, \{(\alpha, H_\alpha)) \mid \alpha < \omega_2\})$ be a $\mathbb{P} * \mathbb{Q}$-generic set and let $p \in \mathbf{P}$ be a condition such that

$$p \Vdash \text{"}\dot{\mathbb{Q}} \text{ is a product Sacks forcing with countable support of length } \omega_2\text{"}.$$

For any set $A \subseteq \omega_2$, we use $\oplus_{\alpha\in A} H_\alpha$ to denote $\{(\alpha, H_\alpha) \mid \alpha \in A\}$. Fix a condition $(p_0, \dot{q}_0) \in (G, \dot{H})$ such that $p_0 \leq p$ and

$$(p_0, \dot{q}_0) \Vdash \dot{x} \in \dot{2^\omega} \setminus \mathcal{M}[\dot{G}].$$

We define a fusion sequence of conditions $\{(p_\sigma, \dot{q}_\sigma)\}_{\sigma\in 2^{<\omega}}$ below $(p_0, \dot{q}_0)$, a sequence of ordinals $\{\alpha_n\}_{n<\omega}$, and a sequence $\{\nu_\sigma\}_{\sigma\in 2^{<\omega}} \subseteq 2^{<\omega}$ with the following properties:

(1) $(\forall\sigma)((p_\sigma, \dot{q}_\sigma) \Vdash \dot{x}\upharpoonright|\nu_\sigma| = \nu_\sigma)$ and $\sigma \preceq \tau \to \nu_\sigma \preceq \nu_\tau \wedge (p_\tau, \dot{q}_\tau) \leq (p_\sigma, \dot{q}_\sigma)$;

(2) For all n, for all $\sigma, \tau \in 2^n$ and all α,

$$\begin{aligned}
&\alpha < \omega_1 \to \\
&((\exists m < n)(\alpha = \alpha_m \wedge \sigma(m) \neq \tau(m)) \to t_{p_\sigma(\alpha)}(\emptyset)|t_{p_\tau(\alpha)}(\emptyset)) \wedge \\
&((\forall m < n)(\alpha = \alpha_m \to \sigma(m) = \tau(m)) \to p_\sigma(\alpha) = p_\tau(\alpha))
\end{aligned}$$

and

$$\begin{aligned}
&p_\sigma = p_\tau \to \\
&(p_\sigma \Vdash \alpha \geq \omega_1 \to \\
&((\exists m < n)(\alpha = \alpha_m \wedge \sigma(m) \neq \tau(m))) \to t_{\dot{q}_\sigma(\alpha)}(\emptyset)|t_{\dot{q}_\tau(\alpha)}(\emptyset)) \wedge \\
&((\forall m < n)(\alpha = \alpha_m \to \sigma(m) = \tau(m))) \to \dot{q}_\sigma(n < \omega\alpha) = \dot{q}_\tau(\alpha)); \text{ and}
\end{aligned}$$

(3) $(\forall\alpha)((\exists\sigma)(p_\sigma(\alpha) \text{ is defined} \vee p_\sigma \Vdash \text{"}\dot{q}_\sigma(\alpha) \text{ is defined"})) \to (\forall n)(\exists m > n)(\alpha = \alpha_m)))$; and

(4) $(\forall n)(\forall\sigma \in 2^n)(\forall\tau \in 2^n)(\exists m < n)(\exists\alpha < \omega_1)(\alpha = \alpha_m \wedge \sigma(m) \neq \tau(m)) \to \nu_\sigma|\nu_\tau)$.

Now for $n < \omega$ and $\alpha < \omega_1$, let

$$p_n(\alpha) = \bigcup_{\sigma \in 2^n} p_\sigma(\alpha),$$

and

$$p_\omega = \bigcap_{n<\omega} p_n.$$

Suppose that $p_\omega \in G$. Then for any n and $m \leq n$, there is a $\sigma_m \in 2^n$ such that if $\alpha_m < \omega_1$, then $G_{\alpha_m} \in [p_{\sigma_m}(\alpha_m)]$. Define $\tau \in 2^n$ so that $\tau(m) = \sigma_m(m)$ for $m \leq n$. By (2), for each such m, $p_\tau(\alpha_m) = p_{\sigma_m}(\alpha_m)$, and for each $\sigma \in 2^n$ and every other α, $p_\tau(\alpha) = p_\sigma(\alpha)$. Thus $G_\alpha \in [p_\tau(\alpha)]$ whenever $p_\tau(\alpha)$ is defined. Hence $p_\tau \in G$. We have the following lemma.

Lemma 9.1.16. *If $p_\omega \in G$, then for each n there is a $\sigma \in 2^n$ such that $p_\sigma \in G$.*

If $\alpha \geq \omega_1$, let $\dot{q}_n$ be a name so that

$$q_n(\alpha) = \bigcup_{\sigma \in 2^n \wedge p_\sigma \in G} q_\sigma(\alpha).$$

By Lemma 9.1.16, $\dot{q}_n$ is well defined. For $\alpha > \omega_1$, let $\dot{q}_\omega$ be a name such that

$$q_\omega(\alpha) = \bigcap_{n<\omega} q_n(\alpha).$$

Lemma 9.1.17. $(p_\omega, \dot{q}_\omega) \Vdash (\forall \sigma)(p_\sigma \in \dot{G} \leftrightarrow v_\sigma = \dot{x} \upharpoonright |v_\sigma|)$.

Proof. By (1), $(p_\omega, \dot{q}_\omega) \Vdash (\forall \sigma)(p_\sigma \in G \to v_\sigma = \dot{x} \upharpoonright |v_\sigma|)$ for any generic set (G, H) with $(p_\omega, q_\omega) \in (G, H)$ and $x \upharpoonright |v_\sigma| = v_\sigma$. By Lemma 9.1.16, there is a $\tau \in 2^{|\sigma|}$ such that $p_\tau \in G$. By (2), $\sigma(m) = \tau(m)$ for any m with $\alpha_m < \omega_1$. By (2) again, $p_\sigma = p_\tau \in G$. □

Obviously $(p_\omega, \dot{q}_\omega) \leq (p_0, \dot{q}_0)$. In fact, for any $(p_0', \dot{q}_0') \leq (p_0, \dot{q}_0)$, there is such a sequence below $(p_0', \dot{q}_0')$. Hence there is a $(p_\omega, \dot{q}_\omega)$ such that $(p_\omega, q_\omega) \in (G, H)$. Now for any $\alpha < \omega_1$, let $z_\alpha = \{(p_\sigma(\alpha), v_\sigma) \mid \sigma \in 2^{<\omega}\}$. Then z_α may be viewed as a real in M. By Lemma 9.1.17, $G_\alpha \leq x \oplus z_\alpha$, proving Lemma 9.1.15.

Next is the construction of the sequences satisfying (1)–(4), of which meeting condition (2) is the key. We need a technical lemma.

Lemma 9.1.18. *If $(p, \dot{q}) \leq (p_0, \dot{q}_0)$, then there exist an n and a condition $(p', \dot{q}') \leq (p, \dot{q})$ such that for any $(p'', \dot{q}') \leq (p', \dot{q}')$, $(p'', \dot{q}') \nVdash \dot{x}(n) = i$ for $i \leq 2$.*

Proof. Suppose not. Then there is a $(p_1, \dot{q}_1) \leq (p_0, \dot{q}_0)$ such that for any n and $(p', \dot{q}') \leq (p_1, \dot{q}_1)$, there exist a $(p'', \dot{q}') \leq (p', \dot{q}')$ and an $i_n \leq 2$ satisfying $(p'', \dot{q}') \Vdash \dot{x}(n) = i_n$. Then the set

$$D_n = \{p \mid p \nleq p_1 \vee (\exists i_n)(p \leq p_1 \wedge (p, \dot{q}_1) \Vdash \dot{x}(n) = i_n)\}$$

is an element of M and is dense in $\mathbb{P}$. Fix a generic set (G', H') such that $p_1 \in G'$. Then $val(\dot{x}, (G', H'))(n) = i_n$ if and only if there exists a $p \in D_n \cap G'$ such that $(p, \dot{q}_1) \Vdash \dot{x}(n) = i_n$. Thus $val(\dot{x}, (G', H')) \in \mathcal{M}[G']$, a contradiction. □

Let $f : \omega \to \omega^3$ be a recursive bijection such that if $f(n) = (n_0, n_1, n_2)$ and $n > 0$, then $n > n_0$. The construction of the sequences proceeds as follows:

Stage 0. Let $(p_\emptyset, \dot{q}_\emptyset) = (p_0, \dot{q}_0)$ and $v_\emptyset = \emptyset$. Let $\{\gamma_i\}_{i<\omega}$ be an enumeration of the countable set

$$\{\alpha < \omega_2 \mid p_0(\alpha) \text{ is defined} \vee p_0 \Vdash \text{“}\dot{q}_0(\alpha) \text{ is defined”}\}.$$

Let $\beta_{(0,i,j)} = \gamma_i$ for $i, j < \omega$. Each γ_i occurs infinitely many times on the list to ensure that it is diagonized infinitely often.

Stage $n+1$. We decompose this into $2^n + 1$-many substages.

At substage $j \le 2^n$, define $\{v^j_\sigma\}_{\sigma \in 2^{n+1}}$ and $\{(p^j_\sigma, \dot{q}^j_\sigma)\}_{\sigma \in 2^{n+1}}$ such that $p^j_{\sigma^\frown 0}(\alpha) = p^j_{\sigma^\frown 1}(\alpha)$ for each $\alpha \neq \beta_{f(n)}$ and $\sigma \in 2^n$.

For $\sigma \in 2^n$ and $i \in 2$, let $v^j_{\sigma^\frown i} = v^j_\sigma$. If $\beta_{f(n)} < \omega_1$, then

$$p^0_{\sigma^\frown i}(\alpha) = \begin{cases} p_\sigma(\alpha) & : \ \alpha \neq \beta_{f(n)}, \\ \{\tau \in p_\sigma(\alpha) \mid \tau \prec t_{p_\sigma(\alpha)}(i) \vee \tau \succeq t_{p_\sigma(\alpha)}(i)\} & : \ \text{Otherwise.} \end{cases}$$

Let $\dot{q}^0_{\sigma^\frown i} = \dot{q}_\sigma$. If $\beta_{f(n)} \ge \omega_1$, then $p^0_{\sigma^\frown i} = p_\sigma$ and $\dot{q}^0_{\sigma^\frown i}$ is a name such that

$$p_\sigma \Vdash \dot{q}^0_{\sigma^\frown i}(\alpha) = \begin{cases} \dot{q}_\sigma(\alpha) & : \alpha \neq \beta_{f(n)}, \\ \{\tau \mid p_\sigma \Vdash \tau \in \dot{q}_\sigma(\alpha) \wedge (\tau \prec t_{\dot{q}_\sigma(\alpha)}(i) \vee \\ \quad \tau \succeq t_{\dot{q}_\sigma(\alpha)}(i))\} & : \text{Otherwise.} \end{cases}$$

Hence $\{p^0_\sigma\}_{\sigma \in 2^{n+1}}$ satisfies (2). There are now two cases to consider.

Case 1. $\beta_{f(n)} \ge \omega_1$. Since (4) only concerns ordinals not less then ω_1, let $(p_{\sigma^\frown i}, \dot{q}_{\sigma^\frown i}) = (p^0_{\sigma^\frown i}, \dot{q}^0_{\sigma^\frown i})$, $\beta_{(n+1,i,j)} = \beta_{(n,i,j)}$, and $v_{\sigma^\frown i} = v_\sigma$ for every $\sigma \in 2^n$ and $i < 2$, and go to the next stage.

Case 2. $\beta_{f(n)} < \omega_1$. Fix an enumeration $\{\sigma_j\}_{1 \le j \le 2^n}$ of 2^n.

Substage $j+1$. By Lemma 9.1.18, there exist an $m > |v^j_{\sigma_j}|$ and a $(p', \dot{q}') \le (p^j_{\sigma_j^\frown 0}, \dot{q}^j_{\sigma_j^\frown 0})$ such that for any $(p'', \dot{q}') \le (p', \dot{q}')$, we have $(p'', \dot{q}') \nVdash \dot{x}(m) = i$. Define

$$p''(\alpha) = \begin{cases} p'(\alpha) & : \ \alpha \neq \beta_{f(n)}, \\ p^j_{\sigma_j^\frown 1}(\alpha) & : \ \text{Otherwise.} \end{cases}$$

Note that $p^j_{\sigma_j^\frown 0}(\alpha) = p^j_{\sigma_j^\frown 1}(\alpha)$ for any $\alpha \neq \beta_{f(n)}$. We have $(p'', \dot{q}^j_{\sigma_j^\frown 1}) \le (p^j_{\sigma_j^\frown 1}, \dot{q}^j_{\sigma_j^\frown 1})$. Let $(r_1, \dot{s}_1) \le (p'', \dot{q}^j_{\sigma_j^\frown 1})$ so that for some $v^{j+1}_{\sigma_j^\frown 1} \succ v^j_{\sigma_j}$ in 2^{m+1}, $(r_1, \dot{s}_1) \Vdash \dot{x}(m) = 1$ and $(r_1, \dot{s}_1) \Vdash \dot{x} \restriction m+1 = v^{j+1}_{\sigma_j^\frown 1}$. Define

$$r'_0(\alpha) = \begin{cases} r_1(\alpha) & : \ \alpha \neq \beta_{f(n)}, \\ p'(\alpha) & : \ \text{Otherwise.} \end{cases}$$

Then $r'_0 \le p'$ and so $(r'_0, \dot{q}') \nVdash \dot{x}(n) = i$ for $i < 2$. Hence there is a $(r_0, \dot{s}_0) \le (r'_0, \dot{q}')$ such that $(r_0, \dot{s}_0) \Vdash \dot{x}(m) = 0$ and $(r_0, \dot{s}_0) \Vdash \dot{x} \restriction m+1 = v^j_{\sigma_j^\frown 0}$ for some $v^j_{\sigma_j^\frown 0} \succ v^j_{\sigma_j}$ in 2^{m+1}. Let $(p^{j+1}_{\sigma_j^\frown 0}, \dot{q}^{j+1}_{\sigma_j^\frown 0}) = (r_0, \dot{s}_0)$. Define

$$p^{j+1}_{\sigma_j^\frown 1}(\alpha) = \begin{cases} r_0(\alpha) & : \ \alpha \neq \beta_{f(n)}, \\ r_1(\alpha) & : \ \text{Otherwise.} \end{cases}$$

Then $(p^{j+1}_{\sigma_j\frown 1}, \dot{s}_1) \le (r_1, \dot{s}_1)$. Let $\dot{q}^{j+1}_{\sigma_j\frown 1} = \dot{s}_1$. Now for $j' \neq j$ and $i < 2$, let $v^{j+1}_{\sigma_{j'}\frown i} = v^{j}_{\sigma_{j'}}$, and

$$p^{j+1}_{\sigma_{j'}\frown i}(\alpha) = \begin{cases} p^{j+1}_{\sigma_j\frown 0}(\alpha) & : \; p^{j}_{\sigma_{j'}\frown i}(\alpha) = p^{j+1}_{\sigma_j\frown 0}(\alpha), \\ p^{j+1}_{\sigma_j\frown 1}(\alpha) & : \; p^{j}_{\sigma_{j'}\frown i}(\alpha) = p^{j+1}_{\sigma_j\frown 1}(\alpha), \\ p^{j}_{\sigma_{j'}\frown i}(\alpha) & : \; \text{Otherwise.} \end{cases}$$

Let $\dot{q}^{j+1}_{\sigma_{j'}\frown i}$ be a name such that

$$\dot{q}^{j+1}_{\sigma_{j'}\frown i}(\alpha) = \begin{cases} \dot{q}^{j+1}_{\sigma_j\frown 0}(\alpha) & : \; p^{j+1}_{\sigma_{j'}\frown i} = p^{j+1}_{\sigma_j\frown 0} \wedge p^{j+1}_{\sigma_{j'}\frown i} \Vdash \dot{q}^{j}_{\sigma_{j'}\frown i}(\alpha) = \dot{q}^{j}_{\sigma_j\frown 0}(\alpha), \\ \dot{q}^{j+1}_{\sigma_j\frown 1}(\alpha) & : \; p^{j+1}_{\sigma_{j'}\frown i} = p^{j+1}_{\sigma_j\frown 1} \wedge p^{j+1}_{\sigma_{j'}\frown i} \Vdash \dot{q}^{j}_{\sigma_{j'}\frown i}(\alpha) = \dot{q}^{j}_{\sigma_j\frown 1}(\alpha), \\ \dot{q}^{j}_{\sigma_{j'}\frown i}(\alpha) & : \; \text{Otherwise.} \end{cases}$$

This completes the construction at substage $j + 1$.

For $\sigma \in 2^n$ and $i < 2$, let $(p_{\sigma\frown i}, \dot{q}_{\sigma\frown i}) = (p^{2^n}_{\sigma\frown i}, \dot{q}^{2^n}_{\sigma\frown i})$ and $v_{\sigma\frown i} = v^{2^n}_{\sigma\frown i}$. Let $\{\gamma_i\}_{i<\omega}$ be an enumeration of the countable set

$$\{\alpha < \omega_2 \mid (\exists\sigma)(\sigma \in 2^{n+1} \wedge (p_\sigma(\alpha) \text{ is defined} \vee p_\sigma \Vdash \text{“}\dot{q}_\sigma(\alpha) \text{ is defined”}))\}.$$

Let $\beta_{n+1,i,j} = \gamma_i$ for $i, j < \omega$. This ends the construction at stage $n + 1$. Finally let $\alpha_n = \beta_{f(n)}$ for each n.

It is immediate that the sequences defined satisfy conditions (1)–(4) and hence the proof of Theorem 9.1.7 is complete. □

9.1.3 Exercises

Exercise 9.1.1. *Prove that there is a partial ordering $\langle 2^\omega, \le_P\rangle$ into which every locally countable partial ordering of size not greater than $2^{\aleph_0}$ is embeddable.*

Exercise 9.1.2. *Prove Theorem 9.1.4.*

Exercise 9.1.3 (Sacks [117]). *A partial ordering $\langle \mathbf{P}, \le_P\rangle$ is locally finite if for any $p \in \mathbf{P}$, $|\{q \mid q \le_P p\}| < \aleph_0$. Prove that every locally finite partial ordering in $\langle 2^\omega, \le_P\rangle$ is embeddable into $\langle \mathscr{D}, \le\rangle$.*

Exercise 9.1.4. *Prove that there is an independent set X of Turing degrees of size $2^{\aleph_0}$.*

Exercise 9.1.5. *Prove Theorem 9.1.6.*

9.2 Embedding locally finite upper semilattices into $\langle \mathscr{D}, \le\rangle$

9.2.1 Embedding an upper semilattice

Definition 9.2.1. A partial ordering $\langle P, \le\rangle$ is an *upper semilattice* if for any $p, q \in P$, there is an $r \in P$ such that r is the $\le$-least upper bound of p and q.

We use $p \cup q$ to denote the least upper bound of p and q. Clearly $\langle \mathcal{D}, \leq \rangle$ is an upper semilattice. The following strengthens Exercise 9.1.3 under an additional hypothesis.

Theorem 9.2.2 (Sacks[117]). *Assume* CH. *Every locally finite upper semilattice $\langle 2^{\omega}, \leq, \cup \rangle$ is embeddable into $\langle \mathcal{D}, \leq \rangle$.*

In [117], Sacks asked whether the conclusion of Theorem 9.2.2 is a theorem of ZFC. Groszek and Slaman provided a negative answer to the question.

Theorem 9.2.3 (Groszek and Slaman [40]). *There is a model $\mathcal{N} \models \mathrm{ZFC} + 2^{\aleph_0} = \aleph_2$ in which there is a locally finite upper semilattice $\langle \omega_2, \leq, \cup \rangle$ not embeddable into $\langle \mathcal{D}, \leq, \cup \rangle$.*

9.2.2 The proof of Theorem 9.2.3

Suppose that $\mathcal{M} \models \mathrm{ZFC} + GCH$. Let $\{X_\beta\}_{\beta<\omega_2}$ be a one-one enumeration of $\mathcal{P}(\omega_1)$ in $\mathcal{M}$. Let $\mathbb{U} = \langle \mathbf{U}, \leq_u, \cup \rangle$ be an upper semilattice in $\mathcal{M}$ such that
(1) $|\mathbf{U}| = \aleph_2$;
(2) $\omega_2 \subseteq \mathbf{U}$, and
(3) $(\forall \alpha < \omega_1)(\forall \beta < \omega_2)(\beta > \omega_1 \to (\omega_1 \leq_u \alpha \cup \beta \leftrightarrow \alpha \in X_\beta))$.

Let $\mathbb{P} = \Pi_{\alpha<\omega_2} \mathbb{P}_\alpha$ be an ω_2-product with finite support of Cohen forcing conditions. By Theorem 6.1.11, we have the following:

Lemma 9.2.4. $\mathbb{P}$ *is c.c.c. and hence preserves cardinals.*

Suppose G is $\mathbb{P}$-generic. Then G may be viewed as an ω_2-sequence $\{G_\alpha\}_{\alpha<\omega_2}$ of reals. By Lemma 6.1.6, we have

Lemma 9.2.5. *For any $\mathbb{P}$-generic set G, $\mathcal{M}[G] \models 2^{\aleph_0} = \aleph_2$.*

Suppose that there is a $\mathbb{P}$-generic set $\mathcal{M}[G] \models (\exists f)$("$f$ is an embedding of $\mathbb{U} \to \langle \mathcal{D}, \leq \rangle$)". Let p_0 be such that

$$p_0 \Vdash \text{"}\dot{f} \text{ is an embedding of } \mathbb{U} \text{ into } \langle \dot{\mathcal{D}}, \leq \rangle\text{"}.$$

Then for $\alpha \leq \omega_1$ and $n < \omega$, there is a condition $p_{\alpha,n} \leq p_0$ in G such that

$$p_{\alpha,n} \Vdash \dot{f}(\alpha)(n) = i$$

for some $i < 2$.

Given a condition p, let $\alpha_p = \min\{\alpha \mid (\forall \beta \geq \alpha)(p(\alpha) \text{ is undefined})\}$. Set

$$\alpha_0 = \sup\{\alpha_{p_{\gamma,n}} \mid \gamma < \omega_1\}.$$

Then $\alpha_0 < \omega_2$. Let $\mathbb{P}_{<\alpha_0} = \Pi_{\beta<\alpha_0} \mathbb{P}_\beta$ and $\mathbb{P}_{\geq\alpha_0} = \Pi_{\beta\geq\alpha_0} \mathbb{P}_\beta$. Then $f(\alpha) \in \mathcal{M}[G_{<\alpha_0}]$ where $G_{<\alpha_0} = \{(\beta, G_\beta) \mid \beta < \alpha_0\}$ is a $\mathbb{P}_{<\alpha_0}$-generic set. Since $\mathbb{P} = \mathbb{P}_{<\alpha_0} \times \mathbb{P}_{\geq\alpha_0}$, by Theorem 6.1.9, $G_{\geq\alpha_0} = \{(\beta, G_\beta) \mid \beta \geq \alpha\}$ is a $\mathbb{P}_{\geq\alpha_0}$-generic set over $\mathcal{M}[G_{<\alpha_0}]$. Obviously both $\mathbb{P}_{<\alpha_0}$ and

$\mathbb{P}_{\geq\alpha_0}$ preserve cardinals. Moreover, $\mathcal{M}[G_{<\alpha_0}] \models$ GCH. Thus without loss of generality, we may assume that $\mathcal{M}[G_{<\alpha_0}] = \mathcal{M}$. In other words, for any $\alpha \leq \omega_1$, there is a real $x_\alpha \in M$ such that $p_0 \Vdash \dot{f}(\alpha) = x_\alpha$.

Now for $\alpha < \omega_1 < \beta < \omega_2$, we have

$$p_0 \Vdash \text{“}x_{\omega_1} \leq_T \dot{f}(\beta) \oplus x_\alpha\text{”} \Leftrightarrow \alpha \in X_\beta.$$

By Proposition 6.1.17 and a rearrangement of p_0 (if necessary), we may assume that $p_0 \in \mathbb{P}_{<\omega_1}$. Let $\alpha_{p_0} = \min\{\alpha \mid (\forall \beta \geq \alpha)(p_0(\alpha) \text{ is undefined})\}$. Then α_{p_0} is a countable ordinal. Since only at most countably many ordinals occur in $\dot{f}(\beta)$, for $\beta > \omega_1$, if $\dot{f}(\beta)$ is not a $\mathbb{P}_{<\omega_1}$-name then there is a permutation i_β switching ordinals in $f(\beta)$ to countable ordinals greater than α_{p_0}. In other words, i_β induces an automorphism of $\mathbb{P}$ so that $i_\beta(p_0) = p_0$ but $i_\beta(\dot{f}(\beta)) \in \mathbb{P}_{<\omega_1}$. By Proposition 6.1.17 again,

$$p_0 \Vdash \text{“}x_{\omega_1} \leq_T i_\beta(\dot{f}(\beta)) \oplus x_\alpha\text{”} \Leftrightarrow \alpha \in X_\beta.$$

Now in $\mathcal{M}$, there are at most $2^{\aleph_0} = \aleph_1$ many $\mathbb{P}_{<\omega_1}$-names for reals. Thus there exist $\beta_0 \neq \beta_1$ such that $i_\beta(\dot{f}(\beta_0)) = i_\beta(\dot{f}(\beta_1))$. Hence $X_{\beta_0} = X_{\beta_1}$, contradicting the assumption.

Let $\mathcal{N} = \mathcal{M}[G]$. Then $\mathcal{N} \models$ “$\mathbb{U}$ is not embeddable into $\langle \mathcal{D}, \leq, \cup\rangle$”. This completes the proof of Theorem 9.2.3.

Remark 9.2.6. Both the conclusions of Theorem 9.1.4 (see [4]) and Theorem 9.2.2 (see [167]) are provable under ZFC + *MA* + ¬CH.

9.2.3 Exercises

Exercise 9.2.1. *Prove that given a model $\mathcal{M}$ of* ZFC, *if both g_1 and g_2 are Cohen generic reals over $\mathcal{M}$, then $\mathcal{M}[g_1] \equiv \mathcal{M}[g_2]$, i.e. elementarily equivalent. Show that this is not true for forcing notions in general.*

Exercise 9.2.2 (Simpson [4]). *Assume* ZFC + *MA* + ¬CH. *Prove that every locally countable partial ordering $\langle 2^\omega, \leq_P\rangle$ is embeddable into $\langle \mathcal{D}, \leq\rangle$.*

9.3 Cofinal chains in $\mathcal{D}$

9.3.1 Existence of a cofinal maximal chain of order type ω_1

Definition 9.3.1. Given a partial ordering $\langle P, <_P\rangle$, a set $A \subseteq P$ is a *cofinal chain* if the members of A are linearly ordered under P and for any $z \in P$, there is an $x \in A$ such that $z \leq_P x$.

It is not difficult to see that under ZFC, there is a cofinal chain of Turing degrees if and only if CH holds. A natural question is whether there exists a cofinal maximal chain of order type ω_1 in $\langle \mathcal{D}, \leq\rangle$ assuming ZFC + CH. The question is considered in this section.

The following fact is immediate.

Lemma 9.3.2. *Assume* ZF. *There exist* x_0 *and* x_1 *of minimal degree such that* $\emptyset'' \equiv_T x_0 \oplus x_1 \equiv_T x_0'' \equiv_T x_1''$.

Lemma 9.3.3. *Assume* ZF. *There is a maximal chain* $\mathbf{0} = \mathbf{c}_0 < \mathbf{c}_1 < \ldots < \mathbf{c}_\omega = \mathbf{0}''$ *in* $[\mathbf{0}, \mathbf{0}'']$.

Proof. Fix an enumeration $(\mathbf{d}_n : n \in \omega)$ of degrees below $\mathbf{0}''$. Let $\mathbf{c}_0 = \mathbf{0}$.

Suppose that $\mathbf{c}_n$ is defined with $\mathbf{c}_n'' = \mathbf{0}''$. Relativize Lemma 9.3.2 to obtain degrees $\mathbf{x}_0$ and $\mathbf{x}_1$ minimal over $\mathbf{c}_n$ such that $\mathbf{x}_0 \cup \mathbf{x}_1 = \mathbf{c}_n'' = \mathbf{0}'' = \mathbf{x}_0'' = \mathbf{x}_1''$. If $\mathbf{d}_n < \mathbf{0}''$ then there is an $i < 2$ with $\mathbf{x}_i \not\leq \mathbf{d}_n$, and let $\mathbf{c}_{n+1} = \mathbf{x}_i$. If $\mathbf{d}_n = \mathbf{0}''$, let $\mathbf{c}_{n+1} = \mathbf{x}_0$.

Finally let $\mathbf{c}_\omega = \mathbf{0}''$. □

Theorem 9.3.4 (Wang, Wu and Yu [168]). *Assume* ZFC. *Then* CH *holds if and only if there is a cofinal maximal chain in* $\mathcal{D}$ *of order type* ω_1.

Proof. ($\Rightarrow$) Assume CH and let $\{\mathbf{d}_\alpha \mid \alpha < \omega_1\}$ be an enumeration of the Turing degrees. We construct inductively a sequence $\{C_\alpha\}_{\alpha<\omega_1}$ of chains such that

(1) For each $\beta < \alpha$ there is a $\mathbf{c} \in C_\alpha$ with $\mathbf{c} \geq \mathbf{d}_\beta$;

(2) C_α has order type $\alpha \times \omega + 1$ if $\alpha = 0$ or is a successor ordinal;

(3) C_α has order type $\alpha \times \omega$ if α is a limit ordinal, and

(4) $C_\beta \subset C_\alpha$ if $\beta < \alpha$.

Let $C_0 = \{\mathbf{0}\}$. Suppose that $\alpha = \beta + 1$ and C_β is defined. Given $\mathbf{d}_\beta$, we apply Lemma 8.3.1 to obtain a minimal cover $\mathbf{a}$ of C_β with $\mathbf{a}'' \geq \mathbf{d}_\beta$. Relativizing Lemma 9.3.3 yields a maximal chain $\mathbf{a} = \mathbf{a}_0 < \mathbf{a}_1 < \ldots < \mathbf{a}_\omega = \mathbf{a}''$ in $[\mathbf{a}, \mathbf{a}'']$. Let $C_\alpha = C_\beta \cup \{\mathbf{a}_\gamma : \gamma \leq \omega\}$. For α a limit ordinal, let $C_\alpha = \bigcup_{\gamma<\alpha} C_\gamma$. Let $C = \bigcup_{\alpha<\omega_1} C_\alpha$. It is straightforward to verify that the order type of C is ω_1, C is a maximal chain, and that for each $\alpha < \omega_1$ the $(\alpha \times \omega + \omega + 1)$-th element of C bounds $\mathbf{d}_\alpha$.

($\Leftarrow$) Since every degree bounds at most countably many degrees, CH follows immediately if there is a maximal chain as described. □

9.3.2 CH without AC

To continue our tour of the ordering of Turing degrees, we recall some set-theoretic facts. CH is the statement that there is no set X with $|\omega| < |X| < |2^\omega|$, where "$|X| \leq |Y|$" means the existence of a surjective map from Y onto X. Hence ZF + CH does not obviously imply the existence of a well-ordering of reals. The same is true of GCH with regard to AC. It turns out that there is a link between these two hypotheses.

Theorem 9.3.5 (Sierpinski [136]). ZF $\vdash$ GCH $\rightarrow$ AC.

A question along this line is whether Theorem 9.3.5 holds level by level. For example, does CH imply the existence of a well-ordering of 2^ω?

Theorem 9.3.6 (Kanamori and Pincus [62]). ZFC + *"There exists an inaccessible cardinal" is consistent if and only if* ZF + CH + *"There is no well-ordering of* 2^ω*" is consistent.*

Proof. Assume that ZFC + "There exists an inaccessible cardinal" is consistent. By Solovay [150], there is a model $\mathcal{M}$ of ZF in which every uncountable set of reals contains a perfect subset. In other words, $\mathcal{M} \models$ ZF + CH. Furthermore, there is no well-ordering of the set of reals in $\mathcal{M}$.

Conversely, suppose that ZF + CH + "There is no well-ordering of 2^ω" is consistent and let $\mathcal{M}$ be a transitive model of this theory. We claim that $(\omega_1)^M$ is inaccessible in L. Clearly $(\omega_1)^M$ is regular in L. It is sufficient to prove that for any $\alpha < (\omega_1)^M$, $|(2^\alpha)^L| < \omega_1$. If not, there exists a countable α such that $|(2^\alpha)^L| = (\omega_1)^M$. Then $(2^\alpha)^L \in M$ and there is a well-ordering of $(2^\alpha)^L$ in M. Since $|\omega| < |(2^\alpha)^L| \leq |2^\omega|$, we have $|(2^\alpha)^L| = |2^\omega|$. In other words, there is a bijection between $(2^\alpha)^L$ and 2^ω in $\mathcal{M}$. It follows that there is a well-ordering of 2^ω in $\mathcal{M}$, which is a contradiction. □

Theorem 9.3.6 says that "ZF + CH + There is no well-ordering of 2^ω" is such a strong theory that an inaccessible cardinal is required to guarantee its consistency. However, for forcing extensions, one has the following:

Proposition 9.3.7. *Let* $\mathcal{M} \models$ ZFC. *If* $\mathcal{N}$ *is an extension of* $\mathcal{M}$ *preserving* $(\omega_1)^M$ *such that* $\mathcal{N} \models$ ZF+*"There is a cofinal chain of Turing degrees", then* $\mathcal{N} \models \neg$CH *if and only if* $\mathcal{N} \models$ *"There is no well-ordering of* 2^ω*".*

Proof. If $\mathcal{N} \models$ "There is a well-ordering of 2^ω", then $\mathcal{N} \models$ CH since $\mathcal{N} \models$ ZF+"There is a cofinal chain of Turing degrees".

Now suppose that $\mathcal{N} \models$ "There is no well-ordering of 2^ω". Take $C \in \mathcal{M}$ to be a subset of $(2^\omega)^M$ such that $<_T$ is a well-ordering of C of order type $(\omega_1)^M$. Since $\mathcal{N}$ is an extension of $\mathcal{M}$ preserving $(\omega_1)^M$, $\mathcal{N} \models |\omega| < |C|$. Since there is no well-ordering of 2^ω in $\mathcal{N}$, we have $\mathcal{N} \models |C| < |2^\omega|$. So $\mathcal{N} \models \neg$CH. □

9.3.3 Cofinal chains versus CH

Pursuing further the link between CH and Turing degrees, one asks if, under ZF, CH is equivalent to the existence of a cofinal maximal chain of Turing degrees of order type ω_1. The answer is negative: By Solovay [150], there exists a model $\mathcal{M}$ of ZF in which every uncountable set of reals contains a perfect subset. Since it is a theorem of ZF that every perfect set contains two reals with incomparable Turing degrees, there is no chain of Turing degrees of order type ω_1.

It is not known if ZF + "There exists a cofinal maximal chain of Turing degrees of order type ω_1" implies CH. However, there is a partial answer:

Theorem 9.3.8 (Wang, Wu and Yu [168]). *If* ZFC *is consistent, then so is* ZF + $\neg$CH + *"There exists a cofinal chain in the Turing degrees of order type* ω_1*".*

The rest of this section is devoted to a proof Theorem 9.3.8. We begin with the ground model $\langle L, \in\rangle$. Let $\mathbb{P}$ be the collection of Cohen forcing conditions. $\mathbb{P}_{\aleph_1}$ is the resulting product forcing with finite support of length $\aleph_1$.

For $\alpha < \omega_1$, let $\mathbb{P}_{<\alpha} = \langle \mathbf{P}_{<\alpha}, \leq\rangle$, where $p \in \mathbf{P}_{<\alpha}$ if and only if $p \in \mathbf{P}_{\aleph_1}$ and $\mathrm{dom}(p) \subseteq \alpha$. The partial order on $\mathbb{P}_{<\alpha}$ is the restriction of that on $\mathbb{P}_{\aleph_1}$ to $\mathbf{P}_{<\alpha}$. Thus $\mathbb{P}_{\aleph_1} = \mathbb{P}_{<\aleph_1}$. Let E be the set of limit ordinals less than ω_1. For $\alpha < \omega_1$, let ω_1^α be the least admissible ordinal greater than α. Then $\mathbb{P}_{<\alpha} \in L_{\omega_1^\alpha}$.

Lemma 9.3.9. *There exist an uncountable set $E_1 \subseteq E$ in L and a sequence of binary relations $\{R_\alpha\}_{\alpha\in E_1} \in L$ such that for each $\alpha \in E_1$,*

- *$R_\alpha \in L$ is a binary relation on $\{2^n \mid n < \omega\}$ that is a well-ordering of order type α;*
- *for each $\beta \leq \alpha$, there is a function $f_\beta \leq_T R_\alpha$ which is an isomorphism between R_β and an initial segment of R_α.*

Proof. We define E_1 and $\{R_\alpha\}_{\alpha\in E_1}$ by induction on $\alpha < \omega_1$. Let $E_{1,0} = \emptyset$. At stage $\alpha + 1$, suppose that $E_{1,\alpha}$ and $\{R_\gamma\}_{\gamma\in E_{1,\alpha}}$ have been defined. Let $R \in L$ be a well-ordering of $\{2^{2n} \mid n < \omega\}$ of order type a limit ordinal greater than that of R_γ for $\gamma \in E_{1,\alpha}$. Then for every $\gamma \in E_{1,\alpha}$, there is a function f_γ which maps R_γ isomorphically onto an initial segment of R. Let $x \in L$ be a real that computes all the f_γ's. Let $E_{1,\alpha+1} = E_{1,\alpha} \cup \{\gamma_1\}$ where $\gamma_1 = \alpha + \omega + \omega$. Let R^1 be a binary relation on $\{2^{2n+1} \mid n < \omega\}$ such that for any $n \neq m$, $(2^{2n+1}, 2^{2m+1}) \in R^1$ if and only if either $n \in x$ and $m \notin x$, or $n \in x \leftrightarrow m \in x$ and $n < m$. So $R^1 \geq_T x$ in L and is a well-ordering of order type $\omega + \omega$ on $\{2^{2n+1} \mid n < \omega\}$. Let

$$R_{\alpha+1} = R \cup R^1 \cup \{(2^{2n}, 2^{2m+1}) \mid n, m \in \omega\}.$$

If α is a limit ordinal, let $E_{1,<\alpha} = \bigcup_{\gamma<\alpha} E_{1,\gamma}$ and repeat the above upon replacing $E_{1,\alpha}$ by $E_{1,<\alpha}$.

Let $E_1 = \bigcup_{\gamma<\omega_1} E_{1,\gamma}$. Then E_1 and $\{R_\alpha\}_{\alpha\in E_1}$ are as required. □

From now on, let E_1 and $\{R_\alpha\}_{\alpha\in E_1}$ in L be as above. Let G be a $\mathbb{P}_{\aleph_1}$-generic set over L. Then G may be viewed as an ω_1-sequence $\{G_\alpha\}_{\alpha<\omega_1}$ such that for every $\alpha < \omega_1$, $G_\alpha = \bigcup_{p\in G} p(\alpha)$ is a real. Let $G_{<\alpha} = \{(\beta, G_\beta) \mid \beta < \alpha\}$.

Definition 9.3.10. For any $\gamma < \alpha \in E_1$ and any sequence $\{z_\beta\}_{\beta<\alpha}$ of reals, define *the join of $\{z_\beta\}_{\beta<\gamma}$ along R_α* to be the set

$$\bigoplus_{R_\alpha}\{z_\beta\}_{\beta<\gamma} = R_\alpha \oplus \{2^n \cdot 3^j \mid (\exists\beta < \gamma)(j \in z_\beta \wedge \beta = \text{order type of } (R_\alpha \restriction n)\}.$$

The following lemma says that the join operator is order preserving.

Lemma 9.3.11. *For any $\beta_1 \leq \beta_2 \leq \alpha_1 \leq \alpha_2 \in E_1$,*

$$\bigoplus_{R_{\beta_2}}\{G_\gamma\}_{\gamma<\beta_1} \leq_T \bigoplus_{R_{\alpha_2}}\{G_\gamma\}_{\gamma<\alpha_1}.$$

Proof. It is immediate that $R_{\beta_2} \leq_T R_{\alpha_2}$. Let f be an R_{α_2}-recursive function which is an isomorphism between R_{β_2} and an initial segment of R_{α_2}. Then for any n, j,

$$2^n \cdot 3^j \in \bigoplus_{R_{\beta_2}}\{G_\gamma\}_{\gamma<\beta_1} \text{ if and only if } 2^{f(n)} \cdot 3^j \in \bigoplus_{R_{\alpha_2}}\{G_\gamma\}_{\gamma<\alpha_1}.$$

Hence $\bigoplus_{R_{\beta_2}}\{G_\gamma\}_{\gamma<\beta_1} \leq_T \bigoplus_{R_{\alpha_2}}\{G_\gamma\}_{\gamma<\alpha_1}$. □

The following lemma is used in the construction of a cofinal chain.

Lemma 9.3.12. *For any real x and $\alpha \in E_1$, if $x \in L[G_{<\alpha}]$ then there is a $\beta \geq \alpha$ in E_1 such that $x \leq_T (\bigoplus_{R_\beta}\{G_\gamma\}_{\gamma<\beta})^{(\beta)}$.*

Proof. Note that for any $\alpha < \omega_1$, $L[G_{<\alpha}] = L[\bigoplus_{R_\alpha}\{G_\gamma\}_{\gamma<\alpha}]$ since $R_\alpha \in L$. Suppose that $x \in L[G_{<\alpha}]$. There is a $\beta_1 \geq \alpha$ in E such that $x \in L_{\beta_1}[\bigoplus_{R_\alpha}\{G_\gamma\}_{\gamma<\alpha}]$. By Lemma 9.3.11, $\bigoplus_{R_\alpha}\{G_\gamma\}_{\gamma<\alpha} \leq_T \bigoplus_{R_{\beta_1}}\{G_\gamma\}_{\gamma<\beta_1}$. Thus $x \in L_{\beta_1}[\bigoplus_{R_{\beta_1}}\{G_\gamma\}_{\gamma<\beta_1}]$. Since $\beta_1 < \omega_1^{\bigoplus_{R_{\beta_1}}\{G_\gamma\}_{\gamma<\beta_1}}$, by relativizing Theorem 3.7.1 to $\bigoplus_{R_{\beta_1}}\{G_\gamma\}_{\gamma<\beta_1}$ we have x to be Δ^1_1 in $\bigoplus_{R_{\beta_1}}\{G_\gamma\}_{\gamma<\beta_1}$. Hence there is a $\beta_2 \geq \beta_1$ below $\omega_1^{\bigoplus_{R_{\beta_1}}\{G_\gamma\}_{\gamma<\beta_1}}$ such that $x \leq_T (\bigoplus_{R_{\beta_1}}\{G_\gamma\}_{\gamma<\beta_1})^{(\beta_2)}$. By Lemma 9.3.11 again, $x \leq_T (\bigoplus_{R_{\beta_2}}\{G_\gamma\}_{\gamma<\beta_2})^{(\beta_2)}$. Then β_2 is the required β. □

In $L[G]$, if $\alpha \in E_1$ let $x^g_\alpha = (\oplus_{R_\alpha}\{G_\gamma\}_{\gamma<\alpha})^{(\alpha)}$, and let τ_α be the canonical $\mathbb{P}_{\aleph_1}$-name of x^g_α. Let

$$A = \{z \mid (\exists \alpha \in E_1)(z \equiv_T x^g_\alpha)\}.$$

Obviously A is a chain of Turing degrees in $L[G]$.

Set

$$\mathcal{N} = (\mathrm{HOD}(\mathrm{TC}(\{A\})))^{L[G]},$$

where $\mathrm{TC}(\{A\}) \in L[G]$ is the transitive closure of $\{A\}$.

Lemma 9.3.13. *For any $\alpha < \omega_1$,*
(i) $\mathbb{P}_{<\alpha}$ preserves cardinals;
(ii) $G_{<\alpha}$ is $\mathbb{P}_{<\alpha}$-generic over L;
(iii) $A \in \mathcal{N}$ is a cofinal chain of Turing degrees of order type ω_1;
(iv) $(2^\omega)^{L[G]} = (2^\omega)^{\mathcal{N}}$.

Proof. (i) follows from Theorem 6.1.11 and (ii) from Theorem 6.1.9, while (iv) is an immediate consequence of (iii). We prove (iii).

Given $x \in L[G]$, there is an ordinal $\alpha < \omega_1$ such that $x \in L[G_{<\alpha}]$. Then there is a countable ordinal $\alpha_1 \geq \alpha$ such that $x \in L_{\alpha_1}[G_{<\alpha}]$. By Lemma 9.3.12, for any $\beta \geq \alpha_1$ in E_1, $x \leq_T (\oplus_{R_\beta}\{G_\gamma\}_{\gamma<\beta})^{(\beta)}$. Thus $x \leq_T x^g_\beta$. □

Hence $\mathcal{N}$ is a cardinal preserving extension of L satisfying ZF+"There exists a cofinal chain of Turing degrees". If we take $\mathcal{N}' = \mathrm{HOD}(\{G_\alpha \mid \alpha < \omega_1\})$, then in $\mathcal{N}'$ there is no well-ordering of 2^ω by the proof of Theorem 6.2.15. However, to apply the argument for $\mathcal{N}$, additional precaution is required to preserve the name of A. For this purpose, we work with a subset of permutations of ω_1.

If π is a permutation of ω_1, let $\mathrm{spt}(\pi) = \{\alpha < \omega_1 \mid \pi(\alpha) \neq \alpha\}$. Let H_0 be the set of all permutations π of ω_1 with $|\mathrm{spt}(\pi)| < \omega$, and set

$$H = \{\pi \in H_0 \mid (\forall \alpha < \omega_1)(\pi(\alpha) < \alpha + \omega)\}.$$

For each $\pi \in H$, let $\bar{\pi}$ be the permutation of $L^{\mathbb{P}_{\aleph_1}}$ induced by π.

Let B be the set of $\mathbb{P}_{\aleph_1}$-names of reals, and define

$$\dot{A} = \{(\tau, p) \in B \times \mathbf{P}_{\aleph_1} \mid (\exists \alpha \in E_1)(p \Vdash (\tau \equiv_T \tau_\alpha))\}.$$

Lemma 9.3.14. *If $\pi \in H$ and $\gamma \in E_1$ then $\Vdash \tau_\gamma \equiv_T \bar{\pi}(\tau_\gamma)$.*

Proof. Note that the canonical name $\check{\gamma}$ of γ is invariant under π, and R_γ is definable in γ as L is the minimal inner model. Then for any $\mathbb{P}_{\aleph_1}$-generic G, τ_γ^G and $\bar{\pi}(\tau_\gamma)^G$ differ on at most finitely many columns, the lemma follows immediately. □

Corollary 9.3.15. *If $\pi \in H$, then $\emptyset \Vdash \dot{A} = \bar{\pi}(\dot{A})$.*

Hence the permutations in H are sufficient to show

$$\mathcal{N} \models \text{“There is no well-ordering of } 2^\omega\text{”}.$$

By Proposition 9.3.7, $\mathcal{N} \models \neg\mathrm{CH}$ and this completes the proof of Theorem 9.3.8.

9.3.4 Exercises

Exercise 9.3.1. *Prove Lemma* 9.3.2.

Exercise 9.3.2 (Kunen). *Prove that if there is a well-ordering of the Turing degrees, then there is a well-ordering of 2^ω.*

Exercise 9.3.3 (Wang, Wu and Yu [168]). *Prove that in the proof of Theorem* 9.3.8, *we may assume $\mathcal{M} \models \mathrm{ZFC} + \mathrm{CH} + \omega_1 = (\omega_1)^L$.*

Exercise 9.3.4. *Prove that there is a model $\mathcal{M}$ of ZF in which no set of reals has exactly one representative from each Turing degree, and in which there is a thin set A where $2^\omega \setminus A$ is also thin.*

9.4 ω-homogeneity of the Turing degrees

Definition 9.4.1. $\langle \mathcal{D}, \leq \rangle$ is *ω-homogeneous* if for any two $n+1$-tuples $(\mathbf{a}_0, \ldots, \mathbf{a}_n)$ and $(\mathbf{b}_0, \cdots, \mathbf{b}_n)$ in $\mathcal{D}$, if $\langle \mathcal{D}, \leq, \mathbf{a}_0, \ldots, \mathbf{a}_n \rangle \equiv \langle \mathcal{D}, \leq, \mathbf{b}_0, \ldots, \mathbf{b}_n \rangle$ then there is an automorphism $\pi : \mathcal{D} \to \mathcal{D}$ such that $\pi(\mathbf{a}_i) = \pi(\mathbf{b}_i)$ for each $i \leq n$.

Slaman and Woodin [144] showed:

Theorem 9.4.2. *The ω-homogeneity of $\langle \mathcal{D}, \leq \rangle$ is independent of* ZFC.

We say that a function $f : 2^\omega \to 2^\omega$ presents an automorphism $\pi : \langle \mathcal{D}, \leq \rangle \to \langle \mathcal{D}, \leq \rangle$ if for every Turing degree $\mathbf{x}$ and real $x \in \mathbf{x}$, $f(x) \in \pi(\mathbf{x})$. The next two results are needed for Theorem 9.4.2 and are stated without proof.

Theorem 9.4.3. *Every automorphism $\pi : \langle \mathcal{D}, \leq \rangle \to \langle \mathcal{D}, \leq \rangle$ is presented by an arithmetical function $f : 2^\omega \to 2^\omega$.*

If R is a degree invariant relation over $(2^\omega)^{n+1}$ for some n, then it induces a relation $\tilde{R}$ on degrees such that $(\mathbf{x}_0, \dots, \mathbf{x}_n) \in \tilde{R} \Leftrightarrow (\exists x_0) \cdots (\exists x_n)(x_0 \in \mathbf{x}_0 \wedge \cdots \wedge x_n \in \mathbf{x}_n \wedge (x_0, \dots, x_n) \in R)$. We identify $\tilde{R}$ with R for simplicity.

Theorem 9.4.4. *For any relation R over $\mathcal{D}$ which is induced by a degree invariant analytical relation over 2^ω, if R is invariant under any automorphism of $\mathcal{D}$, then R is definable in $\langle \mathcal{D}, \leq \rangle$.*

We now prove Theorem 9.4.2.

Lemma 9.4.5. *If $V = L$, then $\langle \mathcal{D}, \leq \rangle$ is ω-homogeneous.*

Proof. By Proposition 4.4.2, $<_L$ is a Δ^1_2 well-ordering of 2^ω. The lexicographic ordering of $n+1$-tuples induced by $<_L$ implies that for each n there is a Δ^1_2 well-ordering $<_{n+1}$ of $(2^\omega)^{n+1}$.

For $m < \omega$ and $k \leq n$, let $R_{m,k} \subseteq (2^\omega)^{n+1}$ be such that

$$\begin{aligned}(x_0, \dots, x_n) \in R_{m,k} \Leftrightarrow k \leq n \wedge \\ (\exists y_0) \cdots (\exists y_n)((y_0, \dots, y_n) \text{ is the } <_{n+1}\text{-least so that} \\ (\exists \pi)(\forall i \leq n)(\pi : \langle \mathcal{D}, \leq \rangle \to \langle \mathcal{D}, \leq \rangle \text{ is an automorphism} \wedge \pi(\mathbf{y}_i) = \mathbf{x}_i \wedge m \in y_k)).\end{aligned}$$

It is straightforward to verify that $R_{m,k}$ is degree invariant and hence induces a relation on $\mathcal{D}$ invariant under any automorphism. By Theorem 9.4.3, $R_{m,k}$ is also analytical. Hence by Theorem 9.4.4, $R_{m,k}$ is definable in $\langle \mathcal{D}, \leq \rangle$.

Suppose that $\langle \mathcal{D}, \leq, \mathbf{a}_0, \dots, \mathbf{a}_n \rangle \equiv \langle \mathcal{D}, \leq, \mathbf{b}_0, \dots, \mathbf{b}_n \rangle$. Then $R_{m,k}(\mathbf{a}_0, \dots, \mathbf{a}_n) \Leftrightarrow R_{m,k}(\mathbf{b}_0, \dots, \mathbf{b}_n)$ for every m and $k \leq n$. For such a k, define y_k so that $m \in y_k$ if and only if $R_{m,k}(\mathbf{a}_0, \dots, \mathbf{a}_n)$. By the definition of $R_{m,k}$, there are automorphisms π_0 and π_1 such that for every $k \leq n$, $\pi_0(\mathbf{y}_k) = \mathbf{a}_k$ and $\pi_1(\mathbf{y}_k) = \mathbf{b}_k$. Then $\pi_1 \circ \pi_0^{-1}$ is an automorphism between $\langle \mathcal{D}, \leq \rangle$ and $\langle \mathcal{D}, \leq \rangle$, and $\pi_1 \circ \pi_0^{-1}(\mathbf{a}_k) = \mathbf{b}_k$, implying that $\langle \mathcal{D}, \leq \rangle$ is ω-homogeneous. □

We complete the proof of Theorem 9.4.2 with the following lemma.

Lemma 9.4.6. *If $V = L[g]$ where g is a Cohen generic real, then $\langle \mathcal{D}, \leq \rangle$ is not ω-homogeneous.*

Proof. Let $g = g_0 \oplus g_1$. We leave it to the reader to verify that (g_0, g_1) is a pair of generic reals over a product Cohen forcing $\mathbb{C} \times \mathbb{C}$. Then by Theorem 9.4.3, there is no automorphism of $\langle \mathcal{D}, \leq \rangle$ sending $\mathbf{g}_0$ to $\mathbf{g}_1$.

Now let φ be a formula such that $\langle L[g], \in\rangle = \langle L[(g_0, g_1)], \in\rangle \models$ "$\langle \mathscr{D}, \leq\rangle \models \varphi(\mathbf{g}_0)$". Then there is a condition $(p_0, p_1) \in 2^{<\omega} \times 2^{<\omega}$ such that $|p_0| = |p_1|$, $(p_0, p_1) \prec (g_0, g_1)$ and $(p_0, p_1) \Vdash$ "$\langle \dot{\mathscr{D}}, \dot{\leq}\rangle \models \varphi(\dot{\mathbf{g}}_0)$". Obviously there is an automorphism $\pi : \mathbb{C} \times \mathbb{C} \to \mathbb{C} \times \mathbb{C}$ so that $\pi((p_0, p_1)) = (p_1, p_0)$ and $\pi_*(\dot{g}_0) = \dot{g}_1$. By Proposition 6.1.17, $(p_1, p_0) \Vdash$ "$\langle \dot{\mathscr{D}}, \dot{\leq}\rangle \models \varphi(\dot{\mathbf{g}}_1)$". Then $\langle L[(g_1, g_0)], \in\rangle \models$ "$\langle \mathscr{D}, \leq\rangle \models \varphi(\mathbf{g}_1)$". However, $L[(g_1, g_0)] = L[(g_0, g_1)] = L[g]$ and thus $\langle L[g], \in\rangle \models$ "$\langle \mathscr{D}, \leq\rangle \models \varphi(\mathbf{g}_1)$". Hence $\langle \mathscr{D}, \leq\rangle$ is not ω-homogeneous. □

9.5 Some general independence results

For a general reference on the subject of Turing degrees, see Lerman [82]. Let $\mathscr{L}(\leq)$ be the language of partial ordering. $\mathscr{L}(\leq)$ is a set definable in the language of set theory. Furthermore, both $\langle \mathscr{D}, \leq\rangle$ and the relation "$\langle \mathscr{D}, \leq\rangle \models \varphi$" may be viewed as definable sets in the language of set theory. We code each formula φ in $\mathscr{L}(\leq)$ by a number $\ulcorner\varphi\urcorner$. Let

$$\mathrm{Th}_{\exists_2}\langle \mathscr{D}, \leq\rangle = \{\ulcorner\varphi\urcorner \mid \varphi \text{ is a } \Sigma_2\text{-sentence in } \mathscr{L}(\leq) \wedge \langle \mathscr{D}, \leq\rangle \models \varphi\}.$$

Lerman and Shore [131] proved:

Theorem 9.5.1. *The set* $\mathrm{Th}_{\exists_2}\langle \mathscr{D}, \leq\rangle$ *is recursive.*

Hence the set $\mathrm{Th}_{\exists_2}\langle \mathscr{D}, \leq\rangle$ is definable by a Σ^0_1-formula ψ_0 and a Π^0_1-formula ψ_1. It follows that for any Σ_2-sentence φ,

$$\mathrm{ZFC} \vdash \text{“}\langle \mathscr{D}, \leq\rangle \models \varphi \Leftrightarrow \psi_0(\ulcorner\varphi\urcorner)\text{”}$$

and

$$\mathrm{ZFC} \vdash \text{“}\langle \mathscr{D}, \leq\rangle \models \neg\varphi \Leftrightarrow \neg\psi_1(\ulcorner\varphi\urcorner)\text{”}.$$

Since for any Σ^0_1-formula $\psi(u)$ and natural number n,

$$\mathrm{ZFC} \vdash \psi(n) \Leftrightarrow \langle \omega, \leq, +, \cdot, 0, 1\rangle \models \psi(n),$$

(see [145]), for any Σ_2-sentence φ either

$$\mathrm{ZFC} \vdash \text{“}\langle \mathscr{D}, \leq\rangle \models \varphi\text{”}$$

or

$$\mathrm{ZFC} \vdash \text{“}\langle \mathscr{D}, \leq\rangle) \models \neg\varphi\text{”}.$$

Hence we have the following conclusion.

Corollary 9.5.2. *For any Σ_2-sentence φ in $\mathscr{L}(\leq)$, either "$\langle \mathscr{D}, \leq\rangle \models \varphi$" or "$\langle \mathscr{D}, \leq\rangle \models \neg\varphi$" is provable in* ZFC.

Clearly the set $\mathrm{Pr} = \{\ulcorner\varphi\urcorner \mid \mathrm{ZFC} \vdash \text{“}\langle \mathscr{D}, \leq\rangle \models \varphi\text{”}\}$ is r.e. The complexity of the theory of the Turing degrees was established by Simpson [139]:

Theorem 9.5.3. ZFC *proves that there is a recursive bijection* $f : \omega \to \omega$ *such that for any* n, $n \in \{\ulcorner\varphi\urcorner \mid (\mathscr{D}, \leq) \models \varphi\}$ *if and only if* $f(n) \in \{\ulcorner\psi\urcorner \mid (\omega, 2^\omega, +, \cdot, \leq, \in) \models \psi\}$, *where* ψ *is a formula in the language of second order arithmetic.*

Thus ZFC proves that the set $\{\ulcorner\varphi\urcorner \mid (\mathscr{D}, \leq) \models \varphi\}$ is not analytical. It follows that for any $\mathcal{M} \models$ ZFC, we have $\mathrm{Pr} \subset \{\ulcorner\varphi\urcorner \mid \mathcal{M} \models \text{“}(\mathscr{D}, \leq) \models \varphi\text{”}\}$. This leads to the following conclusion.

Corollary 9.5.4. *For any* $\mathcal{M} \models$ ZFC, *there is a sentence* φ *such that* $\ulcorner\varphi\urcorner \in \{\ulcorner\psi\urcorner \mid \mathcal{M} \models \text{“}(\mathscr{D}, \leq) \models \psi\text{”}\} \setminus \mathrm{Pr}$.

Let $\mathcal{M} \models$ ZFC and φ be as above. Since ZFC $\nvdash$ “$(\mathscr{D}, \leq) \models \varphi$”, there exists a model $\mathcal{N} \models$ ZFC such that $\mathcal{N} \models$ “$(\mathscr{D}, \leq) \models \neg\varphi$”. Hence the statement “$(\mathscr{D}, \leq) \models \varphi$” is independent of ZFC. This argument implies the existence of a sentence φ not provable in ZFC, but does not provide any information about φ. The following offers more information: By Corollary 4.4.1, the statement “every real is constructible” is Π^1_3. Hence there is a sentence φ in $\mathscr{L}(\leq)$ such that $(\mathscr{D}, \leq) \models \varphi$ if and only if every real is constructible. Thus the statement “$(\mathscr{D}, \leq) \models \varphi$” is independent of ZFC. However, a weakness in the argument is that the sentence φ is highly complicated, not necessarily considered “natural”, and heavily dependent on the coding scheme. It is not known if there is a “natural” sentence in $\mathscr{L}(\leq)$ that is independent of ZFC.

Part III: Hyperarithmetic degrees and perfect set property

10 Rigidity and biinterpretability of hyperdegrees

Let $\langle \mathcal{D}_h, \leq \rangle$ denote the structure of hyperdegrees under the partial ordering $\leq$ for hyperarithmetic reducibility. As for the Turing degrees $\langle \mathcal{D}, \leq \rangle$, the hyperdegrees form an upper semilattice under the join operator $\cup$, where if the hyperdegrees of x and y are respectively $\mathbf{a}$ and $\mathbf{b}$, then $x \oplus y$ has hyperdegree $\mathbf{a} \cup \mathbf{b}$. Note that $\cup$ may be defined in terms of $\leq$. A map of $\mathcal{D}$ or $\mathcal{D}_h$ into itself is an *automorphism* if it is a bijection and preserves $\leq$. A major open problem in recursion theory asks whether the identity map is the only automorphism for the Turing degrees, i.e. whether the Turing degrees are *rigid*. Slaman and Woodin ([141] and [144]) used a sophisticated metamathematical argument to prove a number of remarkable results about $\langle \mathcal{D}, \leq \rangle$, including the theorem that there are at most countably many nontrivial automorphisms for the structure (indeed the degrees $\geq \mathbf{0}'$ are invariant under any automorphism), and that the hyperdegrees are rigid. In this chapter, we present a more "elementary" proof of the latter due to Shore [133]. The biinterpretability of $\langle \mathcal{D}_h, \leq \rangle$ with second order arithmetic follows as an application of rigidity (see § 10.3).

10.1 Embedding lattices into hyperdegrees

Recall that a partially ordered set $\langle \mathcal{C}, \leq \rangle$ is a *lattice* if any two members $c, c' \in \mathcal{P}$ have a least upper bound and a greatest lower bound denoted $c \cup c'$ and $c \cap c'$ respectively. A lattice is recursively presented if its elements (via some effective coding) are enumerated recursively. A partial ordering under $\leq$ is a partial lattice if the relations $\leq, \nleq$ and operators $\cup, \cap$ are partial recursive. The next theorem goes back to Jónsson [61] (see also Lerman [81] and Shore [132]).

Theorem 10.1.1. *Suppose $\mathcal{C} = \{c_i\}_{i<\omega}$ is an enumeration of a recursively presentable partial lattice with least element c_0. There is a uniformly recursive array $\{f_n \subset \omega^\omega \mid n < \omega\}$ of functions such that for all i, j, k, m and n,*
(i) $f_n(0) = 0$;
(ii) $c_i \leq c_j \Rightarrow [f_n(j) = f_m(j) \to f_n(i) = f_m(i)]$;
(iii) $c_i \nleq c_j \Rightarrow [(\exists d, e)(f_d(j) = f_e(j) \wedge f_d(i) \neq f_e(i))]$;
(iv) $c_i \cup c_j = c_k \Rightarrow [f_n(i) = f_m(i) \wedge f_n(j) = f_m(j) \to f_n(k) = f_m(k)]$;
(v) $c_i \cap c_j = c_k \wedge f_n(k) = f_m(k) \Rightarrow (\exists d_1, d_2, d_3)[f_n(i) = f_{d_1}(i) \wedge f_{d_1}(j) = f_{d_2}(j) \wedge f_{d_2}(i) = f_{d_3}(i) \wedge f_{d_3}(j) = f_m(j)]$.

Proof. Define

$$f^*_{2j+1}(i) = \begin{cases} 0 & i = 0, \\ 2^{2j+1} & i > 0. \end{cases}$$

and

$$f^*_{2j}(i) = \begin{cases} 0 & i = 0, \\ 2^{2j+1} & c_i \leq c_j, \\ 2^{2j} & c_i \nleq c_j. \end{cases}$$

We use $\{f^*_j\}_{j<\omega}$ as the basis to generate the array $\{f_n\}_{n<\omega}$. It is straightforward to verify that (i)–(iv) hold for each f^*_j. For any two functions $f, \hat{f}$ that have been generated and satisfy the hypothesis of (v), we define the functions $\hat{f}_{d_1}$, $\hat{f}_{d_2}$ and $\hat{f}_{d_3}$ as follows: If $c_i \leq c_j$, then set $\hat{f}_{d_1} = \hat{f}_{d_2} = \hat{f}_{d_3} = \hat{f}$ and (v) holds. If $c_i \nleq c_j$, let a_1, a_2, a_3, a_4 be numbers not in the range of f and $\hat{f}$. Define

$$\hat{f}_{d_1}(s) = \begin{cases} f(s) & c_s \leq c_i, \\ a_1 & c_s \nleq c_i. \end{cases}$$

$$\hat{f}_{d_2}(s) = \begin{cases} f(s) & c_s \leq c_j, \\ a_2 & c_s \leq c_i \wedge c_s \nleq c_j, \\ a_3 & \text{Otherwise}. \end{cases}$$

$$\hat{f}_{d_3}(s) = \begin{cases} \hat{f}(s) & c_s \leq c_j, \\ a_2 & c_s \leq c_i \wedge c_s \nleq c_j, \\ a_4 & \text{Otherwise}. \end{cases}$$

We leave it to the reader to verify that (v) is satisfied for $f, \hat{f}$, and that (i)–(iv) remain valid with the addition of the three new functions. The collection of functions generated through this process is a recursive array $\{f_n\}_{n<\omega}$ stipulated by the theorem. □

An adaptation of Cohen forcing (§ 5.2) allows one to embed a recursively presented lattice into the hyperdegrees.

Theorem 10.1.2. *Let $\mathfrak{C} = \{c_i\}_{i<\omega}$ be a recursive partial lattice with 0 and let $\{f_n\}_{n<\omega}$ be a recursive representation of $\mathfrak{C}$ as given in Theorem 10.1.1. Let $\mathfrak{A}(\omega_1^{\mathrm{CK}})$ be the structure defined in § 5.1 for the ramified analytical hierarchy. If g is Cohen generic over $\mathfrak{A}(\omega_1^{\mathrm{CK}})$, then there is a sequence of sets $\{h_i\}_{i<\omega}$ such that $h_i \leq_h g$ and the map taking c_i to the hyperdegree of h_i is an embedding of $\mathfrak{C}$ in $\mathcal{D}_h$ preserving $\leq$, $\nleq$, $\cup$ and $\cap$.*

Proof. The notion of Cohen forcing we consider here is defined over the Baire space ω^ω rather than over the Cantor space 2^ω as was defined in § 5.2. A condition is a finite collection of a consistent set of relations of the form $\sigma(m) = n$. The generic set g is then a function that is an element of ω^ω such that $\omega_1^g = \omega_1^{\mathrm{CK}}$. Recall also by Corollary 5.2.6 (ii) that if g is Cohen generic and $h \leq_h g$, then $h \leq_T g \oplus \emptyset^\gamma$ for some recursive ordinal γ.

For each i, let $h_i(n) = f_{g(n)}(i)$. First of all, $h_0(n) = f_{g(n)}(0) = 0$ for all n according to Theorem 10.1.1 (i), so that $0 \in \mathfrak{C}$ is mapped to a recursive set. Suppose $c_i \leq c_j$. For each n, search for the least m such that $f_m(n) = h_j(n) = f_{g(n)}(j)$. Note that such m must exist since $g(n)$ is a candidate. Then $f_m(i) = f_{g(n)}(i) = h_i(n)$. Since $\{f_n\}_{n<\omega}$ is uniformly recursive. We have in fact $h_i \leq_T h_j$, showing that the map preserves order.

Now suppose $c_i \cup c_j = c_k$. To compute $h_k(n)$ which is $f_{g(n)}(k)$, recursively find an m such that $f_m(i) = h_i(n) = f_{g(n)}(i)$ and $f_m(j) = h_j(n) = f_{g(n)}(j)$. By Theorem 10.1.1 (iv), one has $f_m(k) = f_{g(n)}(k) = h_k(n)$. Hence the embedding preserves joint when it is defined.

We use the genericity of g to argue that $\not\leq$ and $\cap$ are preserved under the embedding. Assume $c_i \not\leq c_j$. By Theorem 10.1.1 (iii), there exist m, n such that $f_m(j) = f_n(j)$ and $f_m(i) \neq f_n(i)$. If $h_i \leq_h h_j$, then there is a $k \in \mathcal{O}$, a number e and a condition p such that

$$p \Vdash h_i = \Phi_e^{\hat{H}_k^{h_j}}.$$

Let q be a condition extending p containing d not in the domain of p such that $q \Vdash \dot{g}(\bar{d}) = \bar{n}$, and q decides the value of $\Phi_e^{\hat{H}_k^{h_j}}(\bar{d})$. Let q' be similar to q except that it replaces m for n. Since generic extensions of q and q' which differ in value only at d have the same interpretation for h_j, both conditions force the same value for $\Phi_e^{\hat{H}_k^{h_j}}(\bar{d})$. Then

$$q \Vdash \Phi_e^{\hat{H}_k^{h_j}}(\bar{d}) = f_n(\bar{i}) \quad \text{and} \quad q' \Vdash \Phi_e^{\hat{H}_k^{h_j}}(\bar{d}) = f_m(i)$$

which contradict our choice of m and n. Hence $c_i \not\leq c_j$ implies $h_i \not\leq_h h_j$.

The final case to consider is when $c_i \cap c_j = c_k$. By the preservation of order under embedding, we have $h_i \cap h_j \leq_h h_k$. Assume that $f \leq_h h_i$ and $f \leq_h h_j$ with the aim of showing $f \leq_h h_k$. Let p be a condition extended by g such that

$$p \Vdash f = \Phi_{e_0}^{\hat{H}_m^{h_i}} = \Phi_{e_1}^{\hat{H}_n^{h_j}}$$

for some $m, n \in \mathcal{O}$ and e_0, e_1. For a given u, use h_k to find a condition q extending p such that q forces a (necessarily common) value for $\Phi_{e_0}^{\hat{H}_m^{h_i}}(\bar{u})$ and $\Phi_{e_1}^{\hat{H}_n^{h_j}}(\bar{u})$ so that $h_k(d) = f_{q(d)}(k)$ for every d in the domain of q (note that $f_{g(d)}(k) = f_{q(d)}(k)$ for every d in the domain of q). We claim that this is an algorithm for computing f hyperarithmetically in h_k.

Since forcing relation for a ranked formula is hyperarithmetic, and since the search for a condition to satisfy the second half of the above statement is recursive, q may be computed hyperarithmetically in h_k. We show that the value $\Phi_{e_1}^{\hat{H}_n^{h_j}}(\bar{u})$ is the correct value for $f(u)$. Assume for the sake of contradiction that $v \neq w$ are numbers such that

$$\Phi_{e_0}^{\hat{H}_m^{h_i}}(u) = \Phi_{e_1}^{\hat{H}_n^{h_j}}(u) = w \quad \text{and} \quad q \Vdash \Phi_{e_0}^{\hat{H}_m^{h_i}}(\bar{u}) = \Phi_{e_1}^{\hat{H}_n^{h_j}}(\bar{u}) = \bar{v}.$$

By the genericity of g, there is a condition r stronger than p and extended by g such that

$$r \Vdash \Phi_{e_0}^{\hat{H}_m^{h_i}}(u) = \Phi_{e_1}^{\hat{H}_n^{h_j}}(u) = \bar{w}.$$

By extending the domains if necessary, we may assume that q and r have the same domain. Then $f_{q(d)}(k) = h_k(d) = f_{r(d)}(k)$ for every d in the domain. By Theorem 10.1.1 (v), for each such d there exist numbers $\hat{d}_1, \hat{d}_2$ and $\hat{d}_3$ satisfying the conclusion in (v) for $f_{q(d)}$ and $f_{r(d)}$. For $l = 1, 2, 3$, let q_l be a condition such that $q_l(d) = \hat{d}_l$ and let $g_l(d) = g(d)$ if d is not in the domain of q_l, and equal to $q_l(d)$ otherwise. Then g_l

differs from g on only finitely many values and is hence a generic function extending q_l. Let $h_{l,i}$, $h_{l,j}$ be the functions obtained in place of h_i and h_j upon substituting g with g_l. Then $h_i = h_{1,i}$, $h_{1,j} = h_{2,j}$, $h_{2,i} = h_{3,i}$ and $h_{3,j} = h_j$. The choice of the condition p and the fact that $h_i = h_{1,i}$ imply that

$$\Phi_{e_0}^{\hat{H}_m^{h_i}}(u) = \Phi_{e_0}^{\hat{H}_m^{h_{1,i}}}(u) = v.$$

Then p and the other equalities give

$$\Phi_{e_0}^{\hat{H}_m^{h_{1,i}}}(u) = \Phi_{e_1}^{\hat{H}_n^{h_{1,j}}}(u) = \Phi_{e_1}^{\hat{H}_n^{h_{2,j}}}(u) = \Phi_{e_0}^{\hat{H}_m^{h_{2,i}}}(u) = \Phi_{e_0}^{\hat{H}_m^{h_{3,i}}}(u) = \Phi_{e_1}^{\hat{H}_n^{h_{3,j}}}(u) = \Phi_{e_1}^{\hat{H}_n^{h_j}}(u) = w.$$

Since $v \neq w$, we have a contradiction. □

10.2 The rigidity of hyperdegrees

Let $x \subseteq \omega$. The next step in the proof of rigidity is to define a partial lattice $\mathfrak{C}_x$ with a finite set of parameters that generates a copy of ω. In addition, x will be coded in $\mathfrak{C}_x$. For the intended application, the coding of membership and nonmembership in x will be in terms of positive formulas involving only $\leq$ and $\cup$, i.e. the negation symbol $\neg$ and the meet symbol $\cap$ do not appear in the formula defining an $n \in x$ or $n \notin x$. This is a key device to ensure that in any embedding f of $\mathfrak{C}_x$ into $\mathcal{D}_h$ preserving $\leq$ and $\cup$, one always has x hyperarithmetic in the parameters defining the partial lattice (a consequence of Lemma 10.2.1 (iv) below).

Firstly, let d_0, e_0, e_1, f_0, f_1 be elements of $\mathfrak{C}_x$ which will be the generators of $\{d_n \mid n < \omega\}$ intended as a copy of ω. For each $n < \omega$, let:
- $(d_{2n} \cup e_0) \cap f_1 = d_{2n+1}$
- $(d_{2n+1} \cup e_1) \cap f_0 = d_{2n+2}$.

These conditions guarantee that the d_n's are enumerated recursively. Given $x \subseteq \omega$, we add two additional parameters c_x and $c_{\bar{x}}$, and require $d_n \leq c_x$ for $n \in x$, $d_n \cap c_x = 0$ for $n \notin x$, $d_n \leq c_{\bar{x}}$ for $n \notin x$ and $d_n \cap c_{\bar{x}} = 0$ for $n \in x$. We also set $d_{2n+1} \leq (d_{2n} \cup e_0)$, $d_{2n+1} \leq f_1$ and $d_{2n+1} \neq 0$. To avoid the use of $\neg$, the last inequality may be satisfied by introducing two further parameters c and c' and requiring that in the partial lattice $c' \neq 0$ and $c \cup d_n \geq c'$ for each n. If $u = d_n$ for some n, then $u \neq 0$ is expressed as $c \cup u \geq c'$.

Lemma 10.2.1. *Let $x \subseteq \omega$. There is a recursive set $\{\varphi_n \mid n \in \omega\}$ of positive existential formulas involving only $\leq$, $\cup$ and parameters $c_x, c_{\bar{x}}, c, c', d_0, e_0, e_1, f_0, f_1$ such that in any partial lattice $\mathfrak{C}_x$ defined as above,*
(i) $\varphi_n(u) \Leftrightarrow 0 < u \leq d_n$;
(ii) $n \in x \Leftrightarrow (\exists u)(\varphi_n(u) \wedge u \leq c_x)$;
(iii) $n \notin x \Leftrightarrow (\exists u)(\varphi_n(u) \wedge u \leq c_{\bar{x}})$;
(iv) *In any embedding of $\mathfrak{C}_x$ into $\mathcal{D}_h$ preserving $\cap$, $\cup$ and $\leq$, the formulas are equivalent to Π^1_1-formulas (with the relevant parameters).*

Proof. First set $\varphi_0(u)$ to be $u = d_0$. Recursively define $\varphi_{2n+1}(u)$ to be

$$(\exists w)(\varphi_{2n}(w) \wedge u \le w \cup e_0 \wedge u \le f_1 \wedge c \cup u \ge c'),$$

and let $\varphi_{2n+2}(u)$ to be

$$(\exists w)(\varphi_{2n+1}(w) \wedge u \le w \cup e_1 \wedge u \le f_0 \wedge c \cup u \ge c').$$

Suppose that $\varphi_{2n+1}(u)$ holds for u. Then there is a w as described such that, by induction, $0 < w \le d_{2n}$. Thus $w \cup e_0 \le d_{2n} \cup e_0$ and so $u \le d_{2n} \cup e_0$ and f_1. As $d_{2n+1} = (d_{2n} \cup e_0) \cap f_1$, $u \le d_{2n+1}$. Furthermore, $c' \le u \cup c$ ensures that $u > 0$ as well and so (i) holds for φ_{2n+1}. The argument for φ_{2n+2} is similar.

We leave the verification of (ii) and (iii) to the reader. Finally, (iv) follows from Corollary 3.7.4. □

Remark 10.2.2. Theorem 10.1.2 may be relativized to any real x. The partial lattice $\mathcal{C}_x$ is defined recursively in x from the parameters $d_0, e_0, e_1, f_0, f_1, c_x, c_{\bar{x}}, c, c'$ using $\le$, $\cup, \cap$ as partial functions. Then any embedding of $\mathcal{C}_x$ into the hyperdegrees maps the 0 of $\mathcal{C}_x$ to $\mathbf{x}$.

Now fix a real x. By Theorem 10.1.2 relativized to x, there is an embedding $f : \mathcal{C}_x \to \mathscr{D}_h$ preserving $\le, \not\le, \cup$ and $\cap$ so that $f(0_{\mathcal{C}_x}) = \mathbf{x}$. By Lemma 10.2.1, both x and $\omega \setminus x$ are Π^1_1-definable relative to a representation of

$$f(d_0) \cup f(e_0) \cup f(e_1) \cup f(f_0) \cup f(f_1) \cup f(c_x) \cup f(c_{\bar{x}}) \cup f(c) \cup f(c') \le f(1_{\mathcal{C}_x}).$$

Hence $\mathbf{x} \le f(1_{\mathcal{C}_x})$.

The following result will be used in the proof of the biinterpretability of $\langle \mathscr{D}_h, \le \rangle$ with second order arithmetic.

Lemma 10.2.3. *For any x and y, $y \le_h x$ if and only if there exist $\mathbf{y}_0, \mathbf{y}_1 > \mathbf{x}$ such that $\mathbf{y}_0 \cap \mathbf{y}_1 = \mathbf{x}$ and $\mathcal{C}_y$ is embeddable into both $[\mathbf{x}, \mathbf{y}_0]$ and $[\mathbf{x}, \mathbf{y}_1]$ with minimal element $\mathbf{x}$.*

Proof. Let g_0 and g_1 be Cohen generic reals relative to x so that $(\mathbf{x} \cup \mathbf{g_0}) \cap (\mathbf{x} \cup \mathbf{g_1}) = \mathbf{x}$. Then the "only if" direction follows immediately from Theorem 10.1.2 relativized to x by letting $\mathbf{y}_0 = \mathbf{x} \cup \mathbf{g_0}$ and $\mathbf{y}_1 = \mathbf{x} \cup \mathbf{g_1}$.

For the other direction, note that by Lemma 10.2.1 $\mathbf{y} \le \mathbf{y}_0$ and $\mathbf{y} \le \mathbf{y}_1$. So $\mathbf{y} \le \mathbf{x}$. □

We are now ready to prove the main theorem of this section.

Theorem 10.2.4 (Slaman and Woodin [141], Shore [133]). *If $\pi : \mathscr{D}_h \to \mathscr{D}_h$ is an automorphism, then π is the identity map.*

Proof. Let $\mathbf{x}$ be a hyperdegree containing x as an element. Let $\mathcal{C}_x$ be the lattice generated by d_0, e_0, e_1, f_0, f_1 as described above. Let $\hat{\mathcal{C}}_x$ be a disjoint copy of $\mathcal{C}_x$, and define a partial lattice $\mathcal{P}_x$ which is the union of $\mathcal{C}_x$ and $\hat{\mathcal{C}}_x$ such that $0_{\mathcal{P}_x} = 1_{\mathcal{C}_x} \cap 1_{\hat{\mathcal{C}}_x}$ and $1_{\mathcal{P}_x} = 1_{\mathcal{C}_x} \cup 1_{\hat{\mathcal{C}}_x}$. By Theorem 10.1.2 relativized to x, there is an embedding f of $\mathcal{P}_x$ into the hyperdegrees such that $f(0_{\mathcal{P}_x}) = \mathbf{x}$.

Suppose $\pi : \mathscr{D}_h \to \mathscr{D}_h$ is an automorphism such that $\pi(\mathbf{x}) \neq \mathbf{x}$. Then the coding scheme (see Remark 10.2.2) ensures that $\mathbf{x} \le \pi f(1_{\mathcal{C}_x})$ and $\mathbf{x} \le \pi f(1_{\mathcal{C}_{\hat{x}}})$. Since $\mathbf{x} = f(0_{\mathcal{P}_x}) = f(1_{\mathcal{C}_x}) \cap f(1_{\hat{\mathcal{C}}_x})$, we have $\pi(\mathbf{x}) = \pi(f(1_{\mathcal{C}_x}) \cap f(1_{\hat{\mathcal{C}}_x})) = \pi f(1_{\mathcal{C}_x}) \cap \pi f(1_{\hat{\mathcal{C}}_x})$ so that $\mathbf{x} \le \pi(\mathbf{x})$. A similar argument with π^{-1} shows that $\pi(\mathbf{x}) \le \mathbf{x}$ so that $\pi(\mathbf{x}) = \mathbf{x}$, a contradiction. □

10.2.1 Exercises

Exercise 10.2.1. *Prove Theorem* 10.2.1 (ii) *and* (iii).

Exercise 10.2.2. *Prove that if $x \in L_{\omega_1^x}$ and $\pi : \mathscr{D}_h(\le \mathbf{x}) \to \mathscr{D}_h(\le \mathbf{x})$ is an isomorphism, then π is the identity map.*

10.3 Biinterpretability

Biinterpretability in the Turing degrees says essentially that a relation on $\mathscr{D}$ is definable if and only if it is definable in second order arithmetic.

While the biinterpretability of $\langle \mathscr{D}, \le \rangle$ has yet to be established, we show in this section that it holds for the hyperdegrees.

10.3.1 Slaman–Woodin Coding

Slaman and Woodin [143] showed the definability (with parameters) of any countable relation on the Turing degrees. This was achieved using a technique of what is now called the Slaman–Woodin coding. The technique is applicable to an array of degree structures, including the hyperdegrees, and is key to proving the biinterpretability of $\langle \mathscr{D}_h, \le \rangle$ with second order arithmetic.

Theorem 10.3.1 (Slaman and Woodin [143]). *For each $n < \omega$ there is a formula φ_n in the language of hyperdegrees such that for any countable n-ary relation $R \subseteq \mathscr{D}_h$, there exist $\mathbf{g}_1, \dots, \mathbf{g}_m$ and*

$$(\mathbf{a}_1, \dots, \mathbf{a}_n) \in R \Leftrightarrow \langle \mathscr{D}_h, \le \rangle \models \varphi_n(\mathbf{a}_1, \dots, \mathbf{a}_n, \mathbf{g}_1, \dots, \mathbf{g}_m).$$

We establish this theorem by first showing that any countable antichain of hyperdegrees is definable in $\langle \mathscr{D}_h, \le \rangle$. Here is the setup: Let $A = \{\mathbf{a}_i \mid i < \omega\}$ be an antichain and let a_i be a real of hyperdegree $\mathbf{a}_i$ such that a_i is recursive in each of its infinite subsets (note that every infinite set is Turing equivalent to the set of (codes of) its finite initial segments). Let $a \in \mathbf{a}$ be an upper bound of A.

Define a notion of forcing $\mathbb{P} = \langle \mathbf{P}, \leq \rangle$ such that each $p \in \mathbf{P}$ is a triple (p_1, p_2, p_3) such that $p_1, p_2 \in 2^{<\omega}$ have the same length (called the length of p), $p_3 \in \omega$ and for $p, q \in \mathbf{P}$, $q \leq p$ if and only if

(i) $q_j \succeq p_j$ for $j = 1, 2$;

(ii) $q_3 \geq p_3$;

(iii) for all $k \leq p_3$, for all $m \in a_k$, if $\langle k, m \rangle$ is in the domain of $q_1 \setminus p_1$ (hence in the domain of $q_2 \setminus p_2$), then $q_1(\langle k, m \rangle) = q_2(\langle k, m \rangle)$.

Expand the language of Cohen forcing in second order arithmetic (see § 5.2) relativized to a by introducing ranked variables x^β for $\beta < \omega_1^a$, terms $\dot{a}_i$ for each $i < \omega$, $\dot{a}$ for a and $\dot{g}_1, \dot{g}_2$ for the generic reals g_1 and g_2 to be obtained from $\mathbb{P}$. For a formula φ in the expanded language, define $p \Vdash \varphi$ as in Section 5.2. We leave it to the reader to verify that $\omega_1^{a \oplus g} = \omega_1^a$ for any generic g.

Let g be $\mathbb{P}$-generic with $\mathbf{g}_j$ the hyperdegree of g_j for $j = 1, 2$ where $g_1 = \bigcup\{p_1 \mid (\exists p_2)(\exists p_3)((p_1, p_2, p_3) \in g))\}$ and $g_2 = \bigcup\{p_2 \mid (\exists p_1)(\exists p_3)((p_1, p_2, p_3) \in g))\}$. The plan is to show that A is the set of minimal solutions of

$$\{\mathbf{x} \leq \mathbf{a} \mid (\exists \mathbf{z})((\mathbf{z} \leq \mathbf{x} \cup \mathbf{g}_1) \wedge (\mathbf{z} \leq \mathbf{x} \cup \mathbf{g}_2) \wedge (\mathbf{z} \nleq \mathbf{x}))\}.$$

Lemma 10.3.2. *Let g_1 and g_2 be as above. For each k, there is a real c_k so that $c_k \leq_h a_k \oplus g_j$ for each j and $c_k \nleq_h a_k$.*

Proof. Let $p \prec g$ be a condition such that $|p_1| > k$. Let $c_k = \{m \mid m \in a_k \wedge |p_1| < \langle k, m \rangle \in g_1\}$. Clearly $c_k \leq_T a_k \oplus g_1$. Also by (iii), we have $c_k = \{m \mid m \in a_k \wedge |p_1| < \langle k, m \rangle \in g_2\}$ so that $c_k \leq_T a_k \oplus g_2$ as well. Note that $c_k \nleq_h a_k$: A simple diagonization method shows that c_k is not equal to x for any $x \leq_h a$. So $c_k \nleq_h a$ and so $c_k \nleq_h a_k$. □

Lemma 10.3.3. *For any reals $x, z \leq_h a$, if $z \leq_h x \oplus g_j$ for $j = 1, 2$ and $z \nleq_h x$, then $x \geq_h a_k$ for some k.*

Proof. Fix ranked formulas φ_x, φ_z so that $m \in x \Leftrightarrow \mathfrak{A}(\omega_1^a, a) \models \varphi_x(\dot{m})$ and $m \in z \Leftrightarrow \mathfrak{A}(\omega_1^a, a) \models \varphi_z(\dot{m})$. Also for $j = 1$ and 2, let $\varphi_{x \oplus g_j}$ be such that $m \in z \Leftrightarrow \mathfrak{A}(\omega_1^a, a, g_j) \models \varphi_{x \oplus g_j}(\dot{m}) \Leftrightarrow \mathfrak{A}(\omega_1^a, a, g) \models \varphi_{x \oplus g_j}(\dot{m})$. Note that $\varphi_{x \oplus g_j}$ is expressed in terms of φ_x and g_j. Hence there is a condition $p \prec g$ such that

$$p \Vdash (\forall m)(\varphi_z(m) \leftrightarrow \varphi_{x \oplus g_1}(m)) \wedge (\varphi_z(m) \leftrightarrow \varphi_{x \oplus g_2}(m)).$$

Claim. *There exist infinitely many m's and conditions $q^m, r^m \succ p$ of common length at least m such that $q^m \Vdash \varphi_{x \oplus g_1}(\dot{m})$ and $r^m \Vdash \neg \varphi_{x \oplus g_1}(\dot{m})$.*

Proof of Claim. Otherwise, for all sufficiently large m, for all $q, r \succ p$ with length at least m, q and r do not force different values for $\varphi_{x \oplus g_1}(\dot{m})$. Since $x \leq_h a$, the sentence $\varphi_{x \oplus g_1}(\dot{m})$ has (the same) rank less than ω_1^a for each m and so the relation "$q \Vdash \varphi_{x \oplus g_1}(\dot{m})$" is hyperarithmetic in x. But then this implies that $z \leq_h x$ which contradicts the hypothesis.

Thus let q^m, r^m be chosen as in the Claim. Note that $m \mapsto (q^m, r^m)$ is hyperarithmetic in x.

We start with (the least) $i < |q|$ such that q_1^m and r_1^m disagree at i (if any). Define a condition $s(0) = (s_1(0), s_2(0), s_3(0))$ by first changing the value of q_1^m at the point i to agree with that of r_1^m, $s_2(0) \restriction |q_2^m| = q_2^m$ and $s_3(0) = q_3^m$, and then extending it to decide the value of $\varphi_{x\oplus g_1}(\dot{m})$. Then extend the domains of q_1^m, q_2^m and r_1^m, r_2^m to that of $s_1(0)$ so that the new points in these domains agree with $s_1(0)$ in their values. Call the resulting conditions $q^m(0)$ and $r^m(0)$. Either $s_1(0)$ and $q_1^m(0)$ disagree at exactly one point, $s_2(0) = q_2^m(0)$, and $s(0)$ and $q^m(0)$ force incompatible values for $\varphi_{x\oplus g_1}(\dot{m})$, or $s(0)$ and $r^m(0)$ force incompatible values and have fewer points of disagreement between them than that between q^m and r^m. For q_2^m and r_2^m, we carry out the same construction. By induction, one defines successively $s(0), s(1), \ldots$ and concludes that there is a j so that $s(j)$ and $s(j+1)$ with exactly one point of disagreement in the first two components forcing incompatible values for $\varphi_{x\oplus g_1}(\dot{m})$.

Thus we may assume that q^m and r^m differ exactly at one point $\langle k, l\rangle$ so that $q_1^m(\langle k, l\rangle) \neq r_1^m(\langle k, l\rangle)$. If either $k > p_3$ or $l \notin a_k$, then let $s = (q_1^m, r_2^m, q_3 + r_3)$. Then $s \leq p$. This implies that we can extend p to a generic g having the same values for g_1 and g_2 as those for q^m and r^m respectively at their common points. This, however, contradicts the assumption on p.

It follows that there exist infinitely many pairs of conditions extending p which differ at exactly one $\langle k, l\rangle$ with $k \leq p_3$ and $l \in a_k$, forcing different values as described above. In particular there is a $k \leq p_3$ for infinitely many such l's to satisfy this requirement. Now as noted earlier the forcing relation for sentences of fixed rank considered here is hyperarithmetic in the parameter x, and so one can hyperarithmetically in x enumerate an infinite set of l's contained in a_k. Since a_k is recursive in each of its infinite subsets, this set computes a_k.

Thus $x \geq_h a_k$, proving the lemma. □

One may now convert an arbitrary countable relation on the degrees into an antichain so that the original relation is definable from the antichain using parameters.

Proof of Theorem 10.3.1. Let R be an n-ary relation on the hyperdegrees. Let $\{\mathbf{a}_j \mid j < \omega\}$ be a list of all the hyperdegrees $\leq \mathbf{a}$, with $a_j \in \mathbf{a}_j$ and $a \in \mathbf{a}$. Let $\{g_{i,j} \mid 1 \leq i \leq n \text{ and } j < \omega\}$ be reals which are mutually Cohen generic over a. This can be obtained following §5.2 with the expanded language described earlier, taking $g_{i,j}$ to be the appropriate columns of a single generic g. By Exercise 10.3.1, the following sets are antichains of hyperdegrees:

$$S_i = \{\mathbf{g}_{i,j} \mid j < \omega\};$$

$$T_i = \{\mathbf{a}_j \cup \mathbf{g}_{i,j} \mid j < \omega\};$$

$$U = \{\mathbf{g}_{1,j_1}, \ldots \mathbf{g}_{n,j_n} \mid R(\mathbf{a}_{j_1}, \ldots, \mathbf{a}_{j_n})\}.$$

Then S_i, T_i and U are definable in $\langle \mathscr{D}_h, \leq \rangle$. For any $\mathbf{x}_1, \ldots, \mathbf{x}_n \leq \mathbf{a}$, $R(\mathbf{x}_1, \ldots, \mathbf{x}_n)$ holds if and only if for each $i \in [1, n]$, there is a $\mathbf{y}_i \in S_i$ and $\mathbf{z}_i \in T_i$ such that $\mathbf{y}_1 \cup \cdots \cup \mathbf{y}_n \in U$ and $\mathbf{x}_i \cup \mathbf{y}_i = \mathbf{z}_i$. Hence R restricted to the hyperdegrees below $\mathbf{a}$ is definable, completing the proof of Theorem 10.3.1. □

10.3.2 Coding an expanded standard model of PA

Consider the standard model $\langle \omega, x, <, +, \cdot, 0, 1 \rangle$ of Peano arithmetic expanded by a unary predicate x. We may represent it as a relation on degrees. Let $\{\mathbf{d}_n\}_{n<\omega}$ be a countable sequence of hyperdegrees. We interpret $\mathbf{d}_n$ as the number n and define arithmetical relations and functions on finite subsets of $\{\mathbf{d}_n\}_{n<\omega}$ in the obvious way. From these one may define a countable relation R^x on $\{\mathbf{d}_n\}_{n<\omega}$ turning it into (a copy of) the standard model of arithmetic with a unary predicate representing the real x. By Theorem 10.3.1, there is a finite sequence of degrees $\vec{\mathbf{p}}$ and a formula φ in the language of partial ordering which together define R^x. Fix the formula φ throughout this section. Note that not every $\vec{\mathbf{p}}$ together with φ necessarily code a standard model of arithmetic.

Definition 10.3.4. Given a finite sequence $\vec{\mathbf{p}}$ of hyperdegrees and a real x, we say that $\vec{\mathbf{p}}$ *codes a standard model with* x (as a second order element) if $\vec{\mathbf{p}}$ and φ together define a relation $R^x = \langle R_\omega, R_x, R_<, R_+, R_\cdot, R_0, R_1 \rangle$ which represents the standard model of arithmetic $\mathcal{M} = \langle \omega, x, <, +, \cdot, 0, 1 \rangle$.

If one removes the unary predicate x, then $\vec{\mathbf{p}}$ *codes a standard model* if $\vec{\mathbf{p}}$ and φ define a relation $R = \langle R_M, R_<, R_+, R_\cdot, R_0, R_1 \rangle$ which represents $\mathcal{M} = \langle \omega, <, +, \cdot, 0, 1 \rangle$. Then the relation "$\vec{\mathbf{p}}$ and φ code a standard model" is definable in $\langle \mathscr{D}_h, \leq \rangle$. To see this, we make the following observations:

(i) The model $\mathcal{M}$ represented by R satisfies the system of Robinson arithmetic Q. Since Q is a finite collection of axioms, it can be expressed in one sentence;
(ii) $\mathcal{M}$ is wellfounded. In other words, there is no infinite descending $R_<$-sequence. By Theorem 10.3.1, there is a formula $\psi(\vec{u}, v)$ so that $\mathcal{M}$ is wellfounded if and only if for any finite sequence $\vec{\mathbf{q}}$, the set $\{\mathbf{d} \mid \langle \mathscr{D}_h, \leq \rangle \models \psi(\vec{\mathbf{q}}, \mathbf{d})\}$ is not an infinite descending $R_<$-sequence.

Thus the statement that $\vec{\mathbf{p}}$ and φ code a standard model is definable.

10.3.3 Biinterpretability

For a general definition of biinterpretability, see Hodges [52]. We focus our attention on biinterpretability in the context of hyperdegrees.

Fix a relation R^* so that

$$R^*(\vec{\mathbf{p}}, \mathbf{x}) \text{ if and only if } \vec{\mathbf{p}} \text{ codes a real } x \text{ and } x \in \mathbf{x}.$$

Definition 10.3.5. The degree structure $\langle \mathcal{D}_h, \leq \rangle$ is *biinterpretable* with second order arithmetic if R^* is definable in $\langle \mathcal{D}_h, \leq \rangle$.

The intuition is as follows: It is immediate that $\langle \mathcal{D}_h, \leq \rangle$ can be "interpreted" in the standard model of second order arithmetic $\mathcal{M} = \langle \omega, \mathcal{P}(\omega), <, +, \cdot, 0, 1, \in \rangle$. In other words, any $\langle \mathcal{D}_h, \leq \rangle$-definable relation is definable in $\mathcal{M}$. The converse is not obvious. Slaman and Woodin [144] showed that the biinterpretability of $\langle \mathcal{D}_h, \leq \rangle$ would enable one to interpret $\mathcal{M}$ in $\langle \mathcal{D}_h, \leq \rangle$. In other words, any $\mathcal{M}$-definable degree relation is definable in $\langle \mathcal{D}_h, \leq \rangle$. This is the content of the following theorem. We say that a set of hyperdegrees is analytical if it is the set of hyperdegrees of an analytical set.

Proposition 10.3.6. *If $\langle \mathcal{D}_h, \leq \rangle$ is biinterpretable with second order arithmetic, then every analytical set of hyperdegrees is definable in $\langle \mathcal{D}_h, \leq \rangle$.*

Proof. Suppose that A is analytical. Let $\hat{A} = \{\mathbf{x} \mid (\exists x)(x \in A \wedge x \in \mathbf{x})\}$. We show that $\hat{A}$ is definable in $\langle \mathcal{D}_h, \leq \rangle$. Let ψ be a formula in the language of second order arithmetic such that

$$x \in A \Leftrightarrow \langle \omega, \mathcal{P}(\omega), <, +, \cdot, 0, 1, \in \rangle \models \psi(x).$$

Since quantifying over countable sets of hyperdegrees can be expressed in $\langle \mathcal{D}_h, \leq \rangle$ through the codes, the second order quantifiers over ω are also interpretable in $\langle \mathcal{D}_h, \leq \rangle$ via first order quantifiers over codes for subsets of ω. Then $\mathbf{x} \in \hat{A}$ if and only if there is a finite sequence $\vec{\mathbf{p}}$ such that $R^*(\vec{\mathbf{p}}, \mathbf{x})$ and $\psi(x)$ is satisfied in the model with x coded by $\vec{\mathbf{p}}$ and φ.

Thus $\hat{A}$ is definable in $\langle \mathcal{D}_h, \leq \rangle$. □

The rigidity of the hyperdegrees proved in Theorem 10.2.4 allows one to show the biinterpretability of this structure with second order arithmetic:

Theorem 10.3.7 (Slaman and Woodin [141], Shore [133]). *The structure $\langle \mathcal{D}_h, \leq \rangle$ is biinterpretable with second order arithmetic.*

Proof. Let $\psi(\vec{\mathbf{p}}, \mathbf{x})$ be defined as follows:
- $\vec{\mathbf{p}}$ codes a real x;
- For any coded real $y \leq_h x$, there exist degrees $\mathbf{y}_0, \mathbf{y}_1 > \mathbf{x}$ such that
  - $\mathbf{y}_0 \cap \mathbf{y}_1 = \mathbf{x}$, and
  - $\mathcal{C}_y$ can be embedded into both $[\mathbf{x}, \mathbf{y}_0]$ and $[\mathbf{x}, \mathbf{y}_1]$ with least element $\mathbf{x}$.

By Theorem 10.3.1, ψ may be viewed as a first order formula in the language of partial ordering for $\langle \mathcal{D}_h, \leq \rangle$. If $R^*(\vec{\mathbf{p}}, \mathbf{x})$ holds, then by Lemma 10.2.3 $\langle \mathcal{D}_h, \leq \rangle \models \psi(\vec{\mathbf{p}}, \mathbf{x})$.

If $\langle \mathcal{D}_h, \leq \rangle \models \psi(\vec{\mathbf{p}}, \mathbf{x})$ then by Lemma 10.2.3, the real x stated must have hyperdegree $\mathbf{x}$. So $R^*(\vec{\mathbf{p}}, \mathbf{x})$ is true.

In summary, the structure $\langle \mathcal{D}_h, \leq \rangle$ is biinterpretable with second order arithmetic. □

10.3.4 Exercises

Exercise 10.3.1. *Prove that the sets S_i, T_i and U_i are antichains.*

Exercise 10.3.2. *Prove that if $\pi : \mathscr{D}_h \to \mathscr{D}_h$ is an elementary embedding, then π is the identity map.*

Exercise 10.3.3 (Simpson [139]). *Prove that the first order theory of hyperdegrees is recursively isomorphic to the second order theory of arithmetic.*

Exercise 10.3.4 (Slaman and Woodin [143]). *Prove that Theorem 10.3.1 holds for Turing degrees.*

Exercise 10.3.5. *Prove that under the assumption $V = L$, if $\langle \mathscr{D}_h, \leq, \mathbf{a} \rangle \equiv \langle \mathscr{D}_h, \leq, \mathbf{b} \rangle$ then* $\mathbf{a} = \mathbf{b}$.

11 Basis theorems

In earlier chapters we have seen examples of uncountable sets of reals (e.g. Σ_1^1 and certain Π_1^1-sets) that contain perfect subsets. In this chapter we study basis theorems for uncountable sets of reals as well as basis properties of perfect sets.

11.1 A basis theorem for Δ_1^1-sets of reals

11.1.1 Friedman's conjecture

Friedman [29] conjectured that every uncountable Δ_1^1-set of reals has a member in each hyperdegree greater than or equal to the hyperdegree of $\mathcal{O}$. This conjecture was settled by Martin [94] and Friedman independently.

Theorem 11.1.1 (Martin [94] and Friedman). *If $A \subseteq \omega^\omega$ is an uncountable Δ_1^1-set, then there is a perfect tree $T \leq_T \mathcal{O}$ such that $[T] \subseteq A$ and for any $x \geq_h \mathcal{O}$, there is a $z \in [T]$ with $z \equiv_h x$.*

Suppose that $A \subseteq \omega^\omega$ is an uncountable Δ_1^1-set. By Corollary 2.2.14, there is a recursive function $h : \omega^\omega \to \omega^\omega$ and a Π_1^0-set $B \subseteq \omega^\omega$ such that h maps B onto A and is injective on B. Then for every $x \in B$, $h(x) \equiv_h x$ and for every $y \in A$, there exists an $x \in B$ such that $x \equiv_h h(x) = y$. Hence it suffices to show that every $x \geq_h \mathcal{O}$ satisfies $z \equiv_h x$ for some $z \in B$. Since B is a Π_1^0-class, there is a recursive tree $T \subseteq \omega^{<\omega}$ so that $B = [T] = \{x \mid (\forall n)(x \upharpoonright n \in T)\}$. We shall prove in the next section a slightly stronger result:

Theorem 11.1.2. *Every Σ_1^1-tree T_1 with uncountably many infinite paths has a member in each hyperdegree greater than or equal to that of $\mathcal{O}$.*

11.1.2 Proof of Friedman's conjecture

We prove Theorem 11.1.2 following [172]. Let $T_1 \subseteq \omega^{<\omega}$ be a Σ_1^1-tree with a nonhyperarithmetic path. Then there is a $\Sigma_1(L_{\omega_1^{\mathrm{CK}}})$-enumeration of the nodes of $\omega^{<\omega} \setminus T_1$. We use $T_1[\beta]$ to denote the collection of nodes not yet enumerated up to stage β. Then $T_1 = \bigcap_{\beta<\omega_1^{\mathrm{CK}}} T_1[\beta]$. Fix a nonempty recursive tree $T_0 \subseteq \omega^{<\omega}$ having an infinite path and without a hyperarithmetic infinite path. Let f be the leftmost path of T_0. Then $f \equiv_h \mathcal{O}$.

Lemma 11.1.3. *If $T_1 \cap [\nu] = \{\tau \in T_1 \mid \tau \succ \nu\}$ has no nonhyperarithmetic infinite path, then there is a stage $\beta < \omega_1^{\mathrm{CK}}$ witnessing it. In other words, there is a $\Sigma_1(L_{\omega_1^{\mathrm{CK}}})$ partial function $H : \omega^{<\omega} \to \{0\}$ such that $H(\sigma)$ is undefined if and only if $T_1 \cap [\sigma]$ contains a nonhyperarithmetic infinite path.*

Proof. Suppose that $\sigma \in T_1$ and $T_1 \cap [\sigma]$ has no nonhyperarithmetic infinite path. We perform a Cantor–Bendixon type operation on $T_1 \cap [\sigma]$ as follows: Let $<_{KB}$ be the Kleene–Brouwer ordering on $\omega^{<\omega}$. At stage $\beta < \omega_1^{\mathrm{CK}}$, define a subtree T_1^β. T_1^β is in fact $T_1[\beta]$ except for the removal of those strings σ where $[\sigma]$ is seen at stage β not to contain a nonhyperarithmetic path in T_1.

Stage 0. Let $T_1^0 = T_1[\emptyset]$.

Stage $\beta + 1$. Given $\sigma \in T_1^\beta$ and $\tau \succ \sigma$, if either

(1) there exists a $<_{KB}$ order-preserving function $h : T_1^\beta \cap [\tau] \to \beta$ in L_β (hence there is no infinite path in $T_1^\beta \cap [\tau]$, or

(2) there exists a function $h \in L_\beta$ that enumerates $T_1^\beta \cap [\tau]$ as a unique hyperarithmetic path.

Then let $T_1^{\beta+1} = T_1^\beta \setminus [\tau]$ and declare that τ is removed from T_1^β at stage $\beta + 1$.

By the usual Cantor–Bendixon proof as in Proposition 3.6.10, if $T_1 \cap [\sigma]$ has no nonhyperarithmetic infinite path, then every $\tau \succ \sigma$ is removed at some stage β. Then by the admissibility of ω_1^{CK}, there is a stage $\beta_\sigma < \omega_1^{\mathrm{CK}}$ such that every $\tau \succ \sigma$ is removed at β_σ. At stage β_σ, define $H(\sigma) = 0$. □

Thus H may be viewed as a $\Sigma_1(L_{\omega_1^{\mathrm{CK}}})$-partial enumeration function. A node $\sigma \in \omega^{<\omega}$ is a *splitting node in the tree* T_1 if there exist $i \neq j$ such that both $H(\sigma^\frown i)$ and $H(\sigma^\frown j)$ are undefined. For $\beta < \omega_1^{\mathrm{CK}}$, a node $\sigma \in \omega^{<\omega}$ is a *splitting node in the tree* $T_1[\beta]$ if there exist $i \neq j$ such that both $H(\sigma^\frown i)$ and $H(\sigma^\frown j)$ are undefined at stage β. By a similar argument, there is a $\Sigma_1(L_{\omega_1^{\mathrm{CK}}})$-partial function $K : \omega^{<\omega} \to \{0\}$ such that $K(\sigma)$ is undefined if and only if there exists an infinite path in $T_0 \cap [\sigma]$. Note that K is different from H.

We call a node $\sigma \in \omega^{<\omega}$ a *terminal node in* T_0 if $K(\sigma)$ is defined. For $\beta < \omega_1^{\mathrm{CK}}$, $\sigma \in \omega^{<\omega}$ is a *terminal node in* $T_0[\beta]$ if $K(\sigma)$ is defined at stage β.

Now for any $x \geq_h \mathcal{O}$ in 2^ω, we build a $z \in [T_1]$ such that $x \equiv_h z$:

Stage 0. Let $\sigma_0 = \emptyset$.

Stage $s + 1$. Let $\sigma' \succ \sigma_s$ be the leftmost splitting node of T_1 extending σ_s so that there are $f(s)$ many splitting nodes between σ_s and σ'. Let $\sigma'' \succ \sigma'$ be the second leftmost splitting node extending σ' in T_1. Set $\sigma_{s+1} \succ \sigma''$ to be the $x(s) + 1$-th leftmost splitting node extending σ'' in T_1.

Let $z = \bigcup_{s \in \omega} \sigma_s$.

We prove that $x \in L_{\omega_1^z}[z]$. In fact we show that there is a $\Sigma_1(L_{\omega_1^z}[z])$ (and hence $\Delta_1(L_{\omega_1^z}[z])$)-definable increasing sequence of ordinals $\{\alpha_s\}_{s \in \omega}$ below ω_1^{CK} such that $x \in L_{(\bigcup_s \alpha_s)+1}[z]$. Then $\bigcup_s \alpha_s + 1 < \omega_1^z$, proving $x \in L_{\omega_1^z}[z]$.

The idea is to decode the construction of z in $L_{\omega_1^z}[z]$. For each $n < \omega$ we use z to identify the value of $f(n)$, where f is the leftmost path of T_0. As an example, consider the case $n = 0$. Given $\beta < \omega_1^{\mathrm{CK}}$, we compute, beginning at the root of $T_1[\beta]$, the number l_β of times z "turns left" in $T_1[\beta]$. Then l_β is a possible value of $f(0)$. Of course this guess may be wrong. As an aid to making a better guess, we also determine if l_β is $f(0)$ by asking if $0^\frown l_\beta$ is not a terminal node in $T_0[\beta]$ but $0^\frown l'$ is terminal for every $l' < l_\beta$. If this is not true, proceed to a stage $\beta' > \beta$ where this holds. By the coding procedure, there must be a $\beta' < \omega_1^{CK}$ greater than β such that $l_{\beta'}$ matches these requirements.

Let β_1 be the first such stage. One is not able to conclude that l_{β_1} is indeed the real value of $f(0)$. Both l_{β_1} and the computed value of $f(0)$ (as seen in $T_0[\beta_1]$) may change at a later stage. However, this situation does not occur infinitely often since otherwise either l_β goes to infinity or $l_\beta = l$ infinitely often for some finite number l. In either case, since "σ is a terminal node in T_1" is a Σ_1-statement, z would be the leftmost path in T_1, a contradiction. Hence incorrect guesses occur at most finitely many times and l_β reaches a limit after finitely many changes.

In general, to compute $f(n)$ from z one imitates the guessing procedure for $f(0)$. This will involve a finite injury argument. Thus once the finite sequence $(0, l_\beta^0)^\frown (1, l_\beta^1)^\frown \ldots {}^\frown(n, l_\beta^n)$ matches the computed values of $f \restriction n$ at stage s, we set β_s to be β. The sequence and values of $f \restriction n$ computed at a later stage may change, but changes happen at most finitely often. Hence l_β^n will reach a limit. Let γ be the least upper bound of $\{\beta_s\}_{s\in\omega}$. Clearly $\gamma < \omega_1^z$ and the limit of $\{f_{\beta_s}\}_{s\in\omega}$ is an infinite path in $T_0[\gamma]$. Since $[T_0] = [T_0[\gamma]]$, it is also an infinite path in T_0. Moreover, it is the leftmost path. Hence z hyperarithmetically computes the leftmost path of T_0. Using this, one shows by a decoding method that $z \equiv_h x$.

We now proceed with the formal proof.

Define $\alpha_s, \sigma_s^n, f_s(n)$ as follows. The aim is to ensure that $\lim_{s\to\omega} f_s(n) = f(n)$.

Stage 0. Let $\sigma_0^0 = \emptyset$ and $\beta_0 = f_0(n) = 0$ for every n. We say that 0 receives attention.

Assume that at stage $s \geq 0$, β_s is defined and that if $j \leq s$ receives attention, then $f_s(j)$ and σ_s^j are defined. Then z is said to be *correct up to j at β* if for every $i \leq j$, if $0 < i$ then

(1) $f_s \restriction (i+1)$ is the leftmost nonterminal node in $T_0[\beta]$;

(2) there exists a $\sigma' \prec \sigma_s^i$ which is the leftmost splitting node of $T_1[\beta]$ so that σ' has $f_s(i)$ many splitting nodes extending σ_s^{i-1} in $T_1[\beta]$, and

(3) for the σ' above, there is a node $\sigma'' \prec \sigma_s^i$ extending σ' which is the second leftmost splitting node extending σ' in $T_1[\beta]$ so that σ_s^i is the next splitting node extending σ'' in $T_1[\beta]$.

Stage $s+1$. Let j_{s+1} be the minimum of $j \leq s$ such that z is not correct up to j at $\beta_s + 1$ if such a j exists, and let it be $s+1$ otherwise. Initialize each $j > j_{s+1}$ and let $f_{s+1}(j) = f_s(j)$ and $\sigma_{s+1}^j = \sigma_s^j$ for $j < j_{s+1}$. For $j = j_{s+1}$, there are two cases:

Case 1. j_{s+1} has not received any attention since its last initialization, if any. Let $\beta_{s+1} = \beta_s + 1$, $\sigma_{s+1}^{j_{s+1}} \succ \sigma_s^{j_{s+1}-1}$ be the next splitting node in $T_1[\beta_{s+1}]$ and $f_{s+1}(j_{s+1}) = 0$ and declare that j_{s+1} receives attention at stage $s+1$.

Case 2. Otherwise. Search for a $\beta > \beta_s$ less than ω_1^{CK} such that z is correct up to $j_{s+1} - 1$ at β, and there exists a $\sigma \prec z$ extending $\sigma_s^{j_{s+1}-1}$ for which there is an l satisfying

(i) $f_s \restriction j_{s+1}{}^\frown(j_{s+1}, l)$ is the leftmost nonterminal node in $T_0[\beta]$;

(ii) there exists a $\sigma' \prec \sigma$ which is the leftmost splitting node of $T_1[\beta]$ such that σ' has l-many splitting nodes extending σ_s^{j-1} in $T_1[\beta]$, and

(iii) for the σ' above, there is a node $\sigma'' \prec \sigma$ extending σ' which is the second leftmost splitting node extending σ' in $T_1[\beta]$ so that σ is the next splitting node extending σ'' in $T_1[\beta]$.

Suppose that during the search in Case 2, z is always correct up to $j_{s+1} - 1$. By the definition of z, a β as stipulated exists. Then let $\beta_{s+1} = \beta$, $f_{s+1}(j_{s+1}) = l$ and $\sigma_{s+1}^{j_{s+1}} = \sigma$. Declare that j_{s+1} receives attention at stage $s + 1$.

Otherwise, z is not correct up to $j_{s+1} - 1$ at β, and we replace j_{s+1} by $j_{s+1} - 1$ to do the search. However, this process of replacement occurs only finitely many times since z is always correct up to 0 at any β.

This completes the construction.

Let $\gamma = \bigcup_{s\in\omega} \beta_s$. Then $\{\beta_s\}_{s\in\omega}$ is a $\Delta_1(L_{\omega_1^z}[z])$-sequence and hence $\gamma < \omega_1^z$.

Lemma 11.1.4. *For every j, there exists a stage s_j such that if $s \geq s_j$ then z is correct up to j at β_s.*

Proof. Suppose otherwise. Let j^* be the least j such that for infinitely many s, z is not correct up to j^* at β_s. Then Case 2 in the construction applies to j^*. Let s_{j^*-1} be the least stage such that for every $s \geq s_{j^*-1}$, z is correct up to $j^* - 1$ at β_s.

For $s > s_{j^*-1}$, let σ_s' be the σ' defined in the construction for j^* at stage s. In other words, for some initial segment σ of z, $\sigma_s' \prec \sigma$ is the leftmost splitting node of $T_1[\beta_s]$ such that σ_s' has $f_s(j^*)$-many splitting nodes extending $\sigma_s^{j^*-1}$ in $T_1[\beta_s]$. Note that for any $s \geq s_{j^*-1}$, $f_{s+1}(j^*) \geq f_s(j^*)$. Hence $|\sigma_{s+1}'| \geq |\sigma_s'|$ whenever $s > s_{j^*-1}$. Since for any τ, z is not the leftmost path of $T_1 \cap [\tau]$, $\sigma^{j^*} = \lim_s \sigma_s'$ exists. Let $s_1 > s_{j^*-1}$ be chosen such that $\sigma_s' = \sigma^{j^*}$ for $s \geq s_1$. By the construction, $f_s(j^*) = f_{s_1}(j^*)$ for $s \geq s_1$. It follows that there exists an $s_{j^*} \geq s_1$ such that for all $s \geq s_{j^*}$, $\sigma_s^{j^*} = \sigma_{s_{j^*}}^{j^*} = \sigma^{j^*}$. Then for any $s \geq s_{j^*}$, z is correct up to j^* at β_s. □

By Lemma 11.1.4, we may let $\hat{f}(j) = \lim_{s\to\omega} f_s(j)$. Obviously, $\hat{f} \in L_{\gamma+1}[z]$. Note that for any j, $T_0 \supseteq \{\sigma \mid (\forall k \leq j)(\sigma(k) = \hat{f}(k))\}$. So $\hat{f}$ is an infinite path in T_0. By the construction, $\hat{f} = f$ and hence $f \in L_{\gamma+1}[z]$. Since $f \geq_h \mathcal{O}$, we have $\mathcal{O} \leq_h z$ and $\omega_1^{\mathrm{CK}} < \omega_1^z$. Then the construction may be decoded from z. Thus $z \geq_h x$. If $x \geq_h \mathcal{O}$, then x may also decode the construction so that $x \equiv_h z$.

The construction in the proof above yields a perfect tree $T \leq_T \mathcal{O}$ with $[T] \subseteq A$ in which every infinite path x is hyperarithmetically above $\mathcal{O}$. This completes the proof of Theorem 11.1.2.

11.1.3 Exercises

Exercise 11.1.1. *Prove that there is a Σ_1^1-set A containing a real $x \geq_h \mathcal{O}$ but not every hyperdegree above that of $\mathcal{O}$ has a representative in A.*

Exercise 11.1.2 (Feferman). *Prove that there is a Σ^1_1-set A in which any two distinct members are hyperarithmetically incomparable.*

Exercise 11.1.3. *Prove that if a Σ^1_2-set A contains a member which hyperarithmetically computes every constructible real, then there is a cone of hyperdegrees in which every member has a representative in A.*

Exercise 11.1.4. *Prove that there is a perfect Π^0_1-set $A \subseteq 2^\omega$ in which no member Turing computes $\emptyset'$.*

Exercise 11.1.5. *Prove that the set $\{x \oplus y \mid x \leq_h y\} \subseteq 2^\omega$ is not Borel.*

Exercise 11.1.6 (Kechris [64]). *Prove that* HYP *is the largest Π^1_1-proper subset of 2^ω which is closed under hyperarithmetic reducibility.*

11.2 An antibasis theorem for Π^0_1-sets

11.2.1 An analysis of Theorem 11.1.1

A natural question arising from Theorem 11.1.1 is whether the condition $x \geq_h \mathcal{O}$ in the statement may be weakened to $x^{(\beta)} \geq_T \mathcal{O}$ or even $x^{(\beta)} \geq_T \emptyset^{(\beta+1)}$ for some $\beta < \omega_1^{\mathrm{CK}}$. The proof of the theorem requires at least ω_1^{CK}-many iterates of the Turing jump. A negative answer is given by Harrington:

Theorem 11.2.1 (Harrington [43]). *For any ordinal $\beta < \omega_1^{\mathrm{CK}}$, there is an uncountable Π^0_1-set $A \subseteq \omega^\omega$ such that for each $x \in A$, $x^{(\beta)} \not\geq_T \emptyset^{(\beta+1)}$.*

We begin our discussion with the finite version of Theorem 11.2.1.

Theorem 11.2.2. *There is an uncountable Π^0_1-set $A \subseteq \omega^\omega$ such that for any $x \in A$ and $n < \omega$, $x^{(n)} \not\geq_T \emptyset^{(n+1)}$.*

Harrington's proof uses an ingenious reverse induction construction. The following proof is more complicated than necessary, but the argument is applicable to a more general situation. By Theorem 2.1.4, the set $H = \{(i, \emptyset^{(i)}) \mid i < \omega\} \subseteq \omega \times 2^\omega$ is Π^0_2. Hence there is a recursive predicate $P(i, \sigma, m)$ such that $(i, x) \in H \Leftrightarrow (\forall m)(\exists k)P(i, x \upharpoonright k, m)$. Let $T^\omega \subseteq \omega \times \omega^{<\omega}$ be a recursive tree such that

$$(i, \sigma) \in T^\omega \Leftrightarrow (\exists m)(\exists s_m)(\sigma = \sigma_0 \oplus \sigma_1 \wedge |\sigma_0| = s_m \wedge \sigma_1(|\sigma_0|) = 2^m \cdot 3^{s_m}$$
$$\wedge\, s_m = \min\{s \mid (\forall l \leq m)(\exists k \leq s)(P(i, \sigma_0 \upharpoonright k, l))\}.$$

It is easy to verify that for each $i < \omega$, there is a unique z_i such that $(i, z_i) \in [T^\omega]$ and $z_i = \emptyset^{(i)} \oplus y_i$ for some $y_i \in \omega^\omega$. Hence uniformly each $\emptyset^{(i)}$ is a Π^0_1-singleton (in ω^ω).

We define three sequences of trees $\{T_i\}_{i<\omega}$, $\{T_i^j\}_{i<\omega,j<2}$ in $\omega^{<\omega} \times \omega^{<\omega}$, a sequence of functions $\{\psi_i\}_{i<\omega}$, and constants $\{c_j\}_{j\leq 3}$ in ω such that for each $i < \omega$,

(1) $T_i^1 = \{\sigma \mid (i, \sigma) \in T^\omega\}$;
(2) $(\sigma, \tau) \in T_i \Leftrightarrow (\exists \tau' \succeq \tau)(\tau' \in T_i^1 \wedge \Phi_{c_0}^{\tau'}(\sigma)[|\tau'|] = 1)$;
(3) $T_i^0 = \{\sigma \mid (\exists \tau \prec \emptyset^{(i)})((\sigma, \tau) \in T_i)\}$;
(4) $\psi_i : T_{i+1}^0 \to T_i^0$ is $\Phi_{c_1}^{\emptyset^{(i+1)}}$;
(5) ψ_i is a homeomorphism from $[T_{i+1}^0]$ to $[T_i^0]$ such that

$$(\forall x \in [T_i])(\psi_i^{-1}(x) \oplus \emptyset^{(i+1)} \equiv_T (x \oplus \emptyset^{(i)})')$$

via (c_2, c_3) and
(5a) $\sigma \prec \tau \Leftrightarrow \psi_i(\sigma) \preceq \psi_i(\tau)$;
(5b) If $\sigma \in T_{i+1}^0$, then $|\psi_i(\sigma)| \geq |\sigma|$;
(6) $T_i^0 \cap \omega^{\leq i} = 2^{\leq i}$, and
(7) $(\forall \sigma \in T_{i+1}^0 \cap \omega^{\leq i})(\sigma = \psi_i(\sigma))$.

Assume that these sequences have been defined.

Lemma 11.2.3. *For each $x \in [T_0^0]$ and $i < \omega$,*
(i) $x^{(i)} \equiv x \oplus \emptyset^{(i)}$ *and*
(ii) $x^{(i)} \not\geq_T \emptyset^{(i+1)}$.

Proof. By (4), (5) and an induction on i, (i) is immediate. If $x^{(i)} \geq_T \emptyset^{(i+1)}$, then $x^{(i+1)} \equiv_T x \oplus \emptyset^{(i+1)} \equiv_T x^{(i)}$, a contradiction. Hence (ii) holds. □

Lemma 11.2.4. *$[T_0^0]$ is homeomorphic to 2^ω.*

Proof. We define a bijection f between $[T_0^0]$ and 2^ω as follows: For each $x \in [T_0^0]$, let

$$f(x) = \bigcup_{i<\omega} \psi_i^{-1} \circ \psi_{i-1}^{-1} \circ \cdots \circ \psi_0^{-1}(x) \upharpoonright i.$$

Let $x \in [T_0^0]$. For $i < \omega$, let $\sigma_i = \psi_i^{-1} \circ \psi_{i-1}^{-1} \circ \cdots \circ \psi_0^{-1}(x) \upharpoonright i$. Then $\psi_i^{-1} \circ \cdots \circ \psi_0^{-1}(x) \in [T_{i+1}^0] \cap [\sigma_i]$. Then $\psi_{i+1}^{-1} \circ \cdots \circ \psi_0^{-1}(x) \in [T_{i+2}^0]$. Since ψ_{i+1}^{-1} is one-one, $\psi_{i+1} \circ \psi_{i+1}^{-1} \circ \cdots \circ \psi_0^{-1}(x) \in [T_{i+1}^0] \cap [\sigma_i]$. By (7), $\psi_{i+1} \circ \psi_{i+1}^{-1} \circ \cdots \circ \psi_0^{-1}(x) \upharpoonright (i+1) \succeq \psi_{i+1}^{-1} \circ \cdots \circ \psi_0^{-1}(x) \upharpoonright i+1 = \sigma_{i+1}$. Hence $\sigma_{i+1} \succeq \sigma_i$. By (5b), $f(x) \in 2^\omega$.

Let $y \in 2^\omega$ and $\tau_{n,n+i} = \psi_n \circ \cdots \circ \psi_{n+i}(y \upharpoonright n+i)$ for $n, i < \omega$. By (5a), (6) and (7),

$$y \upharpoonright n \preceq \tau_{n,n} \preceq \tau_{n,n+1} \preceq \tau_{n,n+2} \cdots.$$

Moreover, by (5b), for each i, $|\tau_{n,n+i}| \geq n+i$. Let $x_n = \bigcup_{i\in\omega} \tau_{n,n+i}$. Then $x_n \in [T_n^0] \cap [y \upharpoonright n]$. Since ψ_n^{-1} is one-one, by (7) $\psi_n^{-1} \circ \psi_{n-1}^{-1} \circ \cdots \circ \psi_0^{-1}(x_0) \upharpoonright n = x_n \upharpoonright n = y \upharpoonright n$. By the definition of f, we have $f(x_0) = y$. So f is onto.

Now assume $x_0, x_1 \in [T_0^0]$ such that $f(x_0) = f(x_1)$ and for some n_0, and $i \neq j$, $x_0 \upharpoonright n_0 = x_1 \upharpoonright n_0$, and $x_0 \succ x_0 \upharpoonright n_0{}^\frown i$, $x_1 \succ x_1 \upharpoonright n_0{}^\frown j$. By (5b) and (7), $\psi_{n_0+1}^{-1} \circ \psi_{n_0}^{-1} \circ \cdots \circ \psi_0^{-1}(x_0) \upharpoonright n_0 + 1 = \psi_{n_0+1}^{-1} \circ \psi_{n_0}^{-1} \circ \cdots \circ \psi_0^{-1}(x_1) \upharpoonright n_0 + 1$. Then by (5a) and (5b), $x_0 \upharpoonright n_0 + 1 = x_1 \upharpoonright n_0 + 1$, a contradiction.

Clearly both f and f^{-1} are continuous. Thus $[T_0]$ is homeomorphic to 2^ω. □

Note that $T^0_0 = \{\sigma \mid \Phi^0_{c_0}(\sigma) = 1\}$ is a recursive tree. Hence the set $A = [T_0]$ is an uncountable Π^0_1-set with the desired property.

We turn now to defining the sequence of trees and functions as well as constants that satisfy (1)–(7). Given a tree T, write $T = \Phi^x_i$ if for any σ, $\sigma \in T$ if and only if $\Phi^x_i(\sigma) = 1$.

Lemma 11.2.5. *There is a recursive function $t : \omega \to \omega$ such that for any $x \in 2^\omega$ and tree $T = \Phi^{x'}_e$ in $\omega^{<\omega}$, $\Phi^x_{t(e)}$ is a tree $\hat{T}$ in $\omega^{<\omega}$ and $[\hat{T}] = \{z \oplus h_x \mid z \in [T]\}$, where h_x is an x-recursive function so that $h_x(\langle e, n\rangle)$ is either 0 or $\min\{s + 1 \mid \Phi^{x\restriction s}_e(n)[s] \downarrow\}$.*

Proof. Let

$$\begin{aligned}\hat{T} = \{\sigma \oplus \tau \mid (\forall \langle e, n\rangle)(\langle e, n\rangle < |\tau| \wedge \tau(\langle e, n\rangle) > 0 \to \tau(\langle e, n\rangle) = \\ \min\{s + 1 \leq |\tau| \mid \Phi^{x\restriction s}_e(n)[s] \downarrow\}) \wedge (\exists \nu \in 2^{<|\tau|})(\forall e \leq |\nu|) \\ (\nu(e) = 1 \leftrightarrow \Phi^{x\restriction \tau(\langle e,e\rangle)}_e(e)[\tau(\langle e, e\rangle)] \downarrow) \wedge \Phi^\nu_i(\sigma) = 1)\}.\end{aligned}$$

Then $\hat{T}$ is as required. Since $\hat{T}$ is obtained uniformly, we have a recursive function $t : \omega \to \omega$ such that $\hat{T} = \Phi^x_{t(i)}$. □

Now for $i, j < \omega$ such that $T^0_{i+1} = \Phi^{\emptyset^{(i+1)}}_j$, we define a tree T_i such that T_i satisfies (1)–(7) upon substituting $\{c_j\}_{j\leq 3}$ with parameters depending on T_{i+1}. The construction will be uniform in i and j. This means that there is a recursive function t such that $T^0_i = \Phi^{\emptyset^{(i)}}_{t(j)}$. By the Recursion Theorem, there is a c_0 such that $T^0_i = \Phi^{\emptyset^{(i)}}_{c_0}$. Let $T_i = \{(\sigma, \tau) \mid (\exists \tau' \succeq \tau)((i, \tau') \in T^\omega \wedge \Phi^{\tau'}_{c_0}(\sigma)[|\tau'|] = 1)\}$.

The other parameters c_i for $1 \leq i \leq 3$ are obtained in the same way. Thus we may assume that T^0_{i+1} has properties (1)–(7). T^0_i is defined from T^0_{i+1}.

Lemma 11.2.6. *Given a recursive tree $T_0 \subseteq \omega^{<\omega}$, there is a recursive tree T_1 and a Δ^0_2-homeomorphism $\psi : T_0 \to T_1$ such that for any $x \in [T_1]$, $x' \equiv_T x \oplus \emptyset'$.*

Proof. The proof is a finite injury argument. Given a recursive tree $T_0 \subseteq \omega^{<\omega}$, we define ψ and T_1 by a full approximation construction.

Stage 0. Let $\psi_0(\emptyset) = \emptyset$ and $T_{1,0} = \emptyset$. We declare that all nodes $\sigma \in T_0$ are *fresh*.

Stage $s + 1$. Suppose $\psi_s : T_0 \to T_{1,s} \subseteq 2^{\leq s}$ is partially defined. Search for the least $t \leq s$ with a fresh $\sigma \in 2^t \cap T_0$ for which there is a $\tau \in T_{1,s} \cap [\psi_s(\sigma)] \cap 2^{\leq s}$ so that $\Phi^\tau_t(t)[|\tau|] \downarrow$. For each such σ, define $\psi_{s+1}(\sigma) = \tau' \succeq \tau$ where $|\tau'| = \tau^\frown 0^{s+1-t}$. For all ν's in $T_0 \cap \mathrm{Dom}(\psi_s)$ which are incompatible with such σ's, set $\psi_{s+1}(\nu) = \psi_s(\nu)$. Define $\psi_{s+1}(\nu^\frown i) = \psi_s(\nu)^\frown i$ if $\nu \in \mathrm{Dom}(\psi_s)$ is incompatible with the σ's and $\nu^\frown i \in T_0 \setminus \mathrm{Dom}(\psi_s)$. Declare that σ is not fresh for each such σ and τ is fresh for any τ extending such a σ. Let $T_{1,s+1} = \mathrm{Range}(\psi_{s+1})$.

This completes the construction at stage $s + 1$. Note that for each σ, $|\{s \mid \psi_s(\sigma) \neq \psi_{s+1}(\sigma)\}| \leq |\sigma|$. Hence $\psi(\sigma) = \lim_{s\to\infty} \psi_s(\sigma)$ exists for every σ. Then ψ is Δ^0_2.

Let $T_1 = \bigcup_{s<\omega} T_{1,s}$. By the construction, $\psi : [T_0] \to [T_1]$ is a homeomorphism. Suppose $x \in [T_1]$. Given t, to decide if $\Phi^x_t(t)$ converges, we $\emptyset'$-recursively search for a

$\sigma \in T_1 \cap 2^t$ and a stage $s \geq t$ such that $\psi_s(\sigma) = \psi(\sigma) \prec x$. Then $\Phi_t^x(t)$ is convergent if and only if $\Phi_t^{\psi_s(\sigma)}(t)[s]$. Thus $x' \leq_T x \oplus \emptyset'$. □

Now for $i < \omega$ and $\sigma \in 2^i$, let $\hat{T}_{i+1,\sigma}$ be a $\emptyset^{(i)}$-recursive tree as in Lemma 11.2.5 by setting T as $\{\tau \mid \sigma^\frown\tau \in T^0_{i+1}\}$ and x as $\emptyset^{(i)}$. Relativizing Lemma 11.2.6 to $\emptyset^{(i)}$ by replacing T_0 with $\hat{T}_{i+1,\sigma}$, we have a $\emptyset^{(i)}$-recursive homeomorphism $\psi_\sigma : \hat{T}_{i+1,\sigma} \to T_{i,\sigma}$ where $T_{i,\sigma}$ is T_1 in Lemma 11.2.6. Define T^0_i to be a $\emptyset^{(i)}$-recursive tree so that $\tau \in T^0_i$ if and only if either

- $|\tau| \leq i$, or
- $\tau = (\tau \upharpoonright i)^\frown\tau'$ for some $\tau' \in T_{i,\tau\upharpoonright i}$.

Define $\psi_i \leq_T \emptyset^{(i+1)} : T^0_{i+1} \to T^0_i$ such that

- if $\tau \in 2^{\leq i}$, then $\psi_i(\tau) = \tau$;
- otherwise, let $\tau = (\tau \upharpoonright i)^\frown\tau'$. Define $\psi_i(\tau) = \tau \upharpoonright i^\frown\psi_{\tau\upharpoonright i}(\tau' \oplus (h_{\emptyset^{(i)}} \upharpoonright |\tau'|))$.

Hence for any path $x \in [T^0_i]$,

$$(x \oplus \emptyset^{(i)})' \equiv_T x \oplus \emptyset^{(i+1)} \equiv_T \psi_i^{-1}(x) \oplus \emptyset^{(i+1)}.$$

Since the construction is uniform, it is straightforward to verify that (1)–(7) are satisfied. This completes the proof of Theorem 11.2.2.

11.2.2 The proof of Harrington's Theorem

We follow the presentation of Hjorth [50]. By Theorem 2.6.5, for any recursive ordinal $\beta \geq \omega$, there is an $n \in \mathcal{O}$ such that $|n| = \beta$ and the set $\{m \mid m <_o n\}$ is recursive. We choose such an ordinal β and the corresponding n.

For each $m = 3 \cdot 5^k <_o n$, let $\{i_{j,m}\}_{j<\omega}$ be a sequence of numbers such that

$$i_{0,m} = \Phi_k(m)$$

and

$$i_{j+1,m} = \Phi_k(m + j + 1).$$

By the proof of Theorem 2.6.5, we may assume that $m \leq i_{j,m} < i_{j+1,m}$ for all j. Clearly the set $\{(i_{j,m}, j, m) \mid (\exists k)(m = 3 \cdot 5^k \wedge m <_o n \wedge j \in \omega)\}$ is recursive.

Lemma 11.2.7.
(i) *For any $m = 3 \cdot 5^k <_o n$ and $j < \omega$, $i_{j,m} <_o i_{j+1,m}$;*
(ii) *for any $l <_o n$, the set $\{m \mid (\exists k)(i_{0,m} <_o l <_o m = 3 \cdot 5^k <_o n)\}$ is finite;*
(iii) *the set $\{(m, l) \mid (\exists k)(i_{0,m} <_o l <_o m = 3 \cdot 5^k <_o n)\}$ is recursive;*
(iv) *if $m = 3 \cdot 5^k <_o n$ and $|m| = \lim_{i\to\infty} \beta_i$ where $\{\beta_i\}$ is an increasing sequence of limit ordinals, then for any i there is an $m'' = 3 \cdot 5^{k''} <_o m$ with $|m''| > \beta_i$ such that for any $m' = 3 \cdot 5^{k'} <_o m$ with $|m'| > |m''|$, we have $|i_{0,m'}| > |m''|$.*

Proof. (i) is immediate. (ii) is true since for any number l, there are only finitely many $m \le l$ such that $l <_O m <_O n$. (iii) is true since $<_O$ is a recursive relation over the recursive set $\{i \mid i <_O n\}$.

We prove (iv). If $m = 3 \cdot 5^k <_O n$ and $|m| = \lim_{i\to\infty} \beta_i$ where $\{\beta_i\}$ is an increasing sequence of limit ordinals, then for any i there is an $m'' = 3 \cdot 5^{k''} <_O m$ with $|m''| > \beta_i$ such that whenever $m' = 3 \cdot 5^{k'} <_O m$ with $|m'| > |m''|$, we have $m' > m''$. Hence $|i_{0,m'}| > |m''|$. □

The main difficulty with extending Theorem 11.2.2 to arbitrary levels of recursive ordinals concerns coding at limit ordinal stage. Here, as in the proof of Theorem 11.2.2, at each successor stage one attempts to carry out a stepdown construction that codes information computed at a limit level. There is no difficulty dealing with only one limit ordinal. With more than one limit ordinal, one of them has to be chosen for coding. Lemma 11.2.7 allows the use of nice sequences in the approximation of limit ordinals to facilitate the choice.

For convenience, in the rest of this section let $|n| = \beta_\infty$ and identify each natural number $\le_O n$ with the ordinal $\beta < \beta_\infty$ of which it is an ordinal notation. In particular, we use λ for natural numbers which are ordinal notations of limit ordinals. Also $\beta_{j,\lambda} = |i_{j,m}|$ for $\lambda = |m|$. We write $\emptyset^{(\beta)}$ and $x^{(\beta)}$ for H_i and $\hat{H}^x_i$ where $|i| = \beta$. Finally, β^+ is identified with 2^k where $|k| = \beta$.

For $\beta \le_O n$, let

$$\Lambda(\beta) = \{\lambda \mid \beta <_O \lambda \wedge (\exists j \in \omega)(\beta_{j,\lambda} \le_O \beta)\}.$$

Note that by Lemma 11.2.7, $\Lambda(\beta)$ is finite for any $\beta \le_O n$. Hence the following is well defined:

$$j(\beta) = \max(\{0\} \cup \{j \mid (\exists \lambda \in \Lambda(\beta))(\beta_{j,\lambda} \le_O \beta)\}).$$

For each $a \subseteq \omega$, define the following function "specialized" to $\omega \setminus a$:

$$p_a(\sigma) = \tau \Leftrightarrow (\forall i, j, k)(\tau(\langle j, i\rangle) = k > 0 \leftrightarrow j \notin a \wedge \sigma(\langle j, i\rangle) = k > 0).$$

Then

$$p_a(T) = \{p_a(\sigma) \mid \sigma \in T\},$$

and

$$p_a(\bar{x}) = x \Leftrightarrow (\forall i, j, k(x(\langle j, i\rangle) = k > 0 \leftrightarrow j \notin a \wedge \bar{x}(\langle j, i\rangle) = k > 0).$$

For $\beta <_O \gamma \le_O \beta_\infty$, let

$$(\beta, \gamma] = \{\delta \mid \beta <_O \delta \le_O \gamma\}.$$

Clearly if $\beta_o < \gamma_o < \delta$, then $p_{(\beta,\delta]} = p_{(\beta,\gamma]} \circ p_{(\gamma,\delta]}$. As in the previous section, there is a recursive tree $T \subseteq \beta_\infty \times \omega^{<\omega}$ such that $T(\beta) = \{\sigma \mid (\beta, \sigma) \in T\}$ contains exactly one path $\emptyset^{(\beta)}$. Let

$$T^{\beta_\infty} = \{\sigma \mid (\forall \beta)(\exists i)(\exists \tau)(\sigma(\beta, i) \text{ defined } \to ((\beta, \tau) \in T \wedge (\forall j)(\sigma(\langle \beta, j\rangle) = \tau(j))))\}.$$

Hence T^{β_∞} codes T as a single tree. For $\beta \le_o \beta_\infty$, let y_β be a real such that

$$(\forall i, j, k)(y_\beta(\langle i, j\rangle) = k \leftrightarrow i \le_o \beta \wedge \emptyset^{(|i|)}(j) = k).$$

Note that there are constants d_0, d_1 such that $y_\beta \equiv_T \emptyset^{(\beta)}$ via (d_0, d_1). We build a system consisting of a uniform sequence of r.e. trees $\{T_\beta\}_{\beta \le_o \beta_\infty} = \{\check{T}^0_\beta \times T^1_\beta\}_{\beta \le_o \beta_\infty}$ of $\omega^{<\omega} \times \omega^{<\omega}$, functions $\{\psi_{\gamma,\beta}\}_{\beta \le_o \gamma \le_o \beta_\infty}$ and constants $\{c_i\}_{i \le 3}$ such that

(1) $T_{\beta_\infty} = 2^\omega \times T^{\beta_\infty}$;

(2) $T_\beta \cap \omega^{\le j(\beta)} \times \omega^{\le j(\beta)} = \{(\sigma, \tau) \in \omega^{\le j(\beta)} \times \omega^{\le j(\beta)} \mid (\exists \lambda \in \Lambda(\beta))(\exists j)(\exists \bar{\tau})(\beta_{j,\lambda} \le \beta \wedge (\sigma, \bar{\tau}) \in T_\lambda \cap 2^{\le j(\beta)} \times 2^{\le j(\beta)} \wedge p_{(\beta,\lambda]}(\bar{\tau}) = \tau)\}$;

(3) $T^1_\beta = p_{(\beta,\beta_\infty]}(T^{\beta_\infty})$ and hence $[T^1_\beta] = \{y_\beta\}$. Let $T^0_\beta = \{\sigma \mid (\sigma, y_\beta \upharpoonright |\sigma|) \in T_\beta\} = \Phi^{\emptyset^{(\beta)}}_{c_0}$;

(4) $\psi_{\beta^+,\beta} : T^0_{\beta^+} \to T^0_\beta$ is $\Phi^{\emptyset^{(\beta^+)}}_{c_1}$;

(5) $\psi_{\beta^+,\beta}$ may be viewed as a homeomorphism from $[T^0_{\beta^+}]$ to $[T^0_\beta]$ so that

$$(\forall x \in [T^0_\beta])(\psi^{-1}_{\beta^+,\beta}(x) \oplus \emptyset^{(\beta^+)} \equiv_T (x \oplus \emptyset^{(\beta)})')$$

via (c_2, c_3), and

(5i) For $\sigma, \tau \in T^0_{\beta^+}$, $\sigma < \tau \Leftrightarrow \psi_{\beta^+,\beta}(\sigma) \preceq \psi_{\beta^+,\beta}(\tau)$;

(5ii) If $\sigma \in T^0_{\beta^+}$, then $|\psi_{\beta^+,\beta}(\sigma)| \ge |\sigma|$; and

(6) for $\sigma \in T^0_{\beta^+} \cap \omega^{\le j(\beta)}$, $\psi_{\beta^+,\beta}(\sigma) = \sigma$.

Suppose that we have the sequence of trees as above. By (4) and (5), for every $x \in T^0_0$, one concludes by an induction on $\beta \le_o \beta_\infty$ that

$$x^{(\beta)} \equiv x \oplus \emptyset^{(\beta)} \not\ge_T \emptyset^{(\beta^+)}.$$

It would appear that the requirement set by (2) on $T_\beta \cap \omega^{\le j(\beta)} \times \omega^{\le j(\beta)}$ is a problem since too many nodes from T_λ, for $\lambda \in \Lambda(\beta)$, are put into it. However, the following lemma says that $T_\beta \cap \omega^{\le j(\beta)} \times \omega^{\le j(\beta)}$ is an orderly structure due to Lemma 11.2.7 (iv).

Lemma 11.2.8. *Suppose that $\lambda \in \Lambda(\beta)$ with $\beta_{j,\lambda} \le_o \beta$ for some j. Then $(\sigma, \tau) \in T_\beta \cap \omega^{\le j} \times \omega^{\le j}$ if and only if there exists a $(\sigma, \bar{\tau}) \in T_\lambda$ such that $p_{(\beta,\lambda]}(\bar{\tau}) = \tau$.*

Proof. Suppose not. Then there is a $\lambda \in \Lambda(\beta)$, a j with $\beta_{j,\lambda} \le \beta$ such that for some $(\sigma, \tau) \in T_\beta \cap \omega^{\le j} \times \omega^{\le j}$, there is no $(\sigma, \bar{\tau}) \in T_\lambda$ with $p_{(\beta,\lambda]}(\bar{\tau}) = \tau$.

By (2), there is a $\lambda_1 \in \Lambda(\beta)$ with $\beta_{j_1,\lambda_1} \le_o \beta$ such that $j \le j_1$ and for some $(\sigma, \bar{\tau}_1) \in T_{\lambda_1} \cap \omega^{\le j} \times \omega^{\le j}$, one has $p_{(\beta,\lambda_1]}(\bar{\tau}_1) = \tau$.

We claim that $\lambda_1 <_o \lambda$. Otherwise, $\lambda <_o \lambda_1$ and so $\beta_{j_1,\lambda_1} \le_o \beta <_o \lambda <_o \lambda_1$. In other words, $\lambda_1 \in \Lambda(\lambda)$. By applying (2) to λ, $(\sigma, p_{(\lambda,\lambda_1]}(\bar{\tau}_1)) \in T_\lambda$. Then

$$p_{(\beta,\lambda]}(p_{(\lambda,\lambda_1]}(\bar{\tau}_1)) = p_{(\beta,\lambda_1]}(\bar{\tau}_1) = \tau,$$

a contradiction.

Hence $\lambda \in \Lambda(\lambda_1)$. Note that there is no $(\sigma, \bar{\tau}) \in T_\lambda \cap \omega^{\le j} \times \omega^{\le j}$ such that $p_{(\alpha,\lambda]}(\bar{\tau}) = \bar{\tau}_1$. Repeating the above process upon replacing β and (σ, τ) with λ_1 and $(\sigma, \bar{\tau}_1)$ respectively, we see that there is a $\lambda_2 \in \Lambda(\lambda_1)$ and $\beta_{j_2,\lambda_2} \le_o \lambda_1 <_o \lambda_2$ such that $\lambda_2 <_o \lambda$,

$(\sigma, \bar{\tau}_2) \in T_{\lambda_2}$ and $p_{(\beta,\lambda_2]}(\bar{\tau}_2) = \bar{\tau}_1$. It follows that there is an increasing sequence of limit ordinals $\lambda_1 <_o \lambda_2 <_o \cdots <_o \lambda$ such that $\beta_{j,\lambda} <_o \lambda_i$ for every $i \in \omega$, contradicting Lemma 11.2.7 (iv). □

For $\gamma \geq_o \beta$, define a map $\psi_{\gamma,\beta} : T^0_\gamma \to T^0_\beta$ by induction on $\beta \leq_o \beta_\infty$. If $\beta = 0$, then $\psi_{0,0}$ is the identity map. In the following, $\psi_{\beta,\beta}$ is always the identity map.

If $\gamma^+ >_o \beta$, then $\psi_{\gamma^+,\beta} = \psi_{\gamma,\beta} \circ \psi_{\gamma^+,\gamma}$.

If λ is a limit ordinal, then for j such that $\beta \leq_o \beta_{j,\lambda} <_o \lambda$ and $\sigma \in T^0_\lambda \cap \omega^{\leq j}$, we have $\sigma \in T^0_{\beta_j}$ by (2) and Lemma 11.2.8. Define $\psi_{\lambda,\beta}(\sigma) = \psi_{\beta_{j,\lambda},\beta}(\sigma)$.

Note that in the case $\beta = \beta_{j,\lambda}$, we have $\psi_{\lambda,\beta}(\sigma) = \psi_{\beta_{j,\lambda},\beta}(\sigma) = \sigma$. We show that $\psi_{\lambda,\beta}$ is well-defined.

Lemma 11.2.9. *Let $\delta \leq_o \beta_\infty$,*
(i) *If $\beta \leq_o \delta$, then $\psi_{\delta,\beta}$ is well defined.*
(ii) *For $\beta \leq_o \beta_1 \leq_o \delta$, $\psi_{\delta,\beta} = \psi_{\beta_1,\beta} \circ \psi_{\delta,\beta_1}$.*
(iii) *For $\beta \leq_o \delta$ and $\sigma, \tau \in T^0_\beta$, $\sigma \preceq \tau \Leftrightarrow \psi_{\delta,\beta}(\sigma) \preceq \psi_{\delta,\beta}(\tau)$. Moreover, $|\psi_{\delta,\beta}(\sigma)| \geq |\sigma|$.*
(iv) *There is a recursive function g such that for each $\beta \leq_o \delta$, $\psi_{\delta,\beta} = \Phi^{\emptyset(\delta)}_{g(\delta,\beta)}$.*

Proof. We prove (i) and (ii) simultaneously by induction on δ. By the definition of $\psi_{\delta,\beta}$, the only nontrivial case is when $\delta = \lambda$ for some limit ordinal λ.

For (i), fix a $\beta <_o \lambda$ and $\beta \leq_o \beta_{j_0,\lambda} <_o \beta_{j_1,\lambda} <_o \lambda$ for $j_0 < j_1$. Suppose that $\sigma \in T^0_\lambda \cap \omega^{\leq j_0}$ and j_0 is used to define $\psi_{\lambda,\beta}(\sigma)$. By (2) and Lemma 11.2.8, $\sigma \in T^0_{\beta_{j_0,\lambda}} \cap T^0_{\beta_{j_1,\lambda}}$. Note that for any γ with $\beta_{j_0,\lambda} <_o \gamma \leq_o \lambda$, we have $\lambda \in \Lambda(\gamma)$ and $j(\gamma) \geq j_0$. By (6) and an induction on γ in $(\beta_{j_0,\lambda}, \lambda]$, $\psi_{\gamma,\beta_{j_0,\lambda}}(\sigma) = \sigma$. By the same argument, for γ in $(\beta_{j_1,\lambda}, \lambda]$, we have $\psi_{\gamma,\beta_{j_1,\lambda}}(\sigma) = \sigma$. Hence by induction on β, $\psi_{\lambda,\beta}(\sigma) = \psi_{\beta_{j_0,\lambda},\beta}(\sigma) = \psi_{\beta_{j_0,\lambda},\beta}(\psi_{\lambda,\beta_{j_0,\lambda}}(\sigma)) = \psi_{\beta_{j_0,\lambda},\beta}(\psi_{\beta_{j_1,\lambda},\beta_{j_0,\lambda}}(\psi_{\lambda,\beta_{j_1,\lambda}}(\sigma))) = \psi_{\beta_{j_1,\lambda},\beta}(\psi_{\lambda,\beta_{j_1,\lambda}}(\sigma)) = \psi_{\beta_{j_1,\lambda},\beta}(\sigma)$. Thus $\psi_{\delta,\beta}$ is well defined.

For (ii), given σ let j be such that $|\sigma| < j$ and $\beta \leq_o \beta_1 <_o \beta_{j,\lambda} \leq_o \lambda$. Then $\psi_{\lambda,\beta_{j,\lambda}}(\sigma) = \sigma$. By the definition of $\psi_{\lambda,\beta_{j,\lambda}}$, $\psi_{\lambda,\beta}(\sigma) = \psi_{\beta_{j,\lambda},\beta}(\sigma)$. Then by induction, $\psi_{\beta_j,\beta}(\sigma) = \psi_{\beta_1,\beta}(\psi_{\beta_{j,\lambda},\beta_1}(\sigma))$.

We obtain (iii) by using (5i) and (5ii) and an induction on β. Finally, (iv) follows from (3). □

If $x \in [T^0_\beta]$, let $\psi_{\delta,\beta}(x) = \cup_{m\in\omega}\psi_{\delta,\beta}(x \upharpoonright m)$. By Lemma 11.2.9 (iii), $\psi_{\delta,\beta}$ is well defined.

Lemma 11.2.10. *For any $\beta \leq_o \gamma \leq \beta_\infty$, $\psi_{\gamma,\beta} : [T^0_\gamma] \to [T^0_\beta]$ is a homeomorphism.*

Proof. This is proved by induction on γ. First, (5) implies that the only nontrivial case is when $\gamma = \lambda$, a limit ordinal.

For $x \in [T^0_\beta]$ and $m \in \omega$, let $j_m > m$ be chosen such that $\beta_{j_m,\lambda} > \beta$. By induction, there is a unique $z_m \in [T_{\beta_{j,\lambda}}]$ such that $\psi_{\beta_{j_m,\lambda},\beta}(z_m) = x$. By Lemma 11.2.9 (iii), $\psi_{\beta_{j_m,\lambda},\beta}(z_m \upharpoonright m) \succeq x \upharpoonright m$. Let $\sigma_m = z_m \upharpoonright m$. By (ii) and (iii) of Lemma 11.2.9, there is a $z = \cup_m \sigma_m$. Since $j_m > m$, we have $\sigma_m \in T_\lambda$ for every m. Hence $z \in [T^0_\beta]$. Moreover,

$\psi_{\lambda,\beta}(\sigma_m) = \psi_{\beta_{jm,\lambda},\alpha}(\psi_{\lambda,\beta_{jm,\lambda}}(\sigma_m)) = \psi_{\beta_{jm,\lambda},\beta}(\sigma_m) = x \upharpoonright m$. Thus $\psi_{\lambda,\beta}(z) = x$. In other words, ψ is onto.

Now suppose that $x_0, x_1 \in [T^0_\beta]$, $x_0 \upharpoonright n_0 = x_1 \upharpoonright n_0 = \sigma$, $x_0 \succ \sigma^\frown k_0$, $x_1 \succ \sigma^\frown k_1$ for some $k_0 \neq k_1$. Let $j > |\sigma| + 1$ be such that $\beta_{j,\lambda} > \beta$. Then $\psi_{\lambda,\beta_{j,\lambda}}(x_0 \upharpoonright |\sigma| + 1) = x_0 \upharpoonright |\sigma| + 1 \neq x_1 \upharpoonright |\sigma| + 1 = \psi_{\lambda,\beta_{j,\lambda}}(x_1 \upharpoonright |\sigma| + 1)$. By induction, $\psi_{\beta_{j,\lambda},\beta}(y_0) \neq \psi_{\beta_{j,\lambda},\beta}(y_1)$ for any $y_0 \succ x_0 \upharpoonright |\sigma| + 1$ and $y_1 \succ x_1 \upharpoonright |\sigma| + 1$. By Lemma 11.2.9 (ii), $\psi_{\lambda,\beta}(x_0) \neq \psi_{\lambda,\alpha}(x_1)$.

Hence for $\gamma = \lambda$, $\psi_{\gamma,\beta}$ is a bijection between $[T^0_\gamma]$ and $[T^0_\beta]$. Clearly both $\psi_{\gamma,\beta}$ and $\psi^{-1}_{\gamma,\beta}$ are continuous. Hence $\psi_{\gamma,\beta}$ is a homeomorphism. □

By (3), T^0_0 is a recursive tree. By the Recursion Theorem, one may assume that there is already a system satisfying (1)–(6). We define T_β from T_{β^+}. Let

$$T^1_\beta = p_{(\beta,\beta_\infty]}(T^{\beta_\infty})$$

be a recursive tree. Recursively compute $j(\beta)$.

For $\sigma \in \omega^{j(\beta)}$, let $\hat{T}^0_{\beta^+,\sigma}$ be a $\emptyset^{(\beta)}$-recursive tree obtained from Lemma 11.2.5 upon replacing T with $\{\tau \mid \sigma^\frown\tau \in T^0_{\beta^+}\}$ and x with $\emptyset^{(\beta)}$. Relativizing Lemma 11.2.6 to $\emptyset^{(\beta)}$ (replacing T_0 with $\hat{T}^0_{\beta^+,\sigma}$), we have a $\emptyset^{(\beta)}$-recursive homeomorphism $\psi_\sigma : \hat{T}^0_{\beta^+,\sigma} \to T^0_{\beta,\sigma}$ where $T^0_{\beta,\sigma}$ is T_1 in Lemma 11.2.6. Define T^0_β to be a $\emptyset^{(\beta)}$-recursive tree such that $\tau \in T^0_\beta$ if and only if either

- $(\tau, p^{-1}_{(\beta^+,\lambda]}(y_{\beta^+} \upharpoonright j(\beta))) \in T_\lambda \cap \omega^{\leq j(\beta)} \times \omega^{\leq j(\beta)}$ for some $\lambda \in \Lambda(\beta)$, or
- $\tau = (\tau \upharpoonright j(\beta))^\frown\tau'$ for some $\tau' \in T^0_{\beta,\tau\upharpoonright j(\beta)}$.

Define $\psi_{\beta^+,\beta} \leq_T \emptyset^{(\beta^+)} : T^0_{\beta^+} \to T^0_\beta$ such that

- if $(\tau, p^{-1}_{(\beta^+,\lambda]}(y_{\beta^+} \upharpoonright j(\beta))) \in T_\lambda \cap \omega^{\leq j(\beta)} \times \omega^{\leq j(\beta)}$ and $\lambda \in \Lambda(\beta)$, then $\psi_{\beta^+,\beta}(\tau) = \tau$;
- otherwise, let $\tau = (\tau \upharpoonright j(\beta))^\frown\tau'$. Define $\psi_{\beta^+,\alpha}(\tau) = \tau \upharpoonright j(\beta)^\frown\psi_{\tau\upharpoonright j(\beta)}(\tau' \oplus (h_{\emptyset^{(\beta)}} \upharpoonright |\tau'|))$.

Hence for any $x \in T^0_{\beta}$,

$$(x \oplus \emptyset^{(\beta)})' \equiv_T x \oplus \emptyset^{(\beta^+)} \equiv_T \psi^{-1}_{\beta^+,\beta}(x) \oplus \emptyset^{(\beta^+)}.$$

Note that $T^0_\beta = \Phi^{y_\beta}_d$ (see (3)) for some d which is independent of β. Define T_β such that $(\sigma, \tau) \in T_\beta$ if and only if either

- there exists a $(\sigma, \bar{\tau}) \in T_\lambda \cap \omega^{\leq j(\beta)} \times \omega^{\leq j(\beta)}$ for some $\lambda \in \Lambda(\beta)$ such that $\tau = p_{(\beta,\lambda]}(\bar{\tau})$, or
- there is a $\tau' \succeq \tau$ so that $\tau' \in T^1_\beta$, $\Phi^{\tau'}_d(\sigma)[|\tau'|] = 1$ and $(\sigma \upharpoonright j(\beta), \bar{\tau'} \upharpoonright j(\beta)) \in T_\lambda$ for some $\lambda \in \Lambda(\beta)$ and $\bar{\tau'}$ such that $\tau' = p_{(\beta,\lambda]}(\bar{\tau'})$.

Then the tree T_β is as required. Since the construction is uniform, the rest of (1)–(6) follow. This completes the proof of Theorem 11.2.2.

By Σ^1_1-boundedness, we have the following generalization:

Corollary 11.2.11 (Harrington). *There is an uncountable Π^0_1-set A such that for any recursive ordinal β and $x \in A$, $x^{(\beta)} \not\geq_T \emptyset^{(\beta+1)}$.*

Proof. Suppose not. Let $B \subseteq \omega \times \omega^{<\omega}$ be arithmetical such that $\{[B_e]\}_{e<\omega}$, where $B_e = \{\sigma \in \omega^{<\omega} \mid (e, \sigma) \in B)\}$, is a recursive list of all Π^0_1-subsets of ω^ω. Let $O \subseteq \mathcal{O}$ be as defined in Theorem 2.6.5.

Recall that the system satisfying (1)–(6) was obtained by fixing an ordinal notation n with $|n| \geq \omega$. Let e_0 be such that $|3 \cdot 5^{e_0}| = \omega$. Define a Π^1_1-relation $R \subseteq \omega \times \omega$ such that

$$R(e, n) \Leftrightarrow 3 \cdot 5^{e_0} \leq_o n \wedge n \in \mathcal{O} \wedge$$
$$\text{there is no system satisfying (1)–(6) for } n \text{ and } B_e = T^0_0.$$

Then by Theorem 4.2.1 and the assumption, there is a total Π^1_1, hence Δ^1_1, uniformizing function $f : \omega \to O$ for R. Let $I = \{e \mid |[B_e]| > \aleph_0\}$. By Theorem 1.1.10, I is Σ^1_1. By Corollary 1.5.2, there is an $n \in O$ such that for all $m \in \{f(e) \mid e \in I\}$, we have $m <_o n$ and that contradicts the proof above.

Hence there is an $e \in I$ such that $B_e = T^0_0$ and a system satisfying (1)–(6) exists for each $n \in O$ where $|n| \geq \omega$. By the proof above, $\hat{H}^x_n \not\geq_T H_{2^n}$. □

11.2.3 Exercises

Exercise 11.2.1. *Prove that in Theorem 11.2.1, the set A may be chosen to be a subset of 2^ω.*

Exercise 11.2.2 (Harrington). *There is a Π^1_1-path through $\mathcal{O}$ such that the only hyperarithmetic reals recursive in it are recursive.*

Exercise 11.2.3 (Harrington). *Prove that for any recursive ordinal $\beta < \omega_1^{\mathrm{CK}}$, there are two Π^0_2-singletons x and y in 2^ω such that $x^{(\beta)} \not\leq_T y$ and $y^{(\beta)} \not\leq_T x$.*

Exercise 11.2.4 (Harrington). *Prove that there is a transitive model $\mathcal{N}$ of KP such that $\mathcal{N} \models$"Not every Δ^1_1-real is hyperarithmetic".*

11.3 Perfect sets in L

11.3.1 A question of Prikry

In [98], Prikry posed the following question.

Question 11.3.1. *Suppose that there is a nonconstructible real. Does every perfect set contain a nonconstructible real?*

Friedman [29] also raised a question related to Question 11.3.1:

Question 11.3.2. *Suppose that there is a nonconstructible real. Is $L \cap \mathbb{R}$ a $\underset{\sim}{\Sigma}^1_1$-set?*

By Theorem 1.1.10, a positive answer to Question 11.3.1 would imply a negative solution to Question 11.3.2. The latter was given a negative solution by Velickovic and Woodin:

Theorem 11.3.3 (Velickovic and Woodin [166]). *If the set of constructible reals is $\underset{\sim}{\Sigma}^1_1$, then either $(\omega_1)^L$ is countable or every real is constructible.*

On the other hand, Question 11.3.1 was answered positively by Groszek and Slaman [41] and the solution is discussed in the next section.

Theorem 11.3.4. *If there is a perfect set $A \subseteq (2^\omega)^L$, then every real is constructible.*

11.3.2 The proof of Theorem 11.3.4

Given a tree $T \subseteq 2^{<\omega}$, a set $A \subseteq 2^\omega$ is *dense* in T if for any $\sigma \in T$, $[\sigma] \cap [T] \cap A \neq \emptyset$. A real $x \in [T]$ is *eventually constant* in T if there is an n such that x is the leftmost or rightmost infinite path in $[x \upharpoonright n] \cap [T]$.

Lemma 11.3.5. *Let T be a perfect tree. Suppose that $\{g_i\}_{i<\omega} \in L$ contains a dense subset of T no member of which is eventually constant in T. Then for any z, there exist $x, y \in [T]$ such that $x \oplus y \oplus (\oplus_{i<\omega} g_i) \geq_T z$.*

Assume Lemma 11.3.5. Let T be a perfect tree such that $[T] \subseteq L$. Then $(\omega_1)^L = \omega_1$. Select a countable sequence $\{x_i\}_{i<\omega}$ from $[T]$ which is dense and such that no x_i is eventually constant. Since $(\omega_1)^L = \omega_1$, there is an $x^* \in L$ such that $x_i <_L x^*$ for each $i < \omega$. Let $\{g_i\}_{i<\omega}$ be an enumeration of the set $\{x \mid x <_L x^*\}$. Note that $\{g_i\}_{i<\omega}$ belongs to L. By Lemma 11.3.5, for each z there exist $x, y \in [T]$, and hence in L, such that $x \oplus y \oplus (\oplus_{i\in\omega} g_i) \geq_T z$. Thus $z \in L$.

The idea of the proof is as follows. Suppose z is a given real, T is perfect and $\{g_i\}_{i<\omega} \in L$ is a sequence with a subset dense in T no element of which is eventually constant in T. We define $x \in [T]$ and an index function $h_x : \omega \to \omega$ such that for each i, $g_{h_x(i)}$ is a path through T that is not eventually constant in T. The set x agrees with $g_{h_x(i)}$ on longer and longer initial segments as i increases,, and z is coded into x and $\{g_{h_x(i)}\}_{i<\omega}$ so that $z(i)$ is the value of x at the first place where x differs from $g_{h_x(i)}$. It is easy to see that $z \leq_T x \oplus h_x \oplus (\oplus_{i<\omega} g_i)$. There is no problem to recursively define such an x and h_x.

The challenge is to decode h_x. This is handled by defining another real $y \in [T]$ such that $h_x \leq_T x \oplus y \oplus (\oplus_{i<\omega} g_i)$. To achieve this, we define an additional index function $h_y : \omega \to \omega$ with the same properties (except that z is not coded here), in such a way that h_x and h_y can be computed from x, y and $\oplus_{i<\omega} g_i$.

Proof of Lemma 11.3.5. We now describe the construction of the sets x, y and functions h_x, h_y.

Stage 0. Let σ_0 and τ_0 be the first splitting node of T extending its root, i.e. a splitting node of T such that every node in T extends either σ_0 or τ_0. Let $h_x(0)$ be the least i such that $g_i \in [\sigma_0{}^\frown z(0)] \cap [T]$, and let $h_y(0)$ be the least j such that $g_j \in [\tau_0] \cap T$.

Stage $n+1$. Assume by induction hypothesis that σ_n and τ_n are splitting nodes in T, and that $\sigma_n{}^\frown z(n) \prec g_{h_x(n)}$ and $\tau_n \prec g_{h_y(n)}$.

Substage (a). Define j_0 to be the least index j so that $g_j \in [\sigma_n{}^\frown(1-z(n))] \cap [T]$ is not eventually constant. Let m_0 be the least $m > |\tau_n|$ so that $g_i \upharpoonright m \neq g_{j_0} \upharpoonright m$ for every $i < j_0$. Let m_1 be the least $m > m_0$ such that $g_{h_y(n)} \upharpoonright m$ is a splitting node in T. Let $\tau' = g_{h_y(n)} \upharpoonright m_1{}^\frown(1 - g_{h_y(n)}(m_1))$. Then $g_i \upharpoonright m_1 \neq g_{j_0} \upharpoonright m_1$ for $i < j_0$ and m_1 is the least m such that $\tau'(m) \neq g_{h_y(n)}(m)$.

Substage (b). Let j_1 be the least j such that $g_j \in [\tau'] \cap [T]$. Let m_2 be the least m such that for any $i < j_1$, $g_i \upharpoonright m \neq g_{j_1} \upharpoonright m$. Since g_{j_0} is not eventually constant, we may choose m_3 to be the least $m > m_2$ such that $g_{j_0} \upharpoonright m$ is a splitting node in T and $g_{j_0} \succ g_{j_0} \upharpoonright m_3{}^\frown z(n+1)$. Then $g_i \upharpoonright m_3 \neq g_{j_1} \upharpoonright m_3$ for $i < j_1$. Let $\sigma_{n+1} = g_{j_0} \upharpoonright m_3$, $\tau_{n+1} = g_{j_1} \upharpoonright m_3$, $h_x(n+1) = j_0$ and $h_y(n+1) = j_1$.

Let $x = \bigcup_{n<\omega} \sigma_n$ and $y = \bigcup_{n<\omega} \tau_n$. This completes the construction.

The construction ensures that for every n, $h_x(n+1)$ is the least i such that $g_i \upharpoonright m_1 = x \upharpoonright m_1$ where m_1 is the least m satisfying $y(m) \neq g_{h_y(n)}(m)$. Also, $h_y(n+1)$ is the least i such that $g_i \upharpoonright m_3 = y \upharpoonright m_3$ where m_3 is the least m such that $x(m) \neq g_{h_x(n+1)}(m)$. Hence $h_x \oplus h_y \leq_T x \oplus y \oplus (\oplus_{i<\omega} g_i)$, giving $z \leq_T x \oplus y \oplus (\oplus_{i\in\omega} g_i)$. □

11.3.3 Exercises

Exercise 11.3.1. *Prove that Lemma* 11.3.5 *remains true without the condition "no element is eventually constant in T".*

***Exercise 11.3.2** (Harrington). *Prove that for any Π^1_1-set $A \subseteq 2^\omega$ which is not thin, A contains a hyperarithmetically pointed perfect tree T, i.e. $T <_h x$ for every $x \in [T]$.*

Part IV: **Higher randomness theory**

12 Review of classical algorithmic randomness

We assume familiarity with the classical theory of algorithmic randomness and give a review of the subject in this chapter. For details see [111] and [24].

12.1 Randomness via measure theory

Intuitively, a random real should pass every feasible statistical test. This is the motivation for using measure theory to define randomness. In the language of measure theory, a test is associated with a null set. The following is a summary of the key notions of measure-theoretic randomness.

12.1.1 Kurtz and Schnorr randomness

The notion of (now called) Kurtz randomness was introduced in Kurtz [79].

Definition 12.1.1.
(i) A set $A \subseteq 2^\omega$ is a *Kurtz test* if it is Π^0_1 and $\mu(A) = 0$;
(ii) a real x is *Kurtz random* if $x \notin A$ for every Kurtz test A.

Definition 12.1.2 (Schnorr [125] and [127]).
(i) A sequence of open sets $\{U_n\}_{n<\omega}$ is a *Schnorr test* if $\{(n, \sigma) \mid \sigma \in U_n\}$ is r.e. and $\mu(U_n) = 2^{-n}$ for every $n < \omega$;
(ii) a real x is *Schnorr random* if $x \notin \bigcap_{n<\omega} U_n$ for every Schnorr test $\{U_n\}_{n<\omega}$.

Proposition 12.1.3 (Terwijn and Zambella [163]). *A real x is Schnorr random if and only if $x \notin \bigcap_{n<\omega} V_n$ for every sequence of open sets $\{V_n\}_{n<\omega}$ such that $\{(n, \sigma) \mid \sigma \in V_n\}$ is r.e. and $\mu(V_n) = 2^{-f(n)}$ for some recursive function f.*

12.1.2 Martin-Löf randomness

The notions of Kurtz and Schnorr randomness both turn out to be somewhat restrictive and unsatisfactory. The next notion, due to Martin-Löf, is perhaps the most widely studied to-date.

Definition 12.1.4 (Martin-Löf [96]).
(i) A sequence of open sets $\{U_n\}_{n<\omega}$ is a *Martin-Löf test* if $\{(n, \sigma) \mid \sigma \in U_n\}$ is r.e. and $\mu(U_n) < 2^{-n}$ for every $n < \omega$;
(ii) a real x is *Martin-Löf random* if $x \notin \bigcap_{n<\omega} U_n$ for every Martin-Löf test $\{U_n\}_{n<\omega}$.

The next result says that Martin-Löf randomness is decided by a single test.

Proposition 12.1.5 (Martin-Löf [96]). *There exists a universal Martin-Löf test. Namely, there is a Martin-Löf test* $\{U_n\}_{n<\omega}$ *such that* $\bigcap_{n<\omega} V_n \subseteq \bigcap_{n<\omega} U_n$ *for every Martin-Löf test* $\{V_n\}_{n<\omega}$.

12.1.3 Strong Martin-Löf randomness

Definition 12.1.6 (Kurtz [79]).
(i) A sequence of open sets $\{U_n\}_{n<\omega}$ is a *generalized Martin-Löf test* if $\{(n,\sigma) \mid \sigma \in U_n\}$ is r.e. and $\lim_{n\to\infty} \mu(U_n) = 0$;
(ii) a real x is *strongly Martin-Löf random* if $x \notin \bigcap_{n<\omega\omega} U_n$ for any generalized Martin-Löf test $\{U_n\}_{n<\omega}$.

Strong Martin-Löf randomness has a number of regularity properties not shared by Martin-Löf randomness.

12.1.4 Relativized randomness

Each of the notions above may be relativized to an arbitrary real x. For example, an x-Martin-Löf test is a sequence of open sets $\{U_n\}_{n<\omega}$ such that $\{(n,\sigma) \mid \sigma \in U_n\}$ is x-r.e. and $\mu(U_n) < 2^{-n}$ for every $n \in \omega$. Then z is x-Martin-Löf random if $z \notin \bigcap_{n<\omega} U_n$ for any x-Martin-Löf test $\{U_n\}_{n<\omega}$. In particular, z is *n-random* if z is $\emptyset^{(n-1)}$-Martin-Löf random.

The following is known as van Lambalgen's Theorem.

Theorem 12.1.7 (van Lambalgen [165]). *$x \oplus y$ is Martin-Löf random if and only if x is Martin-Löf random and y is x-Martin-Löf random, and also if and only if y is Martin-Löf random and x is y-Martin-Löf random.*

Martin-Löf randomness seems to be the only randomness notion for which van Lambalgen's Theorem holds.

12.1.5 Relative strengths of randomness notions

The relative strengths of the notions of randomness introduced above may be summarized as follows. Note that the implications do not reverse:

$$\begin{aligned}\text{Strong Martin-Löf randomness} &\Rightarrow \text{Martin-Löf randomness}\\ &\Rightarrow \text{Schnorr randomness}\\ &\Rightarrow \text{Kurtz randomness}\end{aligned}$$

A real x is *recursively dominated* if every function $f \leq_T x$ is dominated by a recursive function. The following result gives a characterization of strong Martin-Löf randomness for recursively dominated reals.

Proposition 12.1.8 (Yu). *If x is recursively dominated, then x is strongly Martin-Löf random if and only if it is Kurtz random.*

Next is a characterization of Martin-Löf randomness in terms of Schnorr randomness.

Theorem 12.1.9 (Nies, Terwijn and Stephan [112]). *For any real x,*
(i) *if x is Schnorr random and $x' \not\geq_T \emptyset''$, then x is Martin-Löf random;*
(ii) *if $x' \geq_T \emptyset''$, then there is a Schnorr random real $y \equiv_T x$ such that y is not Martin-Löf random.*

12.1.6 The computational power of random reals

The following theorem says that random reals may have very strong computational power:

Theorem 12.1.10 (Kučera [76]). *For any real $x \geq_T \emptyset'$, there is a Martin-Löf random real $z \equiv_T x$.*

Rather unexpectedly, strong Martin-Löf randomness does not yield the same conclusion:

Theorem 12.1.11 (Downey, Nies, Weber, Yu [23] and Hirschfeldt, Miller). *A real x is strongly Martin-Löf random if and only if x is Martin-Löf random and Turing computes no nonrecursive Δ^0_2-real.*

In algorithmic randomness theory, there is a thesis which states that *more randomness means less computational power.* Although a random real x contains much information, the information may be computationally useless in the sense that it cannot be used to predicate any set except the set x itself. There is much evidence in the literature to support this thesis. We give some examples.

Definition 12.1.12. Let h be a function. A real x is *h-diagonally nonrecursive*, abbreviated h-DNR, if there is a function $f \leq_T x$ such that $f(e) \neq \Phi_e(e)$ and $f(e) < h(e)$ for every $e < \omega$. We say that x is DNR if there is a function $f \leq_T x$ such that $f(e) \neq \Phi_e(e)$ for $e < \omega$.

The next theorem implies that no incomplete Martin-Löf random real is 2-DNR.

Theorem 12.1.13 (Stephan). *Suppose that $x \in 2^\omega$ is Martin-Löf random. Then x is 2-DNR if and only if $x \geq_T \emptyset'$.*

Theorem 12.1.13 has an interesting consequence. By the Low Basis Theorem (Exercise 1.5.3), there is a real x which is 2-DNR and $x' \equiv_T \emptyset'$. By Proposition 12.1.5, there is a nonempty Π^0_1 set which contains only Martin-Löf random reals. Then there is a Martin-Löf random real $z \leq_T x$ since every 2-DNR real Turing computes a member of any given nonempty Π^0_1-set. By Theorem 12.1.13, there is no random real Turing equivalent to x. Thus we have the following:

Theorem 12.1.14 (Kučera [76]). *The collection of Turing degrees of Martin-Löf random reals is not upward closed.*

On the other hand, the Turing degrees of Martin-Löf random reals may be downward closed, as implied by the following.

Theorem 12.1.15 (Demuth [20]). *Any $y >_T \emptyset$ that is truth-table reducible to a Martin-Löf random real is Turing equivalent to a Martin-Löf random real.*

Clearly every real Turing reducible to any recursively dominated x is tt-reducible to x. Hence the collection of Turing degrees of Martin-Löf random reals is downward closed for random reals that are recursively dominated. The following result gives another piece of evidence for the randomness thesis.

Theorem 12.1.16 (Miller and Yu [103]). *For any z, if $x \leq_T y$ are Martin-Löf random and y is z-Martin-Löf random, then so is x.*

12.1.7 Exercises

Exercise 12.1.1. *Prove that there is no universal Kurtz test.*

Exercise 12.1.2 (Schnorr [127]). *Prove that there is no universal Schnorr test.*

Exercise 12.1.3 (Downey, Nies, Weber and Yu [23]). *Prove that there is no universal generalized Martin-Löf test.*

Exercise 12.1.4 (Merkle et al. [99], Yu [170], and Kjos-Hanssen). *Prove that van Lambalgen's Theorem fails for Schnorr randomness.*

12.2 Randomness via complexity theory

Intuitively, a random real is not predicated by any partial information about itself, and no information about its content is compressible. This is the basis for defining randomness in terms of complexity.

12.2.1 Plain Kolmogorov complexity

Definition 12.2.1 (Solomonoff [146] and Kolmogorov [73]). Given a Turing machine M and a finite string $\sigma \in 2^{<\omega}$, the *M-Kolmogorov complexity* of σ, $C_M(\sigma)$, is $\min\{|\tau| \mid M(\tau) = \sigma\}$.

If U is a universal Turing machine, then for any Turing machine M, there is a constant c_M such that

$$(\forall\sigma)(C_U(\sigma) \leq C_M(\sigma) + c_M).$$

In other words, C_U is optimal up to a constant. We drop the subscript U and use C to denote the plain Kolmogorov complexity function C_U. Then one may define a real x to be random if there is a constant c_x such that $(\forall n)(C(x \upharpoonright n) \geq n - c_x)$. However, for any x and n, $x \upharpoonright \mathrm{Code}(x \upharpoonright n)$ (where $\mathrm{Code}(\sigma)$ is the natural number that codes a string σ) may be obtained from a final segment (say σ) of $x \upharpoonright \mathrm{Code}(x \upharpoonright n)$ of length $\mathrm{Code}(x \upharpoonright n) - n$ via U. This is because once σ is known, one may find a unique string τ such that $|\tau^\frown\sigma|$ codes τ, i.e. $|\tau^\frown\sigma| = \mathrm{Code}(\tau)$. Then $|\tau| = n$ and $\tau^\frown\sigma = x \upharpoonright \mathrm{Code}(x \upharpoonright n)$. Thus we have the following.

Proposition 12.2.2 (Martin-Löf). *For any real x, $\overline{\lim}_n n - C(x \upharpoonright n) = +\infty$.*

Proposition 12.2.2 says that there is no random real in the sense of plain Kolmogorov complexity leading to the next definition introduced by Levin and Chaitin.

Definition 12.2.3.
(i) A set $A \subseteq 2^{<\omega}$ is *prefix-free* if for any $\sigma \prec \tau \in 2^{<\omega}$, either $\sigma \notin A$ or $\tau \notin A$.
(ii) A Turing machine M is *prefix-free* if its domain is a prefix-free set.

Definition 12.2.4 (Levin[83] and Chaitin [9]). Given a prefix-free Turing machine M and a string $\sigma \in 2^{<\omega}$, the *M-Kolmogorov complexity* of σ, $K_M(\sigma)$, is $\min\{|\tau| \mid M(\tau) = \sigma\}$.

It is not difficult to see that there is a universal prefix-free Turing machine U and we use $K(\sigma)$ to denote $K_U(\sigma)$. The following technical theorem points a way to constructing a prefix-free machine.

Theorem 12.2.5 (Levin[83] and Chaitin [9] and Schnorr [126]). *Let $\{(d_i, \tau_i)\}_{i<\omega} \subseteq \omega \times 2^{<\omega}$ be a recursive sequence such that $\sum_{i<\omega} 2^{-d_i} \leq 1$. There exist a prefix-free machine M and strings σ_i of length d_i such that $M(\sigma_i) = \tau_i$ for all i and $\mathrm{Dom}(M) = \{\sigma_i \mid i < \omega\}$. Furthermore, an index for M may be obtained effectively from an index for the sequence.*

The following result summarizes some basic properties of K.

Theorem 12.2.6 (Chaitin).
(i) $\sum_{\sigma\in 2^{<\omega}} 2^{-K(\sigma)} \leq 1$;
(ii) $(\exists c)(\forall\sigma)(K(\sigma) \leq |\sigma| + K(|\sigma|) + c)$;
(iii) $(\exists c)(\forall n)(\forall k)(|\{\sigma \in 2^n \mid K(\sigma) \leq n + K(n) - k\}| \leq 2^{n-k+c})$.

12.2.2 Characterizing randomness via complexity

The equivalence between Martin-Löf randomness and Kolmogorov complexity is given by the following:

Theorem 12.2.7 (Schnorr). *A real x is Martin-Löf random if and only if*

$$(\exists c)(\forall n)(K(x \upharpoonright n) \geq n - c).$$

Chaitin [9] gave a "concrete" example of a Martin-Löf random real:

$$\Omega = \sum_{U(\sigma)\downarrow} 2^{-|\sigma|},$$

and several randomness notions have been characterized in terms of Kolmogorov complexity. The next two theorems characterize Martin-Löf randomness and 2-randomness respectively in terms of plain Kolmogorov complexity.

Theorem 12.2.8 (Miller and Yu [103]). *A real x is Martin-Löf random if and only if for every recursive function g with $\sum_{n<\omega} 2^{-g(n)} < \infty$,*

$$(\exists c)(\forall n)(C(x \upharpoonright n) \geq n - g(n) - c).$$

Theorem 12.2.9 (Miller [101] and Nies, Stephan, Terwijn [112]). *A real x is 2-random if and only if*

$$(\exists c)(\forall n)(\exists m \geq n)(C(x \upharpoonright m) \geq m - c).$$

2-randomness may also be characterized via prefix-free Kolmogorov complexity as follows:

Theorem 12.2.10 (Ding, Downey, and Yu [173] and Miller [102]). *A real x is 2-random if and only if*

$$(\exists c)(\forall n)(\exists m \geq n)(K(x \upharpoonright m) \geq m + K(m) - c).$$

A connection between Martin-Löf randomness and the two notions of complexity is given by the following theorem:

Theorem 12.2.11 (Miller and Yu [103]). *$x \oplus y$ is Martin-Löf random if and only if*

$$(\exists c)(\forall n)(K(x \upharpoonright n) + C(y \upharpoonright n) \geq 2n - c).$$

There is a natural way of comparing two reals in terms of plain and prefix-free complexities:

Definition 12.2.12. Write $x \leq_K y$ (resp. $x \leq_C y$) if and only if

$$(\exists c)(\forall n)(K(x \upharpoonright n) \leq K(y \upharpoonright n) + c)$$

(resp. $(\exists c)(\forall n)(C(x \upharpoonright n) \leq C(y \upharpoonright n) + c)$).

By combining van Lambalgen's Theorem 12.1.7 and Theorem 12.2.11, we have the following conclusion.

Corollary 12.2.13 (Miller and Yu [103]). *Suppose that x, y, z are Martin-Löf random. If $x \leq_K y$ or $x \leq_C y$ and x is z-Martin-Löf random, then so is y.*

It is not known whether strong Martin-Löf randomness is $\leq_K$-upward closed. By Theorem 12.1.10, Corollary 12.2.13 says that higher complexity entails stronger randomness in a set. However, the following two results hint at a more subtle picture.

Theorem 12.2.14 (Miller [102]). *If x is 2-random, then $|\{y \mid x \leq_K y\}| = \aleph_0$ and $|\{y \mid x \leq_C y\}| = \aleph_0$.*

Theorem 12.2.15 (Miller and Yu [104]). *There are two Martin-Löf random reals x and y such that $|\{z \mid x \leq_K z \leq_K y\}| = 2^{\aleph_0}$.*

12.2.3 *K*-triviality

Definition 12.2.16. A real x is *K-trivial* if $(\exists c)(\forall n)(K(x \upharpoonright n) \leq K(n) + c)$.

Thus a K-trivial real is one that is far from being random. It is immediate that every recursive real is K-trivial. However, the class of K-trivial reals includes nonrecursive members all of which are recursive in a fixed real.

Theorem 12.2.17 (Solovay [151]). *There is a nonrecursive K-trivial real.*

Theorem 12.2.18 (Nies [110]). *If x is K-trivial, then $x' \leq_{tt} \emptyset'$.*

Theorem 12.2.19 (Kučera and Slaman [77]). *There is an x with $x' \equiv_T \emptyset'$ which computes every K-trivial real.*

12.2.4 Exercises

Exercise 12.2.1 (Miller and Yu [103]). *Prove that a real x is Martin-Löf random if and only if $\sum_{n<\omega} 2^{n-K(x\upharpoonright n)} < \infty$.*

Exercise 12.2.2. *Prove that neither Kurtz randomness nor Schnorr randomness are $\leq_K$-upward closed.*

Exercise 12.2.3 (Chaitin [10]). *Prove that there is no nonrecursive real x such that $(\exists c)(\forall n)(C(x \upharpoonright n) \leq C(n) + c)$.*

Exercise 12.2.4. *Prove that there is a Kurtz random real which is K-trivial.*

12.3 Lowness for randomness

12.3.1 General lowness notions

Given a notion Γ of randomness, we may relativize it to a real x to obtain Γ^x. Define $\mathsf{Low}(\Gamma) = \{x \in 2^\omega \mid \Gamma \subseteq \Gamma^x\}$. Then every member of $\mathsf{Low}(\Gamma)$ is said to be *low* for Γ.

One may consider a slightly stronger lowness notion. Since each Γ is defined in terms of Γ-tests, we may declare x to be *low for Γ-tests* if every Γ^x test is covered (as a set) by a Γ-test. More details concerning these can be found in [111] and [24].

Hence if $x \in \mathsf{Low}(\Gamma)$ then it does not provide any information regarding randomness for Γ. Nevertheless, the property of lowness seems somewhat mysterious since one has no clue on the complexity of members of $\mathsf{Low}(\Gamma)$. A very important and successful application of classical recursion theory is the characterization of the lowness property for various notions of randomness.

12.3.2 Lowness for Kurtz randomness

Theorem 12.3.1 (Stephan, Yu [160] and Greenberg, Miller [38]). *For any real x, the following are equivalent:*
(i) *x is low for Kurtz randomness;*
(ii) *x is low for Kurtz tests;*
(iii) *x is recursively dominated and non-*DNR.

It is known that there are $2^{\aleph_0}$-many recursively dominated and non-DNR reals. Hence Theorem 12.3.1 says that there is an abundance of reals which do not provide any information about Kurtz randomness.

12.3.3 Lowness for Schnorr randomness

Definition 12.3.2. A real x is *recursively traceable* if for any $f \leq_T x$, there is a recursive function $g : \omega \to \mathscr{P}_{<\omega}(2^{<\omega})$ such that for any e, $f(e) \in g(e)$ and $|g(e)| \leq e$.

It is not difficult to see that every recursively traceable real is recursively dominated and non-DNR. Moreover, the construction of a recursively traceable real is quite similar to that of a set of minimal Turing degree. They are both obtained using Sacks forcing with recursive perfect trees.

Theorem 12.3.3 (Terwijn, Zambella [163] and Kjos-Hanssen, Nies, Stephan [67]). *For any real x, the following are equivalent:*
(i) *x is low for Schnorr randomness;*
(ii) *x is low for Schnorr tests;*
(iii) *x is recursively traceable.*

Hence there are again many reals that are low for Schnorr randomness.

12.3.4 Lowness for Martin-Löf randomness

Definition 12.3.4. Let Γ be a notion of randomness. A real x is a *base* for Γ if there is a Γ^x-random y such that $x \leq_T y$.

The following result is one of the deepest in algorithmic randomness theory.

Theorem 12.3.5 (Nies [110] and Hirschfeldt, Nies, Stephan [48]). *For any x, the following are equivalent:*
(i) *x is K-trivial;*
(ii) *x is a base for Martin-Löf randomness;*
(iii) $(\exists c)(\forall n)(K(n) \leq K^x(n) + c)$
(iv) *x is low for Martin-Löf tests;*
(v) *x is low for Martin-Löf randomness.*

The concept of a randomness base is particularly important in the proof of Theorem 12.3.5. Note that one may view Martin-Löf randomness to be the weakest notion of randomness as it has only countably many bases (see [27]). Following the characterization in Theorem 12.3.5, Nies introduced a new reduction relation $\leq_{LR}$:

Definition 12.3.6. $x \leq_{LR} y$ if every y-Martin-Löf random real is x-Martin-Löf random.

Hence by Theorem 12.3.5, $x \leq_{LR} \emptyset$ if and only if x is K-trivial. This relation is expressed in terms of Kolmogorov complexity as follows:

Theorem 12.3.7 (Kjos-Hanssen, Miller, Solomon [66]). *$x \leq_{LR} y$ if and only if*

$$(\exists c)(\forall n)(K^y(n) \leq K^x(n) + c).$$

Combining Corollary 12.2.13 with van Lambalgen's Theorem (12.1.7) offers additional evidence for the randomness thesis:

Proposition 12.3.8. *For any two Martin-Löf random reals x and y, $x \leq_K y$ implies $y \leq_{LR} x$.*

12.3.5 Lowness for strong Martin-Löf randomness

Theorem 12.3.9 (Downey et al. [23], Nies [111], and Kjos-Hanssen et al. [66]). *For any x, the following are equivalent:*
(i) *x is K-trivial;*
(ii) *x is low for strong Martin-Löf test;*
(iii) *x is low for strong Martin-Löf randomness.*

12.3.6 Exercises

Exercise 12.3.1. *Prove that there are $2^{\aleph_0}$-many recursively dominated and non-*DNR *reals.*

Exercise 12.3.2. *Prove that there is a recursively traceable x for which $x' \not\leq_T x \oplus \emptyset'$.*

Exercise 12.3.3. *Let $\mathcal{M}$ be a model of* ZFC *and let x be Sacks generic over $\mathcal{M}$. Prove that if y is random in $\mathcal{M}$, then y is random in $\mathcal{M}[x]$.*

13 More on hyperarithmetic theory

13.1 Hyperarithmetic measure theory

The application of measure theory in recursion theory was initiated by Spector [155], Sacks [117] and Leeuw, Moore, Shannon, and Shapiro [19]. Randomness theory on the natural numbers may be viewed as such an application, while higher randomness theory is the union of measure theory with hyperarithmetic theory. In this chapter we discuss these connections to set the stage for higher randomness in the next chapter.

13.1.1 The measure of Π^1_1-sets

Lemma 13.1.1 (Sacks [120]). *The set*

$$\{(\ulcorner\varphi\urcorner, p) \mid p \textit{ is rational} \wedge \varphi \textit{ is ranked with } x \textit{ as its only free variable} \wedge \mu(\{x \mid \mathfrak{A}(\omega_1^{\mathrm{CK}}, x) \models \varphi\}) \geq p\}$$

is Π^1_1.

Proof. By the Spector–Gandy Theorem (3.7.1) it is sufficient to prove that the set

$$\{(\ulcorner\varphi\urcorner, p) \mid p \text{ is rational} \wedge \varphi \text{ is ranked with } x \text{ as its only free variable} \wedge \mu(\{x \mid \mathfrak{A}(\omega_1^{\mathrm{CK}}, x) \models \varphi\}) \geq p\}$$

is Σ_1 over $\langle L_{\omega_1^{\mathrm{CK}}}, \in\rangle$. We apply Theorem 3.1.8 by performing induction on the rank of φ.

If φ is atomic, then

$$\{(\ulcorner\varphi\urcorner, p) \mid p \text{ is rational} \wedge \varphi\text{'s only free variable is } x \wedge \mu(\{x \mid \mathfrak{A}(\omega_1^{\mathrm{CK}}, x) \models \varphi\}) \geq p\}$$

is recursive. By induction on the complexity of φ, it is not difficult to see that

$$\{(\ulcorner\varphi\urcorner, p) \mid p \text{ is rational } \wedge \varphi\text{'s only free variable is } x \wedge \varphi \text{ is of rank } 0 \wedge \mu(\{x \mid \mathfrak{A}(\omega_1^{\mathrm{CK}}, x) \models \varphi\}) \geq p\}$$

is arithmetical and hence Σ_1 over $\langle L_{\omega_1^{\mathrm{CK}}}, \in\rangle$.

Now suppose that the set

$$\{(\ulcorner\varphi\urcorner, p) \mid p \text{ is rational} \wedge \varphi\text{'s only free variable is } x \wedge \varphi \text{ is of rank at most } \alpha \wedge \mu(\{x \mid \mathfrak{A}(\omega_1^{\mathrm{CK}}, x) \models \varphi\}) \geq p\}$$

is Σ_1 over $\langle L_{\omega_1^{\mathrm{CK}}}, \in\rangle$.

Let $\varphi(x, x^\beta)$ be a formula of rank at most β so that the only free variables are x and x^β. Then $\mathfrak{A}(\omega_1^{\mathrm{CK}}, x) \models (\exists x^\beta)\varphi(\dot{x}, x^\beta)$ if and only if $\mathfrak{A}(\omega_1^{\mathrm{CK}}, x) \models \varphi(\dot{x}, x^\beta \mid \psi)$ for some formula $\psi(n)$ with rank less than β, where n is the only free variable of ψ.

For simplicity, write $\varphi(x, x^\beta \mid \psi)$ as $\varphi(x, \psi)$. Then

$$\{(\ulcorner(\exists x^\beta)\varphi(x, x^\beta)\urcorner, p) \mid \mu(\{x \mid \mathfrak{A}(\omega_1^{\mathrm{CK}}, x) \models (\exists x^\beta)\varphi(\dot{x}, x^\beta)\}) \geq p \wedge p \text{ is rational}\} =$$
$$\{(\ulcorner(\exists x^\beta)\varphi(x, x^\beta)\urcorner, p) \mid (\exists\, \psi(n) \text{ of rank} < \beta)\ (n \text{ is the only free variable of } \psi) \wedge$$
$$p \text{ is rational} \wedge \mu(\{x \mid \mathfrak{A}(\omega_1^{\mathrm{CK}}, x) \models \varphi(\dot{x}, \psi)\}) \geq p\}.$$

Note that the set of formulas ψ of rank $< \beta$ where n is the only free variable belongs to $L_{\omega_1^{\mathrm{CK}}}$. Hence the set $\{(\ulcorner(\exists x^\beta)\varphi(x, x^\beta)\urcorner, p) \mid p \text{ is rational} \wedge \mu(\{x \mid \mathfrak{A}(\omega_1^{\mathrm{CK}}, x) \models (\exists x^\beta)\varphi(\dot{x}, x^\beta)\}) \geq p\}$ is Σ_1 over $\langle L_{\omega_1^{\mathrm{CK}}}, \in\rangle$.

For formulas of the form $(\forall x^\beta)\varphi(\dot{x}, x^\beta)$ where n is the only free variable, note that uniformly in $\langle L_{\omega_1^{\mathrm{CK}}}, \in\rangle$ there is an ω-enumeration of the set of such formulas. Fix such an enumeration $\{\psi_i(n)\}_{i<\omega}$. Then

$$\mu(\{x \mid \mathfrak{A}(\omega_1^{\mathrm{CK}}, x) \models (\forall x^\beta)\varphi(\dot{x}, x^\beta)\}) \geq p$$

if and only if

$$\mu(\{x \mid \mathfrak{A}(\omega_1^{\mathrm{CK}}, x) \models (\forall i)\varphi(\dot{x}, \psi_i)\}) \geq p$$

if and only if

$$(\forall j)\mu(\{x \mid \mathfrak{A}(\omega_1^{\mathrm{CK}}, x) \models \bigwedge_{i \leq j} \varphi(\dot{x}, \psi_i)\}) \geq p.$$

Hence $\{\ulcorner(\forall x^\beta)\varphi(x, x^\beta)\urcorner, p) \mid p \text{ is rational} \wedge \mu(\{x \mid \mathfrak{A}(\omega_1^{\mathrm{CK}}, x) \models (\forall x^\beta)\varphi(\dot{x}, x^\beta)\}) \geq p\}$ is Σ_1 over $\langle L_{\omega_1^{\mathrm{CK}}}, \in\rangle$.

We leave it to the reader to verify the other cases. Note that the induction is uniform and hence the lemma follows. □

Corollary 13.1.2. *Let $p > 0$ be rational. Suppose that $\varphi(n, y)$ is a ranked formula whose only free unranked set variable is y. If for any $n < \omega$, $\mu(\{x \mid \mathfrak{A}(\omega_1^{\mathrm{CK}}, x) \models (\exists y)\varphi(n, y)\}) \geq p$, then there is a $\beta < \omega_1^{\mathrm{CK}}$ such that $(\forall n)\mu(\{x \mid \mathfrak{A}(\omega_1^{\mathrm{CK}}, x) \models (\exists y^\beta)\varphi(n, y^\alpha)\}) \geq p$.*

Proof. Let m_0 be fixed so that $p > 2^{-m_0}$. By Lemma 13.1.1 and the assumption, there is a function $f : \omega \times (\omega \setminus [0, m_0]) \to \omega_1^{\mathrm{CK}}$ in $L_{\omega_1^{\mathrm{CK}}}$ such that for every n and $m > m_0$,

$$\mu(\{x \mid \mathfrak{A}(\omega_1^{\mathrm{CK}}, x) \models (\exists y^{f(n,m)})\varphi(n, y^{f(n,m)})\}) \geq p - 2^{-m}.$$

Let $g(n) = \bigcup_{m_0<m} f(n, m)$. By the admissibility of $L_{\omega_1^{\mathrm{CK}}}$, g is total in $L_{\omega_1^{\mathrm{CK}}}$. Hence for all n, $\mu(\{x \mid \mathfrak{A}(\omega_1^{\mathrm{CK}}, x) \models (\exists y^{g(n)})\varphi(n, y^{g(n)})\}) \geq p$. Let $\beta = \bigcup_{n<\omega} g(n)$. By admissibility again, $\beta < \omega_1^{\mathrm{CK}}$. Moreover, $(\forall n)\mu(\{x \mid \mathfrak{A}(\omega_1^{\mathrm{CK}}, x) \models (\exists y^\alpha)\varphi(n, y^\beta)\}) \geq p$. □

Theorem 13.1.3 (Sacks [120]). *Let p be rational. The set $\{(\ulcorner\varphi\urcorner, p) \mid \varphi$ is $\Sigma_1^1 \wedge \varphi$'s only free variable is $x \wedge \mu(\{x \mid \mathfrak{A}(\omega_1^{\mathrm{CK}}, x) \models \varphi\}) \geq p\}$ is Π_1^1.*

Proof. This follows immediately from Lemma 13.1.1 and Corollary 13.1.2. □

Theorem 13.1.3 is useful as a basic tool for hyperarithmetic measure theory.

13.1.2 Measure-theoretic uniformity

Theorem 13.1.4 (Sacks [120]). *$\{x \mid \omega_1^x = \omega_1^{\mathrm{CK}}\}$ has measure 1.*

Proof. By Theorem 5.1.5, it is sufficient to prove that the set

$$\{x \mid \mathfrak{A}(\omega_1^{\mathrm{CK}}, x) \models \Delta_1^1\text{-comprehension}\}$$

has measure 1. Let $\{(\varphi_i, \psi_i)\}_{i<\omega}$ be a recursive enumeration of arithmetical formulas. For each i, define the Π_1^1-set

$$A_i = \{x \mid \mathfrak{A}(\omega_1^{\mathrm{CK}}, x) \models (\forall n)(\exists y)(\varphi_i(n,y) \vee \neg\psi_i(n,y))\}.$$

Let

$$A_{i,\beta} = \{x \mid \mathfrak{A}(\omega_1^{\mathrm{CK}}, x) \models (\forall n)(\exists y^\beta)(\varphi_i(n,y^\beta) \vee \neg\psi_i(n,y^\beta))\}.$$

Then $A_{i,\beta} \subseteq A_i$ for all $i < \omega$ and $\beta < \omega_1^{\mathrm{CK}}$. By Corollary 13.1.2, for any $i, j < \omega$, there is a $\beta_{i,j} < \omega_1^{\mathrm{CK}}$ such that $\mu(A_i \setminus A_{i,\beta_{i,j}}) < 2^{-i-j}$. Moreover, for any $j < \omega$ and $x \notin \bigcup_{i<\omega}(A_i \setminus A_{i,\beta_{i,j}})$, we have $\mathfrak{A}(\omega_1^{\mathrm{CK}}, x) \models \Delta_1^1$-comprehension. Then for any j,

$$\mu(\{x \mid \mathfrak{A}(\omega_1^{\mathrm{CK}}, x) \models \Delta_1^1\text{-comprehension}\}) \geq 1 - \sum_{i<\omega} 2^{-i-j} \geq 1 - 2^{-j+1}.$$

Thus

$$\mu(\{x \mid \mathfrak{A}(\omega_1^{\mathrm{CK}}, x) \models \Delta_1^1\text{-comprehension}\}) = 1.$$ □

Theorem 13.1.4 is somewhat counterintuitive since it is much easier to construct a real hyperarithmetically computing the hyperjump. The theorem may also be viewed as one more piece of evidence to support the randomness thesis since the set of reals hyperarithmetically computing $\mathcal{O}$ has measure 0.

It is known (see Sacks [117] and Leeuw, Moore, Shannon, and Shapiro [19]) that for any nonrecursive x, the set $\{y \mid y \geq_T x\}$ is null. The following theorem is a hyperarithmetic analog.

Theorem 13.1.5 (Sacks [123]). *If x is not hyperarithmetic, then $\mu(\{y \mid y \geq_h x\}) = 0$.*

Proof. For any x, the set $\{y \mid y \geq_h x\}$ is $\Pi_1^1(x)$ and hence measurable by Corollary 3.7.10.

Now suppose that $\mu(\{y \mid y \geq_h x\}) > 0$. By Theorem 13.1.4, $\omega_1^x = \omega_1^{\mathrm{CK}}$. Moreover, $\mu(\{y \mid \omega_1^y = \omega_1^{\mathrm{CK}} \wedge y \geq_h x\}) > 0$. By Theorem 5.1.4, there is a ranked formula $\varphi(n)$ such that $\mu(\{y \mid n \in x \Leftrightarrow \mathfrak{A}(\omega_1^{\mathrm{CK}}, y) \models \varphi(n)\}) > 0$. By the Lebesgue Density Theorem, there is a string $\sigma \in 2^{<\omega}$ such that $\mu(\{y \succ \sigma \mid n \in x \Leftrightarrow \mathfrak{A}(\omega_1^{\mathrm{CK}}, y) \models \varphi(n)\}) > \frac{1}{2} \cdot 2^{-|\sigma|}$. Then

$$n \in x \implies \mu(\{y \succ \sigma \mid \mathfrak{A}(\omega_1^{\mathrm{CK}}, y) \models \varphi(n)\}) > \frac{1}{2} \cdot 2^{-|\sigma|}$$

and

$$n \notin x \implies \mu(\{y \succ \sigma \mid \mathfrak{A}(\omega_1^{\mathrm{CK}}, y) \models \neg\varphi(n)\}) > \frac{1}{2} \cdot 2^{-|\sigma|}.$$

By Lemma 13.1.1, x is hyperarithmetic, which is a contradiction. □

It is also known that the statement "$\mu(A) > p$" is r.e. if A is an r.e. set of reals and p is rational. The following corresponding result holds in hyperarithmetic theory.

Proposition 13.1.6 (Sacks [123]). *Let $A \subseteq \omega \times 2^\omega$ be Π_1^1 and $A_n = \{x \mid (n, x) \in A\}$. Then $\{(n, p) \mid p$ is rational $\wedge \mu(A_n) > p\}$ is Π_1^1.*

Proof. Since A is Π_1^1, by Corollary 3.7.4 there is an arithmetical formula $\varphi(n, x, z)$ such that

$$(n, x) \in A \Leftrightarrow (\exists z \leq_h x)\varphi(n, x, z).$$

Let $B = \{(n, x) \mid \mathfrak{A}(\omega_1^{\mathrm{CK}}, x) \models (\exists z)\varphi(n, x, z)\} \subseteq A$. By Theorem 13.1.4 for every n, $\mu(B_n) = \mu(\{x \mid (n, x) \in B\})) = \mu(A_n)$. By Corollary 13.1.2, $\mu(B_n) > p$ if and only if

$$(\exists i)(\mu(B_n) \geq p + 2^{-i})$$

if and only if

$$(\exists i)(\exists m)(m \in \mathcal{O}_1 \wedge \mu(\{x \mid \mathfrak{A}(\omega_1^{\mathrm{CK}}, x) \models (\exists z^{|m|})\varphi(n, \dot{x}, z^{|m|})\}) \geq p + 2^{-i}).$$

By Lemma 13.1.1, it follows that the set $\{(n, p) \mid \mu(A_n) > p \wedge p$ is rational$\}$ is Π_1^1. □

The following result was proved by Kechris in [63] under the assumption of PD. It turns out that the additional assumption can be weakened.

Corollary 13.1.7 (Kechris [63]). *Assume that $\omega_1^L < \omega_1$. Given a Σ_2^1-set $A \subseteq 2^\omega$, both $\{p \in \mathbb{Q} \mid \mu(A) > p\}$ and $\{p \in \mathbb{Q} \mid \mu(A) \geq p\}$ are Σ_2^1.*

Proof. By assumption, almost every real is $\mathbb{R}$-generic over L (see § 6.2.2). Note that for any x that is $\mathbb{R}$-generic over L, $\omega_1^{L[x]} = \omega_1^L$. Since A is Σ_2^1, there is a Π_1^1 relation R such that

$$x \in A \Leftrightarrow (\exists y \in L_{\omega_1^{L[x]}}[x])R(x, y).$$

Then $\mu(A) > p$ if and only if there is an ordinal $\beta < \omega_1^L$ such that $\mu(\{x \mid (\exists y \in L_\beta[x]) R(x, y)\}) > p$. Thus $\mu(A) > p$ if and only if there is a real $r \in L$ coding a well-ordering of order type β such that $\mu(\{x \mid (\exists y \in L_\beta[x])R(x, y)\}) > p$. By Proposition 13.1.6, the latter is a Σ_2^1-statement.

The second part of the corollary follows immediately from the first. □

By the proof of Corollary 13.1.7, we have the following conclusion.

Corollary 13.1.8. *If $\omega_1^L < \omega_1$, then every Σ_2^1-set is measurable.*

Proposition 13.1.9 (Kechris [63]). *Suppose that $\omega_1^L < \omega_1$. For any x, if $\{y \mid x \leq_{\Delta_2^1} y\}$ is not null, then x is Δ_2^1;*

Proof. By the Π^1_1-Uniformization Theorem (4.2.1), there is a Π^1_1-partial function $F : \omega \times 2^\omega \to 2^\omega$ such that for any y and $\Pi^1_1(y)$-singleton z, there is an n such that $F(n, y) = z$. By Corollary 4.2.8, $x \leq_{\Delta^1_2} y$ if and only if there exist n and e such that $x = \Phi_e^{F(n,y)}$. Note that for any pair (n, e) and numbers m and i, the set

$$D_{n,e,m,i} = \{y \mid \Phi_e^{F(n,y)}(m) = i\}$$

is Σ^1_2 and hence, by Corollary 13.1.8, is measurable. Note also that for any n and e, the set

$$D_{n,e} = \{y \mid \Phi_e^{F(n,y)}(m) = x(m)\} = \bigcap_{m\in\omega} D_{n,e,m,x(m)}$$

is measurable. Hence $E = \{y \mid x \leq_{\Delta^1_2} y\}$ is measurable. Now suppose that E is not null. Then there is a pair (n, e) that has $D_{n,e}$ positive measure. Let $\sigma \in 2^{<\omega}$ such that $\mu(D_{n,e} \cap [\sigma]) > 2^{-|\sigma|} \cdot \frac{3}{4}$. Then

$$m \in x \Leftrightarrow \mu(D_{n,e,m,1} \cap [\sigma]) > 2^{-|\sigma|} \cdot \frac{3}{4}.$$

By Corollary 13.1.7, x is Σ^1_2. By the same argument, $\omega \setminus x$ is also Σ^1_2 and hence x is Δ^1_2. □

It should be pointed out that the assumption "$\omega_1^L < \omega_1$" in Proposition 13.1.9 is not provable under ZFC.

13.1.3 A measure-theoretic basis theorem

In measure theory, every measurable set of reals is approximated by closed sets. There is an effective version of this in hyperarithmetic theory.

Lemma 13.1.10. *There is a Π^1_1-function $f : \omega \times \omega \to \omega$ such that for any Gödel number n of a ranked formula,*
(i) *$f(n, m)$ is the Gödel number of a ranked formula whose only free variable is x;*
(ii) *the set $\{x|\mathfrak{A}(\omega_1^{\mathrm{CK}}, x) \models \varphi_{f(n,m)}\}$ is a closed subset of $\{x|\mathfrak{A}(\omega_1^{\mathrm{CK}}, x) \models \varphi_n\}$;*
(iii) *$\mu(\{x|\mathfrak{A}(\omega_1^{\mathrm{CK}}, x) \models \varphi_n\} \setminus \{x|\mathfrak{A}(\omega_1^{\mathrm{CK}}, x) \models \varphi_{f(n,m)}\}) < 2^{-m}$.*

Proof. We construct the function f by induction on the rank of formulas. Let $\beta < \omega_1^{\mathrm{CK}}$ and assume by induction that f is defined for every formula with rank less than β. In $\langle L_{\omega_1^{\mathrm{CK}}}, \in\rangle$, there is a uniform enumeration $\{\psi_i(n)\}_{i<\omega}$ of the set of formulas of rank $< \beta$ with n as the only free variable.

Let $\varphi(x^\beta)$ be a formula of rank $\leq \beta$ whose only free ranked variable is x^β. For the formula $(\exists x^\beta)\varphi(x^\beta)$, $\mathfrak{A}(\omega_1^{\mathrm{CK}}, x) \models (\exists x^\beta)\varphi(x^\beta)$ if and only if there is a j such that $\mathfrak{A}(\omega_1^{\mathrm{CK}}, x) \models \bigvee_{i\leq j} \varphi(\psi_i)$. By Theorem 13.1.3, there is a hyperarithmetic function taking each m to a large enough j such that

$$\mu(\{x \mid \mathfrak{A}(\omega_1^{\mathrm{CK}}, x) \models (\exists x^\beta)\varphi(x^\beta)\}) - \mu(\{x \mid \mathfrak{A}(\omega_1^{\mathrm{CK}}, x) \models (\exists i \leq j)(\varphi(\psi_i))\}) < 2^{-m-1}.$$

By induction hypothesis, there is a formula φ' corresponding to $\bigvee_{i\leq j}\varphi(\psi_i)$ (i.e. $\ulcorner\varphi'\urcorner = f(\ulcorner\bigvee_{i\leq j}\varphi(\psi_i)\urcorner, m+1)$) so that

$$\mu(\{x \mid \mathfrak{A}(\omega_1^{\mathrm{CK}}, x) \models \bigvee_{i\leq j}\varphi(\psi_i)\}) - \mu(\{x \mid \mathfrak{A}(\omega_1^{\mathrm{CK}}, x) \models \varphi'\}) < 2^{-m-1}.$$

Then $\mu(\{x \mid \mathfrak{A}(\omega_1^{\mathrm{CK}}, x) \models (\exists x^\beta)\varphi(x^\beta)\}) - \mu(\{x \mid \mathfrak{A}(\omega_1^{\mathrm{CK}}, x) \models \varphi'\}) < 2^{-m}$. Let $f(\ulcorner(\exists x^\beta)\varphi(x^\beta)\urcorner, m) = \ulcorner\varphi'\urcorner$.

For $(\forall x^\beta)\varphi(x^\beta)$, we have $\mathfrak{A}(\omega_1^{\mathrm{CK}}, x) \models (\forall x^\beta)\varphi(x^\beta)$ if and only if for all $j < \omega$, $\mathfrak{A}(\omega_1^{\mathrm{CK}}, x) \models \bigwedge_{i\leq j}\varphi(\psi_i)$. By induction hypothesis, there is a hyperarithmetic function taking each j to a formula φ_j' corresponding to $\bigwedge_{i\leq j}\varphi(\psi_i)$ (i.e. $\ulcorner\varphi_j'\urcorner = f(\ulcorner\bigwedge_{i\leq j}\varphi(\psi_i)\urcorner, m+j+2)$ so that

$$\mu(\{x \mid \mathfrak{A}(\omega_1^{\mathrm{CK}}, x) \models \bigwedge_{i\leq j}\varphi(\psi_i)\}) - \mu(\{x \mid \mathfrak{A}(\omega_1^{\mathrm{CK}}, x) \models \varphi_j'\}) < 2^{-m-j-2}.$$

Now $\{x \mid (\forall j)\mathfrak{A}(\omega_1^{\mathrm{CK}}, x) \models \varphi_j'\}$ is hyperarithmetic. Hence there is a ranked formula φ' such that for any x,

$$\mathfrak{A}(\omega_1^{\mathrm{CK}}, x) \models \varphi' \Leftrightarrow (\forall j)\mathfrak{A}(\omega_1^{\mathrm{CK}}, x) \models \varphi_j'.$$

Note that the set $\{x \mid \mathfrak{A}(\omega_1^{\mathrm{CK}}, x) \models \varphi'\}$ is closed. Moreover,

$$\mu(\{x \mid (\forall x^\beta)\varphi(x^\beta)\}) - \mu(\{x \mid \mathfrak{A}(\omega_1^{\mathrm{CK}}, x) \models \varphi'\}) \leq \sum_{i<\omega} 2^{-m-i-2} < 2^{-m}.$$

Thus define $f(\ulcorner(\forall x^\beta)\varphi(x^\beta)\urcorner, m) = \ulcorner\varphi'\urcorner$. For the other cases, the definition of f is straightforward. □

Corollary 13.1.11. *If A is Π_1^1 and $\mu(A) = r$ where r is hyperarithmetic, then there is a Δ_1^1-set $B \subseteq A$ such that $\mu(B) = \mu(A)$.*

Proposition 13.1.12 (Kechris [63]). *Assume that $\omega_1^L < \omega_1$. Suppose that $A \subseteq 2^\omega$ is a Σ_2^1-set with positive measure. Then for any n, there is a Δ_2^1 perfect set $B_n \subseteq A$ such that $\mu(A) < \mu(B_n) + 2^{-n}$. Furthermore, if $\mu(A)$ is Δ_2^1, then $\{(n, x) \mid x \in B_n \land n < \omega\}$ is Δ_2^1.*

Proof. Since A is Σ_2^1, there is a Π_1^1 relation R such that

$$x \in A \Leftrightarrow (\exists y \in L_{\omega_1^{L[x]}}[x])R(x, y).$$

Let p be a rational between $\mu(A)$ and $\mu(A) - 2^{-n}$. By an argument similar to that for Corollary 13.1.7, $\mu(A) > p$ if and only if there is an $r \in L$ coding a well-ordering of ω such that $\mu(\{x \mid (\exists y \in L_{|r|}[x])R(x, y)\}) > p$. Let $C = \{r \mid r$ codes a well-ordering of $\omega \land \mu(\{x \mid (\exists y \in L_{|r|}[x])R(x, y)\}) > p\}$. Then C is a nonempty Π_1^1-set. Hence there is an $r_n \in C$ which is Δ_2^1. Let

$$B_n = \{x \mid (\exists y \in L_{|r|}[x])R(x, y)\}.$$

Then $B_n \subseteq A$ is Δ_2^1 and $\mu(B_n) > p > \mu(A) - 2^{-n}$. The other part of the proposition follows from the above. □

The following is a measure-theoretic basis theorem:

Theorem 13.1.13 (Sacks [120]). *If $A \subseteq 2^\omega$ is a Π^1_1-set with positive measure, then A contains a hyperarithmetic real.*

Proof. Suppose that $A \subseteq 2^\omega$ is a Π^1_1-set with positive measure. By Corollary 3.7.4, there is an arithmetical formula $\varphi(x, z)$ such that

$$x \in A \Leftrightarrow (\exists z \leq_h x)\varphi(x, z).$$

Let $B = \{x \mid \mathfrak{A}(\omega_1^{\mathrm{CK}}, x) \models (\exists z)\varphi(x, z)\}$. Then $B \subseteq A$. By Theorem 13.1.4, $\mu(B) = \mu(A) > 0$. By Lemma 13.1.10, there is a closed nonempty set $C \subseteq B$ which is defined by a ranked formula $\psi(\dot{x})$. Let

$$T = \{\sigma \in 2^{<\omega} \mid \mu(\{x \mid \mathfrak{A}(\omega_1^{\mathrm{CK}}, x) \models \psi(x) \wedge x \succ \sigma\}) > 0.$$

Then, by Lemma 13.1.1, $[T] \subseteq C$ is a hyperarithmetic tree. The leftmost infinite path in T is hyperarithmetic. □

13.1.4 Exercises

Exercise 13.1.1. *Let $A \subseteq \omega \times 2^\omega$ be Π^1_1 and let $A_n = \{x \mid (n, x) \in A\}$. Show that the set $\{(n, p) \mid \mu(A_n) \geq p \wedge p$ is rational$\}$ is Π^1_1.*

Exercise 13.1.2. *Prove that if $A \subseteq 2^\omega$ is Π^1_1, then $\mu(A) \leq_T \mathcal{O}$.*

Exercise 13.1.3. *Prove that for any real x, the set $\{y \mid (\forall z)(z \leq_h x \wedge z \leq_h y \rightarrow z \in \mathrm{HYP})\}$ has measure 1.*

Exercise 13.1.4. *Prove Corollary* 13.1.11.

Exercise 13.1.5. *Prove that for any recursive ordinal β, there is a hyperarithmetic set A with measure 1 such that no real in A is recursive in $\emptyset^{(\beta)}$.*

Exercise 13.1.6 (Jockusch). *If $A \subseteq 2^\omega$ is a Π^1_1-set containing no hyperarithmetic real, then the set $\{x \mid (\exists y)(y \in A \wedge y \leq_h x)\}$ is null.*

Exercise 13.1.7. *Assume $V = L$. Prove that for any x, $\{y \mid x \leq_{\Delta^1_2} y\}$ has measure 1.*

13.2 Coding sets above Kleene's $\mathcal{O}$

13.2.1 A join basis theorem for Σ^1_1-sets

The question whether there is an uncountable Σ^1_1-set A such that $x \oplus y \not\geq_h \mathcal{O}$ for any $x, y \in A$ was raised in Friedman [29] and Simpson[138]. A negative answer was given by Harrington (unpublished). The following is a stronger version which is used to study higher randomness theory.

Theorem 13.2.1 (Chong and Yu [14]). *Let A_0 and A_1 be uncountable Σ^1_1-sets. For any $z \geq_h \mathcal{O}$, there exist $x_0 \in A_0$ and $x_1 \in A_1$ such that $x_0 \oplus x_1 \equiv_h z$.*

Proof. Fix a real $z \geq_h \mathcal{O}$ and two uncountable Σ^1_1-sets A_0 and A_1. Then there are two recursive trees $T_0, T_1 \subseteq 2^{<\omega} \times \omega^{<\omega}$ such that for $i \leq 1$, $A_i = \{x \mid \exists f \forall n(x \upharpoonright n, f \upharpoonright n) \in T_i\}$. We may assume that neither A_0 nor A_1 contains a hyperarithmetic real. Also assume that if $(\sigma, \tau) \in T_i$, $i \leq 1$, then $|\sigma| = |\mathrm{Dom}(\tau)|$. Let $T_2 \subseteq \omega^{<\omega}$ be recursive so that $[T_2]$ is uncountable and does not contain a hyperarithmetic infinite path. Let $f_{\mathcal{O}}$ be the leftmost path in T_2. Then $f_{\mathcal{O}} \equiv_h \mathcal{O}$.

For $i \leq 1$, let $[T_i] = \{(x, f) \mid \forall n((x \upharpoonright n, f \upharpoonright n) \in T_i)\}$. Our plan is to define $x_i \in A_i$ such that $z \equiv_h x_0 \oplus x_1$. To this end, a procedure of coding z and $f_{\mathcal{O}}$ into x_0 and decoding them from $x_0 \oplus x_1$ will be introduced. Construction of x_i will be carried out in $L_{\omega_1^{\mathrm{CK}}}[z]$ on the recursive tree T_i, hence hyperarithmetically in z (since $z \geq_h \mathcal{O}$). Since A_i is Σ^1_1, x_{1-i} will also code in the function f_i which is the leftmost path in the second component of $[T_i]$ serving as a witness to x_i being in A_i (i.e. $(x_i, f_i) \in [T_i]$ and for any f, if $(x_i, f) \in [T_i]$ and $f_i \neq f$, then the least n where $f_i(n) \neq f(n)$ satisfies $f_i(n) < f(n)$). Since A_i has no hyperarithmetic member, for any $(\sigma, \tau) \in T_i$, if $\sigma \prec x$ and $\tau \prec f$ for some $(x, f) \in [T_i]$, then there exist incompatible extensions σ' and σ'' of σ, and (possibly compatible) extensions τ', τ'' of τ, so that (σ', τ') and (σ'', τ'') both have extensions in $[T_i]$. This "splitting property" of $[A_i]$ allows the coding of z, f_1 and $f_{\mathcal{O}}$ in x_0 and the coding of f_0 in x_1. More specifically, the branch to be selected by x_0 at a splitting node when $z(s)$ is to be coded (at stage $s+1$ of the construction) will follow the value of $z(s)$, so that a "left turn" is taken if $z(s) = 0$ and a right turn is taken if $z(s) = 1$. The coding at stage $s+1$ of $\tau_{1,s}$, which denotes the initial segment of f_1 defined at the end of stage s, is accomplished by taking $\tau_{1,s}(t)$-many consecutive left turns at splitting nodes, for each $t \in \mathrm{Dom}(\tau_{1,s})$. The coding of $f_{\mathcal{O}}(s)$ at stage $s+1$ of the construction is carried out by taking left turns at $f_{\mathcal{O}}(s)$-many consecutive splitting nodes.

For the purpose of decoding, one has to delineate different types of action taken during the coding phase. Since $\tau_{1,s}$ is a finite function, the end of coding the value $\tau_{1,s}(t)$ and the beginning of coding the value $\tau_{1,s}(t+1)$, for $t < |\mathrm{Dom}(\tau_{1,s})|$, is separated by a right turn at the splitting node between the two codings (of course since the construction is executed stage by stage, one may assume that at the beginning of stage $s+1$, the coding of $\tau_{1,s-1}$ is already completed. This means that at stage $s+1$, one only needs to code the values $\tau_{1,s}(t)$ for $t \in \mathrm{Dom}(\tau_{1,s}) \setminus \mathrm{Dom}(\tau_{1,s-1})$). A "right turn" is chosen at the next splitting node to signify the end of coding $\tau_{1,s}$, and the beginning of the coding of $f_{\mathcal{O}}(s)$. Finally, a right turn is taken at the next splitting node after coding $f_{\mathcal{O}}(s)$ to mark the end of the coding action for x_0 at stage $s+1$. This initial segment of x_0 coded at stage $s+1$ is denoted $\sigma_{0,s+1}$. Then $\tau_{0,s+1} \prec f_0$ is a finite string in $\omega^{<\omega}$ such that $|\mathrm{Dom}(\tau_{0,s+1})| = |\sigma_{0,s+1}|$, $\tau_{0,s+1}$ extends $\tau_{0,s}$, and is the leftmost such string. The coding of the initial segment $\tau_{0,s+1}$ of f_0 in x_1 at stage $s+1$, denoted $\sigma_{1,s+1}$, proceeds in a similar fashion. The definition of $\tau_{1,s+1}$, an initial segment of f_1 constructed at stage $s+1$, follows the same format.

We now describe the construction of x_0 and x_1 and the associated strings in detail. For $i \leq 1$ and $\sigma, \tau \in T_i$, let $T_i(\sigma, \tau)$ be the set of strings in T_i compatible with σ and τ), i.e.

$$T_i(\sigma, \tau) = \{(\sigma', \tau') \in T_i \mid \exists(\sigma'', \tau'') \in T_i(\sigma' \preceq \sigma'' \wedge \tau' \preceq \tau'' \wedge \sigma \preceq \sigma'' \wedge \tau \preceq \tau'')\}.$$

Note that it is unnecessary that $|\sigma| = |\tau|$ in the definition above.

We say that a string (or node) $\sigma^* \in 2^{<\omega}$ is *splitting over* (σ, τ) in T_i if $\sigma^* \succeq \sigma$ and for $j \leq 1$,

$$T_{i,\sigma^{*\frown}j}(\sigma, \tau) = \{(\sigma', \tau') \mid \sigma' \succeq \sigma^{*\frown}j \wedge \tau' \succeq \tau \wedge (\sigma', \tau') \in T_i\}$$

contains an infinite path. $T_{i,\sigma^{*\frown}j}(\sigma, \tau)$ is the subtree of T_i with root $\sigma^{*\frown}j$ in its first component (note that $\sigma^* \succeq \sigma$). Since A_i has no hyperarithmetic path, for each $j \leq 1$, there is a string that splits over some (σ', τ') in $T_{i,\sigma^{*\frown}j}(\sigma, \tau)$. Note that σ^* does not lie on T_i but some pair (σ^*, τ') does and we call (σ^*, τ') a *splitting node on* T_i.

For each $i \leq 1$, we construct a sequence $(\sigma_{i,0}, \tau_{i,0}) \prec (\sigma_{i,1}, \tau_{i,1}) \prec \cdots$ on T_i and let $x_i = \bigcup_j \sigma_{i,j}$. Again, the idea is to apply a "mutual coding" technique so that x_0 codes the leftmost witness function $f_1 = \cup_s \tau_{1,s}$ (in the Σ^1_1-definition) for x_1 and x_1 codes the leftmost witness function $f_0 = \cup_s \tau_{0,s}$ for x_0. In addition, we also assign x_0 the responsibility of coding z as well as $f_{\mathcal{O}}$. More precisely, for each $s \in \omega$ we use $\sigma_{0,s}$ to code $z(s)$, $f_{\mathcal{O}}(s)$ and $\tau_{1,s-1}$, and use $\sigma_{1,s}$ to code $\tau_{0,s}$.

At stage 0, let $(\sigma_{i,0}, \tau_{i,0}) = (\emptyset, \emptyset)$ for $i \leq 1$. Without loss of generality, assume that $(\emptyset, \emptyset)$ is a splitting node in both T_0 and T_1.

The construction at stage $s+1$ proceeds as follows:

Substage (i). First let σ^* be the shortest splitting node over $(\sigma_{0,s}, \tau_{0,s})$ in T_0. Thus $T_{0,\sigma^{*\frown}j}(\sigma_{0,s}, \tau_{0,s})$ contains an infinite path for $j \leq 1$. Let $\sigma^0_{0,s+1}$ be the leftmost splitting node over $(\sigma_{0,s}, \tau_{0,s})$ extending $\sigma^{*\frown}z(s)$ in T_0. Thus $z(s)$ is coded here. Next we code $\tau_{1,s}$. Let $n^0_{s+1} = |\mathrm{Dom}(\tau_{1,s})| - |\mathrm{Dom}(\tau_{1,s-1})|$ (let $|\mathrm{Dom}(\tau_{1,s-1})| = 0$ if $s = 0$). Inductively, for any $k \in [1, n^0_{s+1}]$, let $\sigma^k_{0,s+1}$ be the leftmost splitting node over $(\sigma_{0,s}, \tau_{0,s})$ extending $(\sigma^{k-1}_{0,s+1})^\frown 1$ in T_0 so that there are $\tau_{1,s}(k + |\mathrm{Dom}(\tau_{1,s-1})|)$-many splitting nodes over $(\sigma_{0,s}, \tau_{0,s})$ in T_0 between $\sigma^{k-1}_{0,s+1}$ and $\sigma^k_{0,s+1}$. This completes the coding of $\tau_{1,s}$. To code $f_{\mathcal{O}}(s)$, let $\sigma^{n^0_{s+1}+1}_{0,s+1}$ be the leftmost splitting node extending $(\sigma^{n^0_{s+1}}_{0,s+1})^\frown 1$ over $(\sigma_{0,s}, \tau_{0,s})$ in T_0 so that there are $f_{\mathcal{O}}(s)$-many splitting nodes in T_0 over $(\sigma_{0,s}, \tau_{0,s})$ between $\sigma^{n^0_{s+1}}_{0,s+1}$ and $\sigma^{n^0_{s+1}+1}_{0,s+1}$. For $j \leq 1$, let $\sigma^{n^0_{s+1}+1+j+1}_{0,s+1}$ be the next splitting node in T_0 over $(\sigma_{0,s}, \tau_{0,s})$ extending $(\sigma^{n^0_{s+1}+1+j}_{0,s+1})^\frown 1$. This coding tells us that the action at this substage for the "x_0 side" is completed. Define $\sigma_{0,s+1} = \sigma^{n^0_{s+1}+3}_{0,s+1}$. Let $\tau_{0,s+1}$ extend $\tau_{0,s}$ so that $|\mathrm{Dom}(\tau_{0,s+1})| = |\sigma_{0,s+1}|$ and $\tau_{0,s+1}$ is the leftmost node such that the tree $T_0(\sigma_{0,s+1}, \tau_{0,s+1})$ has an infinite path.

Remark. At stage $s+1$, the coding of x_0 in T_0 by way of $\sigma_{0,s+1}$ applies he method of the proof of 11.1.1, treating the Σ^1_1-set A_0 as a "closed set". This means that one first ignores the second component (the "τ side") and applies the coding construction in the proof of Theorem 11.1.1 to the closed set $X_{\sigma_{0,s},\tau_{0,s}} = \{x \succ \sigma_{0,s} \mid \forall n > |\tau_{0,s}| \exists y \exists f \succ \tau_{0,s}(x \upharpoonright n = y \upharpoonright n \wedge (y, f) \in [T_0])\}$. Once the coding of $\sigma_{0,s+1}$ (an initial segment of x_0) is completed, one pairs it with a finite string τ which has an infinite extension to guarantee that x_0 belongs to A_0. Since x_0 does not know the right τ to select, x_1 is designed to help decode the correct τ. The mutual coding strategy is crucial for this purpose.

Substage (ii). Let $\sigma^0_{1,s+1} = \sigma_{1,s}$ and $n^1_{s+1} = |\mathrm{Dom}(\tau_{0,s+1})| - |\mathrm{Dom}(\tau_{0,s})|$. Inductively for any $k \in [1, n^1_{s+1}]$, let $\sigma^k_{1,s+1}$ be the leftmost splitting node over $(\sigma_{1,s}, \tau_{1,s})$ extending $(\sigma^{k-1}_{1,s+1})^\frown 1$ so that in T_1 there are $\tau_{0,s+1}(k + |\mathrm{Dom}(\tau_{0,s})|)$-many splitting nodes over $(\sigma_{1,s}, \tau_{1,s})$ between $\sigma^{k-1}_{1,s+1}$ and $\sigma^k_{1,s+1}$. Hence $\tau_{0,s+1}$ is coded. For $j \leq 1$, let $\sigma^{n^1_{s+1}+j+1}_{1,s+1}$ be the next splitting node over $(\sigma_{1,s}, \tau_{1,s})$ in T_1 extending $(\sigma^{n^1_{s+1}+j}_{1,s+1})^\frown 1$. This coding tells us that the action of coding $\tau_{0,s+1}$ at this substage for the "x_1 side" is completed. Define $\sigma_{1,s+1} = \sigma^{n^1_{s+1}+2}_{1,s+1}$. Thus we have coded $\tau_{0,s+1}$ into $\sigma_{1,s+1}$. Let $\tau_{1,s+1}$ extend $\tau_{1,s}$ so that $|\mathrm{Dom}(\tau_{1,s+1})| = |\sigma_{1,s+1}|$ and $\tau_{1,s+1}$ is the leftmost finite string such that the tree $T_1(\sigma_{1,s+1}, \tau_{1,s+1})$ has an infinite path.

This ends the construction at stage $s + 1$.

Let $x_i = \bigcup_{s<\omega} \sigma_{i,s}$ for $i \leq 1$. Note that the construction is carried out over $L_{\omega_1^{\mathrm{CK}}+1}[z]$ since $\mathcal{O} \in L_{\omega_1^{\mathrm{CK}}+1}$ and whether $(\sigma, \tau) \in T_i$ has an infinite path extension in T_i is decided by stage $L_{\omega_1^{\mathrm{CK}}+1}$. As $z \geq_h \mathcal{O}$ we have $\omega_1^z > \omega_1^{\mathrm{CK}}$ and so $z \geq_h x_0 \oplus x_1$.

We now use x_0 and x_1 to decode the coding construction and hence hyperarithmetically recover $\mathcal{O}$ and z from $x_0 \oplus x_1$. The decoding is achieved via a finite injury method similar to that used in the proof of Theorem 11.1.1. However, a correct decoding requires use of the witness functions f_0 and f_1. Without these witness functions, x_0 would still code z and would belong to the closure of A_0. The entire construction is then reduced to that in the proof of Martin's theorem 11.1.1. However, in this case one cannot conclude that x_0 belongs to A_0. This difficulty is resolved by a procedure of mutual coding, in which x_0 also codes f_1 and x_1 codes f_0. The coding of z is weaved into the mutual coding strategy in the course of the construction.

We now point out the key decoding steps and leave the details to the reader. As in the proof of Theorem 11.1.1, we may fix a Σ_1-enumeration $\{T_i[\alpha]\}_{i\leq 2,\alpha<\omega_1^{\mathrm{CK}}}$ over $L_{\omega_1^{\mathrm{CK}}}$ such that for $i \leq 2$,

- $T_i[0] = T_i$
- $T_i[\alpha] \subseteq T_i[\beta]$ if $\alpha \geq \beta$
- $T_i[\omega_1^{\mathrm{CK}}] = \bigcap_{\alpha<\omega_1^{\mathrm{CK}}} T_i[\alpha]$
- $T_i[\omega_1^{\mathrm{CK}}]$ has no dead end nodes, and
- $A_i = \{x \mid \exists f \forall n (x \upharpoonright n, f \upharpoonright n) \in T_i[\omega_1^{\mathrm{CK}}]\}$.

Since $[T_i]$ does not contain a hyperarithmetic infinite path, we have $[T_i[\omega_1^{\mathrm{CK}}]] = [T_i]$ to be a perfect tree (which as noted earlier enabled the coding procedure to be implemented). Clearly (x_i, f_i) is an infinite path in $T_i[\alpha]$ for each α. As was the case for T_i, $i \leq 1$, for each $\alpha < \omega_1^{\mathrm{CK}}$ one may define the notion of a string σ' being a splitting node over (σ, τ) in $T_i[\alpha]$. In particular, for $(\sigma, \tau) \in T_i[\alpha]$ such that $\sigma \prec x_i$, it makes sense to say that $x_i \upharpoonright n$ *splits over* (σ, τ) in $T_i[\alpha]$. This means that there are strings $\tau_0^i, \tau_1^i \geq \tau$ such that $(x_i \upharpoonright n^\frown 0, \tau_0^i), (x_i \upharpoonright n^\frown 1, \tau_1^i) \in T_i[\alpha]$. In this case we also say that $x_i \upharpoonright n$ is a *locally splitting node* over (σ, τ) in $T_i[\alpha]$. Here "local" refers to the fact that there is no guarantee that $(x_i \upharpoonright n^\frown j, \tau_j^i)$ has an infinite extension in $T_i[\alpha]$ for $j = 0, 1$.

Since for each $i \leq 1$ and $\alpha < \omega_1^{\mathrm{CK}}$, $\langle x_i, f_i \rangle$ is a path on $T_i[\alpha]$, one may use $x_0 \oplus x_1$ to approximate the values of $f_{\mathcal{O}}(s)$ and $f_i(s)$ by simulating in $T_i[\alpha]$ the construction of x_i. This is achieved by attempting to retrieve the sequences $\{(\sigma_{i,s}, \tau_{i,s})\}_{s<\omega}$, $i \leq 1$, using $x_0 \oplus x_1$. Of course, since it is yet to be established that z and $f_{\mathcal{O}}$ are hyperarithmetic in $x_0 \oplus x_1$, the retrieval is only by way of approximating the original construction. Using $x_0 \oplus x_1$, one may mimic the construction of $\{(\sigma_{i,s}, \tau_{i,s})\}_{s<\omega}$ to define $\{(\sigma_{i,s}[\alpha], \tau_{i,s}[\alpha])\}_{s<\omega}$, for $i \leq 1$, so that $\sigma_{i,s+1}[\alpha]$ is an initial segment of x_i (in ascending order of length) and a local splitting node over $(\sigma_{i,s}[\alpha], \tau_{i,s}[\alpha])$ in $T_i[\alpha]$. Furthermore, for each i, $\tau_{i,s+1}[\alpha]$ is an approximation of $f_i \upharpoonright s$ at stage α. In other words, $\tau_{i,s+1}[\alpha]$ is the leftmost finite string so that $(\sigma_{i,s+1}[\alpha], \tau_{i,s+1}[\alpha]) \in T_i[\alpha](\sigma_{i,s}, \tau_{i,s})$.

Note that for each n and $\alpha < \alpha' < \omega_1^{\mathrm{CK}}$, the number of splitting nodes along $(x_i \upharpoonright n, f_i \upharpoonright n)$ in $T_i[\alpha]$ is greater than or equal to that in $T_i[\alpha']$.

One may use $T_2[\alpha]$ to define a $\Sigma_1(L_{\omega_1^{\mathrm{CK}}})$-approximation $f'_{\mathcal{O}}$ of $f_{\mathcal{O}}$ "from the left": $f'_{\mathcal{O}}(0)[\alpha]^\frown \cdots ^\frown f'_{\mathcal{O}}(n)[\alpha]$ is the leftmost string in $T_2[\alpha]$ (it is possible that there exist only finitely many such n's.). Then for $\alpha < \omega_1^{\mathrm{CK}}$, $f'_{\mathcal{O}}(0)[\alpha] \leq f_{\mathcal{O}}(0)$. Suppose $n > 0$ and $f'_{\mathcal{O}}(m)[\alpha] = f_{\mathcal{O}}(m)$ for all $m < n$. Then $f'_{\mathcal{O}}(n)[\alpha] \leq f_{\mathcal{O}}(n)$, and $\lim_{\alpha \to \omega_1^{\mathrm{CK}}} f'_{\mathcal{O}}(n)[\alpha] = f_{\mathcal{O}}(n)$.

The algorithm we adopt for decoding proceeds as follows. Construct a sequence of ordinals $\{\alpha_s\}_{s<\omega}$ which is Δ_1-definable in $L_{\omega_1^{x_0 \oplus x_1}}[x_0 \oplus x_1]$ so that $\lim_{s \to \omega} \alpha_s = \omega_1^{\mathrm{CK}}$, and use it as a parameter to decode the real z, and thereby conclude that $x_0 \oplus x_1 \geq_h z$.

Let $\hat{g}_{\mathcal{O},0}(0) = f_{\mathcal{O}}(0)$ and $\alpha_0 = 0$.

We say that $x_0 \oplus x_1$ *consistently computes* $f_{\mathcal{O}}$ *up to* j *at stage* α if for every $k < j$,

(1) $\tau_{1,k}[\alpha]$ agrees with that coded in x_0 using $\sigma_{0,k+1}[\alpha]$. In other words, let $\sigma_{k+1}^*[\alpha]$ be the shortest locally splitting node over $(\sigma_{0,k}[\alpha], \tau_{0,k}[\alpha])$ in $T_0[\alpha]$. Then there is a $\sigma_{0,k+1}^0[\alpha]$ which is the leftmost locally splitting node over $(\sigma_{0,k}[\alpha], \tau_{0,k}[\alpha])$ extending $\sigma_{k+1}^*[\alpha]^\frown i$, for some $i \in \{0, 1\}$, Moreover, inductively for each $l \in [1, n_{k+1}^0[\alpha]]$ where $n_{k+1}^0[\alpha] = |\mathrm{Dom}(\tau_{1,k}[\alpha])| - |\mathrm{Dom}(\tau_{1,k-1}[\alpha])|$, let $\sigma_{0,k+1}^l[\alpha]$ be the leftmost locally splitting node over $(\sigma_{0,k}[\alpha], \tau_{0,k}[\alpha])$ extending $(\sigma_{0,k}^{l-1}[\alpha])^\frown 1$ in $T_0[\alpha]$. Then there are $\tau_{1,k}[\alpha](l + |\mathrm{Dom}(\tau_{1,k-1}[\alpha])|)$-many locally splitting nodes over $(\sigma_{0,k}[\alpha], \tau_{0,k}[\alpha])$ in $T_0[\alpha]$ between $\sigma_{0,k+1}^{l-1}[\alpha]$ and $\sigma_{0,k+1}^l[\alpha]$;
(2) The approximation of $f_{\mathcal{O}}$ via T_2 using $f'_{\mathcal{O}}$ at stage α agrees with that via $\{\sigma_{0,l}[\alpha]\}_{l \leq k}$ up to the decoding of $f_{\mathcal{O}} \upharpoonright k$. In other words, let $\sigma_{0,k}^{n_{k+1}^0[\alpha]+1}[\alpha]$ be the leftmost locally splitting node extending $(\sigma_{0,k+1}^{n_{k+1}^0[\alpha]}[\alpha])^\frown 1$ in $T_0[\alpha]$ over $(\sigma_{0,k}[\alpha], \tau_{0,k}[\alpha])$ so that there are $f'_{\mathcal{O}}(k)[\alpha]$-many nodes which are locally splitting over $(\sigma_{0,k}[\alpha], \tau_{0,k}[\alpha])$

in $T_0[\alpha]$ between $\sigma_{0,k+1}^{n^0_{k+1}[\alpha]}[\alpha]$ and $\sigma_{0,k+1}^{n^0_{k+1}[\alpha]+1}[\alpha]$. For $i \leq 1$, let $\sigma_{0,k+1}^{n^0_{k+1}[\alpha]+1+i+1}[\alpha]$ be the next local splitting node in $T_0[\alpha]$ over $(\sigma_{0,k}[\alpha], \tau_{0,k}[\alpha])$ extending $(\sigma_{0,k+1}^{n^0_{k+1}[\alpha]+1+i}[\alpha])^\frown 1$. Then $\sigma_{0,k+1}[\alpha] = \sigma_{0,k+1}^{n^0_{k+1}[\alpha]+3}[\alpha]$, and

(3) $\tau_{0,k+1}[\alpha]$ agrees with that coded in x_1 using $\sigma_{1,k+1}[\alpha]$. Thus inductively for each $l \in [1, n^1_{k+1}[\alpha]]$, where $n^1_{k+1}[\alpha] = |\mathrm{Dom}(\tau_{0,k+1}[\alpha])| - |\mathrm{Dom}(\tau_{0,k}[\alpha])|$, let $\sigma^l_{1,k+1}[\alpha]$ be the leftmost locally splitting node over $(\sigma_{1,k}[\alpha], \tau_{1,k}[\alpha])$ extending $(\sigma^{l-1}_{1,k+1}[\alpha])^\frown 1$ in $T_1[\alpha]$ so that there are $\tau_{0,k+1}[\alpha](l+|\mathrm{Dom}(\tau_{0,k}[\alpha])|)$-many locally splitting nodes over $(\sigma_{1,k}[\alpha], \tau_{1,k}[\alpha])$ in $T_1[\alpha]$ between $\sigma^{l-1}_{1,k+1}[\alpha]$ and $\sigma^l_{1,k+1}[\alpha]$. For $i \leq 1$, let $\sigma_{1,k+1}^{n^1_{k+1}[\alpha]+i+1}[\alpha]$ be the next local splitting node over $(\sigma_{1,k}[\alpha], \tau_{1,k}[\alpha])$ in $T_1[\alpha]$. Then $\sigma_{1,k+1}[\alpha] = \sigma_{1,k+1}^{n^0_{k+1}[\alpha]+2}[\alpha]$.

Note that at any stage α, we may assume that $x_0 \oplus x_1$ always computes $f_{\mathcal{O}}$ consistently up to 0. Furthermore, for each $\alpha < \omega_1^{\mathrm{CK}}$ and $j \in \omega$, there is always an $\alpha' \geq \alpha$ such that $\alpha' < \omega_1^{\mathrm{CK}}$ and $x_0 \oplus x_1$ consistently computes $f_{\mathcal{O}}$ up to j at stage α'.

Suppose that α_{t-1} is defined where $t \geq 1$. Search for the least stage $\alpha > \alpha_{t-1}$ so that $x_0 \oplus x_1$ consistently computes $f_{\mathcal{O}}$ up to t at α. Let $\alpha_t = \alpha$ and $\hat{g}_{\mathcal{O},t}(j) = f'_{\mathcal{O}}(j)[\alpha_t]$ for each $j \leq t$.

This completes the construction at step t.

We verify that the decoding construction yields an algorithm to hyperarithmetically compute z from $x_0 \oplus x_1$.

Let $\gamma = \bigcup_{t \in \omega} \alpha_t$.

Clearly $\gamma \leq \omega_1^{\mathrm{CK}}$ and is a limit ordinal. Furthermore $\gamma < \omega_1^{x_0 \oplus x_1}$. We prove that $\gamma = \omega_1^{\mathrm{CK}}$.

Suppose for the sake of contradiction that $\gamma < \omega_1^{\mathrm{CK}}$. Using the fact that neither x_0 nor x_1 is hyperarithmetic, we will show that for $i \leq 1$, each sequence of parameters $n^i_t[\alpha_t]$, $\sigma_{i,t}[\alpha_t]$ and $\tau_{i,t}[\alpha_t]$ introduced in the decoding procedure above eventually stabilizes as $t \to \omega$. Suppose this is false. Then there is a least j_0 and a corresponding least t_0 such that $f_{\mathcal{O}}$ is consistently computed up to $j_0 - 1$, but not j_0, at α_t for all $t \geq t_0$. We argue that such a situation does not occur.

The proof proceeds by induction in the order of the introduction of the parameters at stage α_{j_0}. For convenience, we only show that $\sigma_{0,j_0}^{n^0_{j_0}[\alpha_t]+1}[\alpha_t]$ reaches a limit after some t (essentially the same argument applies to show that the other sequences of parameters also stabilise). This means that $\hat{g}_{\mathcal{O},t}(j_0) \neq \hat{g}_{\mathcal{O},t+1}(j_0)$ for at most finitely many t's.

Assume that $\sigma_{0,j_0}^{n^0_{j_0}[\alpha_t]}[\alpha_t]$ does not change from stage α_{t_0} onwards ($\sigma_{0,k+1}^{n^0_{k+1}[\alpha_t]}[\alpha_t]$ is as defined above before the decoding of $f_{\mathcal{O}} \upharpoonright k$ in (2)). If $\sigma_{0,j_0}^{n^0_{j_0}[\alpha_t]+1}[\alpha_t]$ changes infinitely often, then $\hat{g}_{\mathcal{O},t}(j_0) \neq \hat{g}_{\mathcal{O},t+1}(j_0)$ for infinitely many t's. Then by the decoding construction, we have the following claim.

Claim 1. *$\hat{g}_{\mathcal{O},t+1}(j_0) \geq \hat{g}_{\mathcal{O},t}(j_0)$ for any $t > t_0 + 1$ and hence $\lim_{t\to\omega} \hat{g}_{\mathcal{O},t}(j_0) = +\infty$.*

Proof. If we assign a new value k to $\hat{g}_{\mathcal{O},t}(j_0)$ at some stage $\alpha_t > \alpha_{t_0+1}$, it must be the case that $\hat{g}_{\mathcal{O},t}(0)^\frown \cdots {}^\frown \hat{g}_{\mathcal{O},t}(j_0-1)^\frown k \in T_2[\alpha_t]$. However, $\hat{g}_{\mathcal{O},t}(0)^\frown \cdots {}^\frown \hat{g}_{\mathcal{O},t}(j_0-1) \in T_2[\alpha_{t_0}]$ by the assumption on j_0, so that $k > \hat{g}_{\mathcal{O},t-1}(j_0)$.

Claim 1 implies that at infinitely many t's, left turns were selected at locally splitting nodes in $T_0[\alpha_t]$ during the decoding phase for the purpose of approximating $f_{\mathcal{O}}(j_0)$.

Let $\bar{\sigma} = \sigma_{0,j_0}^{n_{j_0}^0[\alpha_{t_0}]}[\alpha_{t_0}]$ and $\bar{\tau} = \tau_{0,j_0}[\alpha_{t_0}]$. Note that $|\bar{\sigma}| > |\bar{\tau}|$. Define

$$X = \{x \succ \bar{\sigma} \mid \forall n \forall \beta < \gamma \exists \beta' \exists \tau' \succ \bar{\tau}(\beta \leq \beta' < \gamma \wedge |\tau'| = n \wedge (x \upharpoonright n, \tau') \in T_0[\beta'])\}.$$

If we let $\hat{p}(T) = \{\sigma \mid \exists \tau(\sigma, \tau) \in T\}$ be a "local projection" of a tree $T \subseteq \omega^{<\omega} \times \omega^{<\omega}$, then it is not difficult to see that $\bigcap_{\beta<\gamma} \hat{p}(T_0(\bar{\sigma}, \bar{\tau})[\beta])$ is a tree (Note: $T_0(\bar{\sigma}, \bar{\tau})[\beta]$ denotes the analog of $T_0(\bar{\sigma}, \bar{\tau})$ for $T_0[\beta]$). Moreover,

$$X = [\bigcap_{\beta<\gamma} \hat{p}(T_0(\bar{\sigma}, \bar{\tau})[\beta])].$$

In other words, X consists of all the reals x "potentially" with an accompanying witness f so that (x, f) is an infinite path in $T_0(\bar{\sigma}, \bar{\tau})[\gamma]$. Note that since $\gamma < \omega_1^{\mathrm{CK}}$, X is a Δ_1^1 closed subset of 2^ω.

Claim 2. *x_0 is the leftmost path in X.*

Proof. Suppose not. Then there exist y_0 and σ such that $\bar{\sigma} \prec \sigma$, $y_0 \succ \sigma^\frown 0$, $x_0 \succ \sigma^\frown 1$, and $\forall \beta < \gamma \exists \beta' \exists \tau' \succ \bar{\tau}(\beta \leq \beta' < \gamma \wedge |\tau'| = |\sigma| + 1 \wedge (y_0 \upharpoonright |\sigma| + 1, \tau') \in T_0[\beta'])$.

Now by Claim 1 let $t_1 > t_0$ be such that $\hat{g}_{\mathcal{O},t}(j_0) > |\sigma| + 1$ for all $t \geq t_1$. Then there are τ, τ' of length $|\sigma|$ so that $(x_0 \upharpoonright |\sigma| + 1, \tau), (y_0 \upharpoonright |\sigma| + 1, \tau') \in T_0[\alpha_{t_1+1}]$. However, since y_0 is to the left of x_0, the value of $\hat{g}_{\mathcal{O},t_1+1}(j_0)$ coded by x_0 is at most $|\sigma| + 1$. Thus $\hat{g}_{\mathcal{O},t_1+1}(j_0) \leq |\sigma| + 1$, a contradiction.

Claim 2 implies that x_0 is hyperarithmetic which is a contradiction. Hence there is a t^0 such that for any $t \geq t^0$, $\hat{g}_{\mathcal{O},t}(j_0) = \hat{g}_{\mathcal{O},t+1}(j_0)$ and so $\sigma_{0,j_0}^{n_{j_0}^0[\alpha_t]+1}[\alpha_t]$ is a constant for all $t > t^0$.

We leave it to the reader to verify the stabilization of the other sequences of parameters.

It follows that for each j, there is a t_j such that $\hat{g}_{\mathcal{O},t_j}(j) = \hat{g}_{\mathcal{O},t}(j) = f'_{\mathcal{O}}(j)[\alpha_t]$ for $t \geq t_j$. Define $\hat{g}(j) = \lim_{t\to\omega} \hat{g}_{\mathcal{O},t}(j)$ for $j \in \omega$. Then $\hat{g}(j) = \lim_{t\to\omega} f'_{\mathcal{O}}(j)[\alpha_t]$ and $\hat{g} \in L_{\gamma+1}$. Now $\hat{g}$ is the leftmost path in $T_2[\gamma]$ and so in T_2 as well. Thus $\hat{g} = f_{\mathcal{O}} \in L_{\gamma+1}$. This contradicts the assumption that $\gamma < \omega_1^{\mathrm{CK}}$. Hence $\gamma = \omega_1^{\mathrm{CK}}$ and so $f_{\mathcal{O}} \leq_h x_0 \oplus x_1$.

Using this, one may decode the entire construction in $T_i[\omega_1^{\mathrm{CK}}]$ and conclude that $z \leq_h x_0 \oplus x_1$, completing the proof of the Theorem. □

13.2.2 Exercise

Exercise 13.2.1. *Prove that there exists an uncountable Π^0_1-set A such that $x \oplus y \not\geq_T \emptyset'$ for any $x, y \in A$.*

13.3 Hyperarithmetic computation

In this section, we discuss some notions concerning effective bounding in the context of hyperarithmetic theory. The results will be applied to study higher randomness.

13.3.1 Hyperarithmetic domination

Definition 13.3.1. A real x is Δ^1_1-dominated if for any function $f : \omega \to \omega$ with $f \leq_h x$, there is a hyperarithmetic g such that $(\forall n)(g(n) > f(n))$(written as $g > f$).

The first item to note is that there is an abundance of Δ^1_1-dominated reals.

Theorem 13.3.2 (Chong, Nies and Yu [12]). *$\mu(\{x \mid x$ is Δ^1_1-dominated$\}) = 1$.*

Proof. We prove that for any rational number p,

$$\mu(\{x \mid x \text{ is } \Delta^1_1\text{-dominated}\}) \geq p.$$

This is achieved with the use of a fusion argument.

First we show that for any e, rational r, notation $n \in \mathcal{O}$ and Δ^1_1-set A where $p + r < \mu(A)$, there is a hyperarithmetic function f such that

$$\mu(\{x \mid x \in A \wedge \Phi_e^{\hat{H}^x_n} \text{ is total} \to \Phi_e^{\hat{H}^x_n} < f\}) > p + \frac{r}{2}.$$

Note that the set $\{(x, i, m) \mid \Phi_e^{\hat{H}^x_n}(i) \downarrow \to \Phi_e^{\hat{H}^x_n}(i) < m\}$ is Δ^1_1. Since A is Δ^1_1, Proposition 13.1.6 implies that the set

$$C = \{(i, m, k) \in \omega^3 | \mu(\{x \mid x \in A \wedge (\Phi_e^{\hat{H}^x_n}(i) \downarrow \to \Phi_e^{\hat{H}^x_n}(i) < m)\}) > \mu(A) - \frac{r}{2^{k+2}}\}$$

is Δ^1_1. Note that for each k, there is an m such that $(k, m, k) \in C$. Hence there is a Δ^1_1 total function f such that for all k, $(k, f(k), k) \in C$. Define

$$B_k = \{x \mid x \in A \wedge (\Phi_e^{\hat{H}^x_n}(k) \downarrow \to \Phi_e^{\hat{H}^x_n}(k) < f(k))\}.$$

Then the set $\{(k, x) \mid x \in B_k\}$ is Δ^1_1. Moreover, for every k, $B_k \subseteq A$ and $\mu(B_k) > \mu(A) - \frac{r}{2^{k+2}}$. Hence $B = \bigcap_k B_k$ is Δ^1_1 and

$$\mu(B) \geq \mu(A) - \sum_{k\geq 0} \mu(A - B_k) \geq p + r - \sum_{k\geq 0} \frac{r}{2^{k+2}} = p + \frac{r}{2}.$$

Now for every $x \in B$, if $\Phi_e^{\hat{H}_n^x}$ is total, then $\Phi_e^{\hat{H}_n^x} < f$. Thus we may construct an ω-sequence of Δ_1^1-sets $\{B^{\langle e,n\rangle}\}_{e<\omega,n\in\mathcal{O}}$ such that
(1) if $\langle e,n\rangle > \langle e',n'\rangle$, then $B^{\langle e,n\rangle} \subseteq B^{\langle e',n'\rangle}$;
(2) $\mu(B^{\langle e,n\rangle}) > p$.

Define

$$D = \{x \mid \omega_1^x = \omega_1^{\mathrm{CK}}\} \cap (\bigcap_{e<\omega \wedge n\in\mathcal{O}} B^{\langle e,n\rangle}).$$

By Theorem 13.1.4, $D \subseteq \{x | x \text{ is } \Delta_1^1\text{-dominated}\}$ and $\mu(D) \geq p$. □

It is obvious that if x is Δ_1^1-dominated, then $\omega_1^x = \omega_1^{\mathrm{CK}}$. The converse is however false.

Proposition 13.3.3. *There is an uncountable Σ_1^1-set A in which every x hyperarithmetically computes an f such that $f > g$ for every hyperarithmetic function g.*

Proof. It is immediate that there is a Turing oracle function Φ such that $\Phi^{\mathcal{O}}$ dominates all the hyperarithmetic function. Let

$$A = \{x \mid \Phi^x \text{ is total} \wedge (\forall f)(f \in \mathsf{HYP} \to f < \Phi^x)\}.$$

By Corollary 2.2.12, A is Σ_1^1. Since $\mathcal{O} \in A$, by Lemma 2.5.4, A is uncountable. □

Hence by the Gandy Basis Theorem (2.5.3), we have the following conclusion.

Corollary 13.3.4. *There is an x with $\omega_1^x = \omega_1^{\mathrm{CK}}$ which is not Δ_1^1-dominated.*

13.3.2 Traceability

Definition 13.3.5.
(i) Let h be a nondecreasing unbounded and hyperarithmetic function. A Π_1^1-trace/Δ_1^1-trace with bound h is a Π_1^1/ Δ_1^1-set $T \subseteq \omega \times \omega$ such that $|T_e| \leq h(e)$ for each e where $T_e = \{k \mid (e,k) \in T\}$.
(ii) x is Π_1^1-traceable/Δ_1^1-traceable if there is an $h \in \Delta_1^1$ such that, for each $f \leq_h x$, there is a Π_1^1-trace/Δ_1^1 -trace T with bound h satisfying $f(e) \in T_e$ for each e.

It is not difficult to see that the bounding function h may be chosen to be any hyperarithmetic function, for example $h(e) = e$ for every e.

Proposition 13.3.6 (Chong, Nies and Yu [12]). *If x is Π_1^1-traceable, then x is Δ_1^1-traceable.*

Proof. Firstly we show that $\omega_1^x = \omega_1^{\mathrm{CK}}$. It is sufficient to show that $\mathcal{O}$ is not Π_1^1-traceable. Since each Π_1^1-set is many-one reducible to $\mathcal{O}$, there is a uniformly $\mathcal{O}$-recursive list $(T^e)_{e<\omega}$ of all Π_1^1-traces for the bound $h(e) = e$. Define $f \leq_h \mathcal{O}$ by

$$f(e) = \mu n\, [n \notin T_e^e].$$

Then f does not have a Π^1_1 trace. Now given $f \leq_h x$, there is a Π^1_1-trace $(T_e)_{e<\omega}$ such that $f(e) \in T_e$ for each e. Then there is a recursive function $h : \omega^2 \to \omega$ such that $k \in T_e$ if and only if $h(k, e) \in \mathcal{O}$. Define a $\Pi^1_1(x)$-relation $R \subseteq \omega \times \mathcal{O}$ by

$$(e, n) \in R \Leftrightarrow h(e, f(e)) \in \mathcal{O}_n.$$

Note that for each e, there is a notation $n \in \mathcal{O}$ such that $(e, n) \in R$. By the Π^1_1-Uniformization Theorem (4.2.1), there is a total $\Pi^1_1(x)$ (and hence $\Delta^1_1(x)$) function g uniformizing R. Since $\omega^x_1 = \omega^{\mathrm{CK}}_1$, there exists a notation $n_0 \in \mathcal{O}$ so that $|g(e)| < |n_0|$ for every e. Define a set $\hat{T}_e \subseteq T_e$ as follows:

$$k \in \hat{T}_e \Leftrightarrow h(k, e) \in \mathcal{O}_{n_0}.$$

By the definition of n_0, $f(e) \in \hat{T}_e$ for all $e \in \omega$. Note that the relation $n \in \hat{T}_e$ is Δ^1_1. Hence $(\hat{T}_e)_{e\in\omega}$ is a Δ^1_1-trace for f. It follows that f is Δ^1_1-traceable. □

There is also a plentiful supply of Δ^1_1-traceable reals. In fact every Sacks generic real is Δ^1_1-traceable.

Proposition 13.3.7. *Every Sacks generic real is Δ^1_1-traceable.*

Proof. Suppose x is Sacks generic. Then $\omega^{\mathrm{CK}}_1 = \omega^x_1$. If $f \leq_h x$, then there is an e and a notation $n \in \mathcal{O}$ such that $\Phi_e^{\hat{H}^x_n} = f$. Note that the ranked formula

$$(\exists z^{|n|})(H(n, z^{|n|}, y) \wedge \Phi_e^{z^{|n|}}(i) = j)$$

defines the set $A = \{(y, n, i, j) | \Phi_e^{\hat{H}^y_n}(i) = j\}$. For simplicity, we may view $\Phi_e^{\hat{H}^x_n}$ as a ranked formula. Since $\Phi_e^{\hat{H}^x_n}$ is total, by Lemma 5.3.6, there is a condition $T \Vdash$ "$\Phi_e^{\hat{H}^x_n}$ is total". We show that for each condition $S \subseteq T$, there is a condition $Q \subseteq S$ such that

$$Q \Vdash \text{“}(\exists f)(f \in \Delta^1_1 \wedge (\forall i)(\Phi_e^{\hat{H}^x_n}(i) \in D_{f(i)} \wedge |D_{f(i)}| \leq 2^{i+1}))\text{”}.$$

Then by the definition of forcing, there is a Δ^1_1 function f such that for all i, $\Phi_e^{\hat{H}^x_n}(i) \in D_{f(i)} \wedge |D_{f(i)}| \leq 2^{i+1}$. Since $T \Vdash$ "$\Phi_e^{\hat{H}^x_n}$ is total", S also forces the same sentence. We consider two cases.

Case 1. There is a condition $R \subseteq S$ such that for all i, j_0, j_1 and for all conditions $P_0, P_1 \subseteq R$, if $P_0 \Vdash \Phi_e^{\hat{H}^x_n}(i) = j_0$ and $P_1 \Vdash \Phi_e^{\hat{H}^x_n}(i) = j_1$ then $j_0 = j_1$. In this case define $f(i) = j$ if and only if there exists a condition $P \subseteq R$ such that $P \Vdash \Phi_e^{\hat{H}^x_n}(i) = j$. Then f is a total Π^1_1-function and hence Δ^1_1. This implies that $R \Vdash f = \Phi_e^{\hat{H}^x_n}$.

Case 2. Otherwise. Define a Π^1_1-relation $\mathcal{R}(P, \sigma, i, j_0, j_1, Q_0, Q_1)$ as

$$i \geq |\sigma| \wedge j_0 \neq j_1 \wedge Q_0 \cap Q_1 = \emptyset \wedge (\forall k \leq 1)(Q_k \subseteq P \wedge Q_k \Vdash \Phi_e^{\hat{H}^x_n}(i) = j_k).$$

By the Π^1_1-Uniformization Theorem (4.2.1), there is a Π^1_1-function $F : 2^\omega \times 2^{<\omega} \to (\omega)^3 \times (2^\omega)^2$ such that $\mathcal{R}(P, \sigma, i, j_0, j_1, Q_0, Q_1)$ holds if and only if $\mathcal{R}(P, \sigma, F(P, \sigma))$.

Without loss of generality, we assume that if $P \Vdash \Phi_e^{\hat{H}_n^x}(i) = j_i$ then for all $k \le i$, $P \Vdash \Phi_e^{\hat{H}_n^x}(k) = j_k$ for some j_k. Define by induction on ω a Π_1^1-sequence of conditions $\{P_\sigma\}_{\sigma\in 2^{<\omega}}$ as follows:

Step 0. Let $P_\emptyset = S$.

Step $n+1$. For each $\sigma \in 2^n$, define $P_{\sigma^\frown 0} = Q_0$, $P_{\sigma^\frown 1} = Q_1$ if $F(P_\sigma, \sigma) = (i, j_0, j_1, Q_0, Q_1)$.

Let $G(\sigma) = P_\sigma$. Then G is a total Π_1^1-function and so Δ_1^1. Note that for each σ, $G(\sigma^\frown 0) \cap G(\sigma^\frown 1) = \emptyset$ and if $\sigma \preceq \tau$ then $G(\sigma) \supseteq G(\tau)$. Let

$$Q = \bigcap_n \bigcup_{\sigma\in 2^n} G(\sigma).$$

Then Q is a Δ_1^1-perfect set.

Define $g : \bigcup_{i<\omega} i \times 2^{i+1} \to \omega$ such that $g(i, \sigma) = k$ if $\sigma \in 2^{i+1}$ and $G(\sigma) \Vdash \Phi_e^{\hat{H}_n^x}(i) = k$. Hence g is a total Π_1^1 and therefore Δ_1^1-function. Define $f(i) = j$ if j is the least number such that $D_j = \{g(i, \sigma) \mid \sigma \in 2^{i+1}\}$. Then f is a Δ_1^1-function and $|D_{f(i)}| \le 2^{i+1}$ for all i. Since for all i, $Q \subseteq \bigcup_{\sigma\in 2^n} G(\sigma)$, it is easy to see that $Q \Vdash \Phi_e^{\hat{H}_n^x}(i) \in D_{f(i)}$. It follows that x is Δ_1^1-traceable. □

We will see later (Theorem 14.6.8) that the collection of Δ_1^1-traceable reals is null.

By Theorem 5.3.12, there is an uncountable Σ_1^1-set of weakly-Sacks-generic reals in which every real x has the property $\omega_1^x = \omega_1^{\mathrm{CK}}$. By the proof of Proposition 13.3.7, it is easy to derive the following conclusion.

Corollary 13.3.8. *There is an uncountable Σ_1^1-set of Δ_1^1-traceable reals.*

13.3.3 Semitraceability

Definition 13.3.9.

(i) $x \in 2^\omega$ is Δ_1^1-semitraceable if for each $f \le_h x$, there is a Δ_1^1-function g such that, for infinitely many n, $f(n) = g(n)$. We say that g semitraces f.

(ii) $x \in 2^\omega$ is Π_1^1-semitraceable if for each $f \le_h x$, there is a partial Π_1^1-function p such that for infinitely many n we have $f(n) = p(n)$.

Lemma 13.3.10. *x is Π_1^1-semitraceable if and only if x is Δ_1^1-semitraceable.*

Proof. It is not difficult to see that if x is Π_1^1-semitraceable, then $\omega_1^x = \omega_1^{\mathrm{CK}}$. Since $x \ge_h \mathcal{O}$ otherwise. Hence it suffices to show that $\mathcal{O}$ is not Π_1^1-semitraceable. Let $\{\phi_i\}_{i<\omega}$ be an effective enumeration of partial recursive functions. Let m_j^i be the least k such that $p_j(i, k) \in \mathcal{O}$ if it exists, and let it be 0 otherwise. Define a function $g \le_T \mathcal{O}'$ such that $g(i) = \sum_{j\le i} m_j^i + 1$. Note that for any Π_1^1-partial function p, there is a partial recursive function p_j such that for every pair n, m, $p(n) = m$ if and only if $p_j(n, m) \in \mathcal{O}$. Then by the definition of g, for any $i > j$, $g(k) \ne p(i)$. Hence g is not traced by p.

Suppose that x is Π^1_1-semitraceable, $\omega_1^x = \omega_1^{\mathrm{CK}}$, and $f \le_h x$. Fix a Π^1_1-partial function p for f witnessing the semitraceability of x. Since p is a Π^1_1-function, there is a recursive injection h such that $p(n) = m \Leftrightarrow h(n, m) \in \mathcal{O}$.

Let $R(n, m)$ be a $\Pi^1_1(x)$-relation such that $R(n, m)$ if and only if there exists a k, where $m > k \ge n$, so that $f(k) = p(k)$. Then there is a total function g uniformizing R such that g is $\Pi^1_1(x)$, and hence $\Delta^1_1(x)$. Thus for every n there is an $m \in [g(n), g(g(n)))$ such that $f(m) = p(m)$. Let $g'(0) = g(0)$, and $g'(n + 1) = g(g'(n))$ for all $n < \omega$. Define a $\Pi^1_1(x)$-relation $S(n, m)$ such that $S(n, m)$ if and only if $m \in [g'(n), g'(n + 1))$ and $p(m) = f(m)$. Uniformizing S, we obtain a $\Delta^1_1(x)$-function g''.

Let $H = \{h(m, k) \mid (\exists n)(g''(n) = m \wedge f(m) = k)\}$. Then H is $\Delta^1_1(x)$ and since $\omega_1^x = \omega_1^{\mathrm{CK}}$, $H \subseteq \mathcal{O}_n$ for some $n \in \mathcal{O}$. Since $\mathcal{O}_n$ is Δ^1_1, we may define a Δ^1_1-function $\hat{f}$ by setting $\hat{f}(i) = j$ if $h(i, j) \in \mathcal{O}_n$ and $\hat{f}(i) = 1$ otherwise. Then there are infinitely many i such that $f(i) = \hat{f}(i)$. □

The following clarifies the relationship between traceability and semitraceability.

Proposition 13.3.11 (Kjos-Hassen, Nies, Stephan and Yu [68]). *Every Δ^1_1-traceable real is Δ^1_1-dominated and Δ^1_1-semitraceable.*

Proof. Clearly every Δ^1_1-traceable real is Δ^1_1-dominated. Let x be a Δ^1_1-traceable real and let f be a $\Delta^1_1(x)$-function. Define $g(n) = \langle f(2^n), f(2^n + 2), \dots, f(2^{n+1} - 1)\rangle$. Then there is a Δ^1_1 trace T for g such that $|T_n| \le n$ for all n.

For $2^n + 1 \le m \le 2^{n+1}$, let $\hat{f}(m)$ = the $(m - 2^n)$-th entry of the tuple which is the $(m - 2^n)$-th element of T_n if there exists such a tuple, and let $\hat{f}(m) = 1$ otherwise. It is not difficult to see that for every n there is at least one $m \in [2^n, 2^{n+1})$ such that $f(m) = \hat{f}(m)$. □

Corollary 13.3.12. *A real x is Δ^1_1-traceable if and only if for every $\hat{f}$ hyperarithmetic in x, there is a hyperarithmetic f such that*

$$(\forall n)(\exists m \in [2^n, 2^{n+1}))(f(m) = \hat{f}(m)).$$

Proposition 13.3.13. *For any x, the following are equivalent.*
(i) *x is Δ^1_1-semitraceable and Δ^1_1-dominated;*
(ii) *For every function $g \le_h x$, there exist an increasing Δ^1_1-function f and a Δ^1_1-function $F : \omega \to \mathcal{P}_{<\omega}(\omega)$ with $|F(n)| \le n$ such that*

$$(\forall n)(\exists m \in [f(n), f(n + 1)))(g(m) \in F(m)).$$

Proof. (i) ⇒ (ii): Immediate because $1 \le n$.

(ii) ⇒ (i). Suppose $\hat{g} \le_h x$. Without loss of generality, assume $\hat{g}$ is nondecreasing. Let f and F be the corresponding Δ^1_1-functions. Let $u(n) = \sum_{i \le f(n+1)} \sum_{k \in F(i)} k$ and note that u is a Δ^1_1-function dominating $\hat{g}$. Let

$$h(n) = \langle \hat{g}(2^n + 1), \hat{g}(2^n + 2), \dots, \hat{g}(2^{n+1} - 1)\rangle.$$

Then by assumption there are corresponding Δ^1_1-functions f_h and F_h. For every n and $m \in [2^n, 2^{n+1})$, let $g(m)$ = the $(m-2^n)$-th entry of the $(m-2^n)$-th element in $F_h(n)$ if such an m exists, and let $g(m) = 1$ otherwise. Then g is a Δ^1_1-function semitracing $\hat{g}$. □

To separate Δ^1_1-traceability from the conjunction of Δ^1_1-semitraceability and Δ^1_1-domination, one applies a modified Sacks forcing construction.

Definition 13.3.14.

(i) A Δ^1_1-perfect tree $T \subseteq 2^{<\omega}$ *fully splits at n* if for every $\sigma \in T$ with $|\sigma| \in [2^n, 2^{n+1})$, both $\sigma^\frown 0$ and $\sigma^\frown 1$ are in T. Then n is said to be a *fully splitting level* of T.

(ii) A Δ^1_1-perfect tree $T \subseteq 2^{<\omega}$ is *clumpy* if there are infinitely many n's where T fully splits at n.

Let $\mathbb{F} = \langle \mathbf{F}, \subseteq \rangle$ be the partial order of clumpy trees ordered under inclusion. Given a formula φ in the language $\mathscr{L}(\omega_1^{\mathrm{CK}}, \dot{x})$ of ramified analytical hierarchy (§ 5.1.1), define the forcing relation $T \Vdash \varphi$ as in Sacks forcing:

- If φ is a ranked sentence and $(\forall x \in T)(\mathfrak{A}(\omega_1^{\mathrm{CK}}, x) \models \varphi)$, then $T \Vdash \varphi$.;
- if $\varphi(y)$ is unranked and $T \Vdash \varphi(y \mid \psi)$ for some ψ of rank at most β, then $T \Vdash (\exists y^\beta)\varphi(y^\alpha)$;
- if $T \Vdash (\exists y^\beta)\varphi(y^\alpha)$, then $T \Vdash (\exists y)\varphi(y)$;
- if $\varphi(n)$ is unranked and $T \Vdash \varphi(\bar{m})$ for some numeral $\bar{m}$, then $T \Vdash (\exists n)\varphi(n)$;
- if φ and ψ are unranked, $T \Vdash \varphi$ and $T \Vdash \psi$, then $T \Vdash \varphi \wedge \psi$;
- if φ is unranked and $(\forall P)(P \subseteq T \to P \nVdash \varphi)$, then $T \Vdash \neg\varphi$.

By Theorem 5.1.1, we have the following.

Lemma 13.3.15. *The set*

$$\{(T, \ulcorner\varphi\urcorner) \mid T \text{ is a condition} \wedge \varphi \text{ is } \Sigma^1_1 \wedge T \Vdash \varphi\}$$

is Π^1_1.

Lemma 13.3.16.

(i) *Let* $\{\varphi_i\}_{i<\omega}$ *be a hyperarithmetic sequence of* Σ^1_1*-sentences. Suppose that for every* i *and* $Q \subseteq T$, *there exists an* $R \subseteq Q$ *such that* $R \Vdash \varphi_i$. *Then there exists an* $R \subseteq T$ *such that* $R \Vdash \varphi_i$ *for every* i.

(ii) $(\forall\varphi)(\forall T)(\exists Q \subseteq T)(Q \Vdash \varphi \vee Q \Vdash \neg\varphi)$.

Proof. Using the notation $P \upharpoonright n = \{\tau \in 2^{\leq n} \mid \tau \in P\}$, define a Π^1_1-relation $\mathfrak{R}$ by

$$\begin{aligned}\mathfrak{R}(R, i, \sigma, P) \Leftrightarrow (&\sigma \in R \wedge P \subseteq R \wedge P \Vdash \varphi_i \wedge P \upharpoonright |\sigma| = \{\tau \mid \tau \prec \sigma\} \wedge \\ &\log|\sigma| - 1 \text{ is the } i\text{-th fully splitting level of } R).\end{aligned}$$

Then $\mathfrak{R}$ is uniformized by a partial Π^1_1-function $F : \mathbf{F} \times \omega \times 2^{<\omega} \to \mathbf{F}$. Using F, a hyperarithmetic family $\{P_\sigma \mid \sigma \in 2^{<\omega}\}$ may be defined by recursion on σ.

Let $P_\emptyset = T$. If $\sigma \notin P_\sigma$, let $P_{\sigma^\frown 0} = P_{\sigma^\frown 1} = \emptyset$. If $\log|\sigma| - 1$ is not a fully splitting level of P_σ, set $P_{\sigma^\frown 0}, P_{\sigma^\frown 1} = P_\sigma$. Otherwise, $P_{\sigma^\frown 0} \cap P_{\sigma^\frown 1} = \emptyset$, $P_{\sigma^\frown 0} \cup P_{\sigma^\frown 1} \subseteq P_\sigma$,, $P_{\sigma^\frown 0} \upharpoonright |\sigma|, P_{\sigma^\frown 1} \upharpoonright |\sigma| = \{\tau \mid \tau \prec \sigma\}$ and $P_{\sigma^\frown 0}, P_{\sigma^\frown 1} \Vdash \bigwedge_{j \le i} \varphi_j$ where $\log|\sigma| - 1$ is the i-th fully splitting level of T.
Let $R = \bigcap_n \bigcup_{|\sigma|=n} P_\sigma$. Then $R \in \mathbf{F}$. It is routine to verify that $R \Vdash \varphi_i$ for every i.

The proof of (ii) is similar to the proof of Lemma 5.3.4. □

A real x is *generic* if for every dense set $\mathscr{D} \subseteq \mathbf{F}$ arithmetical in $\mathcal{O}$, there is a $T \in \mathscr{D}$ such that $x \in [T]$. One can show as in the proof of Lemma 5.3.6 that for every sentence φ and generic x,

$$\mathfrak{A}(\omega_1^{\mathrm{CK}}, x) \models \varphi \Leftrightarrow (\exists P)(x \in P \wedge P \Vdash \varphi).$$

Lemma 13.3.17. *If x is generic, then*
(i) *$\mathfrak{A}(\omega_1^{\mathrm{CK}}, x)$ satisfies Δ_1^1-comprehension. Hence $\omega_1^x = \omega_1^{\mathrm{CK}}$;*
(ii) *x is Δ_1^1-dominated and Δ_1^1-semitraceable;*
(iii) *x is not Δ_1^1-traceable.*

Proof. The proof of (i) is the same as that of Lemma 5.3.7 and is left to the reader.

For (ii), by Proposition 13.3.13, it suffices to show that for every function $g \le_h x$, there exist an increasing Δ_1^1-function f and a Δ_1^1-function $F : \omega \to \omega^{<\omega}$ such that
(1) $|F(n)| \le n$;
(2) every interval $[f(n), f(n+1))$ captures an m with $g(m) \in F(m)$.

Since $g \le_h x$ and $\omega_1^x = \omega_1^{\mathrm{CK}}$, there is a ranked formula φ such that for every n, $g(n) = m$ if and only if $\mathfrak{A}(\omega_1^{\mathrm{CK}}, x) \models \varphi(n, m)$. Hence there is a condition $S \Vdash (\forall n)(\exists! m)\varphi(n, m)$. Let $T \subseteq S$. As in the proof of Lemma 13.3.16, one can build a hyperarithmetic sequence $\{P_\sigma\}_{\sigma \in 2^{<\omega}}$ of conditions such that

$$P_{\sigma^\frown i} \Vdash \varphi(|\sigma|, m_{\sigma^\frown i}) \text{ for } i \le 1$$

where $\sigma \in P_\sigma$ and $\log|\sigma| - 1$ is a fully splitting level of P_σ. Let Q be as defined in the proof of Lemma 13.3.16. Let f be the Δ_1^1-function such that $f(0) = 0$, and $f(n+1)$ is the least number $k > f(n)$ so that m_σ is defined for some σ with $f(n) < |\sigma| < k$. Let $F(n) = \{0\} \cup \{m_\sigma \mid |\sigma| = n\}$, and note that F is a Δ_1^1-function. Then

$$Q \Vdash (\forall n)(|F(n)| \le n) \wedge (\forall n)(\exists m \in [f(n), f(n+1)))(\exists i \in F(m))(\varphi(m, i)).$$

This implies that

$$Q \Vdash (\exists F)(\exists f)[(\forall n)(|F(n)| \le n) \wedge (\forall n)(\exists m \in [f(n), f(n+1)))(\exists i \in F(m))(\varphi(m, i))].$$

Since T is an arbitrary condition stronger than S, one has

$$S \Vdash (\exists F)(\exists f)[(\forall n)(|F(n)| \le n) \wedge (\forall n)(\exists m \in [f(n), f(n+1))(\exists i \in F(m))(\varphi(m, i))].$$

Since $x \in S$,

$$\mathfrak{A}(\omega_1^{\mathrm{CK}}, x) \models (\exists F)(\exists f)[(\forall n)(|F(n)| \le n) \wedge (\forall n)(\exists m \in [f(n), f(n+1))(\exists i \in F(m))(\varphi(m, i))].$$

Thus x is Δ_1^1-dominated and Δ_1^1-semitraceable.

For (iii), suppose $f : \omega \to \omega$ is a Δ^1_1-function such that for every n, there is an $m \in [2^n, 2^{n+1})$ with $f(m) = x(m)$. Then there is a ranked formula φ such that $f(n) = m \Leftrightarrow \mathfrak{A}(\omega_1^{\mathrm{CK}}, x) \models \varphi(\bar{n}, \bar{m})$. Moreover, $\mathfrak{A}(\omega_1^{\mathrm{CK}}, x) \models (\forall n)(\exists m \in [2^n, 2^{n+1}))(\varphi(m, x(m)))$. Hence there is a condition T which forces $(\forall n)(\exists m \in [2^n, 2^{n+1}))(\varphi(m, \dot{x}(m)))$ and $x \in T$. Let n be such that T is fully splitting at level n and let $\sigma \in 2^{2^n-1}$ be a string in T. Let ν be a string such that $\nu(m) = 1-f(m+2^n-1)$. Define $S = \{\sigma^\frown\nu^\frown\tau \mid \sigma^\frown\nu^\frown\tau \in T\} \subseteq T$. Then $S \Vdash (\forall m \in [2^n, 2^{n+1}))(\neg\varphi(m, x(m)))$. But S is stronger than T, which is a contradiction. By Corollary 13.3.12, x is not Δ^1_1-traceable. □

We may now separate Δ^1_1-traceability from the conjunction of Δ^1_1-semitraceability and Δ^1_1-dominability.

Theorem 13.3.18 (Kjos-Hanssen, Nies, Stephan and Yu [68]). *There are $2^{\aleph_0}$ many Δ^1_1-dominated and Δ^1_1-semitraceable reals which are not Δ^1_1-traceable.*

Proof. This is immediate from Lemma 13.3.17. Note that there are $2^{\aleph_0}$ many generic reals. □

By exactly the same argument as in the proof of Theorem 5.3.12, we have the following.

Proposition 13.3.19. *There exists an uncountable Σ^1_1-set which contains only Δ^1_1-dominated and Δ^1_1-semitraceable reals.*

13.3.4 Exercises

Exercise 13.3.1. *Prove that no Cohen generic real is Δ^1_1-dominated.*

Exercise 13.3.2. *Prove that every Cohen generic real is Δ^1_1-semitraceable.*

Exercise 13.3.3. *Prove Proposition 13.3.19.*

14 The theory of higher randomness

Higher randomness theory studies algorithmic complexity problems of randomness using techniques and ideas from hyperarithmetic theory and set theory. Since second order arithmetic provides a broader, though sometimes coarser, view of the reals, the picture one gets of randomness is quite different from that for first order theory.

14.1 Higher Kurtz randomness

14.1.1 Δ_1^1 and Π_1^1-Kurtz randomness

Definition 14.1.1. Let $x \in 2^\omega$. Given a class Γ of sets of reals, x is *Γ-Kurtz random* if it is not a member of any closed nullset in Γ.

We focus on Δ_1^1, Σ_1^1 and Π_1^1-Kurtz randomness. Recall (§ 2.7.1) that BC denotes the set of Borel codes and $\{A_z \mid z \in \mathsf{BC}\}$ is the set of reals generated by the codes.

Proposition 14.1.2.
(i) *x is Δ_1^1-Kurtz random if and only if x does not belong to any $\Pi_1(\mathsf{HYP})$-nullset.*
(ii) *The set of Δ_1^1-Kurtz random reals is Σ_1^1 but not Δ_1^1.*

Proof. Note that (i) follows immediate from Theorem 2.7.4. We give a self-contained proof here.

By Theorem 2.7.2, every $\Pi_1(\mathsf{HYP})$-set is a Δ_1^1-set. Hence no Δ_1^1-Kurtz random real belongs to a $\Pi_1(\mathsf{HYP})$-nullset. Conversely, let P be a Δ_1^1-nullset. Then $U = \{\sigma \mid (\forall z)(z \in P \to z \nsucc \sigma)\}$ is Π_1^1. Since P is closed and the set of hyperarithmetic reals is dense, we have

$$\sigma \in U \Leftrightarrow (\forall z \in \mathsf{HYP})(z \succ \sigma \to z \notin P).$$

By Corollary 2.2.12, U is also Σ_1^1 and hence Δ_1^1. Thus $P \in \Pi_1(\mathsf{HYP})$. It follows that x does not belong to a $\Pi_1(\mathsf{HYP})$-nullset, and hence $x \notin P$. In other words, x is Δ_1^1-Kurtz random.

To prove (ii), observe that the set

$$C = \{(\langle n, e\rangle, y) \mid n \in \mathcal{O} \wedge \Phi_e^{H_n} \in \Pi_1 \cap \mathsf{HYP} \wedge y \in A_{\Phi_e^{H_n}}\}$$

is Π_1^1, where A_z is as defined in § 2.7.1. By Proposition 13.1.6, the set

$$D = \{(\langle n, e\rangle, y) \mid \mu(\{y \mid (\langle n, e\rangle, y) \in C\}) = 0\}$$

is Π_1^1. Hence the set

$$E = \{y \mid (\exists m)((m, y) \in D)\}$$

is a Π_1^1-nullset. Moreover by (i),

$$(\forall x)(x \in E \leftrightarrow x \text{ is not } \Delta_1^1\text{-Kurtz random}).$$

Hence the set of Δ^1_1-Kurtz random reals is Σ^1_1. Now $\mu(E) = 1$, and since no hyperarithmetic real is Δ^1_1-Kurtz random, by Theorem 13.1.13 one concludes that E is not Δ^1_1. □

Theorem 14.1.3 (Kjos-Hanssen, Nies, Stephan and Yu [68]). *Π^1_1-Kurtz randomness $\subset$ Σ^1_1-Kurtz randomness $=$ Δ^1_1-Kurtz randomness.*

Proof. It is obvious that Π^1_1-Kurtz randomness $\subseteq$ Σ^1_1-Kurtz randomness $\subseteq$ Δ^1_1-Kurtz randomness. We show that the first inclusion is proper. Let

$$R(\langle e, n\rangle, z) \Leftrightarrow n \in \mathcal{O} \wedge \Phi_e^{H_n} \in \Sigma_1 \cap \mathsf{HYP} \wedge z \text{ is finite } \wedge z \subseteq \Phi_e^{H_n} \wedge (\mu(A_z) > 1 - 2^{-\langle e,n\rangle - 2}).$$

By Proposition 13.1.6, R is Π^1_1 and by Theorem 4.2.1, there is a Π^1_1-function $f : \omega \to 2^{<\omega}$ uniformizing R. Let

$$C = \{y \mid (\forall m)(f(m)\downarrow \to y \in A_{f(m)})\}.$$

Then C is a closed Σ^1_1-set and $\mu(C) \geq 2^{-1}$. Note that if $z_1 \in \Sigma_1 \cap \mathsf{HYP}$ and $\mu(A_{z_1}) > p$ for some p, then there is a finite $z \subseteq z_1$ in $\Sigma_1 \cap \mathsf{HYP}$ such that $\mu(A_z) > p$. By Proposition 14.1.2, every real in C is Δ^1_1-Kurtz random. Since C is closed and Σ^1_1, it can be presented by a Σ^1_1-tree. By Theorem 11.1.2, there is an $x \in C$ such that $x \equiv_h \mathcal{O}$. Let e and $n \in \mathcal{O}^{\mathcal{O}}$ be such that $x = \Phi_e^{H_n^{\mathcal{O}}}$. Then

$$z \in \{x\} \Leftrightarrow (\exists y \leq_h x)(y \in \{\mathcal{O}\} \wedge x = \Phi_e^{H_n^y}).$$

In other words, x is a Π^1_1-singleton which is therefore not Π^1_1-Kurtz random.

We now show that every Δ^1_1-Kurtz random real is Σ^1_1-Kurtz random. Let A be a Π^1_1-open set of measure 1. Define $x = \{\sigma \in 2^{<\omega} \mid (\forall y)(y \succ \sigma \to y \in A)\}$. Then x is a Π^1_1-real coding A (i.e. $y \in A$ if and only if there is a $\sigma \in x$ such that $y \succ \sigma$). Hence there is a recursive function $f : 2^{<\omega} \to \omega$ such that $\sigma \in x$ if and only if $f(\sigma) \in \mathcal{O}$. Define a Π^1_1-relation $R \subseteq \omega \times \omega$ so that $(k, n) \in R$ if and only if $n \in \mathcal{O}$ and $\mu(\bigcup\{[\sigma] \mid (\exists m \in \mathcal{O}_n)(f(\sigma) = m)\}) > 1 - \frac{1}{k}$. Clearly R is a Π^1_1-relation which can be uniformized by a Π^1_1-function f. Since $\mu(A) = 1$, f is a total function. Thus the range of f is bounded by a notation $n \in \mathcal{O}$. Define $B = \{y \mid (\exists\sigma)(y \succ \sigma \wedge f(\sigma) \in \mathcal{O}_n)\}$. Then $B \subseteq A$ is a Δ^1_1-open set with measure 1. It follows that every Π^1_1-open co-nullset has a Δ^1_1-open co-null subset. Hence Σ^1_1-Kurtz randomness $=$ Δ^1_1-Kurtz randomness. □

The following result clarifies the connection between Δ^1_1 and Π^1_1-Kurtz randomness.

Proposition 14.1.4. *If $\omega_1^x = \omega_1^{\mathrm{CK}}$, then x is Π^1_1-Kurtz random if and only if x is Δ^1_1-Kurtz random.*

Proof. Suppose that $\omega_1^x = \omega_1^{\mathrm{CK}}$ and x is Δ^1_1-Kurtz random. If A is a Π^1_1-closed nullset such that $x \in A$, then by the Spector–Gandy Theorem 3.7.3, there is an arithmetic formula $\varphi(z, y)$ such that the formula $(\exists z \in \mathsf{HYP})\varphi(z, y)$ defines A. Since $\omega_1^x = \omega_1^{\mathrm{CK}}$, there is a recursive ordinal β such that

$$x \in B = \{y \mid \mathfrak{A}(\omega_1^{\mathrm{CK}}, y) \models (\exists z^\beta)\varphi(z^\beta, y)\} \subseteq A\}.$$

Define $T = \{\sigma \in 2^{<\omega} \mid (\exists y \in B)(y \succ \sigma)\}$. Clearly $B \subseteq [T]$. Since B is Δ^1_1, $[T]$ is Σ^1_1. Since A is closed, we have $[T] \subseteq A$. Also A is null implies that $[T]$ is null as well. By Lemma 13.1.10 and a routine application of Π^1_1-uniformization, there is a total Δ^1_1-function $f : \omega \to \mathsf{HYP}$ such that for any k, $f(k) \in \Sigma_1 \cap \mathsf{HYP}$, $A_{f(k)} \subseteq 2^{\omega} \setminus [T]$ and $\mu(A_{f(k)}) > 2^{-k}$. Let

$$C = \{z \mid (\forall k)(z \notin A_{f(k)})\}.$$

Then C is a Δ^1_1-closed nullset. However, $x \in C$ and this is a contradiction. □

14.1.2 Base of a cone of Kurtz random reals

From the proof of Theorem 14.1.3, one sees that every hyperdegree above that of $\mathcal{O}$ contains a Δ^1_1-Kurtz random real. This fails for Π^1_1-Kurtz randomness. We say that a hyperdegree **a** is a *base of a cone for Γ-Kurtz randomness* if for every hyperdegree $\mathbf{b} \geq \mathbf{a}$, **b** contains a Γ-Kurtz random real.

The hyperdegree of $\mathcal{O}$ is a base of a cone for Δ^1_1-Kurtz randomness as proved in Theorem 14.1.3. Not every nonzero hyperdegree is a base of a cone for Δ^1_1-Kurtz randomness (see Lemma 14.6.1). Does there exist a base of a cone for Π^1_1-Kurtz randomness? If so, such a base, say **b**, is not hyperarithmetically reducible to a Π^1_1-singleton. This means that a hyperdegree which serves as a base must be fairly complicated.

Lemma 14.1.5. *For any reals x and $z \geq_T x'$, there is an x-Kurtz-random real $y \equiv_T z$.*

Proof. Let $z \geq_T x'$ and fix an enumeration of x-r.e. open sets $\{U^x_n\}_{n<\omega}$. We construct an increasing sequence $\{\sigma_s\}_{s<\omega}$ of strings as follows.

Let $\sigma_0 = \emptyset$.

Stage $s+1$. Let $l_0 = 0$, $l_1 = |\sigma_s|$, and $l_{n+1} = 2^{l_n}$ for $n > 1$. For each such n, let

$$A_n = \{\sigma \in 2^{l_n - 1} \mid (\exists m < n)(\forall i)(\forall j)(l_m \leq i, j < l_{m+1} \implies \sigma(i) = \sigma(j))\}.$$

Then

$$|A_n| \leq 2 \cdot 2^{l_{n-1}}.$$

In other words,

$$\mu(B_n = \bigcup\{[\sigma] \mid \sigma \succeq \sigma_s \wedge \sigma \notin A_n\}) \geq 2^{-l_1} \cdot (1 - 2^{l_n + 1 - l_{n+1}}).$$

Case 1. There is an $m > l_1 + 1$ such that $|\{\sigma \succeq \sigma_s \mid \sigma \in 2^m \wedge [\sigma] \subseteq U^x_s\}| > 2^{m - l_1 - 1}$. Let $n = m+1$. Then $l_{n+1} - 1 - l_n > 2$ and $l_n > m$. Then there is a $\sigma \in 2^{l_n - 1} \setminus A_n$ and a $\tau \preceq \sigma$ such that $[\tau] \subseteq U^x_s$ and $\tau \in 2^m$.

Let $\sigma_{s+1} = \sigma^\frown(z(s))^{l_n - 1}$.

Case 2. Otherwise. Let $\sigma_{s+1} = \sigma_s^\frown(z(s))^{l_1 - 1}$.

This completes the construction at stage $s+1$. Let $y = \bigcup_s \sigma_s$. Obviously the construction is recursive in z and so $y \leq_T z$. Moreover, if U^x_s is of measure 1, then Case 1 holds at stage $s+1$. Hence y is x-Kurtz random.

Let $l_0 = 0$ and let $l_n = 2^{l_n}$ for $n > 0$. To compute $z(n)$ from y, we y-recursively find the n-th l_m such that for all i, j with $l_m \le i < j < l_{m+1}, y(i) = y(j)$. Then $z(n) = y(l_m)$. □

We call $A \subseteq \omega \times 2^\omega$ a universal Π^1_1-closed set if (i) it is Π^1_1, (ii) for every n, $A_n = \{x \mid (n, x) \in A\}$ is a Π^1_1-closed set, and (iii) every Π^1_1-closed set is A_n for some n. By Theorems 13.1.3 and 13.1.4, the real $x_0 = \{n \mid \mu(A_n) = 0\}$ is Σ^1_1. Let

$$c = \{(n, \sigma) \mid n \in x_0 \wedge (\exists x)((n, x) \in A \wedge \sigma \prec x)\} \subseteq \omega \times 2^{<\omega}.$$

Then c may be viewed as a Σ^1_2-real. Since every Π^1_1-null closed set is $\Pi^0_1(c)$, every c-Kurtz-random real is Π^1_1-Kurtz random.

Theorem 14.1.6. *The hyperdegree of c is the base of a cone for Π^1_1-Kurtz randomness.*

Proof. Lemma 14.1.5 implies that the Turing degree of any $y \ge_T c'$ contains a member which is c-Kurtz random and therefore Π^1_1-Kurtz random. Thus c is a base for Π^1_1-Kurtz randomness. □

We analyze the complexity of c. Since every Π^1_1-singleton is recursive in c, $c \notin L_{\delta^1_2}$. Moreover, the set $\{x \mid x \text{ is a } \Pi^1_1\text{-singleton}\}$ is $\Delta^1_1(c)$. In other words, $\{x \mid x \text{ is a } \Pi^1_1\text{-singleton}\} \in L_{\omega_1^c}[c]$. Hence one has $\omega_1^c > \delta^1_2$. Applying the same argument as that in Proposition 4.4.4, the reals in $L_{\delta^1_2}$ are precisely those that are Δ^1_2 and so c is not Δ^1_2. Moreover, since c is Σ^1_2, it is Σ_1 definable over $L_{\delta^1_2}$ and hence $c \in L_{\delta^1_2+1}$. Indeed, all Σ^1_2-reals belong to $L_{\delta^1_2+1}$. Thus:

c has the largest hyperdegree among all Σ^1_2 reals.

14.1.3 Exercises

Exercise 14.1.1. *Prove that neither the set of Δ^1_1 nor Π^1_1-Kurtz random reals are $\mathbf{\Sigma}^0_2$.*

Exercise 14.1.2. *Prove that there is a Π^1_1-Kurtz random real which is not Schnorr random.*

Exercise 14.1.3. *Prove that the set of Π^1_1-Kurtz random reals is not Π^1_1.*

Exercise 14.1.4. *Prove that $\langle L_{\delta^1_2}, \in\rangle \prec_{\Sigma_1} \langle L, \in\rangle$.*

14.2 Δ^1_1 and Π^1_1-Martin-Löf randomness

14.2.1 Δ^1_1-randomness

Definition 14.2.1. Let Γ be a definable class of sets of reals. We say that x is *Γ-random* if it does not belong to any Γ-nullset.

In this section, we study the key properties of Δ_1^1-randomness.

Proposition 14.2.2. *Suppose that A is a Δ_1^1-nullset. There is a Δ_1^1-set $V \subseteq \omega \times 2^{<\omega}$ such that $(\forall n)(\mu(V_n) = 2^{-n})$ and $A \subseteq \bigcap_{n<\omega} V_n$, where $V_n = \{\sigma \mid (n, \sigma) \in V)\}$ is a Δ_1^1-open set.*

Proof. As before, $\{A_z\}$ is the collection of sets of reals generated by Borel codes z (§ 2.7.1). Since A is a Δ_1^1-nullset, by Lemma 13.1.10 there is a Δ_1^1-function $f : \omega \to \Sigma_1 \cap \mathrm{HYP}$ such that for every $n < \omega$,

- $\mu(A_{f(n)}) < 2^{-n}$, and
- $A \subseteq A_{f(n)}$

Then $A_{f(n)} = [U_n]$ for some Δ_1^1-set $U_n \subseteq 2^{<\omega}$. It now suffices to show that $\{U_n\}_{n<\omega}$ may be expanded to a Δ_1^1-collection $\{V_n\}_{n<\omega}$ such that $U_n \subseteq V_n$ and $\mu(V_n) = 2^{-n}$. Since the set $U = \{(n, \sigma) \mid \sigma \in U_n\}$ is Δ_1^1, there is an $m \in \mathcal{O}$ such that $U \leq_T H_m$. Note that the set $\{(x, e, p) \in 2^\omega \times \omega \times \mathbb{Q} \mid \mu(\{[\sigma] \mid (\exists i)(\Phi_e^x(i) = \sigma)\}) > p\}$ is r.e. Then by a standard technique in classical randomness theory relativized to H_{2^m}, there is a $V \subseteq \omega \times 2^{<\omega}$ recursive in H_{2^m} such that for every n, $\mu(V_n) = 2^{-n}$ and $U_n \subseteq V_n$. Then $A \subseteq \bigcap_{n<\omega} V_n$. □

By Proposition 14.2.2, Δ_1^1-randomness may be viewed as a higher analog of Schnorr randomness.

Proposition 14.2.3 (Martin-Löf [97]). *The set of Δ_1^1-random reals is Σ_1^1 but not Δ_1^1.*

Proof. Define a Σ_1^1-set

$$A = \{x \mid (\forall e)(\forall n)((n \in \mathcal{O} \wedge (\forall m)\mu(\{[\sigma] \mid (\exists i)(\Phi_e^{H_n}(\langle m, i\rangle) = \sigma)\}) \leq 2^{-m}) \to (\exists m)(\forall i)(x \not\succ \Phi_e^{H_n}(\langle m, i\rangle)))\}.$$

By Proposition 14.2.2, A is precisely the set of Δ_1^1-random reals. Since A does not contain a hyperarithmetic member, by Theorem 13.1.13, it is not Δ_1^1. □

14.2.2 Π_1^1-Martin-Löf randomness

A sequence of open sets $\{U_n\}_{n<\omega}$ is a Martin-Löf test (ML test) if $\mu(U_n) < 2^{-n}$ for all n. Given a class Γ of sets of reals (e.g. Π_1^1), $\{U_n\}_{n<\omega}$ is a Γ-ML test if $\{(n, \sigma) | \sigma \in 2^{<\omega} \wedge [\sigma] \in U_n\} \in \Gamma$.

Definition 14.2.4. Let Γ be a class of sets of reals. Then x is Γ-ML random if $x \notin \bigcap_{n<\omega} U_n$ for any Γ ML test $\{U_n\}_{n<\omega}$.

Proposition 14.2.5 (Hjorth and Nies [51]). *There is a universal Π_1^1-ML test $\{U_n\}_{n<\omega}$, i.e. for every Π_1^1-ML test $\{V_n\}_{n<\omega}$, $\bigcap_{n<\omega} V_n \subseteq \bigcap_{n<\omega} U_n$.*

Proof. Let $\hat{U} \subseteq \omega \times 2^{<\omega}$ be a universal Π^1_1-set, i.e. $\hat{U}$ is Π^1_1 and every Π^1_1-set $\hat{V} \subseteq 2^{<\omega}$ is $\hat{U}_n = \{\sigma \mid (n, \sigma) \in \hat{U}\}$ for some n. Then there is a recursive function $f : \omega \times 2^{<\omega} \to \omega$ such that $(n, \sigma) \in \hat{U} \Leftrightarrow f(n, \sigma) \in \mathcal{O}$. Let

$$\hat{V} = \{(n, \sigma) \mid (n, \sigma) \in \hat{U} \wedge \sum_{(\exists k \in \mathcal{O}_{f(n,\sigma)})(f(n,\tau)=k)} 2^{-|\tau|} < 2^{-n}\}.$$

In other words, (n, σ) is enumerated into $\hat{V}$ if the measure of $\hat{U}_n$ up to stage $f(n, \sigma)$ is less than 2^{-n}. Let

$$U = \{(n, \sigma) \mid (\exists m)(m \geq n + 1 \wedge (m, \sigma) \in \hat{V})\}.$$

Then U is Π^1_1 as well. Setting $U_n = \{\sigma \mid (n, \sigma) \in U\}$, we have

$$\mu(U_n) < \sum_{m \geq n+1} 2^{-n} = 2^{-n}.$$

Hence $\{U_n\}_{n<\omega}$ is a Π^1_1-ML test. By the universality of $\hat{U}$, $\{U_n\}_{n<\omega}$ is a universal Π^1_1-ML test □

Corollary 14.2.6 (Hjorth and Nies [51]).
(i) *There is a Σ^1_1-closed set of positive measure of Π^1_1-ML random reals.*
(ii) *For any $z \geq_h \mathcal{O}$, there is a Π^1_1-ML random $x \equiv_h z$.*

Proof. (i) follows immediately from Proposition 14.2.5, while (ii) is a consequence of (i) and Theorem 11.1.2. □

The following results says that every Σ^1_1-closed set of positive measure essentially contains all the Π^1_1-ML random reals.

Proposition 14.2.7. *Suppose that $T \subseteq 2^{<\omega}$ is a Σ^1_1-tree with $\mu([T]) > 0$. Then for any Π^1_1-ML random x, there is a $z \in [T]$ such that $z =^* x$, i.e. $(\exists m)(\forall n \geq m)(x(n) = z(n))$.*

Proof. Let $T \subseteq 2^{<\omega}$ be as given. For any $n \geq 0$, let

$$U^n = \{\sigma_0{}^\frown\sigma_1{}^\frown \dots {}^\frown\sigma_n \mid (\forall i \leq n)(\sigma_i \notin T)\}.$$

Then $\{U^n\}_{n<\omega}$ is uniformly Π^1_1. Moreover, $\mu(U^0) = \mu(2^\omega \setminus [T]) = 1 - \mu([T])$. For any n, $\mu(U^{n+1}]) = \mu(U^n) \cdot (1 - \mu([T])) = (1 - \mu([T]))^{n+2}$. Let $V_n = \bigcup_{m \geq n} U^m$. Then $\mu(V_n) \leq \sum_{m \geq n}(1 - \mu([T]))^{m+1}$ and hence $\{V_n\}_{n<\omega}$ can be viewed as a Π^1_1-ML test. It follows that there is a least n such that $x \notin V_n$. If $n = 0$, then $x \in [T]$. Otherwise there is a $\sigma_0{}^\frown\sigma_1{}^\frown \dots {}^\frown\sigma_{n-1} \prec x$ so that $x = \sigma_0{}^\frown\sigma_1{}^\frown \dots {}^\frown\sigma_{n-1}{}^\frown z$ for some $z \in [T]$. Hence $x =^* z$. □

14.2.3 Separating Π_1^1-Martin-Löf randomness from Δ_1^1-randomness

Define a notion of forcing $\mathbb{D} = \langle \boldsymbol{D}, \leq \rangle$ as follows:
(i) $P \in \boldsymbol{D}$ if and only if $P \subseteq 2^\omega$ is Δ_1^1-closed and of positive measure;
(ii) for any $P, Q \in \boldsymbol{D}$, $P \leq Q$ if and only if $P \subseteq Q$.

Suppose $U \subseteq \omega \times 2^{<\omega}$ is Δ_1^1 and $U_n = \{\sigma \mid (n, \sigma) \in U\}$. Assume $\lim_{n\to\infty} \mu(U_n) = 0$ and let

$$\mathscr{D}_U = \{P \mid P \in \boldsymbol{D} \wedge P \cap (\bigcap_{n<\omega} U_n) = \emptyset\}.$$

Lemma 14.2.8. *$\mathscr{D}_U$ is dense.*

Proof. Let $P \in \boldsymbol{D}$. There is an n such that $\mu(U_n) < \mu(P)/2$. Now the complement $P_0 = 2^\omega \setminus U_n$ has measure greater than or equal to $1 - (\mu(P)/2)$. Then $P_0 \cap P$ is a Δ_1^1-closed set and has measure greater than or equal to $\mu(P)/2$. Thus $P \cap P_0 \in \boldsymbol{D}$. □

For any $\underset{\sim}{\Pi}{}_1^0$-set G, let $\mathscr{D}_G = \{P \mid P \in \boldsymbol{D} \wedge P \cap G = \emptyset\}$.

Lemma 14.2.9. *If G is a $\underset{\sim}{\Pi}{}_1^0$-set consisting only of Δ_1^1-random reals, then $\mathscr{D}_G$ is dense in $\mathbb{D}$.*

Proof. Let $P \in \boldsymbol{D}$. Then there is a hyperarithmetic x such that P is $\Pi_1^0(x)$. Without loss of generality, we may assume that for any σ, if $[\sigma] \cap P \neq \emptyset$, then $\mu([\sigma] \cap P) > 0$ (since we may assume that P contains only 1-x-random reals). By Theorem 13.1.13, there is a hyperarithmetic z in P. Hence $z \notin G$. Since G is closed, there is an n such that $[z \upharpoonright n] \cap G = \emptyset$. Let $Q = [z \upharpoonright n] \cap P$. Then $Q \in \mathscr{D}_G$. □

Theorem 14.2.10 (Chong, Nies and Yu [12]). *Π_1^1-ML randomness $\subset \Delta_1^1$-randomness.*

Proof. By Proposition 14.2.2, Π_1^1-ML randomness $\subseteq \Delta_1^1$-randomness. By Proposition 14.2.5, let $\{U_n\}_{n<\omega}$ be a universal Π_1^1-ML test. For $n < \omega$, let

$$\mathscr{D}_{U_n} = \{P \mid P \in \boldsymbol{D} \wedge P \subseteq U_n\}.$$

By Lemma 14.2.9, $\mathscr{D}_{U_n}$ is dense. By Lemma 14.2.8, there is a Δ_1^1-random x such that for any n, there is a $P \in \mathscr{D}_{U_n}$ with $x \in P$. Hence $x \in \bigcap_{n<\omega} U_n$, i.e. x is not Π_1^1-ML random. □

We remark that $\mathbb{D}$ is an analog of random forcing $\mathbb{R}$ discussed in § 6.2.2. However, $\mathbb{R}$ is c.c.c. and so preserves all cardinals while $\mathbb{D}$ is not c.c.c. in $L_{\omega_1^{\mathrm{CK}}}$. By the proof of Theorem 14.2.10 and Corollary 14.3.2 below, any $\mathbb{D}$-generic x collapses ω_1^{CK} and hence does not preserve admissible ordinals.

14.2.4 Strong Π^1_1-Martin-Löf randomness

A sequence of open sets $\{U_n\}_{n<\omega}$ is a *generalized Martin-Löf test* (ML test) if $\lim_{n\to\infty}\mu(U_n) = 0$. Given a class Γ of sets of reals (e.g. Π^1_1), $\{U_n\}_{n<\omega}$ is a generalized Γ-ML test if $\{(n,\sigma) \mid \sigma \in 2^{<\omega} \wedge \sigma \in U_n\} \in \Gamma$.

Definition 14.2.11. Given a class Γ of sets of reals, x is strongly Γ-ML random if $x \notin \bigcap_{n<\omega} U_n$ for any generalized Γ-ML test $\{U_n\}_{n<\omega}$.

Obviously every strongly Π^1_1-ML random real is Π^1_1-ML random.

Lemma 14.2.12. *If x is the leftmost path of a Σ^1_1-closed set of reals, then x is not strongly Π^1_1-ML random.*

Proof. Let P be a Σ^1_1-closed set. Then there is a Σ^1_1-tree $T \subseteq 2^{<\omega}$ such that $P = [T]$. By the Gandy–Spector Theorem (3.7.1), there is a Σ_0-formula $\varphi(\sigma,\beta)$ such that

$$\sigma \in T \Leftrightarrow \langle L_{\omega_1^{\mathrm{CK}}}, \in\rangle \models (\forall\beta)\varphi(\sigma,\beta).$$

For $\beta < \omega_1^{\mathrm{CK}}$, let

$$T[\beta] = \{\sigma \mid \langle L_{\omega_1^{\mathrm{CK}}}, \in\rangle \models (\forall\gamma \le \beta)\varphi(\sigma,\gamma)\}.$$

For $n < \omega$ and $\beta < \omega_1^{\mathrm{CK}}$, let

$$U_{n,\beta} = \{\sigma \mid (\exists z)(z \text{ is the leftmost path in } T[\beta] \cap [\sigma] \wedge \sigma = z \restriction n+1)\}.$$

Define

$$U_{n,<\beta} = \bigcup_{\gamma<\beta} U_{n,\gamma}$$

and

$$U_n = \bigcup_{\beta<\omega_1^{\mathrm{CK}}} U_{n,\beta}.$$

The following facts are straightforward to verify: For $n < \omega$ and $\beta < \omega_1^{\mathrm{CK}}$,

(1) $U_{n+1,\beta} \subseteq U_{n,\beta}$;
(2) $\mu(U_{n,\beta}) < 2^{-n}$;
(3) $x \in \bigcap_{n<\omega} U_n$;
(4) for any z, if $z \in U_{n,<\beta} \setminus U_{n,\beta}$, then $z \notin U_{n,\gamma}$ for any $\gamma \ge \beta$.

Now suppose that $\mu(\bigcap_{n<\omega} U_n) > 0$. Then there is a σ_0 such that

$$\mu(\bigcap_{n<\omega} U_n \cap [\sigma_0]) > \frac{3}{4}\cdot 2^{-|\sigma_0|}.$$

Let $n_0 = |\sigma_0| + 2$. Then there is a $\beta_0 < \omega_1^{\mathrm{CK}}$ such that

$$\mu(U_{n_0,\le\beta_0} \cap [\sigma_0]) > \frac{3}{4}\cdot 2^{-|\sigma_0|}.$$

By (2),

$$\mu(U_{n_0,<\beta_0} \cap [\sigma_0]) > \frac{1}{2} \cdot 2^{-|\sigma_0|}.$$

By (1) and (4),

$$\mu(\bigcap_{n>n_0} U_{n,<\beta_0} \cap [\sigma_0]) > \frac{1}{4} \cdot 2^{-|\sigma_0|}.$$

Hence there is a $\sigma_1 \succ \sigma_0$ such that

$$\mu(\bigcap_{n>n_0} U_{n,<\alpha_0} \cap [\sigma_1]) > \frac{3}{4} \cdot 2^{-|\sigma_1|}.$$

Let $n_1 = |\sigma_1| + 2$. There is a $\beta_1 < \beta_0$ such that

$$\mu(U_{n_1,\leq\beta_1} \cap [\sigma_1]) > \frac{3}{4} \cdot 2^{-|\sigma_1|}.$$

Repeating the process, one obtains an infinite descending sequence $\beta_0 > \beta_1 > \cdots$ of ordinals which is a contradiction. It follows that $\{U_n\}_{n<\omega}$ is a generalized Π^1_1-ML test and $x \in \bigcap_{n<\omega} U_n$. □

Theorem 14.2.13 (Chong, Nies and Yu). *There is a real which is Π^1_1-ML random but not strongly Π^1_1-ML random.*

Proof. By Corollary 14.2.6, there is Σ^1_1-closed set P of positive measure consisting only of Π^1_1-ML random reals. By Lemma 14.2.12, there is an $x \in P$ which is not strongly Π^1_1-ML random. □

14.2.5 Exercises

Exercise 14.2.1 (Sacks [123]). *Prove that every Δ^1_1-random real is Σ^1_1-random.*

Exercise 14.2.2. *Prove that for any Δ^1_1-random x and recursive ordinal β, $x^{(\beta)} \equiv_T x \oplus \emptyset^{(\beta)}$.*

Exercise 14.2.3 (Yu [171]). *Prove that the set of Δ^1_1-random reals is not $\utilde{\Sigma}^0_3$.*

Exercise 14.2.4. *Prove that for any Δ^1_1-random x and recursive function $f : 2^\omega \to 2^\omega$, if f is almost everywhere differentiable, then f is differentiable at x.*

Exercise 14.2.5 (Bienvenu, Greenberg and Monin). *Prove that for any x which is the leftmost path of a Σ^1_1-closed set of reals, there is a generalized Π^1_1-Martin-Löf test $\{U_n\}_{n<\omega}$ such that $x \in \bigcap_{n<\omega} U_n$ and $\bigcap_{n<\omega} U_n$ is countable.*

Exercise 14.2.6 (Chong and Yu). *Prove that for any $x \geq_h \mathcal{O}$, there is a $z \equiv_h x$ which is Π^1_1-Martin-Löf random but not strongly Π^1_1-Martin-Löf random.*

14.3 Π^1_1-randomness

14.3.1 The largest Π^1_1-nullset

The notion of Π^1_1-randomness was introduced in Hjorth and Nies [51] and is unique to the study of higher randomness. There is no analog of this notion in the subject of classical randomness theory.

Theorem 14.3.1 (Kechris [64]; Stern [161]; Hjorth and Nies [51]). *There is a largest Π^1_1-nullset.*

Proof. Define

$$\mathcal{P} = \{(\ulcorner\varphi\urcorner, x) \mid \varphi \text{ is a ranked formula} \wedge \mathfrak{A}(\omega_1^{\mathrm{CK}}, x) \models \varphi(\dot{x}) \wedge \\ \mu(\{x \mid \mathfrak{A}(\omega_1^{\mathrm{CK}}, x) \models \neg\varphi(\dot{x})\}) \geq 1)\}$$

and

$$\mathcal{Q}_{\ulcorner\varphi\urcorner} = \{x | (\ulcorner\varphi\urcorner, x) \in \mathcal{P}\}.$$

Define

$$\mathcal{Q} = \bigcup_{n<\omega} \mathcal{Q}_n \cup \{x \mid \omega_1^x > \omega_1^{\mathrm{CK}}\}.$$

We show that $\mathcal{Q}$ is the largest Π^1_1-nullset. By Theorem 13.1.1, $\mathcal{Q}$ is Π^1_1. Moreover, $\mu(\mathcal{Q}_n) = 0$ for all $n < \omega$. By Theorem 13.1.4, $\mu(\mathcal{Q}) = 0$.

If A is a Π^1_1-nullset, then, by Corollary 3.7.4, there is an arithmetic formula $\varphi(\dot{x}, z)$ such that if $\omega_1^x = \omega_1^{\mathrm{CK}}$, then $x \in A \Leftrightarrow \mathfrak{A}(\omega_1^{\mathrm{CK}}, x) \models (\exists y)\varphi(x, y)$. Hence if $\omega_1^x = \omega_1^{\mathrm{CK}}$, then $x \in A \Leftrightarrow \mathfrak{A}(\omega_1^{\mathrm{CK}}, x) \models (\exists y^\beta)\varphi(x, y^\beta)$ for some $\beta < \omega_1^{\mathrm{CK}}$. Since A is null, we have

$$A \cap \{z \mid \omega_1^z = \omega_1^{\mathrm{CK}}\} \subseteq \bigcup_{\beta<\omega_1^{\mathrm{CK}}} \mathcal{Q}_{\ulcorner(\exists y^\beta)\varphi(\dot{x},y^\beta)\urcorner}.$$

Thus $A \subseteq \mathcal{Q}$. □

Corollary 14.3.2. *For any $x \in 2^\omega$, the following are equivalent:*
(i) *x is Π^1_1-random;*
(ii) *x is strongly Π^1_1-ML random and $\omega_1^x = \omega_1^{\mathrm{CK}}$;*
(iii) *x is Π^1_1-ML random and $\omega_1^x = \omega_1^{\mathrm{CK}}$;*
(iv) *x is Δ^1_1-random and $\omega_1^x = \omega_1^{\mathrm{CK}}$.*

Proof. Suppose that x is Π^1_1-random. Since $\{z \mid \omega_1^z > \omega_1^{\mathrm{CK}}\}$ is a Π^1_1-nullset, we have $\omega_1^x = \omega_1^{\mathrm{CK}}$ and hence (i) ⇒ (ii) ⇒ (iii).

By Theorem 14.2.10, (iii) implies (iv). We now show (iv) ⇒ (i). By the proof of Theorem 14.3.1, if $\omega_1^x = \omega_1^{\mathrm{CK}}$ and $x \in \mathcal{Q}$, then $x \in \mathcal{Q}_n$ for some n. Since $\mathcal{Q}_n$ is a Δ^1_1-nullset, if $\omega_1^x = \omega_1^{\mathrm{CK}}$ and x is Δ^1_1-random, then it is Π^1_1-random. □

The equivalence between (i) and (iv) was also obtained by Stern [161].

14.3.2 Domination and Π_1^1-randomness

Lemma 14.3.3. *If x is Π_1^1-random, then it is Δ_1^1-dominated.*

Proof. By the proof of Theorem 13.3.2, for each $e \in \omega$ and $n \in \mathcal{O}$, the set

$$A_{e,n} = \{x \mid (\exists f \in \mathsf{HYP})(\Phi_e^{\hat{H}_n^x} \text{ is total} \rightarrow \Phi_e^{\hat{H}_n^x} < f)\}$$

has measure 1. Note that $A_{e,n}$ is Π_1^1. By Corollary 13.1.11, there is a Δ_1^1-set $\hat{A}_{e,n} \subseteq A_{e,n}$ such that $\mu(A_{e,n}) = 1$. Hence if x is Π_1^1-random, then $x \in A_{e,n}$. By Corollary 14.3.2, $\omega_1^{\mathrm{CK}} = \omega_1^x$. Thus if $g \leq_h x$, then $g = \Phi_e^{H_n^x}$ for some e and $n \in \mathcal{O}$, proving the lemma. □

Proposition 14.3.4 (Kjos-Hanssen, Nies, Stephan and Yu [68]). *For any $x \in 2^\omega$, the following are equivalent:*
(i) *x is Π_1^1-random;*
(ii) *x is strongly Π_1^1-ML random and Δ_1^1-dominated;*
(iii) *x is Π_1^1-ML random and Δ_1^1-dominated;*
(iv) *x is Δ_1^1-random and Δ_1^1-dominated;*
(v) *x is Π_1^1-Kurtz random and Δ_1^1-dominated.*
(vi) *x is Δ_1^1-Kurtz random and Δ_1^1-dominated.*

Proof. By Corollary 14.3.2 and Lemma 14.3.3, we have (i) ⇒ (ii) ⇒ (iii) ⇒ (iv) ⇒ (v) ⇒ (vi).

To prove (vi) ⇒ (i), it is sufficient to prove (vi) ⇒ (iv) according to Corollary 14.3.2. Suppose x is Δ_1^1-dominated and Δ_1^1-Kurtz random. Let $\{U_n\}_{n<\omega}$ be a Δ_1^1-ML test. Then there is a recursive function $f : \omega \times 2^{<\omega} \rightarrow \omega$ such that for any pair (n, σ), $\sigma \in U_n$ if and only if $f(n, \sigma) \in \mathcal{O}$. Since x is Δ_1^1-dominated, $\omega_1^x = \omega_1^{\mathrm{CK}}$. Define a $\Pi_1^1(x)$ relation $R \subseteq \omega \times \omega$ such that

$$R(n, m) \Leftrightarrow (\exists \sigma)(m \in \mathcal{O} \wedge f(n, \sigma) \in \mathcal{O}_m \wedge \sigma \prec x),$$

where $\mathcal{O}_m = \{i \in \mathcal{O} \mid |i| < |m|\}$. By the Π_1^1-Uniformization Theorem (4.2.1), there is a partial function p uniformizing R. Since $x \in \bigcap_n U_n$, p is a total function and so $\Delta_1^1(x)$. Since $\omega_1^x = \omega_1^{\mathrm{CK}}$, there is an $m_0 \in \mathcal{O}$ such that $p(n) \in \mathcal{O}_{m_0}$ for every n. Define a Δ_1^1-Martin-Löf test $\{\hat{U}_n\}_{n<\omega}$ so that $\sigma \in \hat{U}_n$ if and only if $f(n, \sigma) \in \mathcal{O}_{m_0}$. Then $x \in \bigcap_n \hat{U}_n$. Let

$$\hat{f}(n) = \min\{l \mid (\exists \sigma \in 2^l)(\sigma \in \hat{U}_n \wedge x \in [\sigma])\}$$

be a $\Delta_1^1(x)$-function. Then there is a Δ_1^1-function f dominating $\hat{f}$. For $n < \omega$, define

$$V_n = \{\sigma \mid \sigma \in 2^{\leq f(n)} \wedge \sigma \in \hat{U}_n\}.$$

Then $P = \bigcap_{n<\omega} V_n$ is a Δ_1^1-closed set and $x \in P$. Hence x is not Δ_1^1-Kurtz random, a contradiction. □

14.3.3 Hyperdegrees of random reals

By Corollary 14.2.6, we know that the hyperdegrees of the set of Δ^1_1-random reals include the cone of hyperdegrees with base $\mathcal{O}$. This raises the natural question of whether the hyperdegrees of Δ^1_1-random reals are closed upwards.

Proposition 14.3.5. *If x is Δ^1_1-random and $\omega_1^x = \omega_1^{\mathrm{CK}}$, then there is a $y \geq_h x$ such that no $z \equiv_h y$ is Δ^1_1-random.*

Proof. Suppose that x is Δ^1_1-random and $\omega_1^x = \omega_1^{\mathrm{CK}}$. Let

$$F(x) = \{y \mid y \geq_T x \wedge (\exists f \leq_T y)(\forall g \leq_h x)(g < f)\}.$$

Then $F(x)$ is $\Sigma^1_1(x)$. Since $\mathcal{O}^x \in F(x)$, $F(x)$ is not empty. By Gandy's Basis Theorem relativized to x, there is a $y \in F(x)$ with $\omega_1^y = \omega_1^x = \omega_1^{\mathrm{CK}}$. By Corollary 14.3.2 and Proposition 14.3.4, no $z \equiv_h y$ is Δ^1_1-random. □

On the other hand, the hyperdegrees of Π^1_1-random reals are closed downwards.

Lemma 14.3.6. *If x is Π^1_1-random and $y \leq_h x$, then there is a recursive ordinal γ such that $y \leq_T x \oplus \emptyset^{(\gamma)}$.*

Proof. Since x is Π^1_1-random, we have $\omega_1^x = \omega_1^{\mathrm{CK}}$. Let $y \leq_h x$. Then there is a formula $\varphi(\dot{x}, \bar{n})$ with rank at most $\beta_0 < \omega_1^{\mathrm{CK}}$ such that

$$n \in y \Leftrightarrow \mathfrak{A}(\omega_1^{\mathrm{CK}}, x) \models \varphi(x, \bar{n}).$$

By Lemma 13.1.1, there is a recursive ordinal $\beta > \beta_0$ such that

$$A_{\beta_0} = \{\ulcorner\varphi\urcorner \mid \varphi \text{ has rank at most } \beta_0 \wedge \mu(\{z \mid \mathfrak{A}(\omega_1^{\mathrm{CK}}, z) \models \varphi\}) > 0\}$$

is recursive in $\emptyset^{(\beta)}$. By Lemma 13.1.10, there is a recursive ordinal $\beta_1 \geq \beta$ such that for any number i and formula ψ with rank β, there is a formula ψ' with rank at most β_1 such that $\{z \mid \mathfrak{A}(\omega_1^{\mathrm{CK}}, z) \models \psi'\}$ is a $\Pi^0_1(\emptyset^{(\beta_1)})$-subset of $\{z \mid \mathfrak{A}(\omega_1^{\mathrm{CK}}, z) \models \psi\}$ with difference in measure less than 2^{-i}. Note that the search procedure for β_1 from β_0 is effective in $L_{\omega_1^{\mathrm{CK}}}$. In this way one obtains a Δ_1-sequence of ordinals $\beta_0 < \beta_1 < \cdots$ in $L_{\omega_1^{\mathrm{CK}}}$. Let $\gamma = \bigcup_{i<\omega} \beta_i$. Then for any $\beta < \gamma$,

(i) the set

$$A_{\beta} = \{\ulcorner\varphi\urcorner \mid \varphi \text{ has rank at most } \beta \wedge \mu(\{z \mid \mathfrak{A}(\omega_1^{\mathrm{CK}}, z) \models \varphi\}) > 0\}$$

is recursive in $\emptyset^{(\gamma)}$, and

(ii) for any number i and formula ψ with rank β, there is a formula ψ' with rank less than γ so that $\{z \mid \mathfrak{A}(\omega_1^{\mathrm{CK}}, z) \models \psi'\}$ is a $\Pi^0_1(\emptyset^{(\beta')})$-subset of $\{z \mid \mathfrak{A}(\omega_1^{\mathrm{CK}}, z) \models \psi\}$ for some $\beta' < \gamma$ with difference in measure less than 2^{-i}.

Note that for any ranked formula ψ, if $x \in P_\psi = \{z \mid \mathfrak{A}(\omega_1^{\mathrm{CK}}, z) \models \psi\}$, then P_ψ has positive measure.

Since x is Δ^1_1-dominated, there is a hyperarithmetic function $f : \omega \to \omega$ such that for any e and $n \in \mathcal{O}$ with $|n| < \gamma$ such that $\Phi_e^{H_n}$ computes a tree $T_{e,n}$, if $x \notin [T_{e,n}]$ then $x \upharpoonright f(\langle e, n\rangle) \notin T_{e,n}$.

Now recursively in $x \oplus \emptyset^{(\gamma)} \oplus f$, search for a formula ψ_0 with rank less than γ such that $S_0 = \{z \mid \mathfrak{A}(\omega_1^{\mathrm{CK}}, z) \models \psi_0\}$ contains x, has positive measure, and is a closed subset of either $\{z \mid \mathfrak{A}(\omega_1^{\mathrm{CK}}, z) \models \varphi(z, \bar{0})\}$ or $\{z \mid \mathfrak{A}(\omega_1^{\mathrm{CK}}, z) \models \neg\varphi(z, \bar{0})\}$. Since x is Π^1_1-random, by (ii) there is such a ψ_0. Note that one needs $x \oplus \emptyset^{(\gamma)} \oplus f$ to decide if $x \in S_0$. Inductively, for each n, search recursively in $x \oplus \emptyset^{(\gamma)} \oplus f$ for a formula ψ_{n+1} with rank less than γ such that $S_{n+1} = \{z \mid \mathfrak{A}(\omega_1^{\mathrm{CK}}, z) \models \psi_{n+1}\}$ contains x, has positive measure, and is a closed subset of either $S_n \cap \{z \mid \mathfrak{A}(\omega_1^{\mathrm{CK}}, z) \models \varphi(z, \overline{n+1})\}$ or $S_n \cap \{z \mid \mathfrak{A}(\omega_1^{\mathrm{CK}}, z) \models \neg\varphi(z, \overline{n+1})\}$. Since x is Π^1_1-random, by (ii), such a ψ_{n+1} exists. This implies that $y \leq_T x \oplus \emptyset^{(\gamma)} \oplus f$.

Since f is hyperarithmetic, there is a $\gamma_0 < \omega_1^{\mathrm{CK}}$ such that $f \leq_T \emptyset^{(\gamma_0)}$. Without loss of generality, assume $\gamma_0 \leq \gamma$ (replacing γ with $\gamma + \gamma_0$ if necessary). Then $y \leq_T x \oplus \emptyset^{(\gamma)}$. □

Corollary 14.3.7. *If x is Π^1_1-random and $y \leq_h x$, then there is a recursive ordinal β, a function $f \leq_T \emptyset^{(\beta)}$ and a recursive functional Ψ such that for every n, $y(n) = \Psi^{x \oplus \emptyset^{(\beta)} \upharpoonright f(n)}(n)[f(n)]$.*

Proof. Suppose that x is Π^1_1-random and $y \leq_h x$. By Lemma 14.3.6, there is a recursive ordinal γ and a recursive function Φ such that for every n, $y(n) = \Phi^{x \oplus \emptyset^{(\gamma)}}(n)$. Let $g <_h x$ be a function such that for every n, $y(n) = \Phi^{x \oplus \emptyset^{(\gamma)} \upharpoonright g(n)}(n)[g(n)]$. Since x is Π^1_1-dominated, there is a hyperarithmetic function h so that for all n, $h(n) > g(n)$.

Hence there is a sufficiently large recursive ordinal β such that both x and h are many-one reducible to $\emptyset^{(\beta)}$. It is not difficult to define a function $f \leq_T \emptyset^{(\beta)}$ and a recursive function Ψ so that for every n, $y(n) = \Psi^{x \oplus \emptyset^{(\beta)} \upharpoonright f(n)}(n)[f(n)]$. □

Theorem 14.3.8. *If x is Π^1_1-random and $y \leq_h x$ is not hyperarithmetic, then there is a Π^1_1-random $z \equiv_h y$.*

Proof. Suppose x is Π^1_1-random and $y \leq_h x$ is not hyperarithmetic. By Corollary 14.3.7, there is a recursive ordinal β, a nondecreasing function $f \leq_T \emptyset^{(\beta)}$ and a recursive function Ψ such that $\lim_{n\to\infty} f(n) = \infty$ and for every n,

$$y(n) = \Psi^{x \upharpoonright f(n) \oplus \emptyset^{(\beta)} \upharpoonright f(n)}(n)[f(n)].$$

For each $u \in 2^{<\omega}$, let

$$l(u) = \sum_{\tau \in 2^{|u|} \wedge \tau < u} \Bigg(\sum_{\sigma \in 2^{f(|u|)} \wedge \Psi^{\sigma \oplus \emptyset^{(\beta)} \upharpoonright f(|u|)}[f(|u|)] \upharpoonright |u| = \tau} 2^{-|\sigma|} \Bigg)$$

and

$$r(u) = l(u) + \sum_{\sigma \in 2^{f(|u|)} \wedge \Psi^{\sigma \oplus \emptyset^{(\beta)} \upharpoonright f(|u|)}[f(|u|)] \upharpoonright |u| = u} 2^{-|\sigma|},$$

where $\tau < u$ means, as strings, τ is to the left of u.

One may view $\sum_{\sigma \in 2^{f(|u|)} \wedge \Psi^{\sigma \oplus \emptyset^{(\beta)} \upharpoonright f(|u|)}[f(|u|)] \upharpoonright |u| = \tau} 2^{-|\sigma|}$ as a "measure" for τ. For each n, let

$$l_n = l(y \upharpoonright n), \text{ and } r_n = r(y \upharpoonright n).$$

Then $l_n \leq l_{n+1} \leq r_{n+1} \leq r_n$. Since y is not hyperarithmetic, it is not difficult to see that $\lim_{n \to \infty} r_n = 0$. Hence there is a unique

$$z = \bigcap_{n<\omega} (l_n, r_n).$$

Clearly $z \leq_T y \oplus \emptyset^{(\beta)}$. We leave it to the reader to verify that $y \leq_T z \oplus \emptyset^{(\beta)}$. Thus $z \equiv_h y$.

Suppose that z is not Δ_1^1-random. Then there is a recursive ordinal $\gamma < \beta$ and a $\emptyset^{(\gamma)}$-ML-test $\{V_n\}_{n \in \omega}$ such that $z \in \bigcap_{n<\omega} V_n$. Let

$$\begin{aligned} \hat{V}_n = \{u \mid (\exists v)(&v \text{ is the } k\text{-th string in } V_n \wedge \\ &(\exists p \in \mathbb{Q})(\exists q \in \mathbb{Q})(p < l(u) < r(u) < q \wedge [v] \subseteq (p, q) \wedge q - p < 2^{-|v|} - 2^{-n-k-2})\}. \end{aligned}$$

Intuitively, $\{\hat{V}_n\}_{n<\omega}$ is a Δ_1^1-Martin-Löf test relative to the measure λ, where $\lambda(\sigma) = \sum_{\tau \in C(\sigma)} 2^{-|\tau|}$ for any $\sigma \in 2^{<\omega}$.

Since $z \in V_n$, we have $y \in \hat{V}_n$ for every n. Note that $\{\hat{V}_n\}_{n<\omega}$ is $\emptyset^{(\gamma+1+\beta)}$-r.e.

Let

$$U_n = \{\sigma \mid (\exists \tau \in \hat{V}_n)(|\sigma| = f(|\tau|) \wedge \Phi^{\sigma \oplus \emptyset^{(\beta)} \upharpoonright f(|\tau|)}[f(|\tau|)] \succeq \tau)\} = \bigcup_{u \in \hat{V}_n} C(u).$$

Then $\{U_n\}_{n<\omega}$ is $\emptyset^{(\gamma+1+\beta)}$-r.e and $x \in \bigcap_{n<\omega} U_n$. Note that for every n,

$$\mu(U_n) \leq \mu(V_n) + \sum_{k<\omega} 2^{-n-k-2+1} < 2^{-n} + 2^{-n} = 2^{-n+1}.$$

Then $\{U_{n+1}\}_{n<\omega}$ is a $\emptyset^{(\gamma+1+\beta)}$-ML-test. Hence x is not a Δ_1^1-random, a contradiction. □

14.3.4 Exercises

Exercise 14.3.1 (Hjorth and Nies [51]). *Prove that $x \oplus y$ is Π_1^1-random if and only if x is Π_1^1-random and y is $\Pi_1^1(x)$-random.*

Exercise 14.3.2 (Yu [171]). *For any $z \geq_h \mathcal{O}$, there is a Δ_1^1-random $x \equiv_h z$ which is not Π_1^1-ML random.*

Exercise 14.3.3. *Prove that if x is Π_1^1-random, then for any Σ_1^1-set A of reals containing x,*

$$\lim_{n \to \infty} \frac{\mu(A \cap [x \upharpoonright n])}{2^{-n}} = 1.$$

Exercise 14.3.4. *Prove that there is a Σ^1_1-set A and a Π^1_1-ML random $x \in A$ such that* $\lim_{n\to\infty} \frac{\mu(A\cap[x\upharpoonright n])}{2^{-n}} = 0$.

Exercise 14.3.5. *Prove that every $x \notin$ HYP hyperarithmetically reducible to a Π^1_1-random real is Π^1_1-random relative to a continuous measure.*

Exercise 14.3.6 (Kjos-Hanssen and Yu). *Let Φ be a universal partial Π^1_1-function from ω to 2, and assume that $x(e) \neq \Phi(e, e)$ for each e. Show that x hyperarithmetically computes some member of a given nonempty Σ^1_1-closed subset of 2^ω.*

Exercise 14.3.7 (Greenberg, Montalbán and Slaman [39]; Kalimullin and Nies). *Suppose that $\mathcal{M}$ is a structure such that the set $A = \{x \mid (\exists \mathcal{N} \leq_T x)(\mathcal{N} \cong \mathcal{M})\}$ has measure* 1. *Prove that every Π^1_1-random real is in A.*

Exercise 14.3.8 (Kripke [120]). *If A is a nullset of hyperdegrees such that* **0** $\notin A$*, then the set $\mathcal{U}_h(A) = \{x \mid (\exists y \in A)(x \geq_h y)\}$ is also null.*

14.4 Δ^1_2 and Σ^1_2-randomness

14.4.1 Δ^1_2-randomness

Since there is a Δ^1_2-real which is Π^1_1-random, Δ^1_2-randomness is strictly stronger than Π^1_1-randomness. By Corollary 13.1.7 and Proposition 13.1.12, we have the following conclusion.

Proposition 14.4.1. *Assume that $\omega^L_1 < \omega_1$. Then Δ^1_2-randomness = Δ^1_2-ML randomness = Π^1_2-randomness.*

14.4.2 Σ^1_2-ML randomness

Let $\{\varphi_n(\sigma)\}_{n<\omega}$ be a recursive enumeration of Σ^1_2-formulas. There is a Σ_1-formula $\psi(n, \sigma)$ (in set theory) such that

$$\varphi_n(\sigma) \Leftrightarrow (\exists x)(x \text{ is a } \Pi^1_1\text{-singleton} \wedge \langle L_{\omega^x_1}, \in\rangle \models \psi(n, \sigma)).$$

Let $(n, \sigma) \in U$ if and only if

$$(\exists x)(x \text{ is a } \Pi^1_1\text{-singleton} \wedge \langle L_{\omega^x_1}, \in\rangle \models \psi(n, \sigma)) \wedge$$
$$\mu(\{z \mid (\exists m)(\langle L_{\omega^x_1}, \in\rangle \models \psi(n, z\upharpoonright m)\}) < 2^{-n}.$$

One may view $(n, \sigma) \in U$ as an enumeration of strings σ in U up to any stage coded by a Π^1_1-singleton where the measure is less than 2^{-n}. We leave it to the reader to verify that U is Σ^1_2. Then the set $\{U_n\}_{n<\omega}$, where $U_n = \{\sigma \mid (n, \sigma) \in U\}$, is a universal Σ^1_2-ML-test.

Hence there is a Π^1_2-closed set P which contains only Σ^1_2-ML random reals. It follows that the leftmost path in P is Δ^1_2-equivalent to a Σ^1_2-real and so belongs to L.

By Proposition 4.4.6, every $x \in P \cap L$ satisfies $\delta^{1,x}_2 > \delta^1_2$. This applies in particular to the leftmost path x in P. Hence if one stays within L, the situation that one normally sees in standard randomness theory no longer applies. Note that by applying a method similar to that in the proof of Theorem 14.2.10, one may separate Σ^1_2-ML randomness from Δ^1_2-randomness.

Proposition 14.4.2. *If $V = L$, then every nonzero Δ^1_2-degree contains a Σ^1_2-ML random real.*

Proof. First observe that the set P above can be presented by a Π^1_2-tree T. By Proposition 4.4.6, T is Δ^1_2 in every $x \in [T]$ and T has the least nontrivial Δ^1_2-degree. Hence the proposition follows. □

14.4.3 Σ^1_2-randomness

Definition 14.4.3. We say that x is *L-random* if for any Martin-Löf test $\{U_n\}_{n<\omega}$ in L, $x \notin \bigcap_{n<\omega} U_n$.

L-randomness was essentially introduced by Solovay [150]. The following fact is immediate.

Proposition 14.4.4. *The followings are equivalent:*
(i) *x is L-random;*
(ii) *x does not belong to any Borel nullset coded in L;*
(iii) *x is $\mathbb{R} = \langle \mathbf{T}, \leq\rangle$-generic over L, where $\mathbb{R}$ is the notion of random forcing as defined in § 6.2.2.*

Proof. Obviously (iii) ⇒ (ii) ⇔ (i). By Proposition 6.2.6, $\mathbb{R}$ is c.c.c. Then it follows that for any dense subset $\mathscr{D}$, $\bigcup_{p\in\mathscr{D}} p$ is a Borel co-nullset with a Borel code in L. Hence (ii) ⇒ (iii). □

Theorem 14.4.5 (Solovay [150]). *If the set of L-random reals has measure 1, then every Σ^1_2-set is measurable.*

Proof. Suppose that A is a Σ^1_2-set of reals defined by a formula $\varphi(x)$ in the language of set theory so that

$$x \in A \Leftrightarrow \langle L[x], \in\rangle \models \varphi(x).$$

Let $\mathscr{D}$ be a maximal antichain in L so that for any $p \in \mathscr{D}$, $p \Vdash \varphi(\dot{G})$. Then $\mathscr{D}$ is countable. Let $R = \bigcup_{p\in\mathscr{D}} p$. Then R is a Borel set with a Borel code in L. If x is L-random, then x is an $\mathbb{R}$-generic real over L. Hence if x is L-random in R, then $\langle L[x], \in\rangle \models \varphi(x)$ and so x belongs to A. Note that $\mathscr{D}' = \{r \mid (\exists p \in \mathscr{D})(r \leq p) \vee (\forall p)(\forall q)(p \in \mathscr{D} \wedge q \leq p \to q \not\leq r)\}$ is a dense set. Thus if $x \in A$, then $x \in R$ so that for any L-random real x, $x \in A$ if and only if $x \in R$. Since the set of L-random reals has measure 1, A is equal to R modulo a nullset. It follows that A is measurable. □

Theorem 14.4.6 (Stern [161]). *For any real x, x is Δ^1_2-random and $\delta^{1,x}_2 = \delta^1_2$ if and only if x is L-random.*

Proof. Let x be Δ^1_2-random such that $\delta^{1,x}_2 = \delta^1_2$. If x is not L-random, then let

$$A = \{r \mid r \text{ codes a well-ordering} \wedge x \text{ is not } L_{|r|}\text{-random}\}$$

be a nonempty $\Delta^1_2(x)$-set. Then there is an $r \leq_{\Delta^1_2} x$ so that $x \in L_{\delta^1_2}$, a contradiction.

Conversely, let x be L-random. Then x is Δ^1_2-random. Suppose that $\delta^{1,x}_2 > \delta^1_2$. Then there are Σ^1_2-formulas $\varphi(x, m, n)$ and $\psi(x, m, n)$ defining an $r \in L$ coding a well-ordering greater than δ^1_2. By Corollary 4.2.9, $r \in L_{\delta^{1,x}_2}[x]$. By (iii) in Proposition 14.4.4, there is a condition $p \in \mathbf{T} \cap L$ such that

$$p \Vdash (\forall m)(\forall n)((\langle m, n\rangle \in \check{r} \leftrightarrow \varphi(\dot{x}, m, n)) \wedge (\langle m, n\rangle \in \check{r} \Leftrightarrow \psi(\dot{x}, m, n)).$$

Then for any L-random real $z \in p$, $r \leq_{\Delta^1_2} z$. Applying Levy collapse $\mathbb{L}((\aleph_2)^L)$ over V, in $V[G]$ the set of L-random reals is of measure 1. It follows that in $V[G]$, the set

$$A_r = \{z \mid r \leq_{\Delta^1_2} z\}$$

has positive measure. By Theorem 14.4.5 and Proposition 13.1.9, r is Δ^1_2 in $V[G]$ and so in V, a contradiction. □

Note that the set of L-random reals is Π^1_2. Thus if the measure of non-L-random reals is 0, then every Σ^1_2-random real is L-random.

Theorem 14.4.7 (Stern [161]). *Assume that the set of L-random reals has measure 1. Then the set of non-L-random reals is the largest null Σ^1_2-set and so Σ^1_2-randomness is equivalent to L-randomness.*

Proof. For a contradiction, suppose that A is a null Σ^1_2-set containing an L-random real x. Then there is a formula φ defining A such that $\langle L[x], \in\rangle \models \varphi(x)$. By Proposition 14.4.4, there is a condition $p \in \mathbf{T} \cap L$ such that $p \Vdash \varphi(\dot{x})$. Hence for any random real $y \in p$, we have $\langle L[y], \in\rangle \models \varphi(y)$. Thus $\varphi(y)$ holds for any $y \in p$. Since the set of L-random reals has measure 1, it follows that $\mu(A) > 0$, a contradiction. □

Hence if $(\omega_1)^L < \omega_1$, then Σ^1_2-randomness is strictly stronger than Σ^1_2-ML randomness. Indeed, the conclusion is provable in ZFC.

Proposition 14.4.8. *There is a Σ^1_2-ML random real which is not Σ^1_2-random.*

Proof. Let $T \subseteq 2^{<\omega}$ be a Π^1_2-tree containing only Σ^1_2-ML random reals. Then by a method similar to that used in the proof of Lemma 14.2.12, replacing ω_1^{CK} with δ^1_2, one is able to show that the leftmost path of T is covered by a Σ^1_2-nullset. We leave it to the reader to fill in the details. □

One may also introduce the notion of strong Σ^1_2-ML randomness by considering reals that avoid all generalized Σ^1_2-ML tests. Then Proposition 14.4.8 says that strong Σ^1_2-ML randomness is different from Σ^1_2-ML randomness. Note that every generalized Σ^1_2-ML test $\{U_n\}_{n<\omega}$ is a member of L, and by Corollary 13.1.7, $\langle L, \in\rangle \models \mu(\bigcap_{n<\omega} U_n) = 0$. Hence there is a real in L which is strongly Σ^1_2-ML random. It follows that if the set of constructible reals is null, then strong Σ^1_2-ML randomness is strictly weaker than Σ^1_2-randomness.

14.4.4 Exercises

Exercise 14.4.1. *Separate Σ^1_2-ML randomness from Δ^1_2-randomness.*

Exercise 14.4.2. *Every Σ^1_2-set is measurable if and only if almost every real is L-random.*

Exercise 14.4.3. *Assume $V = L$. Prove that there is no largest Σ^1_2-nullset.*

Exercise 14.4.4. *Complete the proof of Proposition 14.4.8.*

Exercise 14.4.5 (Stern [161]). *Prove that it is consistent with ZFC that $(\omega_1)^L = \omega_1$ and almost every real is L-random.*

Exercise 14.4.6. *Assume $V = L$. Show that every nontrivial Δ^1_2-degree contains a strongly Σ^1_2-ML random real.*

Exercise 14.4.7. *Let r be random over L. Show that in $L[r]$, there is no real which is Cohen generic over L.*

Exercise 14.4.8. *Let g be Cohen generic over L. Show that in $L[g]$, the set of constructible reals is null but no real is L-random.*

14.5 Kolmogorov complexity and randomness

14.5.1 The Kolmogorov complexity of Δ^1_1-random reals

Proposition 14.5.1. *The following are equivalent:*
(i) *x is Δ^1_1-random.*
(ii) *For any $z \in$ HYP which is Martin-Löf random,*

$$(\exists c)(\forall n)(K(x\upharpoonright n) \geq 2n - C(z\upharpoonright n) - c).$$

(iii) *For any $z \in$ HYP which is Martin-Löf random,*

$$(\exists c)(\forall n)(C(x\upharpoonright n) \geq 2n - K(z\upharpoonright n) - c).$$

Proof. (i) $\Leftrightarrow$ (ii). By Proposition 14.2.2, x is Δ^1_1-random if and only if for any hyperarithmetic z, x is Martin-Löf random relativized to z. By Theorem 12.1.10, this is equivalent to stating that for any hyperarithmetic Martin-Löf random z, x is Martin-Löf random relativized to z. By van Lambalgen's Theorem 12.1.7, this is in turn equivalent to saying that for every hyperarithmetic Martin-Löf random z, $x \oplus z$ is Martin-Löf random which, by Theorem 12.2.11, is equivalent to "for any hyperarithmetic Martin-Löf random z, $(\exists c)(\forall n)(K(x \upharpoonright n) \geq 2n - C(z \upharpoonright n) - c)$".

One proves similarly that (i) $\Leftrightarrow$ (iii). □

Corollary 14.5.2. *Δ^1_1-randomness is upward closed for both K- and C-degrees.*

14.5.2 The Kolmogorov complexity of Π^1_1-random reals

Proposition 14.5.3. *For any Π^1_1-random x, if either $x \leq_K y$ or $x \leq_C y$, then y is Π^1_1-random.*

Proof. Suppose that x is Π^1_1-random and $x \leq_K y$ (resp. $x \leq_C y$). By Corollary 14.5.2, y is Δ^1_1-random. Note that $\{z \mid x \leq_K z\}$ (resp. $\{z \mid x \leq_C z\}$) is $\Delta^1_1(x)$. By Theorem 12.2.14 and Lemma 2.5.4 relativized to x, $y \leq_h x$ and so $\omega_1^y = \omega_1^{\mathrm{CK}}$. Then by Corollary 14.3.2, y is Π^1_1-random. □

One may also show that Δ^1_2-randomness is both K- and C-upward closed. By Theorem 14.4.7, if $(\omega_1)^L < \omega_1$, then Σ^1_2-randomness is upward closed for both $\leq_K$ and $\leq_C$. It is not known if Π^1_1-ML randomness is upward closed for either of these reducibilities.

14.5.3 Exercise

Exercise 14.5.1. *Prove that there is no function f such that x is Δ^1_1-random if and only if $(\exists c)(\forall n)(K(x \upharpoonright n) \geq n + f(n) - c)$.*

14.6 Lowness for randomness

Lowness notions for higher randomness generally parallel those for randomness over the natural numbers. In this section, we omit proofs about lowness properties which are minor modifications of those for the latter. The reader is referred to the relevant books or papers for details.

14.6.1 Lowness for higher Kurtz randomness

Lemma 14.6.1. *If x is Δ^1_1-dominated and Δ^1_1-semitraceable, then x is low for Δ^1_1-Kurtz tests.*

Proof. Suppose x is Δ^1_1-dominated and Δ^1_1-semitraceable. Let U be a $\Delta^1_1(x)$ open set with measure 1. Then there is a $y \le_h x$ such that U is $\Sigma^0_1(y)$. Let Φ be a Turing reduction so that the domain of Φ^y is U. Write U^z for the domain of Φ^z for any z. Note that $U = U^y$.

Define a $\Delta^1_1(x)$ function $\hat{f}$ by: $\hat{f}(n)$ is the shortest string $\sigma \prec y$ with $\mu(U^\sigma[\sigma]) > 1 - 2^{-n}$. By the hypothesis of the theorem, there is an increasing Δ^1_1-function g and a Δ^1_1-function f such that for every n, there is an $m \in [g(n), g(n+1))$ with $f(m) = \hat{f}(m)$. Without loss of generality, we may assume that $\mu(U^{f(m)}[m]) > 1 - 2^{-m}$ for every m.

Define a Δ^1_1-open set V such that $\sigma \in V$ if and only if there exists an n so that $[\sigma] \subseteq \bigcap_{g(n)\le m<g(n+1)} U^{f(m)}[m]$. By the property of f and g, $V \subseteq U^y = U$. But for every n,

$$\mu(\bigcap_{g(n)\le m<g(n+1)} U^{f(m)}[m]) > 1 - \sum_{g(n)\le m<g(n+1)} 2^{-m} \ge 1 - 2^{-g(n)+1}.$$

Hence

$$\mu(V) \ge \lim_n \mu(\bigcap_{g(n)\le m<g(n+1)} U^{f(m)}[m]) = 1.$$

This implies that x is low for Δ^1_1-Kurtz tests. □

Lemma 14.6.2. *If x is low for Δ^1_1-Kurtz randomness, then x is Δ^1_1-dominated.*

Proof. We first show that if x is low for Δ^1_1-Kurtz tests, then x is Δ^1_1-dominated.

Suppose $f \le_h x$ is an increasing function. Let $S_f = \{z \mid (\forall n)(z(f(n)) = 0)\}$. Obviously S_f is a $\Delta^1_1(x)$-closed nullset. Hence there is a Δ^1_1-closed nullset $[T] \supseteq S_f$ where $T \subseteq 2^{<\omega}$ is a Δ^1_1-tree. Define

$$g(n) = \min\{m \mid \frac{|\{\sigma \in 2^m \mid \sigma \in T\}|}{2^m} < 2^{-n}\} + 1.$$

Since $\mu([T]) = 0$, g is Δ^1_1. We claim that g dominates f.

For every n, $S_{f(n)} = \{\sigma \in 2^{f(n)} \mid (\forall i \le n)(\sigma(f(i)) = 0)\}$ has cardinality $2^{f(n)-n}$. However, if $g(n) \le f(n)$ then applying $S \subseteq [T]$ one has

$$|S_{f(n)}| \le 2^{f(n)-g(n)} \cdot |\{\sigma \in 2^{g(n)} \mid \sigma \in T\}| < 2^{f(n)-g(n)} \cdot 2^{g(n)-n} = 2^{f(n)-n}.$$

This is a contradiction. Hence x is Δ^1_1-dominated.

Now suppose x is not Δ^1_1-dominated witnessed by an $f \le_h x$. Then S_f is not contained in any Δ^1_1-closed nullset. Indeed it is not difficult to see that for any σ with $[\sigma] \cap S_f \neq \emptyset$, $[\sigma] \cap S_f$ is not contained in any Δ^1_1-closed nullset (otherwise, as proved above, one shows that f is dominated by a Δ^1_1-function). Then by induction, one constructs a Δ^1_1-Kurtz random $z \in S_f$ as follows:

Let $P_0, P_1, \dots$ be an enumeration of the Δ^1_1-closed nullsets. First choose the least τ such that $[\tau] \cap S_f \neq \emptyset$ but $[\tau] \cap S_f \cap P_0 = \emptyset$. Let $l_0 = |\tau|$ and $z \upharpoonright l_0 = \tau$. Suppose that at stage $n+1$, $z \upharpoonright l_n$ is defined so that $[z] \upharpoonright l_n] \cap S_f \neq \emptyset$. Then there is a $\tau \succ z \upharpoonright l_n$ such that $[\tau] \cap S_f \neq \emptyset$ but $[\tau] \cap S_f \cap P_n = \emptyset$. Choose the least such τ and let $l_{n+1} = |\tau|$ and $z \upharpoonright l_{n+1} = \tau$. Then $z \in S_f$ is Δ^1_1-Kurtz random. Thus x is not low for Δ^1_1-Kurtz randomness. □

The proof of the following lemma is left to the reader.

Lemma 14.6.3 (Kučera [76]). *Suppose $T \subseteq 2^{<\omega}$ and $P \subseteq \omega^{<\omega}$ are Σ^1_1-trees with $\mu([T]) > 0$. Then there is a Σ^1_1-tree $S \subseteq T$ and a constant c such that for any e, if $[P_e] \subseteq S$ is not empty, then $\mu([P_e]) > 2^{-e-c}$, where $P_e = \{\sigma \mid (e, \sigma) \in P\}$.*

The following lemma is essentially due to Joseph Miller.

Lemma 14.6.4. *For any non-Δ^1_1-semitraceable real x and Σ^1_1-tree $T \subseteq 2^{<\omega}$ with $\mu([T]) > 0$, there exist $z \in [T]$ and $y \leq_h x$ such that $y \subseteq z$ and $|\{n \mid y(n) = 1\}| = \aleph_0$.*

Proof. Let x be non-Δ^1_1-semitraceable and let $T \subseteq 2^{<\omega}$ be a Σ^1_1-tree with $\mu(T) > 2^{-k}$ for some $k > 0$. By Lemma 13.3.10, there is a function $f \leq_h x$ such that for any partial Π^1_1-function $p : \omega \times 2^{<\omega} \to \omega$, there are at most finitely many (n, σ)'s satisfying $f(n, \sigma) = p(n, \sigma)$.

Let $P \subseteq \omega \times 2^{<\omega}$ be a universal Σ^1_1-tree, i.e. P is a Σ^1_1-tree and for any Σ^1_1-tree $Q \subseteq 2^{<\omega}$, there is an e such that $P_e = Q$. We may assume that there is a recursive function h such that for any $\langle e_0, e_1\rangle$, $P_{h(\langle e_0,e_1\rangle)} = P_{e_0} \cap P_{e_1}$.

Let $S \subseteq T$ be as in Lemma 14.6.3. Let $S = P_{e_0}$ for some e_0. We may also assume that there is a recursive function $g : 2^{<\omega} \to \omega$ such that $P_{g(\sigma)} = \{\tau \mid \tau \in P_{e_0} \wedge (\forall n)(\sigma(n) = 1 \to \tau(n) = 1)\}$. For $\sigma \in 2^{<\omega}$, let

$$A_\sigma = \{n \mid [S] \cap \{y \in 2^\omega \mid y(n) = 1 \wedge (\forall m)(\sigma(m) = 1 \to y(m) = 1)\} = \emptyset\}.$$

Then the set $\{(\sigma, A_\sigma) \mid \sigma \in 2^{<\omega}\}$ is a Π^1_1-set. Moreover, if $[P_{g(\sigma)}] \neq \emptyset$, then

$$2^{-h(g(\sigma))-c} \leq \mu([P_{g(\sigma)}]) \leq \mu(\{y \mid A_\sigma \cap y = \emptyset\}) = 2^{-|A_\sigma|}.$$

Thus $|A| \leq h(i(\sigma)) + c$. It follows that there is a partial Π^1_1-function $p : \omega \times 2^{<\omega} \to \omega$ such that for any n, if $p(n, \sigma) = k$ then

- $k \in A_\sigma$, and
- $(\forall m < n)(\exists k')(k' \neq k \wedge p(m, \sigma) = k')$.

Hence $\{p(n, \sigma)\}_{n \leq h(i(\sigma))+c}$ is a one-one enumeration of A_σ. Note that

$$|\{n \mid p(n, \sigma) \text{ is defined}\}| \leq h(i(\sigma)) + c.$$

Hyperarithmetically in x compute a k such that $k \notin A_\sigma$ and $k > |\sigma|$ as follows. Fix a recursive bijection $q : \omega \to \omega^{<\omega}$. We may assume that $f(n, \sigma) \neq q(p(n, \sigma))(n)$ for all n and σ. Define $g_1 : 2^{<\omega} \to \omega^{<\omega}$ hyperarithmetically in x such that $g_1(\sigma)$ is a finite string in $\omega^{h(i(\sigma))+c}$ and

$$(\forall n \leq 2^{h(i(\sigma))+c})(g_1(\sigma)(n) = f(n, \sigma)).$$

Hence if $[P_{i(\sigma)}] \neq \emptyset$, then $g_1(\sigma) \neq q(p(n, \sigma))$ for all $n \leq h(i(\sigma))$. For such a σ, $q^{-1}(g_1(\sigma)) \neq p(n, \sigma)$ for all n. Thus $q^{-1}(g_1(\sigma)) \notin A_\sigma$ and so $P_{g(\sigma \cup \{q^{-1}(g_1(\sigma))\})} \neq \emptyset$. Without loss of generality, we may assume that $q^{-1}(g_1(\sigma)) > |\sigma|$.

Using $q^{-1} \circ g_1$, starting with $\sigma_0 = \emptyset$, hyperarithmetically in x construct a sequence $\sigma_0 \prec \sigma_1 \prec \cdots$ of strings such that for each n,

- $P_{g(\sigma_n)} \neq \emptyset$ and
- $\sigma_{n+1} = \sigma_n \cup \{q^{-1}(g_1(\sigma_n))\}$.

Hence $\bigcap_{n<\omega} P_{g(\sigma_n)} \neq \emptyset$ and for any $z \in \bigcap_{n<\omega} P_{g(\sigma_n)}$, the real $y = \bigcup_{n<\omega} \sigma_n$ is a subset of z. □

Corollary 14.6.5. *If x is not Δ^1_1-semitraceable, then x is not low for Δ^1_1-Kurtz randomness.*

Proof. Suppose that x is not Δ^1_1-semitraceable. Let T be a Σ^1_1-tree so that $\mu([T]) > 0$ and $[T]$ contains only Π^1_1-ML random reals. Then by Lemma 14.6.4, there is a $z \in [T]$ and a $y \leq_h x$ such that z is a member of the $\Delta^1_1(x)$-closed nullset $\{r \mid y \subseteq r\}$. Hence x is not low for Δ^1_1-Kurtz randomness. □

This leads us to the following theorem.

Theorem 14.6.6 (Kjos-Hanssen, Nies, Stephan and Yu [68]). *For any $x \in 2^\omega$, the following are equivalent:*
(i) *x is low for Δ^1_1-Kurtz tests;*
(ii) *x is low for Δ^1_1-Kurtz randomness;*
(iii) *x is Δ^1_1-dominated and Δ^1_1-semitraceable.*

It is not known if there exists a nonhyperarithmetic real which is low for Π^1_1-Kurtz randomness. However, we can prove the following set inclusion.

Proposition 14.6.7 (Kjos-Hanssen, Nies, Stephan and Yu [68]). *If x is low for Π^1_1-Kurtz randomness, then x is low for Δ^1_1-Kurtz randomness.*

Proof. Assume that x is low for Π^1_1-Kurtz randomness, y is Δ^1_1-Kurtz random and there is a $\Delta^1_1(x)$-closed nullset A with $y \in A$. By Proposition 14.1.2, the set

$$B = \bigcup\{C \mid C \text{ is a } \Delta^1_1\text{-closed nullset}\}$$

is a Π^1_1-nullset. Then $A \setminus B$ is $\Sigma^1_1(x)$. Since y is Δ^1_1-Kurtz random, $y \notin B$. Hence $y \in A \setminus B$ and so $A \setminus B$ is $\Sigma^1_1(x)$ and nonempty. Thus there is a $z \in A \setminus B$ with $\omega^z_1 = \omega^x_1 = \omega^{\mathrm{CK}}_1$. Since $z \notin B$, z is Δ^1_1-Kurtz random. By Proposition 14.1.4, z is Π^1_1-Kurtz random. This contradicts the fact that x is low for Π^1_1-Kurtz randomness. □

14.6.2 Lowness for Δ^1_1-randomness

Theorem 14.6.8. *The following are equivalent for a real x.*
(i) *x is Δ^1_1-traceable;*
(ii) *Every $\Delta^1_1(x)$-nullset is contained in a Δ^1_1-nullset;*

(iii) *x is low for Δ^1_1-randomness;*
(iv) *Every Π^1_1-ML random real is $\Delta^1_1(x)$-random.*

Proof. (i) → (ii): Assume that x is Δ^1_1-traceable. Let $\mathcal{S}$ be a $\Delta^1_1(x)$-nullset. By Proposition 14.2.2 relativized to x, there is a $\Delta^1_1(x)$-ML test $\{U_n\}_{n<\omega}$ such that $\mu(U_n) = 2^{-n}$ for each n and $\mathcal{S} \subseteq \bigcap U_n$. Let $\{S_e\}_{e<\omega}$ be a recursive enumeration of all finite subsets of $2^{<\omega}$. There is a function $f \leq_h x$ such that $[S_{f(\langle n,s\rangle)}] = U_{n,s}$ satisfies $U_{n,s} \subseteq U_{n,s+1}$, $U_n = \bigcup_{s<\omega} U_{n,s}$, and $\mu(U_{n,s}) > 2^{-n}(1-2^{-s})$.

Let $T = (T_e)_{e<\omega}$ be a Δ^1_1-trace for f. We may choose T such that in addition $|T_e| \leq e$ for each $e > 0$.

We now define a subtrace $\hat{T}$ of T, i.e. $\hat{T}_{\langle n,s\rangle} \subseteq T_{\langle n,s\rangle}$ for each n, s. The objective is to define open sets V_n via $\hat{T}$ (in a way to be specified) small enough to give us a Δ^1_1-nullset $\mathcal{V} = \bigcap_{n<\omega} V_n$, yet large enough as to keep all "relevant" reals out of $T_{\langle n,s\rangle} - \hat{T}_{\langle n,s\rangle}$, so that $\bigcap_{n<\omega} U_n \subseteq \mathcal{V}$.

Let $\hat{T}_{\langle n,s\rangle}$ be the set of $e \in T_{\langle n,s\rangle}$ such that $2^{-n}(1-2^{-s}) \leq \mu([S_e]) \leq 2^{-n}$ and $[S_e] \supseteq [S_d]$ for some $d \in \hat{T}_{\langle n,s-1\rangle}$ (where $\hat{T}_{\langle n,-1\rangle} = \omega$). Note that $f(\langle n,s\rangle) \in \hat{T}_{\langle n,s\rangle}$. Let

$$V_n = \bigcup \left\{[S_e] \mid e \in \hat{T}_{\langle n,s\rangle},\ s \in \omega\right\}.$$

Then $\mu(V_n) \leq 2^{-n}|\hat{T}_{\langle n,0\rangle}| + \sum_{s<\omega} 2^{-s}2^{-n}|\hat{T}_{\langle n,s\rangle}|$. Since $|\hat{T}_{\langle n,s\rangle}| \leq |T_{\langle n,s\rangle}| \leq \langle n,s\rangle$ for $\langle n,s\rangle \neq 0$, and $\langle n,s\rangle$ has only polynomial growth in n and s, it is clear that $\lim_n \sum_{s<\omega} 2^{-s}2^{-n}|\hat{T}_{\langle n,s\rangle}| = 0$, and hence $\lim_n \mu(V_n) = 0$. Then $\mathcal{V} = \bigcap_n V_n$ is a Δ^1_1-nullset and $\bigcap U_n \subseteq \mathcal{V}$.

(ii) ⇒ (iii) and (iii) ⇒ (iv) are immediate.

(iv) ⇒ (i). The proof is identical to the proof in [67]. We leave it to the reader. □

By Theorem 13.3.18, lowness for Δ^1_1-randomness is different from that for Δ^1_1-Kurtz randomness.

14.6.3 Lowness for Π^1_1-ML randomness

Theorem 14.6.9 (Hjorth and Nies [51]). *Every real low for Π^1_1-ML randomness is hyperarithmetic.*

Proof. We give a sketch. Suppose that x is low for Π^1_1-ML randomness. Then it is not difficult to verify that $\omega_1^x = \omega_1^{\mathrm{CK}}$ and that there is a $\Pi^1_1(x)$-ML random $y \geq_h x$. In fact, one may use a Π^1_1-version of Turing reduction, termed $\leq_{\text{fin-h}}$ and introduced in [51], to argue that $x \leq_{\text{fin-h}} y$. In other words, there is a consistent Π^1_1-partial map $\Phi : 2^{<\omega} \to 2^{<\omega}$ such that $X = \cup_{n<\omega}\Phi(Y \upharpoonright n)$. Applying the method in [48] it can be shown that x is $K_{\Pi^1_1}$-trivial, where $K_{\Pi^1_1}$ is an analog of Kolmogorov complexity in the Π^1_1-setting. It follows as in [51] that every $K_{\Pi^1_1}$-trivial real is either hyperarithmetic or is hyperarithmetically equivalent to $\mathcal{O}$. Since $\omega_1^x = \omega_1^{\mathrm{CK}}$, x has to be hyperarithmetic. □

14.6.4 Lowness for Π^1_1-randomness and strong Π^1_1-ML randomness

Theorem 14.6.9 raises the natural question of characterizing reals that are low for Π^1_1-randomness – in particular whether there is one that is not hyperarithmetic. The answer to the latter is in the negative. Proposition 14.6.10 gives a characterization of such reals. We say that x is *non-Π^1_1-random cuppable* if $x \oplus y \not\geq_h \mathcal{O}$ for all Π^1_1-random y. It is immediate that if x is low for Π^1_1-randomness then $\omega_1^{\mathrm{CK}} = \omega_1^x$.

Proposition 14.6.10 (Harrington, Nies and Slaman [12]). *A real x is low for Π^1_1-randomness if and only if x is low for Δ^1_1-randomness and non-Π^1_1-random cuppable.*

Proof. If x is low for Π^1_1-randomness, then every Π^1_1-random real is $\Pi^1_1(x)$-random. Note that by relativizing Theorem 13.1.4, the $\Pi^1_1(x)$-set $\{y \mid y \oplus x \geq_h \mathcal{O}\}$ is null since $x \not\geq_h \mathcal{O}$. Thus x is non-Π^1_1-random cuppable.

Suppose y is Δ^1_1-random. If y is not $\Delta^1_1(x)$-random, then there is a $\Delta^1_1(x)$-nullset A containing y. By Proposition 14.2.3, the set $B = \bigcup\{C \subset 2^\omega \mid \mu(C) = 0 \wedge C \text{ is } \Delta^1_1\}$ is a Π^1_1-nullset. Now $y \in A \setminus B$. Hence $A \setminus B$ is a nonempty $\Sigma^1_1(x)$-set. By the Gandy Basis Theorem (2.5.3), there is a $z \in A \setminus B$ such that $\omega_1^z = \omega_1^x = \omega_1^{\mathrm{CK}}$. By Corollary 14.3.2, z is Π^1_1-random but not $\Pi^1_1(x)$-random, a contradiction.

Conversely, suppose x is low for Δ^1_1-randomness and non-Π^1_1-random cuppable. Then $x \not\geq_h \mathcal{O}$. Let z be a Π^1_1 random real. Relativizing the proof of Theorem 14.3.1, the largest $\Pi^1_1(x)$-nullset $\mathcal{Q}(x)$ is a union of countably many $\Delta^1_1(x)$-null sets $\mathcal{Q}_n(x)$ and the $\Pi^1_1(x)$-nullset $\{y \mid y \oplus x \geq_h \mathcal{O}\}$. Since x is low for Δ^1_1-randomness, $z \notin \bigcup_n \mathcal{Q}_n(x)$. Since x is non-Π^1_1-random cuppable, $z \oplus x \not\geq_h \mathcal{O}$. Hence z is $\Pi^1_1(x)$-random. We conclude that x is low for Π^1_1-randomness. □

Proposition 14.6.11. *There is a Δ^1_1-traceable x which is Π^1_1-random cuppable.*

Proof. By Corollary 13.3.8, there is an uncountable Σ^1_1-set A containing only nonhyperarithmetic Δ^1_1-traceable reals. Let $\mathcal{Q}$ be the largest Π^1_1-nullset. By Theorem 13.2.1, there exist $x \in A$ and $y \notin \mathcal{Q}$ such that $x \oplus y \geq_h \mathcal{O}$. □

Corollary 14.6.12.
(i) *Lowness for Π^1_1-Kurtz randomness is different from Δ^1_1-Kurtz randomness;*
(ii) *Lowness for Π^1_1-randomness is different from Δ^1_1-randomness.*

Proof. To prove (i), note that by Theorem 14.6.8 and 14.6.6, every Δ^1_1-traceable real is low for Δ^1_1-Kurtz randomness. By Proposition 14.6.11, there is an x which is low for Δ^1_1-Kurtz randomness for which there is a Π^1_1-random y satisfying $x \oplus y \equiv_h \mathcal{O}$. We claim that x is not low for Π^1_1-Kurtz randomness. Let $e < \omega$ and $n \in \mathcal{O}^{\mathcal{O}}$ such that $y = \Phi_e^{H_n^{\mathcal{O}}}$. Then

$$z \in \{y\} \Leftrightarrow (\exists r \leq_h x \oplus z)(r \in \{\mathcal{O}\} \wedge z = \Phi_e^{H_n^r}).$$

Hence y is a $\Pi^1_1(x)$-singleton and so non-$\Pi^1_1(x)$-random. Note that (ii) follows immediately from Proposition 14.6.10 and 14.6.11. □

We now show that no nonhyperarithmetic real is low for Π^1_1-random. Let $\mathbb{P} = \langle \mathbf{P}, \leq \rangle$ be a notion of forcing, where

$$\mathbf{P} = \{P \mid P \text{ is nonempty and } \Sigma^1_1\text{-closed containing only } \Pi^1_1\text{-ML random reals}\}.$$

For $P, Q \in \mathbf{P}$, define $P \leq Q$ if and only if $P \subseteq Q$. Note that $\mu(P) > 0$ for each P.

Let $\{(\varphi_i, \psi_i)\}_{i<\omega}$ be a recursive enumeration of arithmetical formulas. For each i, let

$$A_i = \{x \mid \mathfrak{A}(\omega_1^{\mathrm{CK}}, x) \models (\forall k)(\exists y)(\varphi_i(k, y) \vee \neg\psi_i(k, y))\}.$$

Define an approximation of A_i at j as

$$A_{i,j} = \{x \mid \mathfrak{A}(\omega_1^{\mathrm{CK}}, x) \models (\forall k \leq j)(\exists y)(\varphi_i(k, y) \vee \neg\psi_i(k, y))\}$$

and for $n \in \mathcal{O}$, set

$$A_{i,j,n} = \{x \mid \mathfrak{A}(\omega_1^{\mathrm{CK}}, x) \models (\forall k \leq j)(\exists y^{|n|})(\varphi_i(k, y^{|n|}) \vee \neg\psi_i(k, y^{|n|}))\}.$$

Suppose $x \in A_i$. Then for each j, there is an n such that $x \in A_{i,j,n}$. Uniformly for each $j < \omega$ and $n \in \mathcal{O}$, apply a "dual form" of Lemma 13.1.10 obtained by replacing m, n there respectively with l and the Gödel number of "$\neg(\forall k \leq j)(\exists y^{|n|})(\varphi_i(k, y^{|n|}) \vee \neg\psi_i(k, y^{|n|}))$". We then have a Π^1_1-function $p_i : \omega^2 \times \mathcal{O} \to \omega$ such that for $j, l < \omega$ and $n \in \mathcal{O}$,

$$U_{i,j,l,n} = \{x \mid \mathfrak{A}(\omega_1^{\mathrm{CK}}, x) \models \neg\varphi_{p_i(j,l,n)}\}$$

is a Δ^1_1-open cover of $A_{i,j,n}$, with $\mu(U_{i,j,l,n} \setminus A_{i,j,n}) < 2^{-j-n-l}$. Then $U_{i,j,l} = \bigcup_{n\in\mathcal{O}} U_{i,j,l,n}$ is a Π^1_1-open set. Note that $A_{i,j} \subseteq U_{i,j,l}$ and

$$\mu(U_{i,j,l} \setminus A_{i,j}) < \sum_{n<\omega} 2^{-j-n-l} \leq 2^{-j-l+1}.$$

Let $G_i \supseteq A_i$ be the Π^1_1-set defined as $\bigcap_{j,l} U_{i,j,l}$. Let

$$\mathcal{D}_1 = \{P \in \mathbb{P} \mid P \subseteq G_i \vee P \cap G_i = \emptyset\}.$$

Lemma 14.6.13. *$\mathcal{D}_1$ is dense for each i.*

Proof. Let P be a condition. There are two cases to consider:

Case 1. $\mu(P \setminus G_i) > 0$. Then there is a k such that $\mu(P \setminus \bigcap_{\langle j,l\rangle \leq k} U_{i,j,l}) > 0$. Let $Q = P \setminus \bigcap_{\langle j,l\rangle\leq k} U_{i,j,l}$. Then $Q \in \mathcal{D}_1 \cap \mathbf{P}$.

Case 2. $\mu(P \setminus G_i) = 0$. By Lemma 13.1.10, there exists a Σ^1_1-closed set $F \subseteq G_i$ such that $\mu(F \cap P) > 0$. Let $Q = P \cap F$. Then $Q \in \mathbf{P}$. □

Theorem 14.6.14 (Monin [105]). *If g is a $\mathbb{P}$-generic real meeting all $\mathcal{D}_i$'s, then $\omega_1^g = \omega_1^{\mathrm{CK}}$.*

Proof. Let g be $\mathbb{P}$-generic. It is sufficient to prove that $\mathfrak{A}(\omega_1^{\mathrm{CK}}, g) \models \Delta_1^1$-comprehension.

Let P be a condition such that $g \in P$. If $P \cap G_i = \emptyset$, then $P \cap A_i = \emptyset$. It follows that there is a k such that $\mathfrak{A}(\omega_1^{\mathrm{CK}}, g) \models (\forall y)(\neg\varphi_i(k, y) \wedge \psi_i(k, y))$.

Otherwise $P \subseteq G_i$. By (topological) compactness, for any j and l there is an n such that $P \subseteq U_{i,j,l,n}$. It follows that there is a total Π_1^1 – hence Δ_1^1 – function $f : \omega^2 \to \mathcal{O}$ such that for any j, l, $f(j, l) \in \mathcal{O}$ and $P \subseteq U_{i,j,l,f(j,l)}$. Hence there is an $m \in \mathcal{O}$ such that $|f(j, l)| < |m|$ for any j, l. Then $C_i = \bigcup_{j<\omega} \bigcap_{l<\omega} (U_{i,j,l,f(j,l)} \setminus A_{i,j,f(j,l)})$ is a Δ_1^1-nullset. Since g is Π_1^1-ML-random, $g \notin C_i$. Then for any j, there is an l such that $g \in A_{i,j,f(j,l)}$. Thus $\mathfrak{A}(\omega_1^{\mathrm{CK}}, g) \models (\forall k)(\exists y^{|m|})(\varphi_i(k, y^{|m|}) \vee \neg\psi_i(k, y^{|m|}))$ and so $\mathfrak{A}(\omega_1^{\mathrm{CK}}, g) \models \Delta_1^1$-comprehension. □

Theorem 14.6.15 (Greenberg and Monin). *Every low for Π_1^1-random real is hyperarithmetic.*

Proof. Suppose that x is not hyperarithmetic. By Theorem 14.6.9, x is not low for Π_1^1-ML randomness. Let $\{U_n^x\}_{n<\omega}$ be a universal $\Pi_1^1(x)$-ML test. We claim that for any n and $P \in \mathbf{P}$, $U_n^x \cap P \neq \emptyset$. Otherwise, $P \cap U_n^x = \emptyset$ for some n. Then every real $z \in P$ is $\Pi_1^1(x)$-ML random. However, for any Π_1^1-ML random y, by Proposition 14.2.7, there is a $z \in P$ such that $y =^* z$. Hence y is also $\Pi_1^1(x)$-ML random. In other words, x is low for Π_1^1-ML random, a contradiction.

Since P contains only Π_1^1-ML random reals, $P \cap U_n^x \neq \emptyset$ implies that there is a condition $Q \subseteq P \cap U_n^x$.

Let $\mathcal{D}^n = \{P \in \mathbf{P} \mid P \subseteq U_n^x\}$. Then $\mathcal{D}^n$ is dense for each n. If g is $\mathbb{P}$-generic meeting each $\mathcal{D}_i$ and $\mathcal{D}^n$, then g is Π_1^1-random and $g \in \bigcap_{n<\omega} U_n^x$. In other words, g is not $\Pi_1^1(x)$-ML random and hence not $\Pi_1^1(x)$-random. □

By the proof of Theorem 14.6.15, for any x if every Π_1^1-random real is $\Pi_1^1(x)$-ML random, then x must be hyperarithmetic.

Corollary 14.6.16. *A real x is low for strongly Π_1^1-ML random if and only if it is hyperarithmetic.*

Proof. Obviously every hyperarithmetic real is low for strongly Π_1^1-ML random. Now suppose that x is low for strongly Π_1^1-ML random. Then every Π_1^1-random real is strongly $\Pi_1^1(x)$-ML random and so $\Pi_1^1(x)$-ML random. So x must be hyperarithmetic. □

14.6.5 Lowness for Δ_2^1- and Σ_2^1-randomness

One may also introduce the notion Δ_2^1-traceability as for Δ_1^1-traceability. However, this notion is vacuous under $V = L$ in view of Proposition 14.4.2. Thus we have the following conclusion

Proposition 14.6.17. *If $V = L$, then*

$$\text{Lowness for } \Sigma^1_2\text{-ML randomness} = \text{Lowness for } \Delta^1_2\text{-randomness}.$$

In the case of lowness for Σ^1_2-randomness, one may apply Theorem 14.4.7 to obtain a characterization. Finally, it is possible to introduce the notion of constructible traceability in the obvious way and use it to study lowness property for Σ^1_2-randomness under additional set-theoretic assumptions. This will involve the use of some sophisticated techniques in set theory. The interested reader may consult [2].

14.6.6 Exercises

Exercise 14.6.1. *Prove Lemma* 14.6.3.

Exercise 14.6.2. *Prove that x is Δ^1_1-semitraceable if and only if every Π^1_1-ML random real is $\Delta^1_1(x)$-Kurtz random.*

Exercise 14.6.3. *Prove that there exist two uncountable Σ^1_1-sets A, B such that for any $x \in A$, $y \in B$ and recursive ordinal β, $(x \oplus y)^{(\beta)} \not\geq_T \mathcal{O}$.*

Exercise 14.6.4 (Bienvenu, Greenberg and Monin). *Let $\{G_i\}_{i\in\omega}$ be as defined as before. Prove that a Δ^1_1-random x is Π^1_1-random if and only if for any i, either $x \notin G_i$ or $x \in \bigcup\{P \in \mathbf{P} \mid P \subseteq G_i\}$.*

Exercise 14.6.5 (Monin [105]). *Prove that the set of Π^1_1-random reals is $\underset{\sim}{\Pi}^0_3$ but not $\underset{\sim}{\Sigma}^0_3$.*

Exercise 14.6.6 (Greenberg and Monin). *Prove that if x is low for Δ^1_1-random but not hyperarithmetic, then x is Π^1_1-random cuppable.*

A Open problems

A.1 Hyperarithmetic theory

We first list a question which has remained open for almost 40 years.

Question A.1.1 (Sacks [122]). *Does every countable set of hyperdegrees have a minimal upper bound?*

The next question concerns a version of the Posner–Robinson Theorem in the hyperarithmetic setting.

Question A.1.2. *Let* $x \notin \mathsf{HYP}$ *and let* $A \subseteq 2^\omega$ *be* Σ^1_1 *and uncountable. Is there a* $y \in A$ *such that* $x \oplus y \geq_h \mathcal{O}$*?*

A.2 Set-theoretic problems in recursion theory

The most important problem in this area is probably Martin's Conjecture (6.6.9). We also have the following conjecture.

Conjecture A.2.1. *Assume* ZF + AD + $V = L(\mathbb{R})$. *Then Martin's Conjecture is true for order-preserving functions.*

Question A.2.2 (Sacks [117]). *Is every locally countable partial ordering* $\langle 2^\omega, \leq_P \rangle$ *embeddable into* $\langle \mathcal{D}, \leq \rangle$*?*

The following is motivated by Theorem 9.3.8.

Question A.2.3. *Is it true, under* ZF + DC, *that the existence of a cofinal maximal chain of Turing degrees of order type* ω_1 *implies the existence of a well-ordering of the reals?*

More mysterious is perhaps the structure of antichains of Turing degrees.

Question A.2.4. *Is there a* $\underset{\sim}{\Sigma}^1_1$*-maximal antichain of Turing degrees?*

While a number of examples in recursion theory independent of ZF were presented in Chapter 9, it is not known if there is a "natural" first order sentence φ in the language of partial ordering such that "$\langle \mathcal{D}, \leq \rangle \models \varphi$" is independent of ZF.

Question A.2.5. *Is there a natural first order sentence* φ *in the language of partial ordering such that "*$\langle \mathcal{D}, \leq \rangle \models \varphi$*" is independent of* ZF*?*

In [128], it is proved that if every set of reals is measurable, then ω_1 is inaccessible in L.

Question A.2.6. *Identify a set $\mathscr{A}$ of Turing degrees with the set of reals whose degrees belong to $\mathscr{A}$. Is it true that if every set of Turing degrees is measurable, then ω_1 is inaccessible in L?*

A.3 Higher randomness theory

Question A.3.1. *Is every Π^1_1-random real strongly Π^1_1-ML random?*

By Corollary 14.3.4, Question A.3.1 can be reduced to the question of the existence of a strongly Π^1_1-ML random $x \geq_h \mathcal{O}$.

It is not known if there is a nonhyperarithmetic real which is not Π^1_1-random cuppable, though we know that lowness for Π^1_1-randomness coincides with hyperarithmeticity.

In Chapter 14, the study of algorithmic randomness was generalized to the Π^1_1-level. It was also shown that one would not obtain a satisfactory theory by generalizing notions of randomness to the Σ^1_2-level under the assumption $V = L$. Indeed under this assumption, any generalization of randomness notions to the analytical hierarchy beyond Π^1_1 would result in a theory that is essentially trivial. The key reason is the existence of a Δ^1_2 well-ordering of 2^ω within L. Hence one requires other set-theoretic axioms for a viable theory of algorithmic randomness at higher levels of the analytical hierarchy. There is a "regularity" phenomenon in descriptive set theory where all Π^1_{2n+1} and Σ^1_{2n}-sets are considered analogs of r.e. sets under PD. We are interested in seeing whether this applies to the theory of higher randomness.

Question A.3.2. *Can one extend higher randomness theory to the entire analytical hierarchy under certain large cardinal assumptions?*

Kechris, Martin and Solovay [65] extended hyperarithmetic theory to Q-theory under PD. One may then introduce the theory of higher randomness in Q-theory. However, the modern development of descriptive set theory has embarked on a route somewhat different from that taken previously. Before the 1980s, most descriptive set-theoretic results were obtained using ingenious constructions, such as those associated with games and AD. Today, large cardinals and inner model theory feature prominently in the investigation of descriptive set theory. Even results that may appear to be "straightforward generalizations" of their low level counterparts were established this way. As recursion theory is grounded on the precept of effective computability, one is interested in having a "nice L-like model" to perform recursion-theoretic constructions as those in L. In such a model, the study of higher randomness can be reasonably carried out.

B An interview with Gerald E. Sacks

Chi Tat Chong and Yue Yang[1]

Gerald E. Sacks is Emeritus Professor of Mathematical Logic at Harvard University. He received his PhD degree in 1961 from Cornell University and taught at Cornell until 1967 when he took up an appointment at MIT. He was on the faculty of Harvard and MIT from 1972 and retired from MIT in 2006 and from Harvard in 2012. Professor Sacks's work in recursion theory has had a profound influence on the development of the subject over the past 50 years. This interview was conducted in June 2009 at the Graduate Center of the City University of New York.

Question: Thanks for granting this interview. We thought it would be a good idea to spend some time talking and hearing from you about your work and views on recursion theory and logic in general. They will be an inspiration to students of logic and an important source of information to professional logicians as well as historians of the subject. We think people would be interested to get a sense of the development of major topics and ideas in recursion theory from your perspective. Such an account would be very illuminating.

Sacks: I find that hard to believe but I am willing to try.

(*laughter*)

Q: I think it is good for us, for the purpose of this interview, to start at the beginning. I understand that you come from Brooklyn.

S: Brooklyn, New York. Yes.

Q: You went to a place called Brooklyn Technical High School. Did you already decide then that you wanted to be a mathematician?

S: No. No. I was interested in math when I was around 10 or 11. But my interest faded when I became 12 or 13. I think what I noticed was that girls were not interested in math, or boys who talked about math all the time. But they were interested in literature and art. So my interest shifted. (*laughter*) I became less interested in math and more interested in literature, poetry and arts, that sort of thing.

Q: But when you went to college, did you then . . .

S: When I went to college, I thought I was going to be an engineer. I have no idea why. That would have been a terrible mistake. But fortunately before I finished engineering I went to the army. In the army, where I was for three years, I became interested in math again. Someone gave me a copy of *Introduction to Metamathematics* by S. C. Kleene . . .

Q: A mathematician, I presume, who gave it to you?

S: Yes, yes and I started reading it and I thought it was very interesting.

1 Department of Mathematics, National University of Singapore, Singapore 119076

This was before I graduated. I was sort of half way through the engineering program. I interrupted it to go into the army. I was amazed by, for example, his use of the principle of mathematical induction. Back in the days when I was so interested in math I learned about mathematical induction. But here Kleene used it to prove something about formulas. So I found that amazing. I thought you have to do induction, you know, on integers. But here, he was proving the deduction theorem and related theorems by induction. I think just for the propositional calculus. To me that was amazing.

Q: Before that, you had not taken any courses in logic? So this was your first introduction to the subject?

S: The math courses at Brooklyn Tech were just routine. You know, the old fashioned curriculum. No calculus, just algebra, plane geometry and solid geometry – you know, all that stuff. I mean there were four years of math there but kind of elementary, compared to what's now available at school.

Q: So did you go to Brooklyn College for your engineering?

S: No. I was at Cornell University.

I returned to Cornell and I decided the simplest thing was to finish up the engineering course. But I took every possible elective course in the math department. I was now interested in math. But I still wasn't certain about giving up engineering. By the time I finished the engineering course, I was now committed to math and one of the reasons was Cornell had a famous logician, Barkley Rosser, who had agreed to be my advisor. I was now committed to math and in particular to logic.

Q: But you never took a logic course from him before that.

S: No. But I had read most of Kleene's book. So I knew a little bit about logic. I fought my way through the Incompleteness Theorem, and he had me in my first year explain to him the proof in great detail – that was his method of teaching. I had to stand in front of a blackboard telling him the proof of Gödel's Incompleteness Theorem in detail. He wanted it in detail which I got some pleasure out of at that time, but I never forced it upon any of my students.

Q: Were you the only logic student at that time in Cornell?

S: Yeah, I was his only student. Roughly speaking the logic seminar consisted of him and me. He also spoke. He explained various things to me. But he had me explain – let me see – the Incompleteness Theorem, Gödel's proof of the Generalized Continuum Hypothesis in L. So we just kept going.

Q: Which year was that?

S: That would be the academic year of 1958–59. There's that orange book (*The Consistency of the Axiom of Choice and of the Generalized Continuum Hypothesis with the Axioms of Set Theory*).

Q: One of the earliest volumes in the Princeton *Annals of Mathematical Studies* series.

S: Rosser was of course familiar with it. But you are right though, there was very little available. For example, Rosser knew about Tarski's decision procedure for the reals, and the real closed fields and so on, but all he had was a mimeographed manuscript. There was no book in which it was discussed. There were very few books on logic. I mean, though Gödel's *Continuum* monograph was out, I would not really call that a book. It is just a form of argument. There is not a single intuitive remark in it.

Q: It's a monograph, certainly not something that...

S: So I worked my way through it. I don't claim I really understood what was really going on, but I certainly understood it line by line. I was not yet familiar with the downward Skolem–Löwenheim Theorem[2] and no one had heard of fine structure or anything like that. I mean none of that existed. But I was able to give myself intuitive explanations as well as checking all the formal things. Gödel's monograph tells you every detail, you just have to check it and the argument sort of made sense. I guess sort of. But the real understanding of it came later. Thanks to a number of people, one of whom was Dana Scott.

Q: You met Scott in the late fifties?

S: No. I met him, I guess, in '61. I got my PhD in '61 and I think I met him somewhere around Princeton, maybe the fall of '60. Somewhere in the academic year '61 to '63, he must have been passing through. Let me see. I went to Princeton on a postdoctoral fellowship '61 to '62 and I used it at the Institute for Advanced Study. Of course a lot of people came there and gave talks. He must have been one of them. Kreisel, for example, came through a couple of times. Church was on leave for a semester or two. So he was replaced by Abraham Robinson, whom I met at that time.

Q: Robinson was visiting...

S: Robinson was a visiting professor, I guess, for the academic year beginning 1961. I met him then and he was very encouraging. He seemed eager to speak with me. He certainly gave me a big lift.

Q: Maybe I can bring things back a little, to a few years before that. So you were in the army and you got this book by Kleene. But when you decided to do mathematics – switch to mathematics – how did you decide logic was the way to go, rather than algebra, geometry or whatever?

S: I am not sure. Certainly the idea of a recursive (computable) function was very appealing. But it's hard to explain. Certainly I was very impressed by Gödel's *Continuum* monograph, the ideas involved and that it is possible to prove something like that – its relative consistency – that seems interesting and unusual. It's quite different from, say, real analysis or complex analysis, algebraic topology. I mean, math was interesting, but these ideas somehow seemed more interesting.

Q: Did you see the methods and ideas Gödel introduced in his monograph as an illustration of an infinite Turing machine at work?

---

2 The Downward Löwenheim–Skolem Theorem states that every model of a countable theory has an elementary countable submodel.

S: Oh, no, no. I was just amazed that you could take a question like "Is the Axiom of Choice consistent with ZF" and give an answer yes. I couldn't conceive how anyone could come up with that.

Q: Did you at that time think of the possibility of some independence results?

S: No. Not really. The first year I didn't know what I was going to do. Rosser tried to push me toward set theory, but I resisted, because his interests in set theory were different. He was still concerned with some open problems with respect to new foundations...

Q: Quine's *New Foundations*?

S: Yes. I looked into that. He cleaned it up. His version of new foundations was much simpler than Quine's and equivalent to his. Rosser was also interested in more conventional set-theoretic properties. He was very interested in the question: "Do there exist inaccessible cardinals?" So for example he toyed with the idea trying to prove as a theorem of set theory there were none.

But I couldn't see how to even begin something like that. Clearly you would assume there is an inaccessible and try to get a contradiction, try to do some combinatorial thing and he had some grand schemes for doing that. But I shied away from that...

Q: But he knew that it wasn't provable under ZF – he knew about that?

S: He knew you couldn't prove the existence of an inaccessible. Oh yeah, he knew that. But he was trying to prove there are no inaccessibles. So that was not proven. I mean, nothing to this day, as far as I know, rules that out, unless you assume something stronger. Meanwhile there were papers coming out, more and more papers coming out, in recursive function theory. The first year all we had available was the Kleene–Post paper. I thought that was a good paper. He made me explain the whole paper to him. There was a theorem at the end of that paper that I liked very much, which was, that there exists essentially a minimal pair over the degrees of arithmetical sets: There exist degrees **a** and **b** such that a degree is recursive in both of them if and only if it is arithmetical.

Q: Yes, that's a very nice result?

S: The construction there intrigued me and then the second year Rosser was away. He was replaced for that year by Anil Nerode and Simon Kochen.

Q: They had just joined Cornell.

S: So Kochen taught me a little bit of Model Theory which of course was still in a very primitive state. But Nerode was very eager to discuss recursion theory. Now we had available several more papers: there were two papers by Spector, one on degrees, another on hyperdegrees of Π^1_1-sets.

Also Shoenfield had a paper on degrees and I think that really convinced me. In Shoenfield's paper there was also an open problem at the end that convinced me to try to write a thesis on recursion theory. I wanted to try those questions. And we also had Friedberg's solution of Post Problem. But Shoenfield's paper made me think that Friedberg's method – really the Friedberg–Muchnik method – could be developed a lot further. Certainly the solution to Post Problem really fascinated me. I couldn't get over it. That seemed quite different from what I'd ever seen or thought about. I mean it's trivial now, of course, but at that time, there was nothing like that.

Q: When did you first hear of the Friedberg solution?

S: In the fall of the second year, Nerode taught a course in recursion theory. This would be the academic year '59–'60 and he did the solution. He explained it.

Q: I see. Rosser didn't make you do a seminar and talk about...

S: I guess if he had been there that year, he would have made me do it.

But we kept busy that first year. He knew about Friedberg, when he got to it. We touched on a lot different things. Then Nerode made me aware of these other papers by Shoenfield, Spector and the like. I was also intrigued, Spector gave a measure theoretic argument then, that there exists incomparable hyper-degrees. So again that is intriguing. Again that was an unexpected argument: What made him think like that? At the end of the second year, the beginning of the summer I started my thesis. Of course, I have memories of that time, some of them maybe false, who knows? We always rewrite the past. But Shoenfield had a theorem in his paper: Here is $0'$, there exist degrees **a** and **b** both less than $0'$, and their union is $0'$. He splits $0'$ into two smaller degrees **a** and **b**. So I wondered why can't **a** and **b** be degrees of r.e. sets? I thought about that during the summer. I finally saw, if I took an r.e. set which is nonrecursive... Oh, first I took one of degree $0'$, I finally saw I could split it into two incomparables. Then I realized for any nonrecursive r.e., the argument would still work. So I had a theorem[3] and I was rolling...

That was the summer of, I guess, 1960. That's the end of my second year. Oh yes, I forgot, it was Friedberg's Splitting Theorem. I think Nerode brought that to my attention. Friedberg had split a nonrecursive r.e. set into two nonrecursive r.e. sets – that was in my head. So the question was to get them of lower degrees. So that was an influence. The proof of the Splitting Theorem used a finite injury argument, but there was a further twist.

Q: Yes, we now call it the Sacks preservation strategy. The method is by now classic in recursion theory.

S: I can't explain how my thinking was. I just fooled around with the problem day by day. But it finally worked out. Now Shoenfield had another open problem at the end of that paper, a minimal degree less than $0'$. So I thought about that and didn't get anywhere. At the end of the summer, although I was still a student, I went down to Princeton to become a visiting student at Princeton for my third and final year of graduate study, because Rosser was down there. When he left at the end of the first year, he decided he was going to stay two years in Princeton. So I was reunited with him. I kept thinking about the Shoenfield problem, minimal degree below $0'$, because this minimal degree construction was yet another one that intrigued me immensely.

3 Now known as the Sacks Splitting Theorem.

I read that over and over. I guess a year earlier Nerode had me explain the proof to him. It seems a common teaching method. (*laughter*) I guess, both sides benefit.

Q: So was it before or after you met Spector when you started thinking about the problem?

S: I had not met Spector. I met him after I got my degree. I stayed on in Princeton a second year but now as an NSF PostDoc fellow in the institute. He was just finishing such a year. He was about to leave but he was there over the summer. Unfortunately during that summer he died. He was only 31.

Q: It was a tragic loss . . .

S: I thought about the minimal degree below $\mathbf{0}'$ problem. I looked at his argument. He used Cantor sets, I decided fairly early on, I could use somewhat arbitrary trees, but I wasn't sure just what to do with them. Meanwhile I also thought about a bunch of other problems, such as embedding partial orderings in the degrees. So I played around a lot with that. Finally I guess – I can't remember exactly – maybe February 1961, while I was waiting in line in the supermarket near Princeton, I saw how I could do the minimal degree below $0'$.

Q: Waiting in line at the supermarket helps!

S: The idea is simple enough. In other words, look at the Spector argument and use trees of a suitable type and each tree was essentially coded by an r.e. set of numbers. So I realized you could change your mind. If you change your mind finitely often then things would settle down. I realized that could do it. In other words, $0'$ suffices to be the oracle. I don't need a $0''$ oracle. I could do all this if I knew the story of $0'$.

Q: I guess you wouldn't advise a graduate student to spend time at the supermarket when stuck with a math problem?

S: I don't know. I had an experience a few years later of a similar nature with the Density Theorem.

Q: It was getting to be a habit . . .

S: I worked on that one for quite a long time, maybe for close to two years, off and on.

I graduated in June 1961 at the Institute. I was still trying to do recursion theory. So again in the Shoenfield paper, there was a theorem that says if you take a degree between $0'$ and $0''$, which is r.e. in $0'$, then it is a jump of a degree below $0'$. So I said to myself, could that degree be r.e.? That would be a much prettier theorem for obvious reasons. I started working on it and – he is not here to defend himself – but Shoenfield was one of the people I met during that year in the Institute when he too passed through, and we chatted and I told him I was trying to improve his result in this way. He said he didn't believe that was possible.

I tried to find out why he felt that way; why was he so sure? This is a reconstruction. His argument was a finite injury argument and he felt he had used the strongest method available in recursion theory. Since there is no stronger method, that would be no stronger result, something like that. Yeah, that was the beginning of a very pleasant relationship with him. He was always in a kind of gruff way very encouraging. I met him very often after that. Over the years he always said "What are you doing now?" and I would say "Well, I was doing this and I have done that" or "I was trying to do such and such". He would tell me that's the wrong thing to do; you should be doing this one. But he was the one that pushed for the Density Theorem. He thought it was true. Until it was done, he would bump into me: "Have you done the Density Theorem? What's holding it up?" (*laughter*) I didn't realize for a long time that he was fond of me. I thought he was always being so critical, but in later years I realized that... So he didn't have the personal warmth that someone like Abraham Robinson had. He had the same good feelings.

Q: About the Density Theorem: One has to have a strong belief in the order of the r.e. degrees to work on the problem. There were very few clues on the structure available then.

S: I took an awful chance putting all that time and energy. Well, with the Splitting Theorem, you have a little bit of evidence. Alright, what else? I succeeded with the Jump Theorem[4] that was an improvement of Shoenfield's result. In order to do that, I made use of infinite injury argument.

Q: Was this the first time when infinite injury...

S: Well, that was for me, but I found out several years later that Shoenfield had done an infinite injury argument and published a paper.

I thought I had a method that seemed to deal with some of the difficulties that came up with the Density Theorem, not all of them. There was one, for example, I fooled around with sort of to warm up: If I had r.e. sets $A <_T B$, and I assumed that $A' <_T B'$, then I could do it. I could get an r.e. set C strictly between A and B. So that was a warm up but it avoided the real problem. The problem itself took a while. But I was waiting in line at the supermarket. So this was back in the spring of 1962. I had an appointment with the person I just had lunch with.

Q: You were in Princeton at that time?

S: No. I came to visit him in New York City. I was supposed to meet him in front of his apartment building at a certain time and he was late. So I was sort of thinking about this problem one more time. How could one define an r.e. set C to outwit any computation using A as an oracle that tries to compute C? And I suddenly saw how it could be done.

I tried in the paper to say something intuitive about the idea behind the construction, but I failed. I said, well, we have a scheme here for sort of pretending to embed B into C, but then erasing it if necessary. Erasing it is intuitively reasonable here, if you look at it, but I don't think it helps very much in the exposition.

Q: But your early papers were pretty formal.

S: But that was... Whom can I blame that on? (*laughter*) I complained to Spector that his paper was terribly formal. His excuse was that Kleene was his thesis advisor

4 The Jump Theorem states that every degree that is r.e. in and above $\mathbf{0}'$ is the jump of an r.e. degree.

and Kleene was also the referee of that paper. So the referee forced him to write in that language.

Q: To be 100 % accurate, I suppose.

S: Yes, I thought I was a bit less formal than that, but not much.

Q: I think in your orange book *Degrees of Unsolvability* there are pages and paragraphs in which you did try to motivate the reader.

S: Yes, but it's still terrible. I guess Rosser also encouraged that approach. He didn't seem to need intuitive remarks. (*laughter*) I mean he could just read this stuff and run through it. He was actually a very strong mathematician. He was quicker than I was. I could present him the mass of equations. He could just run right through.

Q: Who was his teacher?

S: Church. He was a student of Church. That's interesting. He and Kleene were students of Church at the same time. Rosser was thought to be the quicker one. Again Kleene is not here to defend himself. But I noticed that Kleene was more persistent. That did pay off. Rosser was less persistent. He knew lots more mathematics. He was quicker. But...

Q: His interest seemed to be pretty wide, not just in mathematics, also in all sort of things.

S: Rocket ballistics, analytic number theory: he went from one thing to another.

Q: Kleene was probably more focused.

S: Yes, very focused, focused to a degree that you cannot imagine. I once asked him about my favorite theorem of his, which is simply this: The one section of ${}^2E$ consists of exactly the hyperarithmetic sets.[5] I was very intrigued by that. But it's not a put-up job. Both of the things were defined by him – years apart! (*laughter*) One day he realized they were equal. So I think that's the theorem in question. He didn't do that in one day. No, no, I think the thing he was most persistent about was: If a set of integers is Δ^1_1, then it is HYP – it's hyperarithmetic. So I learned that one because when I was an assistant professor at Cornell, Nerode asked me to explain that. (*laughter*) That was a challenge. That paper was totally formal. There is not a single intuitive remark. I read it over and over and over and over. He used the Fixed Point Theorem to do effective transfinite recursion, but the phrase effective transfinite recursion does not occur! He just does it.

Q: The line of thought was there.

S: I didn't know that was what he was doing. He just made very, to me, surprising use of the Fixed Point Theorem. So I was able to present it formally, but I didn't really understand it intuitively. I mean I was able to verify everything. So I once asked him about how he got that theorem. He also had no knowledge of classical descriptive set

5 An object of type 0 is a nonnegative integer. An object of type $n+1$ is a set of objects of type n. ${}^{n+1}E$ is the characteristic function of the set of objects of type n. In Kleene's theory of recursion in objects of finite type, ${}^{n+1}E$ is recursive if and only if $n = 0$. He defined the 1-section of ${}^{n+1}E$ to be the set of type 1 objects recursive in ${}^{n+1}E$. He proved that the 1-section of ${}^2E$ is HYP.

theory. If he had, that would help, because he could sort of use that as a guide. But he knew nothing. He just did it from nothing. It took a while, but he said he was very conscious about the fact that you can't work on a problem forever. You have to draw some limit, otherwise you just destroy yourself. So he had chosen a limit, a time limit on this problem, a certain day, 5 o'clock, and he would no longer work on that problem. So the months passed and the day approached. Around 4:30 that day, he figured it out!

I'm not making this up. He told me this story. So I think I learned from him. This is quite a digression you are asking me why my papers were formal, those early papers. I think, it's partly his influence. He was too formal even for me. I had no love with formalism and he loved it. But I guess some pressure from him and Rosser made me write that way. I finally made Rosser admit that he had various intuitive pictures in his mind, though some of the things he did I don't see how that was possible. You see, he did something in combinatoric logic and in λ-calculus, where I don't see how there are any possible intuitive pictures. It's truly formal and you are stuck with that. But he can do that. Maybe these men were more able than you or I to perceive formally (*laughter*) to a degree that I find impossible even to contemplate. It's when I was working in classical recursion theory, I was constantly making pictures with dots, crosses, lines on pieces of paper or on the blackboard, none of which made it to the write-up. (*laughter*)

Q: You were already using Church's thesis very freely, at least internally?

S: Of course, of course. The write-ups had nothing to do with the way that I thought about those things which, looking back now, was senseless. I would advise people to try to write things up the way they actually did it, in so far as it is possible.

Q: That's right. So it seems very soon after you did those two or three major results – the Splitting Theorem, the Jump Inversion Theorem, the Density Theorem – you...

S: I was also very proud of the minimal degree below $0'$...

Q: Yes, I forgot to mention that one.

S: ... because that was a finite priority argument on trees. In other words, the whole tree changes its mind to what its definition was finitely often.

Q: So by the year 1963 or even some time before that, you started thinking about higher recursion theory, I mean, going beyond the natural numbers?

S: Yes, I guess, I owe that initially to Nerode. In the fall of '62, my first year as an assistant professor at Cornell, he asked me to give a seminar for as long as it would take. He had a bunch of logic students. Now Cornell had, it must have been, four or five logic students, which was an amazing change. Rosser had like one logic student every seven years or something. (*laughter*) But the times had changed. So Nerode thought people should learn HYP theory, if they were interested in logic. That started my interests in things higher up.

Q: So you did a seminar on that, a year long seminar?

S: That set me up to try to understand what Kreisel was talking about. Kreisel was talking in an incomprehensible way about generalizations of recursion theory. He had worked out something that, of course, was unnecessarily complex and his way of explaining it was very hard to penetrate or comprehend. But finally it began to make sense to me. His initial formulations were based on HYP theory and that became what he and I called metarecursion theory. This was recursion theory on recursive ordinals.

Q: That was '65 or '66, when it appeared?

S: But we started talking about it in summer '63. So in the spring of '63, I finished the proof of the Density Theorem, and during the summer of '63, I wrote it up (in a very unappealing way) and I started speaking with him about recursion theory on the recursive ordinals, metarecursion theory. So it's the summer of '63. In the fall of '63 we had the basic results.

Q: How did the collaboration happen?

S: I usually went to visit him. He went to Princeton a lot, I went down there a few times. Of course there was no email or anything. I tried to visit him as much as I could. And he wanted me to do Post's Problem on recursive ordinals in his formulation. So I did. He suggested some other problems and so we wrote up a paper. We presented it, I think, in the spring of '64. There was an AMS meeting when Kleene was the Chairman. Kreisel said I should give the talk and he would sit in the front row and maybe answer questions. So I gave the talk. Of course, it's an extension of HYP theory. Kleene was very interested and when the clock ran out, I stopped. So he said: Oh, you can talk longer if you'd like; there's plenty of time. (*laughter*) He was very very cordial.

Q: From then on you never returned to classical recursion theory. You just kept moving on …

S: Yeah, I got completely caught up in Kreisel's program and he had a student Platek, whom he put to work on what became known as admissible ordinals. Gaisi Takeuti first noted that constructible cardinals were admissible. The formal notion of an admissible ordinal, and sets in general, were introduced by Richard Platek and Saul Kripke. Jensen's fine structure theory of L has been fruitfully exploited by Sacks and others to study problems in α-recursion theory. I met him and I spoke with him and I kept talking to Kreisel, who didn't want to write any more joint papers. He said he would supply me with questions, but just I should write the papers. I said alright. He didn't want to be coauthor anymore. He was very generous I thought, in this respect. This is the side of him he'd shown to very few people. (*laughter*) His generous side, it was there. He wanted Post's Problem for every admissible ordinal. I worked on that. So I did the initial result, kind of trivial, that there is always an intermediate degree – α-degree – for every admissible α. But I didn't get two incomparables, sort of left out for the moment.

Q: That would be around '65 or '66?

S: This was around '65 and '66, because I got interested in forcing. I wanted to think about forcing, and forcing with perfect sets and some arguments with measure, getting Cohen's result just with a simple measure-theoretic argument. But then I came back to α-recursion theory and in fact, I owe that to Shore, Richard Shore. I was also getting interested in model theory and I wanted to think about that a bit. But Richard

Shore was a student at the time at MIT and I came back after I was on leave for a year. So in the fall of 1970, he said how about Post Problem for every admissible ordinal α? Why hasn't that been done? I said well I thought a student of Rogers had done it. Let's call them back. Rogers had a student who claimed to have done it in his thesis. So Shore said he had looked at that, but that was wrong. You know the thesis had been accepted sometime earlier, I said really? He said it's completely wrong. There was nothing. So I was willing to believe him, I mean, I knew him. His first year as a graduate student I had been around and I'd spoken to him and if he said it was wrong, it was wrong. I started talking about it and as I talked I went to the board and said: Well, suppose we look at this case and I said that doesn't seem too bad. So I said I think this probably can be done. But I decided to share it with Steve Simpson. I got together with him and we did it as a joint result. He was a more advanced student.

Q: You didn't give this problem to Simpson to work on? He did write his thesis on α-recursion.

S: No, he was in his final year and yes, he was thinking about α-recursion theory. But I decided to push him a little harder, because he just began his, what has to be his, final year. I said I divided this thing up to cases, and I showed him the case I'd already done. I said only this case remains, and I had a vague idea how it should go. He finished that and he wrote the paper. He wrote it very nicely I thought.

Q: So that's the Sacks–Simpson paper.

S: Actually we had several proofs of that. But I decided the way he did it in the paper was a nice way to write it up. I developed a proof which uses much more fine structure than necessary. So he got rid of some of it and Manny Lerman got rid of all of it. (*laughter*)

Q: Lerman didn't like fine structure?

S: I discussed fine structure with him and he knew something about fine structure. But I was certainly interested in the idea of not using any fine structure at all. He nailed it.

Q: I always think incorporating methods and ideas of fine structure theory in recursion theory is a beautiful thing. You see a very nice interplay between recursion theory and set theory, and model theory as well.

S: Certainly. In Shore's proof of the Density Theorem for all admissible ordinals, there's some use of fine structure. In fact, he proved a new fact about fine structure which had not come up in the set theory world. So there is some fine structure there, I mean, I can't prove it, but I would claim it's absolutely necessary. Once you get deep enough in α-recursion theory, you certainly need fine structure. So you might as well have it from the start. So I got very caught up in α-recursion theory...

Q: Then you decided that admissibility was not really necessary, and you moved to β-recursion theory.

S: I did a little bit on the inadmissible case. Then Sy Friedman dived in, and got some terrific results.

Q: After this you thought, well, maybe E-recursion[6] was the way to go?

S: I still wasn't there yet. Somewhere in there I looked at Kleene's papers on higher types, recursion in objects of finite type. I guess maybe Shoenfield got me interested in that. He simplified Kleene's definitions a bit. He looked at recursion in type-two objects; and he wrote a very interesting paper explaining exactly what happened there. Suddenly at least for that case everything was crystal clear, which hadn't been in Kleene. And Moschovakis[7] was working on ^3E and he had done something that looked very interesting. I got interested in the problem of sections, 1-sections, k-sections...

Q: I remember in 1970, at the International Congress of Mathematicians, the title of your talk was k-section of a type n object, something along that...

S: Yes, I got interested in that. I had a conjecture about sections. Of course I asked somebody who is an expert in that area, and he told me that couldn't possibly be true. (*laughter*) It's amazing how easy it is to believe firmly that such and such is the case, but it's not! I think we all made that mistake, but I can tell you I guess I have a whole list of things that I worked on that some recursion theorists told me that couldn't be true. I succeeded every time (*laughter*).

Q: Your intuition told you otherwise?

S: Yes, of course. At least that's the way I wanted it to be. I don't think my intuition is any better than theirs. The way I wanted was prettier than what they wanted.

Q: Coming back to minimal degrees. One of the major threads of your work seems to revolve around ideas and techniques connected to forcing with perfect sets: From minimal degree below $0'$ to minimal hyperdegree, minimal degree of constructibility, and minimal upper bounds for HYP degrees and so forth. Was the intuition to see how far you could push the minimal degree construction?

S: Yes, I wrote this paper with Robin Gandy ("A minimal hyperdegree"). Again, that's perfect sets and that was around '65. So now forcing was involved; it was really needed. That was a forcing argument with perfect sets. Of course it made sense in set theory as well.

Q: But into the 1980s you were still looking at problems that were related to or making use of techniques that came from there, for example the problem of minimal upper bounds for a collection of hyperdegrees.

S: Yes yes, I guess I finally said my goodbye (*laughter*) to perfect set forcing, I don't know what year it was, but the theorem is on upper bounds for countable sets of hyperdegrees. That's the name of the paper. That took a while. I'd previously done it for a countable initial segment of L. But that used very heavily the very nice effective well-ordering of L. The admissible set now has no well-ordering yet satisfies Σ_1-dependent

6 Kleene developed a computation theory of higher type objects generated by n-th iterates of the power set operation on $\mathbb{N}$. A unified approach from the set-theoretic point of view, E-recursion, was introduced by Dag Normann in 1967.

7 Yiannis Moschovakis.

choice. So there is something in there, an ugly face triple forcing. I didn't write that one as clearly as I might have. I was certainly more intuitive by that time. That's kind of funny. The paper was read by several students, all seemed to understand it readily, even though there are gaps in the argument, but they, a number of students, seemed to understand it without any problem. Bob Lubarsky was one of them, Leo Harrington was another – of course he could understand anything – and Fred Abramson.

Q: When did you first meet Jensen? How did the idea of using fine structure of L for α-recursion theory come about?

S: My first meeting with Jensen, I have to think about that for a moment. I guess Carol Karp told me about him. They wrote an important joint paper which goes by the title...

Q: "Primitive recursive set functions"? It appeared in an AMS Summer Institute volume on set theory.

S: I think the phrase "condensation lemma" occurred for the first time there. So they tried to clean up, what was now becoming folklore or common knowledge, in other words, they tried and succeeded in saying in print exactly what had to be said. I thought it was overly formal at that time, but I think I was wrong. Someone had to do that and they did. She got him interested in that, I think, and he carried on with it and of course he developed fine structure. Now he paid a visit to Cambridge (Massachusetts), I think, in 1967 and we talked. I think I met him earlier than this. There was this big Set Theory meeting at UCLA in 1967. He came to that. I don't remember talking to him much at that time. He was kind of shy. He avoided situations where people might be abrasive to each other. At that time, for instance, a number of people sort of challenged each other. I think they were less civil than they would be now. He didn't like anything like that. I was game for it, and he didn't care for it. When he came to Cambridge, I understood him better then so we got along. And I think he was somewhat low at that time. I guess he was working on some hard problems and it was just before his many breakthroughs. This was in the late '60s. I must have had some inkling about what he was doing with fine structure, before this Sacks–Simpson thing, which was done pretty much in the fall of 1970. Simpson finished in 1971. So we must have done that in the fall of 1970. So fine structure had been around for a while. No?

Q: Jensen's paper itself only appeared in '72.

S: What paper is that?

Q: "The fine structure of the constructible universe". The preprint of the paper was of course already available before that.

S: Well, he must have given talks. In other words, I was certainly not an expert, but I knew a little bit about it. It was easy to get started, having experience with recursive ordinals, I mean, the key there was the projection, right? I certainly knew about the Σ_1-projectum. I could see that made sense: For any α, you can define its Σ_1-projectum. I certainly knew about the condensation method, I guess, from the Karp–Jensen paper.

Q: But the study of fine structure, at least for the reals, actually started with Hilary Putnam.

S: Oh, yes, yes, right. That's another source, Boolos and Putnam.

Q: That was before Jensen. Did you take a look at that paper and think about it?

S: Yes. I was aware of that, but of course what I needed was essentially the fact that the two different definitions of say Σ_n-projectum are the same. I only needed that for Σ_1. In the case of ω_1^{CK}, that was sort of trivial. For Σ_1 at least, it was very easy to prove by a simple condensation argument. Of course, the Boolos–Putnam thing was a condensation argument.

Q: Also, the notion of a master code and its characterization was already in their paper. It's embedded somewhere referred to as "arithmetical copy".

S: Oh, yes. But it's just that the motivations were so much different. In other words, what I needed was exactly that fact, the two definitions of Σ_1-projectum. In other words, that a very small, so to speak, α-r.e. set had to be α-finite. That was the key. But I spent lots of time with higher type objects, I don't know why. I became quite interested again for a while, in recursion in objects of finite type. I mean that was what Harrington wrote his thesis on, and Ted Slaman wrote his thesis on that. That stayed in my mind for quite a while until, I mean there were various problems I'd thought about, I was just unable to do.

Q: About the minimal α-degree problem,[8] first of all, do you think the question has a positive answer?

S: Oh, yes, the minimal α-degree problem for arbitrary admissible α. Yes, I thought about it, but I got nowhere, absolutely nowhere. I mean, of course I find myself thinking about ω_ω in L. Maybe that's too tough (*laughter*).

Q: How about ω_{ω_1}?

S: Maybe it would be wise to try that. It could be that ω_ω is the toughest case of all and one should leave that for the next generation or something, or the next century.

Q: That long?

S: No. I'm joking. But, yes, ω_{ω_1} maybe a sensible person may try that.

Q: But do you still think the answer is yes?

S: I'm tempted, I mean, I have no good reason. I guess that is the way I wanted it to be.

Q: Consistent with your past intuition about other problems?

S: No no. My intuition on this is worth nothing. It would just be, the argument would have to be interesting, right? If one could do that, it would be interesting almost even to a set theorist. I mean one might need an insight into fine structure that is new. It's hard to believe that there is anything left to say about fine structure but maybe there is. I mean, before Jensen came along, I didn't think there was anything new to say about it, right? So he corrected that. Maybe there is more. I guess, some

---

8 The problem asks if there is a minimal α-degree for every admissible ordinal α.

expert would say no, we know all about fine structure. Well, maybe not. It's hard to see. Everyone I have ever talked to in detail about fine structure of L seems to have a slightly different intuitive picture. I mean, the way, (Menachem) Magidor say, talks about fine structure, or the way Sy Friedman does, or the way Jensen does, it's all different. Magidor is much more intuitive and then he brushes off the J-hierarchy. You can be fairly crude, fairly intuitive and it would work if you just push it hard enough. But certainly Jensen's presentation is very beautiful. His great insight that there are these master codes which are not merely oracles but which are preserved by condensation arguments. Now I never needed that. Maybe that had a role to play in α-recursion theory. I think Sy Friedman used it in the inadmissible case. But I haven't looked at it in a long time. It doesn't seem to come up in the admissible case. Maybe I'm wrong.

Q: So, looking ahead, how would you like to see recursion theory develop?

S: Well, I tended over the years to keep predicting the immediate end of recursion theory (*laughter*). But I have been wrong every time...

Q: By "end" you mean people lose interested in the subject?

S: Yeah. You know they just all stop working on it. But it hasn't happened yet. This whole randomness phenomena is a big surprise to me. I mean, it has been around – these notions have been around for a long time.

Q: You even mentioned that in your book (*Higher Recursion Theory*), hidden in pages here and there.

S: Really? I mentioned randomness?

Q: Yes, in some exercises.

S: But that's very weak use of it. People have gotten very subtle now with the study of randomness. There is an immense amount of activity and piles of papers that go on and on. I also thought forcing would be over by now but there are still people doing very complicated forcing arguments of a highly combinatoric nature. I thought set theory had sort of died out, but there were AD people, led by Hugh Woodin. I think in a way he is becoming more and more isolated. Fewer and fewer people know what he is doing.

Q: Getting harder and harder to understand.

S: And he is going further and further. The amount of material, I don't see how a student could go to graduate school and hope to write a thesis in that area.

Q: It would also be challenging for a student in the area to have publications upon graduation to land an academic job.

S: Well, it's probably not a good reason to produce so many academic mathematicians. I think the world needs more mathematicians, but they don't have to be all professors. They don't have to all have academic positions. They can work elsewhere...

We have to try to infiltrate all walks of life around the world with mathematicians. People who routinely think the way mathematicians think. I think there is a shortage of such people. I don't think the phrase "applied mathematics" is right, but they can simply use mathematical thinking or a mathematical approach in whatever they do in their work. It can be molecular biology or it can be even in the legal field. I think the mathematical approach is very powerful. Gödel certainly thought so. Gödel certainly thought that's the way to approach almost any problem.

Q: Could you say something about Gödel? When did you first meet him?

S: I met him, I guess in the Fall of '61. In other words, my visit to the Institute was under his auspices. One has to apply, even though there is money provided by the National Science Foundation, and they have to accept you as a temporary member. Obviously it's up to him (Gödel).

Q: So he was your host at the Institute?

S: Yes, but of course, I was afraid of him and never sought him out, essentially never spoke with him. But at another year in the Institute – I think it was 1974/75 – when we had a lot of contact, maybe we met once a month.

Q: Was he still interested in mathematics at that time?

S: Oh yes. I remember asking him one day what he thought the outstanding problems were in logic and he gave me a short list right there in person. One of them was the $P \stackrel{?}{=} NP$ problem.

Q: He was aware of that?

S: Yes, he was one of the very first, if not the first to draw attention to that problem. He wrote a letter to John von Neumann when von Neumann was on his deathbed, and he got a reply. Von Neumann said, roughly speaking, this is a very interesting problem, I would like to think about it sometime. (*laughter*)

This correspondence, I think, was shown to me by Martin Davis. That's a long time ago. Von Neumann died in the '50s. Another problem which is not entirely hopeless is consistency of analysis – proof of the consistency of analysis. There are many ways to interpret that. So you could try to find an ordinal which does it. The ordinal bears the same relation to analysis that ε_0 bears to formal number theory. What's the ordinal? And there are people who ... well, I have met one person who thought this was doable, Michael Rathjen. But he also seemed to think it might be insane to pursue it, because it is so complicated. In other words, the description of the ordinal would be book length. But it would be possible to give a sort of a clear description of that ordinal and then demonstrate it's the correct one.

And Gödel had one more problem: Look, what's now called computable functions – he called them general recursive functions, total functions – there is an intuition that this one is more complicated than that one. Find the corresponding formal concept.

Q: Not in terms of definability?

S: No. Find the true hierarchy of the complexity of computable functions. Of course, no one has managed to do this and I think lots of people think it can be done. It seemed to him people readily said this was more complicated than that one, so there is evidence that the intuitive concept exists. And if that exists, there's got to be a formal concept that captures that intuitive concept. That's how his mind worked. It might also be a little complicated. (*laughter*)

Q: What about set theory? Did he have any comments on set theory?

S: I don't recall. I mean, I tried to get him interested in AD, but he wouldn't listen. I understood his objections to AD, but I thought I could say something that he might find of interest. But he would not tolerate.

Q: Not even Projective Determinacy?

S: I tried to work up from, say Borel, since for Borel it's true. I mean, I had a pep talk prepared. He said AD is obviously false; for it falsifies the Axiom of Choice which is obviously true. He even made a nasty remark which is unusual for him. He said: I suppose sometimes it is interesting in seeing what a man can do with one hand tied behind his back. I mean, that was how the study of AD was to him. I think he was hasty there. I think there is something to talk about even if you believe AD is false, because it might be true in a very restricted setting. It might be true in $L(\mathbb{R})$ or it may be true for the projective sets, or something. I mean, there is something fruitful there, you are not wasting your time even if AD is false, even if AD is dramatically false. There is something going on there.

Q: But he never believed that $V = L$, right? I guess he only saw L as a tool.

S: Oh, I teased him about that. I teased him because if you go back far enough, he thought $V = L$ might be true; and it might be the completion of ZF – this completed it and in other words, it settled every reasonable problem in set theory and so on. Then some time passed, he seemed to say that Continuum Hypothesis was false. It's just definitely false. He said so in a paper.

So he tried to settle the matter. He tried to find the right cardinal. He seemed to think it's $\aleph_2$, because he spent lots of time trying to show that the continuum was $\aleph_2$, and he even developed some false proofs. But what he was calling a proof was not from ZF. He knew ZF couldn't do it. So he tried to write down some axioms about the real numbers, just true statements about real numbers which he found intuitively appealing, and these kept changing over time. But every once a while he would use that trying to prove the continuum was $\aleph_2$, and he would succeed. Then he would find a mistake. So then more time passed. Now it's near the end of his life. Now he told me he thought the continuum was $\aleph_1$.

I said – I was teasing him and said – I thought you've said it's equal to $\aleph_2$. So he looked at me and said, I was misled (*laughter*). Apparently the people who told him all those examples, weird examples that you derive from Continuum Hypothesis misled him. So in the end, he tried to prove it with new axioms that were plausible. So for most of his life he is working on $\aleph_2$.

Q: Thank you very much for spending time here.

S: Oh, I enjoyed the conversation very much. Thanks for your patience.

C Notations and symbols

- A real x is a member of 2^ω or ω^ω. A string is a member of either $2^{<\omega}$ or $\omega^{<\omega}$. Strings are denoted σ, τ etc.
- If x is a real, then $x \upharpoonright n$ is a finite string σ of length n such that for all $i < n$, $\sigma(i) = x(i)$.
- If σ is a finite string, then $|\sigma|$ is its length.
- $x \succ \sigma$ means $x \upharpoonright |\sigma| = \sigma$.
- $[\sigma] = \{x \mid x \in \omega^\omega \wedge x \succ \sigma\}$.
- If f is a function, then $\mathrm{Dom}(f)$ denotes the domain of f, and $\mathrm{Rang}(f)$ its range. $\mathrm{Dom}(\sigma)$ and $\mathrm{Rang}(\sigma)$ carry the obvious meaning where σ is considered as a finite function.
- If T is a tree, then $[T] = \{x \mid x \in \omega^\omega \wedge (\forall n)(x \upharpoonright n \in T)\}$.
- Given a set A, $\mathscr{P}(A) = \{B \mid B \subseteq A\}$ is the power set of A.
- $\leq_T$ denotes Turing reduction.
- x' is the Turing jump of x.
- $x \oplus y$ is the real z such that $z(2n) = x(n)$ and $z(2n + 1) = y(n)$.
- $\langle \cdot, \cdot \rangle$ is a recursion bijection from ω^2 to ω.
- For any n and $x \in 2^\omega$, $x^{[n]} = \{m \mid \langle n, m \rangle \in x\}$.
- If A is a set, then $\mathscr{P}(A)$ denotes the power set of A.
- $A \subseteq 2^\omega$ or ω^ω is $\utilde{\Pi}^0_n$ or $\utilde{\Pi}^1_n$ (resp. $\utilde{\Sigma}^0_n$ or $\utilde{\Sigma}^1_n$) if it is Π^0_n or Π^1_n (resp. Σ^0_n or Σ^1_n) definable with real number parameters.
- Given a partial ordering $\langle \mathbf{P}, \leq \rangle$, $p|q$ means p and q are incomparable, i.e. $p \nleq q$ and $q \nleq p$.

Bibliography

[1] J. W. Addison, Separation principles in the hierarchies of classical and effective descriptive set theory, *Fund. Math.* **46** (1959), 123–135.

[2] T. Bartoszyński and H. Judah, *Set Theory: On the structure of the real line*, A K Peters Ltd., Wellesley, MA, 1995.

[3] J. Barwise, *Admissible Sets and Structures*, Springer-Verlag, Berlin, 1975.

[4] J. Barwise, *Handbook of Mathematical Logic. With the cooperation of H. J. Keisler, K. Kunen, Y. N. Moschovakis, A. S. Troelstra. 2nd printing*, Studies in Logic and the Foundations of Mathematics, Vol. 90, Amsterdam, New York, Oxford: North-Holland Publishing Company. XI, 1978.

[5] J. E. Baumgartner, Sacks forcing and the total failure of Martin's axiom, *Topology Appl.* **19** (1985), 211–225.

[6] J. E. Baumgartner and R. Laver, Iterated perfect-set forcing, *Ann. Math. Logic* **17** (1979), 271–288.

[7] A. Blass, Complexity of winning strategies, *Discrete Math.* **3** (1972), 295–300.

[8] G. Boolos and H. Putnam, Degrees of unsolvability of constructible sets of integers, *J. Symbolic Logic* **33** (1968), 497–513.

[9] G. J. Chaitin, A theory of program size formally identical to information theory, *J. Assoc. Comput. Mach.* **22** (1975), 329–340.

[10] G. J. Chaitin, Information-theoretic characterizations of recursive infinite strings, *Theoret. Comput. Sci.* **2** (1976), 45–48.

[11] C. T. Chong, A recursion-theoretic characterization of constructible reals, *Bull. London Math. Soc.* **9** (1977), 241–244.

[12] C. T. Chong, A. Nies and L. Yu, Lowness of higher randomness notions, *Israel J. Math.* **166** (2008), 39–60.

[13] C. T. Chong, W. Wang and L. Yu, The strength of the projective Martin conjecture, *Fund. Math.* **207** (2010), 21–27.

[14] C. T. Chong and L. Yu, Randomness in the higher setting, *To appear.*

[15] C. T. Chong and L. Yu, Maximal chains in the Turing degrees, *J. Symbolic Logic* **72** (2007), 1219–1227.

[16] C. T. Chong and L. Yu, Thin maximal antichains in the Turing degrees, *Computation and Logic in the Real World*, Lecture Notes in Comput. Sci. 4497, pp. 162–168, Springer, Berlin, 2007.

[17] C. T. Chong and L. Yu, A Π^1_1-uniformization principle for reals, *Trans. Amer. Math. Soc.* **361** (2009), 4233–4245.

[18] P. J. Cohen, *Set Theory and the Continuum Hypothesis*, W. A. Benjamin, Inc., New York-Amsterdam, 1966.

[19] K. de Leeuw, E. F. Moore, C. E. Shannon and N. Shapiro, Computability by probabilistic machines, *Automata Studies*, Annals of Mathematics Studies, no. 34, pp. 183–212, Princeton University Press, Princeton, N. J., 1956.

[20] O. Demuth, Remarks on the structure of tt-degrees based on constructive measure theory, *Comment. Math. Univ. Carolin.* **29** (1988), 233–247.

[21] K. I. Devlin and R. B. Jensen, Marginalia to a theorem of Silver, *ISILC Logic Conference* (Proc. Internat. Summer Inst. and Logic Colloq., Kiel, 1974), pp. 115–142, Springer, Berlin, 1975. Lecture Notes in Math., Vol. 499.

[22] K. J. Devlin, *Constructibility*, Perspectives in Mathematical Logic, Springer-Verlag, Berlin, 1984.

[23] R. Downey, A. Nies, R. Weber and L. Yu, Lowness and Π^0_2 nullsets, *J. Symbolic Logic* **71** (2006), 1044–1052.
[24] R. G. Downey and D. R. Hirschfeldt, *Algorithmic Randomness and Complexity*, Theory and Applications of Computability, Springer, New York, 2010.
[25] S. Feferman, Some applications of the notions of forcing and generic sets, *Fund. Math.* **56** (1964/1965), 325–345.
[26] S. Feferman and C. Spector, Incompleteness along paths in progressions of theories, *J. Symbolic Logic* **27** (1962), 383–390.
[27] J. N. Y. Franklin, F. Stephan and L. Yu, Relativizations of randomness and genericity notions, *Bull. Lond. Math. Soc.* **43** (2011), 721–733.
[28] R. M. Friedberg, Two recursively enumerable sets of incomparable degrees of unsolvability (solution of Post's problem, 1944), *Proc. Nat. Acad. Sci. U.S.A.* **43** (1957), 236–238.
[29] H. Friedman, One hundred and two problems in mathematical logic, *J. Symbolic Logic* **40** (1975), 113–129.
[30] H. Friedman, Uniformly defined descending sequences of degrees, *J. Symbolic Logic* **41** (1976), 363–367.
[31] H. M. Friedman and R. Jensen, Note on admissible ordinals, *Lecture Notes in Mathematics* **72** (1968), 77–79.
[32] H. Gaifman, *Elementary embeddings of models of set-theory and certain subtheories*, Axiomatic set theory (Proc. Sympos. Pure Math., Vol. XIII, Part II, Univ. California, Los Angeles, Calif., 1967), Amer. Math. Soc., Providence R.I., 1974, pp. 33–101.
[33] R. O. Gandy, On a problem of Kleene's, *Bull. Amer. Math. Soc.* **66** (1960), 501–502.
[34] R. O. Gandy, Proof of Mostowski's conjecture, *Bull. Acad. Polon. Sci. Sér. Sci. Math. Astronom. Phys.* **8** (1960), 571–575.
[35] R. O. Gandy, G. Kreisel and W. W. Tait, Set existence. II, *Bull. Acad. Polon. Sci. Sér. Sci. Math. Astronom. Phys.* **9** (1961), 881–882.
[36] R. O. Gandy and G. E. Sacks, A minimal hyperdegree, *Fund. Math.* **61** (1967), 215–223.
[37] K. Gödel, *The Consistency of the Continuum Hypothesis*, Annals of Mathematics Studies, no. 3, Princeton University Press, Princeton, N.J., 1940.
[38] N. Greenberg and J. S. Miller, Lowness for Kurtz randomness, *J. Symbolic Logic* **74** (2009), 665–678.
[39] N. Greenberg, A. Montalbán and T. A. Slaman, The Slaman–Wehner theorem in higher recursion theory, *Proc. Amer. Math. Soc.* **139** (2011), 1865–1869.
[40] M. J. Groszek and T. A. Slaman, Independence results on the global structure of the Turing degrees, *Trans. Amer. Math. Soc.* **277** (1983), 579–588.
[41] M. J. Groszek and T. A. Slaman, A basis theorem for perfect sets, *Bull. Symbolic Logic* **4** (1998), 204–209.
[42] D. Guaspari, A note on the Kondo–Addison theorem, *J. Symbolic Logic* **39** (1974), 567–570.
[43] L. A. Harrington, McLaughlin's Conjecture, *Handwritten notes* (1976).
[44] L. A. Harrington, Long projective wellorderings, *Annals of Mathematical Logic* **12** (1977), 1–24.
[45] L. A. Harrington, Analytic determinacy and $0^\sharp$, *J. Symbolic Logic* **43** (1978), 685–693.
[46] L. A. Harrington and A. S. Kechris, A basis result for Σ^0_3 sets of reals with an application to minimal covers, *Proc. Amer. Math. Soc.* **53** (1975), 445–448.
[47] P. G. Hinman, Recursion-theoretic Hierarchies, *Perspectives in Mathematical Logic*, Springer-Verlag, Berlin, 1978.
[48] D. R. Hirschfeldt, A. Nies and F. Stephan, Using random sets as oracles, *J. Lond. Math. Soc. (2)* **75** (2007), 610–622.
[49] G. Hjorth, *Continuous Images of Coanalytic sets*, Logic Colloquium '98 (Prague), Lecture Notes in Logic 13, Assoc. Symbol. Logic, Urbana, IL, 2000, 226–237.

[50] G. Hjorth, An argument of due to Leo Harrington, *Online draft* (2004).

[51] G. Hjorth and A. Nies, Randomness via effective descriptive set theory, *J. Lond. Math. Soc. (2)* **75** (2007), 495–508.

[52] W. Hodges, *Model Theory*, Encyclopedia of Mathematics and its Applications 42, Cambridge University Press, Cambridge, 1993.

[53] T. Jech, *Set Theory*, Springer Monographs in Mathematics, Springer-Verlag, Berlin, 2003.

[54] R. B. Jensen, Admissible sets, *Online draft* (1969).

[55] R. B. Jensen, The fine structure of the constructible hierarchy, *Ann. Math. Logic* **4** (1972), 229–308; erratum, ibid. 4 (1972), 443.

[56] C. G. Jockusch, Jr., Recursiveness of initial segments of Kleene's O, *Fund. Math.* **87** (1975), 161–167.

[57] C. G. Jockusch, Jr. and R. A. Shore, Pseudojump operators. II. Transfinite iterations, hierarchies and minimal covers, *J. Symbolic Logic* **49** (1984), 1205–1236.

[58] C. G. Jockusch, Jr. and S. G. Simpson, A degree-theoretic definition of the ramified analytical hierarchy, *Ann. Math. Logic* **10** (1976), 1–32.

[59] C. G. Jockusch, Jr. and R. I. Soare, Minimal covers and arithmetical sets, *Proc. Amer. Math. Soc.* **25** (1970), 856–859.

[60] C. G. Jockusch, Jr. and R. I. Soare, Π^0_1 classes and degrees of theories, *Trans. Amer. Math. Soc.* **173** (1972), 33–56.

[61] B. Jónsson, On the representation of lattices, *Math. Scand* **1** (1953), 193–206.

[62] A. Kanamori and D. Pincus, *Does GCH imply AC locally?*, Paul Erdős and his mathematics, II (Budapest, 1999), Bolyai Soc. Math. Stud. 11, pp. 413–426, János Bolyai Math. Soc., Budapest, 2002.

[63] A. S. Kechris, Measure and category in effective descriptive set theory, *Ann. Math. Logic* **5** (1972/73), 337–384.

[64] A. S. Kechris, The theory of countable analytical sets, *Trans. Amer. Math. Soc.* **202** (1975), 259–297.

[65] A. S. Kechris, D. A. Martin and R. M. Solovay, Introduction to Q-theory, *Cabal Seminar 79–81*, Lecture Notes in Math. 1019, pp. 199–282, Springer, Berlin, 1983.

[66] B. Kjos-Hanssen, J. Miller and R. Solomon, Lowness notions, measure and dominations, *J. Lond. Math. Soc.* **85** (2012), 869–888.

[67] B. Kjos-Hanssen, A. Nies and F. Stephan, Lowness for the class of Schnorr random reals, *SIAM J. Comput.* **35** (2005), 647–657.

[68] B. Kjos-Hanssen, A. Nies, F. Stephan and L. Yu, Higher Kurtz randomness, *Ann. Pure Appl. Logic* **161** (2010), 1280–1290.

[69] S. C. Kleene, Arithmetical predicates and function quantifiers, *Trans. Amer. Math. Soc.* **79** (1955), 312–340.

[70] S. C. Kleene, Hierarchies of number-theoretic predicates, *Bull. Amer. Math. Soc.* **61** (1955), 193–213.

[71] S. C. Kleene, On the forms of the predicates in the theory of constructive ordinals. II, *Amer. J. Math.* **77** (1955), 405–428.

[72] S. C. Kleene, Quantification of number-theoretic functions, *Compositio Math.* **14** (1959), 23–40.

[73] A. N. Kolmogorov, Three approaches to the quantitative definition of information, *Internat. J. Comput. Math.* **2** (1968), 157–168.

[74] M. Kondô, Sur l'uniformisation des complémentaires analytiques et les ensembles projectifs de la seconde classe, *Japan. J. Math.* **15** (1938), 197–230 (French).

[75] S. Kripke, Transfinite recursions on admissible ordinals, I, II, *J. Symbolic Logic* **29** (1964), 161–162.

[76] A. Kučera, Measure, Π^0_1-classes and complete extensions of PA, *Recursion Theory Week* (Oberwolfach, 1984), Lecture Notes in Math. 1141, pp. 245–259, Springer, Berlin, 1985.

[77] A. Kučera and T. A. Slaman, Low upper bounds of ideals, *J. Symbolic Logic* **74** (2009), 517–534.

[78] K. Kunen, *Set Theory: An introduction to independence proofs*, Studies in Logic and the Foundations of Mathematics 102, North-Holland Publishing Co., Amsterdam, 1980.

[79] S. Kurtz, *Randomness and genericity in the degrees of unsolvability*, Ph.D. Dissertation, University of Illinois, Urbana, 1981.

[80] A. H. Lachlan, Uniform enumeration operations, *J. Symbolic Logic* **40** (1975), 401–409.

[81] M. Lerman, Initial segments of the degrees of unsolvability, *Ann. of Math. (2)* **93** (1971), 365–389.

[82] M. Lerman, *Degrees of Unsolvability: Local and global theory*, Perspectives in Mathematical Logic, Springer-Verlag, Berlin, 1983.

[83] L. A. Levin, Laws on the conservation (zero increase) of information, and questions on the foundations of probability theory, *Problemy Peredači Informacii* **10** (1974), 30–35.

[84] A. Lévy, Definability in axiomatic set theory. I, *Logic, Methodology and Philos. Sci.* (Proc. 1964 Internat. Congr.), pp. 127–151, North-Holland, Amsterdam, 1965.

[85] A. Louveau, A separation theorem for Σ^1_1 sets, *Trans. Amer. Math. Soc.* **260** (1980), 363–378.

[86] N. Lusin, Sur la classification de *M. Baire*, *C. R.* **164** (1917), 91–94 (French).

[87] R. Mansfield, Perfect subsets of definable sets of real numbers, *Pacific J. Math.* **35** (1970), 451–457.

[88] R. Mansfield and G. Weitkamp, *Recursive Aspects of Descriptive Set Theory*, Oxford Logic Guides 11, The Clarendon Press Oxford University Press, New York, 1985, With a chapter by Stephen Simpson.

[89] W. Markwald, Zur Theorie der konstruktiven Wohlordnungen, *Math. Ann.* **127** (1954), 135–149.

[90] D. A. Martin and R. M. Solovay, Internal Cohen extensions, *Ann. Math. Logic* **2** (1970), 143–178.

[91] D. A. Martin, The axiom of determinateness and reduction principles in the analytical hierarchy, *Bull. Amer. Math. Soc.* **74** (1968), 687–689.

[92] D. A. Martin, Measurable cardinals and analytic games, *Fund. Math.* **66** (1969/1970), 287–291.

[93] D. A. Martin, Borel determinacy, *Ann. of Math. (2)* **102** (1975), 363–371.

[94] D. A. Martin, Proof of a conjecture of Friedman, *Proc. Amer. Math. Soc.* **55** (1976), 129.

[95] D. A. Martin and J. R. Steel, A proof of projective determinacy, *J. Amer. Math. Soc.* **2** (1989), 71–125.

[96] P. Martin-Löf, The definition of random sequences, *Information and Control* **9** (1966), 602–619.

[97] P. Martin-Löf, On the notion of randomness, *Intuitionism and Proof Theory* (Proc. Conf., Buffalo, N.Y., 1968), pp. 73–78, North-Holland, Amsterdam, 1970.

[98] A. R. D. Mathias, Surrealist landscape with figures (a survey of recent results in set theory), *Period. Math. Hungar.* **10** (1979), 109–175.

[99] W. Merkle, J. S. Miller, A. Nies, J. Reimann and F. Stephan, Kolmogorov–Loveland randomness and stochasticity, *Ann. Pure Appl. Logic* **138** (2006), 183–210.

[100] A. W. Miller, Infinite combinatorics and definability, *Ann. Pure Appl. Logic* **41** (1989), 179–203.

[101] J. S. Miller, Every 2-random real is Kolmogorov random, *J. Symbolic Logic* **69** (2004).

[102] J. S. Miller, The K-degrees, low for K-degrees, and weakly low for K sets, *Notre Dame J. Form. Log.* **50** (2009), 381–391 (2010).

[103] J. S. Miller and L. Yu, On initial segment complexity and degrees of randomness, *Trans. Amer. Math. Soc.* **360** (2008), 3193–3210.

[104] J. S. Miller and L. Yu, Oscillation in the initial segment complexity of random reals, *Adv. Math.* **226** (2011), 4816–4840.

[105] B. Monin, Higher randomness and forcing with closed sets, in: *31st International Symposium on Theoretical Aspects of Computer Science (STACS 2014)* (E. W. Mayr and N. Portier, eds.), Leibniz International Proceedings in Informatics (LIPIcs) 25, pp. 566–577, Schloss Dagstuhl, Leibniz-Zentrum für Informatik, Dagstuhl, Germany, 2014.

[106] Y. N. Moschovakis, *Elementary Induction on Abstract Structures*, Studies in Logic and the Foundations of Mathematics, Vol. 77, North-Holland Publishing Co., Amsterdam, 1974.

[107] Y. N. Moschovakis, *Descriptive Set Theory*, second ed, Mathematical Surveys and Monographs 155, American Mathematical Society, Providence, RI, 2009.

[108] A. Mostowski, An undecidable arithmetical statement, *Fund. Math.* **36** (1949), 143–164.

[109] A. A. Mučnik, On the unsolvability of the problem of reducibility in the theory of algorithms, *Dokl. Akad. Nauk SSSR (N.S.)* **108** (1956), 194–197.

[110] A. Nies, Lowness properties and randomness, *Adv. Math.* **197** (2005), 274–305.

[111] A. Nies, *Computability and Randomness*, Oxford Logic Guides 51, Oxford University Press, Oxford, 2009.

[112] A. Nies, F. Stephan and S. A. Terwijn, Randomness, relativization and Turing degrees, *J. Symbolic Logic* **70** (2005), 515–535.

[113] R. Platek, Foundations of recursion theory, Ph.D thesis, Stanford University, 1966.

[114] D. B. Posner and R. W. Robinson, Degrees joining to $\mathbf{0}'$, *J. Symbolic Logic* **46** (1981), 714–722.

[115] H. Putnam, A note on constructible sets of integers, *Notre Dame J. Formal Logic* **4** (1963), 270–273.

[116] H. Rogers, Jr., *Theory of Recursive Functions and Effective Computability*, McGraw-Hill Book Co., New York, 1967.

[117] G. E. Sacks, *Degrees of Unsolvability*, Princeton University Press, Princeton, N.J., 1963.

[118] G. E. Sacks, Post's problem, admissible ordinals, and regularity, *Trans. Amer. Math. Soc.* **124** (1966), 1–23.

[119] G. E. Sacks, On a theorem of Lachlan and Martin, *Proc. Amer. Math. Soc.* **18** (1967), 140–141.

[120] G. E. Sacks, Measure-theoretic uniformity in recursion theory and set theory, *Trans. Amer. Math. Soc.* **142** (1969), 381–420.

[121] G. E. Sacks, Forcing with Perfect Closed Sets, *Axiomatic Set Theory* (Proc. Sympos. Pure Math., Vol. XIII, Part I, Univ. California, Los Angeles, Calif., 1967), pp. 331–355, Amer. Math. Soc., Providence, R.I., 1971.

[122] G. E. Sacks, Countable admissible ordinals and hyperdegrees, *Adv. Math* **19** (1976), 213–262.

[123] G. E. Sacks, *Higher Recursion Theory*, Perspectives in Mathematical Logic, Springer-Verlag, Berlin, 1990.

[124] G. E. Sacks, *Selected Logic Papers*, World Scientific Series in 20th Century Mathematics 6, World Scientific Publishing Co., Inc., River Edge, NJ, 1999.

[125] C.-P. Schnorr, A unified approach to the definition of random sequences, *Math. Systems Theory* **5** (1971), 246–258.

[126] C.-P. Schnorr, Process complexity and effective random tests, *J. Comput. System Sci.* **7** (1973), 376–388, Fourth Annual ACM Symposium on the Theory of Computing (Denver, Colo., 1972).

[127] C.-P. Schnorr, *Zufälligkeit und Wahrscheinlichkeit. Eine algorithmische Begründung der Wahrscheinlichkeitstheorie*, Lecture Notes in Mathematics, Vol. 218, Springer-Verlag, Berlin, 1971.

[128] S. Shelah, Can you take Solovay's inaccessible away?, *Israel J. Math.* **48** (1984), 1–47.

[129] J. R. Shoenfield, The problem of predicativity, *Essays on the Foundations of Mathematics*, pp. 132–139, Magnes Press, Hebrew Univ., Jerusalem, 1961.

[130] J. R. Shoenfield, Unramified forcing, *Axiomatic Set Theory* (Proc. Sympos. Pure Math., Vol. XIII, Part I, Univ. California, Los Angeles, Calif., 1967), pp. 357–381, Amer. Math. Soc., Providence, R.I., 1971.

[131] R. A. Shore, On the ∀∃-sentences of α-recursion theory, *Generalized Recursion Theory, II* (Proc. Second Sympos., Univ. Oslo, Oslo, 1977), Stud. Logic Foundations Math. 94, pp. 331–353, North-Holland, Amsterdam, 1978.

[132] R. A. Shore, Finitely generated codings and the degrees r.e. in a degree $\mathbf{d}$, *Proc. Amer. Math. Soc.* **84** (1982), 256–263.

[133] R. A. Shore, Rigidity and biinterpretability in the hyperdegrees, *Computational Prospects of Infinity.* Part II. Presented talks, Lecture Notes Ser. Inst. Math. Sci. Natl. Univ. Singapore 15, pp. 299–312, World Sci. Publ., Hackensack, NJ, 2008.

[134] R. A. Shore and T. A. Slaman, Defining the Turing jump, *Math. Res. Lett.* **6** (1999), 711–722.

[135] W. Sierpiński, Sur une classe d'ensembles., *Fundamenta* **7** (1925), 237–243 (French).

[136] W. Sierpiński, L'hypothèse généralisée du continu et l'axiome du choix, *Fund. Math.* **34** (1947), 1–5.

[137] J. H. Silver, Some applications of model theory in set theory, *Ann. Math. Logic* **3** (1971), 45–110.

[138] S. G. Simpson, Minimal covers and hyperdegrees, *Trans. Amer. Math. Soc.* **209** (1975), 45–64.

[139] S. G. Simpson, First-order theory of the degrees of recursive unsolvability, *Ann. of Math. (2)* **105** (1977), 121–139.

[140] T. A. Slaman, On a question of Sierpiński, *Fund. Math.* **159** (1999), 153–159.

[141] T. A. Slaman, Global properties of the Turing degrees and the Turing jump, *Computational Prospects of Infinity.* Part I. Tutorials, Lecture Notes Ser. Inst. Math. Sci. Natl. Univ. Singapore 14, pp. 83–101, World Sci. Publ., Hackensack, NJ, 2008.

[142] T. A. Slaman and J. R. Steel, Definable functions on degrees, *Cabal Seminar 81–85*, Lecture Notes in Math. 1333, pp. 37–55, Springer, Berlin, 1988.

[143] T. A. Slaman and W. H. Woodin, Definability in the Turing degrees, *Illinois J. Math.* **30** (1986), 320–334.

[144] T. A. Slaman and W. H. Woodin, Definability of degrees structures, *Online draft* (2005).

[145] R. M. Smullyan, *Gödel's Incompleteness Theorems*, Oxford Logic Guides 19, The Clarendon Press Oxford University Press, New York, 1992.

[146] R. J. Solomonoff, A formal theory of inductive inference, *Information and Control* **7** (1964), 1–22, 224–254.

[147] R. M. Solovay and S. Tennenbaum, Iterated Cohen extensions and Souslin's problem, *Ann. of Math. (2)* **94** (1971), 201–245.

[148] R. M. Solovay, A nonconstructible Δ^1_3 set of integers, *Trans. Amer. Math. Soc.* **127** (1967), 50–75.

[149] R. M. Solovay, On the cardinality of Σ^1_2 sets of reals, *Essays on the Foundations of Mathematics* (Symposium Commemorating Kurt Gödel, Columbus, Ohio, 1966), pp. 58–73, Springer, New York, 1969.

[150] R. M. Solovay, A model of set-theory in which every set of reals is Lebesgue measurable, *Ann. of Math. (2)* **92** (1970), 1–56.

[151] R. M. Solovay, Draft of a paper (or series of papers) on Chaitin's work, (1975), Unpublished notes, 215 pages.

[152] R. M. Solovay, Hyperarithmetically encodable sets, *Trans. Amer. Math. Soc.* **239** (1978), 99–122.

[153] C. Spector, Recursive well-orderings, *J. Symb. Logic* **20** (1955), 151–163.

[154] C. Spector, On degrees of recursive unsolvability, *Ann. of Math. (2)* **64** (1956), 581–592.

[155] C. Spector, Measure-theoretic construction of incomparable hyperdegrees, *J. Symb. Logic* **23** (1958), 280–288.
[156] C. Spector, Hyperarithmetical quantifiers, *Fund. Math.* **48** (1959/1960), 313–320.
[157] J. R. Steel, Descending sequences of degrees, *J. Symbolic Logic* **40** (1975), 59–61.
[158] J. R. Steel, Forcing with tagged trees, *Ann. Math. Logic* **15** (1978), 55–74.
[159] J. R. Steel, A classification of jump operators, *J. Symbolic Logic* **47** (1982), 347–358.
[160] F. Stephan and L. Yu, Lowness for weakly 1-generic and Kurtz-random, *Theory and Applications of Models of Computation*, Lecture Notes in Comput. Sci. 3959, pp. 756–764, Springer, Berlin, 2006.
[161] J. Stern, Some measure theoretic results in effective descriptive set theory, *Israel J. Math.* **20** (1975), 97–110.
[162] M. Suslin, Sur une definition des ensembles measurables B sans nombres transinis, *Compts rendus Acad. Science* **164** (1917), 88–91.
[163] S. A. Terwijn and D. Zambella, Computational randomness and lowness, *J. Symbolic Logic* **66** (2001), 1199–1205.
[164] F. van Engelen, A. W. Miller and J. Steel, Rigid Borel sets and better quasi-order theory, *Logic and Combinatorics* (Arcata, Calif., 1985), Contemp. Math. 65, pp. 199–222, Amer. Math. Soc., Providence, RI, 1987.
[165] M. van Lambalgen, The axiomatization of randomness, *J. Symbolic Logic* **55** (1990), 1143–1167.
[166] B. Velickovic and W. H. Woodin, Complexity of reals in inner models of set theory, *Ann. Pure Appl. Logic* **92** (1998), 283–295.
[167] W. Wang, Martin's axiom and embeddings of upper semi-lattices into the Turing degrees, *Ann. Pure Appl. Logic* **161** (2010), 1291–1298.
[168] W. Wang, L. Wu and L. Yu, Cofinal maximal chains in the Turing degrees, *Proc. Amer. Math. Soc.* **142** (2014), 1391–1398.
[169] W. H. Woodin, A tt version of the Posner–Robinson theorem, *Computational Prospects of Infinity*. Part II. Presented talks, Lect. Notes Ser. Inst. Math. Sci. Natl. Univ. Singap. 15, pp. 355–392, World Sci. Publ., Hackensack, NJ, 2008.
[170] L. Yu, When van Lambalgen's theorem fails, *Proc. Amer. Math. Soc.* **135** (2007), 861–864.
[171] L. Yu, Descriptive set theoretical complexity of randomness notions, *Fund. Math.* **215** (2011), 219–231.
[172] L. Yu, A new proof of Friedman's conjecture, *Bull. Symbolic Logic* **17** (2011), 455–461.
[173] L. Yu, D. Ding and R. Downey, The Kolmogorov complexity of random reals, *Ann. Pure Appl. Logic* **129** (2004), 163–180.

Index

De Gruyter Series in Logic and Its Applications

Volume 7
Itay Neeman
The Determinacy of Long Games, 2008
ISBN 978-3-11-018341-2, e-ISBN 978-3-11-020006-5, Set-ISBN 978-3-11-916429-0

Volume 6
Godehard Link (Ed.)
One Hundred Years of Russell´s Paradox, 2008
ISBN 978-3-11-017438-0, e-ISBN 978-3-11-019968-0, Set-ISBN 978-3-11-916554-9

Volume 5
Martin Zeman
Inner Models and Large Cardinals, 2011
ISBN 978-3-11-016368-1, e-ISBN 978-3-11-085781-8, Set-ISBN 978-3-11-179317-7

Volume 4
Rod Downey, Denis R. Hirschfeldt (Eds.)
Aspects of Complexity, 2011
ISBN 978-3-11-016810-5, e-ISBN 978-3-11-088917-8, Set-ISBN 978-3-11-180439-2

Volume 3
Sy. D. Friedman
Fine Structure and Class Forcing, 2011
ISBN 978-3-11-016777-1, e-ISBN 978-3-11-080911-4, Set-ISBN 978-3-11-174922-8

Volume 2
Marat N. Arslanov, Steffen Lempp
Recursion Theory and Complexity, 2014
ISBN 978-3-11-016587-6, e-ISBN 978-3-11-080748-6, Set-ISBN 978-3-11-175948-7